SCIENCE
A CLOSER LOOK

Macmillan
McGraw-Hill

Program Authors

Dr. Jay K. Hackett
Professor Emeritus of Earth Sciences
University of Northern Colorado
Greeley, CO

Dr. Richard H. Moyer
Professor of Science Education and
    Natural Sciences
University of Michigan–Dearborn
Dearborn, MI

Dr. JoAnne Vasquez
Elementary Science Education Consultant
NSTA Past President
Member, National Science Board
    and NASA Education Board

Mulugheta Teferi, M.A.
Principal, Gateway Middle School
Center of Math, Science, and Technology
St. Louis Public Schools
St. Louis, MO

Dinah Zike, M.Ed.
Dinah Might Adventures LP
San Antonio, TX

Kathryn LeRoy, M.S.
Executive Director
Division of Mathematics and Science Education
Miami-Dade County Public Schools, FL
Miami, FL

Dr. Dorothy J. T. Terman
Science Curriculum Development Consultant
Former K–12 Science and Mathematics Coordinator
Irvine Unified School District, CA
Irvine, CA

Dr. Gerald F. Wheeler
Executive Director
National Science Teachers Association

Bank Street College of Education
New York, NY

Contributing Authors

Dr. Sally Ride
Sally Ride Science
San Diego, CA

Lucille Villegas Barrera, M.Ed.
Elementary Science Supervisor
Houston Independent School District
Houston, TX

American Museum of Natural History
New York, NY

Contributing Writer

Ellen Grace
Albuquerque, NM

RFB&D learning through listening Students with print disabilities may be eligible to obtain an accessible, audio version of the pupil edition of this textbook. Please call Recording for the Blind & Dyslexic at 1-800-221-4792 for complete information.

C

The McGraw·Hill Companies

Macmillan/McGraw-Hill

Send all inquiries to:
Macmillan/McGraw-Hill
8787 Orion Place
Columbus, OH 43240-4027

FOLDABLES™ is a trademark of The McGraw-Hill Companies, Inc.

ISBN: 978-0-02-284138-6
MHID: 0-02-284138-5

Printed in the United States of America

4 5 6 7 8 9 079/043 11 10 09 08

Be a Scientist

AMERICAN MUSEUM ᴼᶠ NATURAL HISTORY

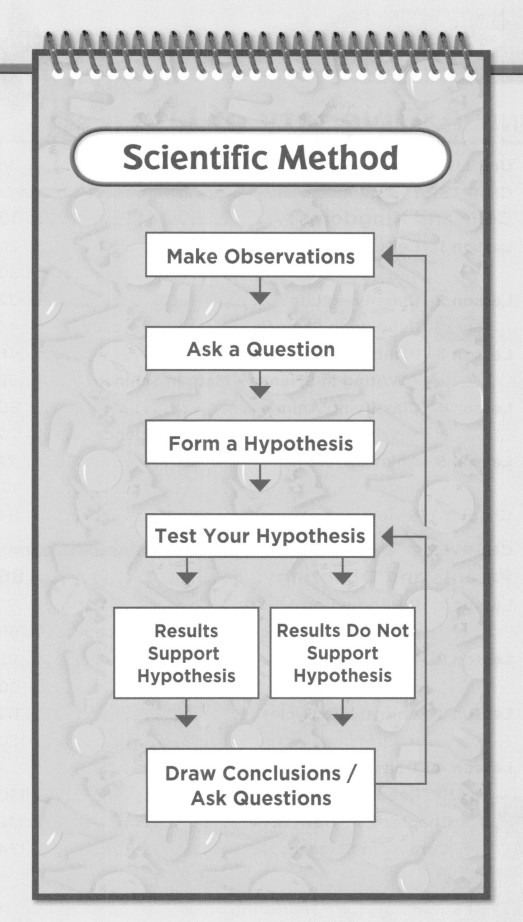

Scientific Method

Make Observations

↓

Ask a Question

↓

Form a Hypothesis

↓

Test Your Hypothesis

Results Support Hypothesis

Results Do Not Support Hypothesis

Draw Conclusions / Ask Questions

Life Science

UNIT A Diversity of Life

UNIT B Ecosystems

Earth Science

UNIT C Earth and Its Resources

UNIT D Weather and Space

Physical Science

UNIT E Matter

UNIT F Forces and Energy

Activities and Investigations

Life Science

Earth Science

Activities and Investigations

Physical Science

Be a Scientist

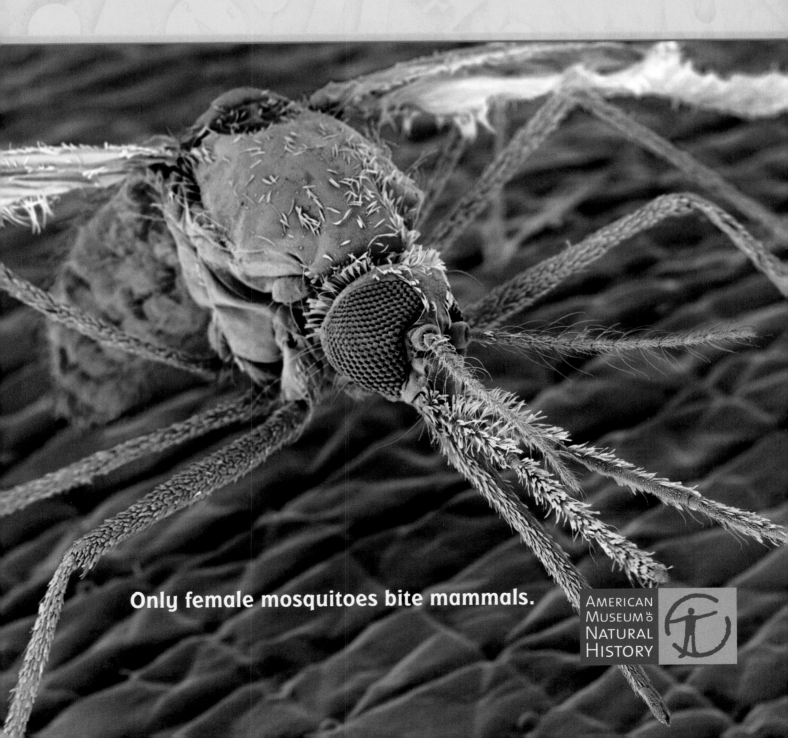

Only female mosquitoes bite mammals.

AMERICAN
MUSEUM OF
NATURAL
HISTORY

The Scientific Method

Look and Wonder

One way diseases spread is by mosquitoes. They lay their eggs in waters like these in tropical Costa Rica. How might these pesky insects affect people who live nearby? What are other ways that diseases are spread?

What do you know about disease?

How do people get sick? Do other animals get sick too? What are some of the diseases that both people and other animals get? How do scientists study diseases?

Biologists are curious about the natural world and everything that lives in it. Susan Perkins and Liliana Dávolos are biologists at the American Museum of Natural History in New York City. They investigate organisms by looking at their cells under a microscope and analyzing them in the laboratory.

Liliana Dávolos

Susan Perkins

Scientific Method

Make Observations

↓

Ask a Question

↓

Form a Hypothesis

↓

Test Your Hypothesis

↓

| Results Support Hypothesis | Results Do Not Support Hypothesis |

↓

Draw Conclusions / Ask Questions

What do scientists do?

Malaria is a serious disease that kills more than one million people every year. Malaria is caused by a parasite that infects red blood cells. A parasite is an organism that lives in or on a host.

Scientists have studied malaria in humans for many years, but humans aren't the only animals that get malaria. Birds, lizards, and other mammals also get malaria. Scientists are now studying malaria in different animals. "The more we understand about the organism that causes malaria, the more tools we have to fight the disease," Susan Perkins explains.

She and Liliana Dávolos use the scientific method to learn more about the malaria parasite. The scientific method is a process that scientists use to investigate and answer questions. This method helps them explain how things happen in the natural world.

Scientists don't always follow the steps of the scientific method in order, but they do make sure that they and others can reliably repeat their procedures. This way, the work can be checked by other scientists.

Fence lizards can catch malaria from sand flies.

Female mosquitoes need a blood meal before they can lay their eggs.

Susan and other scientists know that humans get malaria when they are bitten by mosquitoes that have the organism that causes malaria in them. They have also observed that lizards get the disease when they are bitten by sand flies with malaria parasites.

Susan has a question: "Does malaria transmitted between different organisms behave the same way inside different hosts?" She predicts the cells of malaria parasites in lizards and in sand flies will be similar to the malaria parasites that scientists have observed in mosquitoes and in humans. That is her hypothesis.

There are different types of variables in Susan's hypothesis. Her independent variable, the factor that changes, is the type of species. Her dependent variable, a factor that varies based on the independent variable, is the genes in the cells of the parasites. Controlled variables are factors that are not tested and remain constant.

Striped manakins can also catch malaria.

How do scientists test their hypothesis?

Scientists have made an interesting observation by studying certain genes from malaria parasites. The genes they are studying have directions for making ribosomes. Ribosomes are "factories" within cells that make proteins.

Most multicellular organisms have only one kind of ribosome. But some malaria parasites have two kinds of ribosomes. One kind of ribosome is observed when malaria parasites are living inside the mosquito. A different kind of ribosome is made once the malaria parasite enters a human.

This makes malaria passed between mosquitoes and humans very unusual! "This is an interesting pattern," Susan observes. "Can we find it in other animals?"

Susan's hypothesis is that the cells of malaria parasites in lizards and sand flies will show a similar pattern to the malaria parasites in humans and mosquitoes. However, she needs evidence to test her hypothesis. Evidence is data that scientists collect in different ways.

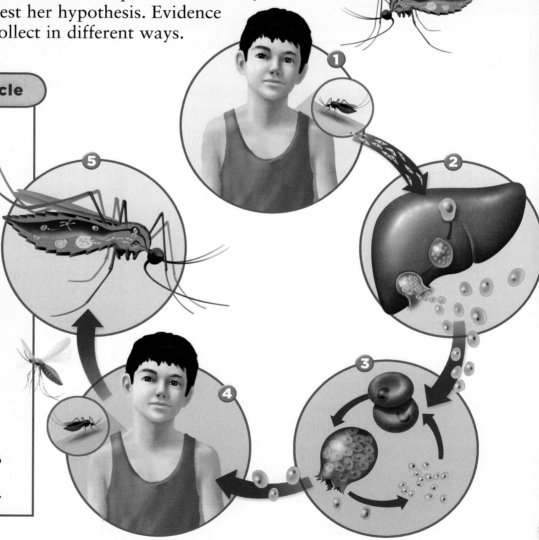

mosquito infected with parasite

Malaria Parasite Cycle

Stage 1 A mosquito infected with parasite bites a healthy person.

Stage 2 The parasite travels to the liver and makes more parasites.

Stage 3 The parasites now infect the red blood cells, make more parasites, and the disease spreads throughout the body.

Stage 4 An uninfected mosquito bites the infected human and becomes infected.

Stage 5 That mosquito can now spread the disease to other people.

In this example, Susan traveled to California to get her evidence. She took two blood samples from fence lizards, which are common in California and easy to catch. One sample is used to see whether the lizard has malaria by looking for parasites under the microscope. If it does, she uses the second sample to look at the parasites' genes.

Back in the lab, Liliana helps analyze the blood samples Susan collected. They remove the parasites' genes from the lizards' blood so they can compare them to genes from parasites in other hosts. They also use powerful microscopes and computers to examine the malaria parasites.

Test Your Hypothesis

❶ Think about the different kinds of data that could be used to test the hypothesis.

❷ Choose the best method to collect this data:

- **Perform an experiment** (in the lab)
- **Observe the natural world** (in the field)
- **Make a model** (on a computer)

❸ Plan a procedure to collect this data.

▶ **Make sure the procedure can be repeated.**

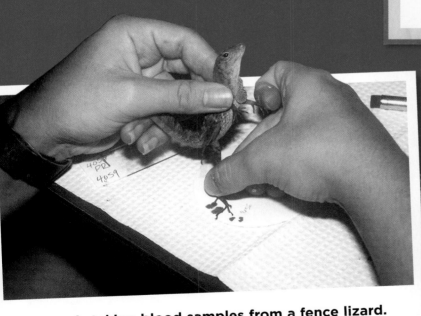

Susan is taking blood samples from a fence lizard.

How do scientists analyze data?

Part of testing a hypothesis is arranging the data to look for patterns. Susan and Liliana collect genes from the malaria parasites found in lizards. They also collect genes from malaria parasites found in rodents. Just like humans and some other mammals, rodents are studied because they get malaria from mosquitoes.

Single genes are too small to see. So scientists use a special chemical process to make copies of them. Then they run the copied genes through a piece of gelatin, called a gel. The gel has tiny openings of different sizes called pores.

Susan and Liliana prepare the genes to run through the gel.

Susan and Liliana collect genes from the blood.

Liliana is testing the gene samples in the gel to determine their patterns.

Analyze the Data

❶ Organize the data as a chart such as a table, graph, diagram, map, or group of pictures.

❷ Look for patterns in the chart that show connections between important variables in the hypothesis being tested.

▶ Make sure to check the data by comparing it to data from other sources.

The pores separate pieces of the gene based on how large they are. The pieces of the gene form bands on the gel.

These bands form patterns that can be used to compare the genes. "It's a delicate process that can take several weeks," says Liliana.

Susan and Liliana found that all of the genes of the malaria parasite in the lizards have the same pattern. This is evidence that malaria parasites that infect lizards have only one kind of ribosome.

However, the genes of the malaria parasite in the rodent show different patterns. This is evidence that the malaria parasites that infect other organisms have different kinds of ribosomes.

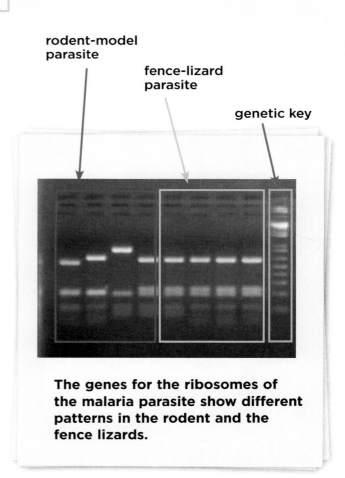

rodent-model parasite

fence-lizard parasite

genetic key

The genes for the ribosomes of the malaria parasite show different patterns in the rodent and the fence lizards.

How do scientists draw conclusions?

Now Susan and Liliana see if their evidence supports their hypothesis. Do the malaria parasites transmitted between different groups of organisms behave the same way?
No, they do not! The evidence shows that malaria parasites that infect mammals have two kinds of ribosomes and malaria parasites that infect lizards have only one.

The results of Susan and Liliana's work do not support their hypothesis. "We thought all malaria parasites would behave the way it does in humans. It came out differently," says Liliana.

Susan and Liliana look forward to investigating new questions.

The next animals to be studied may be deer in India, lemurs in Madagascar, or penguins in Chile.

When it comes to scientific research, disproving a hypothesis is as good as confirming one—in fact, it is often better! Susan and Liliana carefully check their data and procedures. Then they write up their results so that other scientists who study malaria can learn from their work.

Their results lead biologists to ask new questions. Is the way that malaria infects humans and other mammals unique? Why do those parasites have different kinds of ribosomes? How can these results lead to new tools to fight this disease? New questions are important because they lead to new hypotheses.

Liliana continues research on malaria parasites in the field.

Focus on Skills

Scientists use many skills as they work through the scientific method. Skills help them gather information and answer questions they have about the world around us. Here are some skills they use:

Observe Use your senses to learn about an object or event.

Form a Hypothesis Make a statement that can be tested to answer a question.

Communicate Share information with others.

Classify Place things with similar properties into groups.

Use Numbers Order, count, add, subtract, multiply, and divide to explain data.

Make a Model Make something to represent an object or event.

Use spring scales to measure an object's weight.

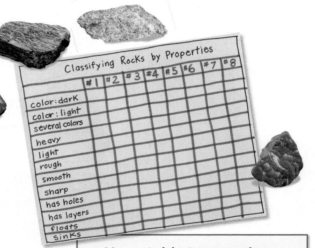

Use a table to organize and interpret data.

Scientists communicate with other scientists by writing down their observations.

Use a microscope to observe very small items.

Scientists use experiments to test a hypothesis.

Use Variables Identify things that can control or change the outcome of an experiment.

Interpret Data Use information that has been gathered to answer questions or solve a problem.

Measure Find the size, distance, time, volume, area, mass, weight, or temperature of an object or event.

Predict State possible results of an event or experiment.

Infer Form an idea or opinion from facts or observations.

Experiment Perform a test to support or disprove a hypothesis.

Inquiry Skill Builder

In each chapter of this book, you will find an Inquiry Skill Builder. These features will help you build the skills you need to be a scientist.

Scientists use models to explain how an event happens.

Safety Tips

In the Classroom

- Read all of the directions. Make sure you understand them. When you see "⚠ **Be Careful**," follow the safety rules.

- Listen to your teacher for special safety directions. If you do not understand something, ask for help.

- Wash your hands with soap and water before an activity.

- Be careful around a hot plate. Know when it is on and when it is off. Remember that the plate stays hot for a few minutes after it is turned off.

- Wear a safety apron if you work with anything messy or anything that might spill.

- Clean up a spill right away, or ask your teacher for help.

- Dispose of things the way your teacher tells you to.

- Tell your teacher if something breaks. If glass breaks, do not clean it up yourself.

- Wear safety goggles when your teacher tells you to wear them. Wear them when working with anything that can fly into your eyes or when working with liquids.

- Keep your hair and clothes away from open flames. Tie back long hair, and roll up long sleeves.

- Keep your hands dry around electrical equipment.

- Do not eat or drink anything during an experiment.

- Put equipment back the way your teacher tells you to.

- Clean up your work area after an activity, and wash your hands with soap and water.

In the Field

- Go with a trusted adult—such as your teacher, or a parent or guardian.

- Do not touch animals or plants without an adult's approval. The animal might bite. The plant might be poison ivy or another dangerous plant.

Responsibility
Treat living things, the environment, and one another with respect.

Diversity of Life

The golden butterfly fish can leap out of the water to catch flying insects.

Coral reef in the Red Sea, Egypt

from *Ranger Rick*

Adventures in Eating

You just take a bite, chew, swallow, and that's that. Right? Well, dinner isn't always as cooperative as that. Some animals, like giraffes and anteaters, have special adaptations that help them succeed in their eating adventures!

How about dinner with a view? A giraffe uses its extra-long neck to stretch to the treetops. There, it munches on leaves—up to 34 kilograms (75 pounds) of them each day! Giraffes even eat the leaves of thorny acacia trees. The giraffe's long, flexible tongue weaves past the thorns, curls around a leaf, and tugs it free. If it grabs a thorn by mistake, thick, gooey saliva inside the giraffe's mouth and throat protects it from the sharp spines.

Giant anteaters have all the right tools to help them eat ants. Nose to the ground, the anteater sniffs out an ant nest. It makes a hole with a sharp claw, pokes in its long snout, and sticks out its tongue. This is no ordinary tongue. It is 2 feet long and covered with tiny spines and sticky saliva. It flicks in and out more than 150 times a minute, slithering through the tunnels where ants live and slurping as many as 30,000 of them a day.

 ## Write About It

Response to Literature This article tells about different adaptations for eating. Research two more animals that have interesting adaptations. Write a report that explains how these adaptations help the animals eat. Compare these adaptations to the ones you read about in the article.

 **LOG ON** e-**Journal** Write about it online at **www.macmillanmh.com**

Cells and Kingdoms

The Big Idea

How are living things similar?

Euglena

Key Vocabulary

organism
any living thing that can carry out its life on its own (p. 22)

cell
the smallest unit of living matter (p. 22)

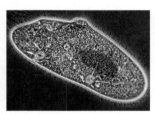

unicellular
one-celled organism (p. 23)

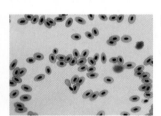

multicellular
many-celled organism (p. 23)

species
a group of similar organisms in a genus that can reproduce more of their own kind (p. 34)

photosynthesis
the food-making process in green plants that uses sunlight (p. 54)

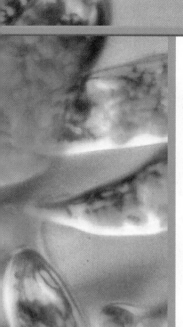

Cells

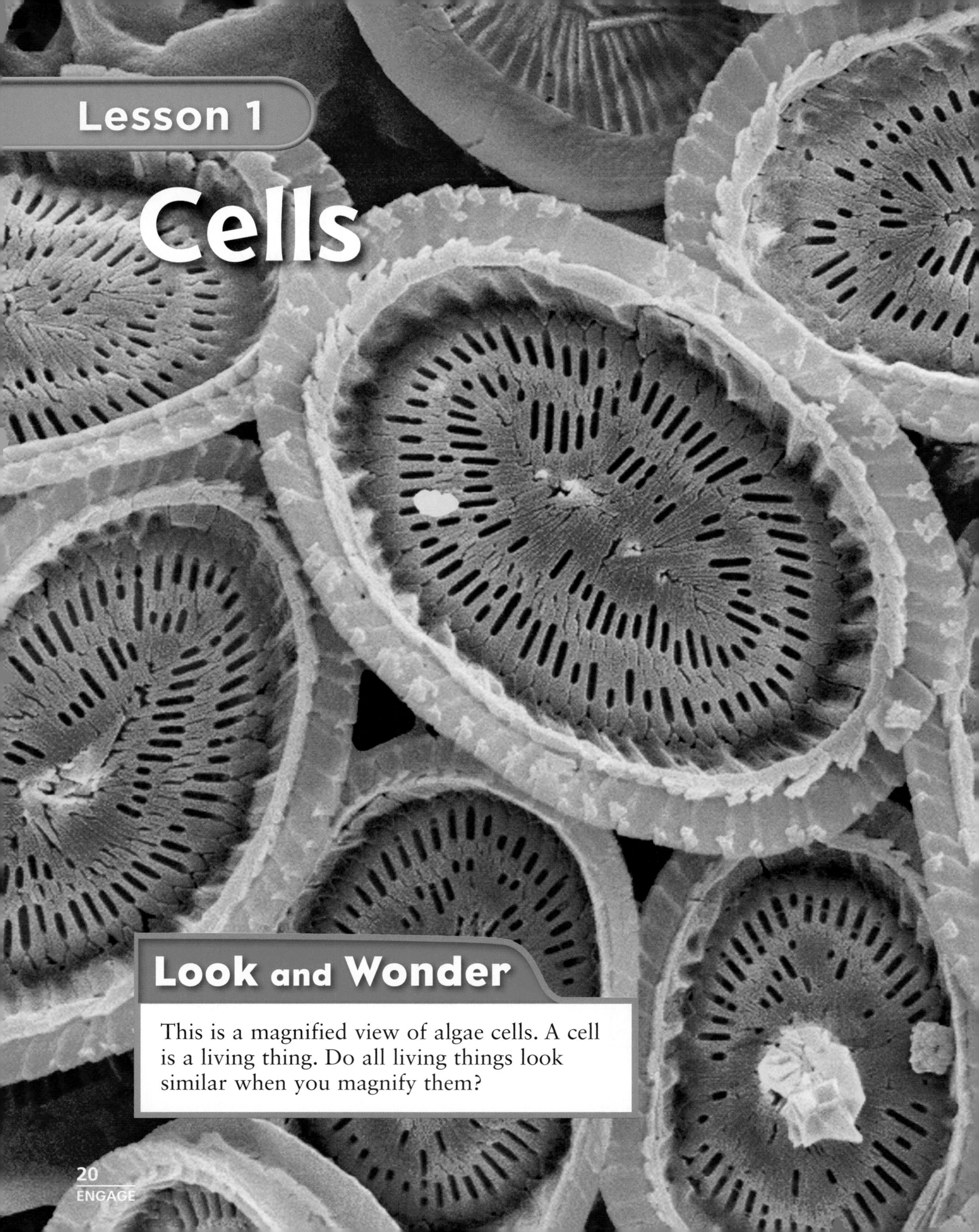

Look and Wonder

This is a magnified view of algae cells. A cell is a living thing. Do all living things look similar when you magnify them?

What are plants and animals made of?

Make a Prediction

Plants and animals are living things. Think about a plant and an animal you have seen. Do you think they are made of similar or different parts?

Test Your Prediction

1 Observe Look at the prepared slide of a plant leaf under the microscope. For help using the microscope, ask your teacher and look at page R5.

2 Draw what you see.

3 Look at the prepared slide of animal blood under the microscope.

4 Draw what you see. Compare your drawings.

Draw Conclusions

5 Interpret Data How were the plant slide and animal slide alike? How were they different?

6 Communicate Write a report explaining whether or not your observations supported your prediction.

Explore More

Examine the drawings you made and think about the living things they came from. Mushrooms are also living things. What do you think a mushroom slide looks like? Make a prediction and plan an experiment to test it.

Materials

- microscope
- prepared slides of plant-leaf cells
- prepared slides of animal-blood cells

Step **1**

Step **2**

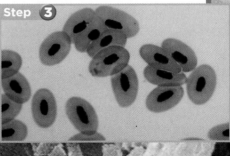

Step **3**

▶ Main Idea
Living things are all made of the same basic building blocks—cells.

▶ Vocabulary
organism, p. 22

cell, p. 22

unicellular, p. 23

multicellular, p. 23

chlorophyll, p. 27

tissue, p. 28

organ, p. 28

organ system, p. 28

LOG ON e-Glossary

at **www.macmillanmh.com**

▶ Reading Skill ✔
Compare and Contrast

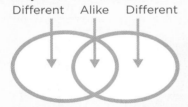

Different   Alike   Different

▶ Technology
Explore the levels of organization from cells to organs with Team Earth.

What are cells?

Earth is home to many different living things— big, small, strange, beautiful, and everything in between. You might think a microscopic amoeba and an 18.3-meter-long (60 foot) giant squid could have nothing in common. However, if you look closely through a microscope you will see how similar they are. A giant squid and an amoeba are both organisms (OR•guh•niz•uhmz). An **organism** is a living thing. They are made of the same tiny building blocks. From the smallest organism to the largest, they are all made of cells. A **cell** is the smallest unit of living things that can carry out the basic processes of life.

Where do cells come from? The simple answer is that cells come from other cells! Every cell in every living thing on Earth originally came from another cell. A cell divided, or split into two new cells, and so did the cell before that, and so on.

In 1665, English scientist Robert Hooke looked at a slice of cork in his microscope and saw many "little boxes" like these that he called cells.

Unicellular and Multicellular Organisms

amoeba cell

▲ frog cells

Read a Photo

How is a frog cell similar to the cell of an amoeba?

Clue: Compare the cells in the photos. How are they alike? How are they different?

A **unicellular** (ew•nuh•SEL•yuh•luhr), or one-celled, organism is made of a single cell that carries out its life processes. Life processes include growing, responding to an environment, reproducing, and getting food. **Multicellular** (mul•ti•SEL•yuh•luhr), or many-celled, organisms are made of more than one cell. Multicellular organisms include frogs, trees, and you!

In multicellular organisms, every cell carries out its own life process. The cells also work together to take care of different functions for the organism. For example, all of your heart muscle cells carry out their own life processes, and they work together to keep your heart beating.

How plentiful are unicellular and multicellular organisms? More than $1\frac{1}{2}$ million kinds of organisms have been identified. That number is small compared to the estimated number of unicellular organisms that exist and have not been identified. Scientists estimate that there are more than 1 billion kinds of unicellular organisms!

 Quick Check

Compare and Contrast How are unicellular and multicellular organisms similar and different?

Critical Thinking Why do you think there are more unicellular organisms than multicellular organisms?

What is inside an animal cell?

All organisms are made of cells. Your own body has more than 200 different kinds of cells. Plant and animal cells have several basic structures, called *organelles* (OR•guh•nelz), that help them perform life processes. Organelles have functions that help keep the cell alive.

Animal Cell

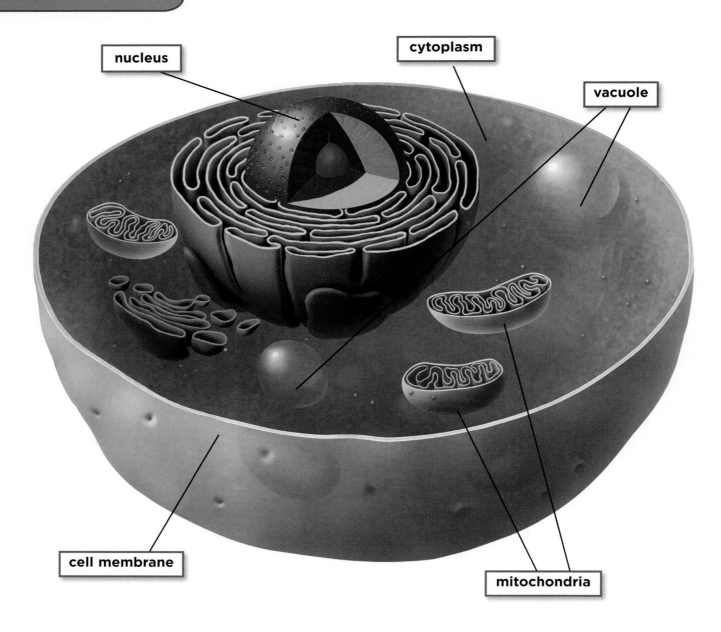

nucleus

cytoplasm

vacuole

cell membrane

mitochondria

Cell Membrane

Animal cells are surrounded by a flexible wrapping called a *cell membrane*. The cell membrane is a layer around the outside of the cell. It wraps around the cell in somewhat the same way your skin wraps around you. It gives the cell its shape.

The cell membrane controls what materials move into and out of the cell. Only certain substances are able to enter and leave the cell.

Cytoplasm

The cell membrane is filled with a gel-like liquid called *cytoplasm* (SYE•tuh•plaz•uhm). It occupies the region from the nucleus to the cell membrane. Cytoplasm is made mostly of water. A variety of organelles float in the cytoplasm.

The cytoplasm supports all the cell's structures. It is constantly moving through the cell in a stream-like motion. Some of the cell's life processes take place in the cytoplasm.

Nucleus

The *nucleus* (NEW•clee•uhs) is the cell's control center. It is a large, round organelle usually found in the center of the cell. It has a membrane with pores, or openings, that allow certain materials to pass in and out.

The nucleus contains the master plans for all the cell's activities. It sends signals to all other parts of the cell. Cells grow, move, and at some point may divide. These functions are controlled by a cell's nucleus.

Mitochondria

Mitochondria (mye•tuh•KON•dree•uh) are oval, membrane-covered organelles that supply energy for the cell. Each mitochondrion is a tiny power plant. They break down food, which releases energy for the cell to use.

Some cells are more active than others and require more energy. Cells that require a lot of energy, such as muscle cells, usually have a great many mitochondria.

Vacuoles

A *vacuole* (VAK•yew•ohl) is a membrane-covered structure used for storage. It can store water, food, and wastes. The nucleus can signal a vacuole to release whatever it is storing. Some animal cells have many small vacuoles and some may not have any vacuoles.

 Quick Check

Compare and Contrast How is a mitochondrion similar to a tiny power plant?

Critical Thinking Do you think a cell would function without a nucleus? Explain your answer.

What is inside a plant cell?

Plant cells have many of the same structures and organelles as animal cells. However, plant cells often have a box-like shape and are a bit larger than animal cells. They also have some additional organelles that animal cells do not have.

Plant Cell

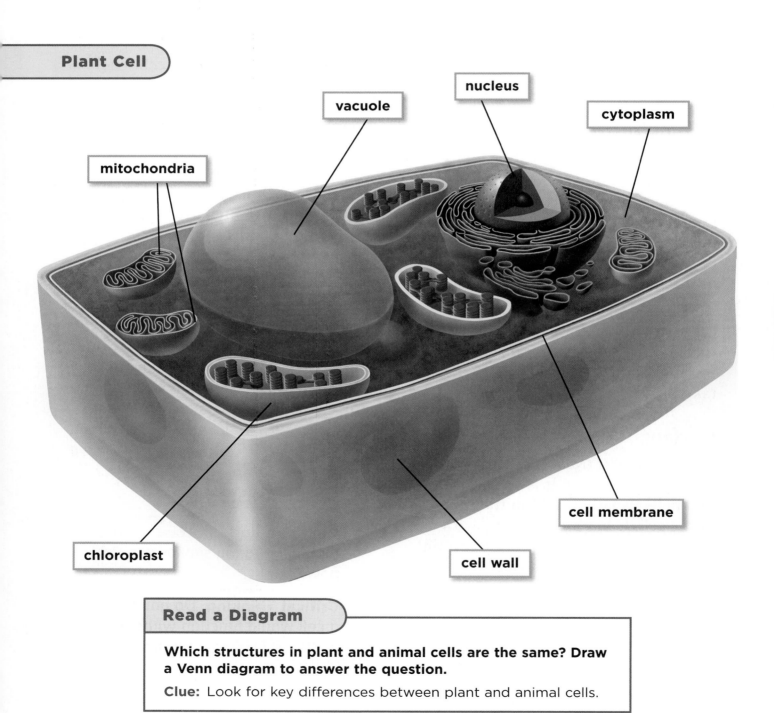

mitochondria

vacuole

nucleus

cytoplasm

cell membrane

chloroplast

cell wall

Read a Diagram

Which structures in plant and animal cells are the same? Draw a Venn diagram to answer the question.

Clue: Look for key differences between plant and animal cells.

Cell Wall

Plant cells have an additional outer covering around the outside of the cell. This layer is called the *cell wall*. The cell wall is a stiff structure outside the cell membrane. It provides the plant cell with strength and extra support.

Vacuole

Unlike animal cells, plant cells usually have one large, central vacuole. In plant cells this vacuole stores excess water and provides extra support. The extra water in the vacuoles of plant cells keeps the plant from drying out. When a plant needs extra water the vacuoles release the water they have stored into the cells.

Chloroplast

Plants make their own food in structures inside their cells called *chloroplasts* (KLOR•uh•plastz). A chloroplast is a green structure where the energy from sunlight is used to produce food for the plant. Chloroplasts are green because they contain a chemical called chlorophyll (KLOR•uh•fil). **Chlorophyll** is able to use the energy in sunlight.

Many plant cells are green because of the chlorophyll in their chloroplasts. Plant cells that lack chloroplasts are not green. Chloroplasts are mainly found in the cells of leaves and stems of plants.

Quick Lab

Plant and Animal Cells

1. **Make a Model** Put one plastic bag in a storage container. This is a plant cell. Use another plastic bag as an animal cell.

2. Using a spoon, carefully put gelatin in both bags until the bags are almost full.

3. Choose vegetables that look the most like the plant-cell and animal-cell organelles.

4. Place the vegetables that you have picked into the appropriate container and seal the bags.

5. Try to stack your models. How well do the plant cells stack compared to the animal cells?

6. **Communicate** Discuss with your classmates which vegetables you selected for your organelles and explain why.

✓ Quick Check

Compare and Contrast Which cell has a stronger outer covering—a plant cell or an animal cell?

Critical Thinking A plant cell has a thick cell wall and large vacuoles. However, it does not seem to have chloroplasts. What part of the plant might this cell be from?

How are cells organized?

For unicellular organisms, organization is simple. The organism has only one cell that performs all life functions. Multicellular organisms are more specialized. Your own body contains many different cell types that have specific functions. Muscle cells, for example, specialize in movement. Red blood cells, on the other hand, carry oxygen to other cells.

In a complex organism like a salamander, organization starts at the cell level. Cells are the building blocks of the body. Similar cells working together at the same job, or function, form a **tissue** (TISH•ew). A group of tissues that work together to perform a specific function form an **organ** (OR•guhn). The salamander's heart, liver, brain, and skin are organs.

Organs that work together to perform a certain function make up an **organ system**. For example, the salamander's circulatory system includes its heart, blood, and blood vessels. These work together to bring food, oxygen, and other materials to the salamander's cells. Some organ systems work together with other organ systems. The digestive system sends food to the circulatory system. The blood vessels in the circulatory system bring this food to the salamander's cells. All of the cells, tissues, organs, and organ systems form an organism.

 Quick Check

Compare and Contrast How do organs compare to organ systems?

Critical Thinking How are complex organisms organized?

From Cells to Organisms

cell

tissue

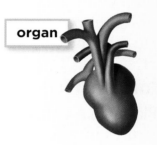
organ

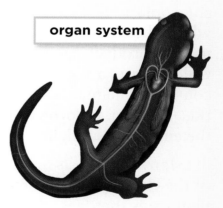
organ system

Lesson Review

Visual Summary

Cells are the basic building blocks of all living things.

Animal cells and plant cells share some **organelles**, but animal cells do not have cell walls, chloroplasts, or large vacuoles.

Organisms can exist as single cells, or they can be organized into **tissues, organs,** and **organ systems**.

Make a FOLDABLES™ Study Guide

Make a Three-Tab Book. Use the titles shown. Tell about the topic on the inside of each tab.

Cells

Organelles

Tissues, organs, and organ systems

Think, Talk, and Write

1 **Main Idea** What is the main difference between the ways unicellular and multicellular organisms are organized?

2 **Vocabulary** The cell's power plants are the _____.

3 **Compare and Contrast** How can you tell the difference between a typical plant cell and a typical animal cell?

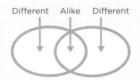
Different   Alike   Different

4 **Critical Thinking** Do bigger organisms have bigger cells? What kind of test could you do to answer this question?

5 **Test Prep** Which of the following exists in both plant and animal cells?
   A chloroplast
   B cell wall
   C mitochondrion
   D chlorophyll

6 **Test Prep** Which of the following is the cell's control center?
   A cytoplasm
   B nucleus
   C cell membrane
   D vacuole

Math Link

Dividing Bacteria
A single unicellular bacterium divides every half hour. How many bacteria cells will exist after 3 hours?

Social Studies Link

The Plague
The Plague was a disease that killed many in the Middle Ages. Write a report on this disease. What kind of organism caused plague—unicellular or multicellular?

Inquiry Skill: Experiment

All living things are made up of cells. Every cell has a cell membrane. The cell membrane is a layer around the cell that lets substances in and out.

Substances move in or out of a cell depending on their concentrations, or amounts. Substances move from areas where they are crowded to areas where they are less crowded. For example, if a cell has a higher concentration of water than its environment, water will flow out of the cell until the concentration on the inside and outside is balanced. One way to learn more about how cell membranes work is by doing an **experiment**.

▶ Learn It

An **experiment** is a test that supports or does not support a hypothesis. To carry out an experiment you need to perform a test that examines the effects of one variable on another using controlled conditions. You can then use your data to draw a conclusion about whether or not the hypothesis has been supported.

In the following experiment, you will test the effects of various substances on a cell membrane. You will gather and analyze data to support or disprove the following hypothesis: If the concentration of a substance is higher outside the membrane, then the substance will move inside the membrane to balance the concentration.

▶ Try It

Materials 2 eggs, balance, 2 glass jars with lids, vinegar, spoon, 2 beakers, water, corn syrup

1 Measure two eggs using a balance. Record the measurements in a chart.

2 Pour 200 mL of vinegar into two jars with lids. Carefully lower the two eggs into the jars of vinegar. Tighten the lids and leave the eggs inside for two days.

3 Use a spoon to carefully remove the eggs. Rinse the eggs under water.

4 Measure each egg and record the data in your chart.

5 Pour 200 mL of water into a beaker and 200 mL of corn syrup into another beaker. Carefully lower an egg into each beaker. Leave the eggs inside for one day.

6 Use the spoon to carefully remove the eggs. Rinse the eggs under water.

7 Measure each egg and record the data in your chart.

▶ Apply It

1 Now it is time to analyze your data and observations. Use your chart to compare the masses of the eggs.

2 Did the mass of both eggs change? Explain why the masses changed.

3 Did this **experiment** support or disprove the hypothesis?

	First Measurement	Second Measurement	Third Measurement
Egg #1			
Egg #2			

Classifying Life

Lionfish swimming over a coral reef in the Red Sea, Egypt

Look and Wonder

Scientists have identified and grouped about two million types of organisms. How do they organize all these living things?

How can living things be classified?

Purpose

Scientists group, or classify, organisms with certain similarities together. Compare specimens and classify them based on their characteristics.

Procedure

1 **Observe** Look at the specimens your teacher has given you.

2 Examine the specimens two at a time and compare them. How are they alike? How are they different? Record your findings in a chart.

3 **Classify** Find ways to group the specimens based on characteristics. For example, you might group the specimens based on whether they move from place to place or whether they take in food or make their own food.

4 **Communicate** Compare your classification chart with a classmate's chart. How did your classification methods compare?

Draw Conclusions

5 **Infer** Why do you think classifying organisms helps scientists? Explain.

6 Which of the items you classified are more similar, or more closely related, to each other?

Explore More

What other organisms or items can you classify? Observe organisms near your house or your school. Classify them into one of the groups.

Materials

- **plant specimens**
- **rock specimens**
- **fungal specimens**
- **animal specimens**

Step 2

Step 3

► **Main Idea**

All organisms can be classified into six separate kingdoms.

► **Vocabulary**

classification, p. 34
kingdom, p. 34
species, p. 34
vertebrate, p. 36
invertebrate, p. 36
vascular, p. 38
nonvascular, p. 38

LOG ON ℮-Glossary
at www.macmillanmh.com

► **Reading Skill** ✔

Classify

How are organisms classified?

There are millions of different organisms on Earth. Scientists organize these organisms by sorting, or classifying, them into groups according to shared characteristics. **Classification** (klas•uh•fi•KAY•shuhn) has been called the science of finding patterns. Classifying helps scientists identify, study, group, and name organisms.

One classification system used today divides all organisms into six major groups called kingdoms. The broadest group an organism is classified into is a **kingdom**. Rather than outward characteristics, such as color, kingdoms are grouped by internal form and structure. Scientists classify organisms into kingdoms by carefully comparing their cells, tissues, organs, and organ systems.

Kingdoms are very broad groups. For example, horses and spiders seem to have little in common, yet both are members of the animal kingdom. Clearly, more and smaller groups are needed to fully classify an organism. Scientists use six subgroups to classify within kingdoms. This allows scientists to separate organisms into smaller groups that have the most characteristics in common. These subgroups include *phylum* (FYE•luhm), *class, order, family, genus,* and *species* (SPEE•sheez).

The narrowest group an organism can be classified into is a **species**. A species only contains organisms that are very closely related. A horse and zebra share many traits, but they are not even similar enough to be the same species. However, a horse and a pony are so similar that they are the same species.

A scientific name is created using an organism's genus and species. The scientific name of this horse is *Equus caballus.*

Kingdom

Phylum

Class

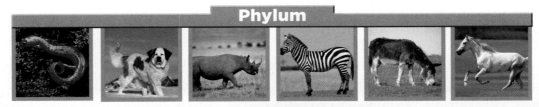

Order

Family

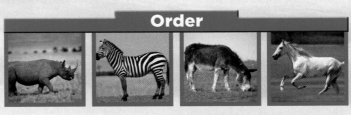

Genus

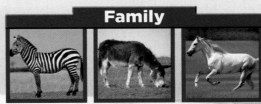

Species

Read a Chart

Is a horse more closely related to a spider or a dog?

Clue: Which groups do the spider and the dog share with the horse?

 ### Quick Check

Classify What is the broadest classification group? The narrowest group?

Critical Thinking Why do scientists use only the genus and species names to identify an organism and not all of the classification groups?

FACT Classification is based on the relationships between organisms.

What are animals?

What makes organisms in the animal kingdom different from organisms in the other kingdoms? First, animals are completely multicellular. Plants and animals are the only kingdoms that include only multicellular organisms. Two other kingdoms—fungi and protists—have both unicellular and multicellular species. Second, animals do not make their own food. Plants can make their own food, but animals must consume other organisms to get energy.

A third difference is with the structure of animal cells. Plants have a cell wall. Unlike plants, animal cells do not have cell walls. Finally, many animals are capable of moving from place to place. Plants only move when carried by wind, water, or animals.

The animal kingdom is one of the largest kingdoms. The animal kingdom features 11 phyla that belong to two major groups, vertebrates (VER•tuh•brayts) and invertebrates (in•VER•tuh•brayts). A **vertebrate** is an animal with a backbone. An animal without a backbone is an **invertebrate**.

Invertebrate phyla include the *mollusks* (MOL•uhsks), *echinoderms* (i•KYE•nuh•duhrmz), and *arthropods* (ARTH•ruh•podz). Mollusks are mainly shelled animals, like clams and snails. Echinoderms include sea stars, sea cucumbers, and sea urchins.

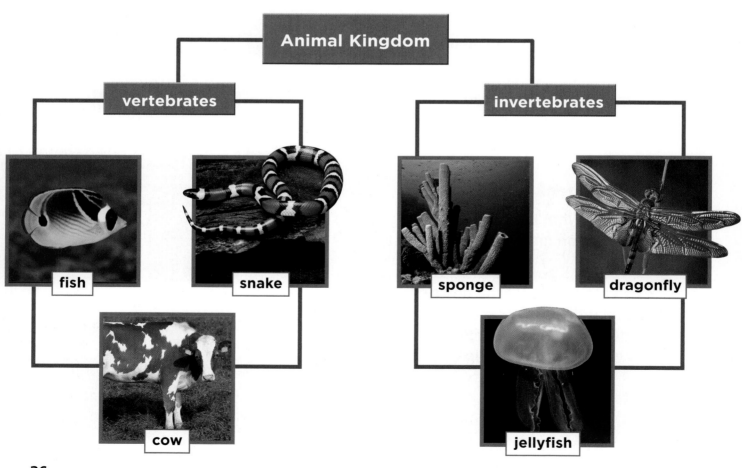

Animal Kingdom

vertebrates

fish

snake

cow

invertebrates

sponge

dragonfly

jellyfish

Number of Animal Species

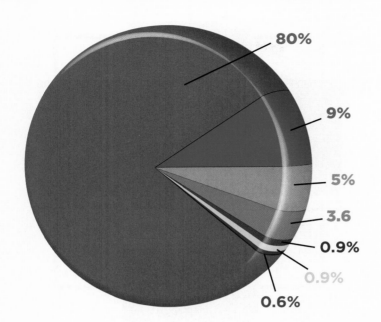

80%
9%
5%
3.6
0.9%
0.9%
0.6%

Animal Phyla Key

- Arthropods
- Mollusks
- Worms
- Chordates
- Porifera
- Cnidaria
- Echinoderms

Read a Graph

Which animal phylum contains the most species? Which contains the fewest species?

Clue: Compare the number of species for each different phylum.

The arthropods are the largest invertebrate phylum. In fact, there are more arthropod species than there are species in all the other animal phyla. Most arthropods are insects, though arthropods also include spiders, crabs, and lobsters.

All vertebrates belong to the phylum *Chordata* (KAWR•day•tuh). Animals in this phylum are called *chordates*. They are animals that have a supporting rod that runs most of the length of their body for at least part of their lives. All vertebrates have a backbone, a nervous system, and a brain. Within the vertebrate group, there are five main classes: fish, amphibians, reptiles, birds, and mammals.

Mammals are the most familiar vertebrate group. They include dogs, horses, kangaroos, and humans. There are about 5,000 identified mammals. Mammals make up about 10 percent of the entire vertebrate group, which has about 50,000 identified species.

 Quick Check

Classify How would you classify a dog and a butterfly—invertebrate or vertebrate?

Critical Thinking An organism looks like an animal but it does not move. How could you determine whether it was an animal?

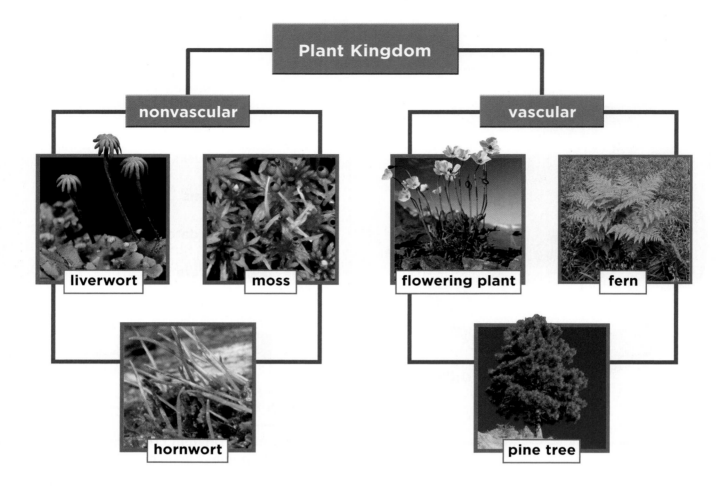

Plant Kingdom

nonvascular

liverwort

moss

hornwort

vascular

flowering plant

fern

pine tree

What are plants and fungi?

Cell walls are a key feature for identifying members of certain kingdoms. Organisms from only two kingdoms, plants and *fungi* (FUN•ji), always have cell walls. Plants and fungi cannot move from place to place and do not have true sense organs.

Plant Kingdom

The plant kingdom includes about 350,000 plant species. All plants are multicellular and make their own food. Plants are organized into two major groups: vascular (VAS•kyuh•luhr) and nonvascular (non•VAS•kyuh•luhr) plants. The word **vascular** means "contains tubes or vessels." Vascular plants have a system of vessels that run up and down the body of the plant. Vascular tissue carries water and nutrients up from the plant's roots to its leaves. It also moves sugars made in the leaves to other parts of the plant.

Nonvascular plants do not have vascular tissue. For this reason, nonvascular plants tend to be much smaller than vascular plants. A tree, for example, can grow tall because it has vascular tubes that carry materials up and down its trunk. Nonvascular plants remain small and close to the ground, where they soak up water directly. Nonvascular plants include mosses, hornworts, and liverworts.

Fungus Kingdom

Fungi differ from plants in the way they obtain energy. Plants can make their own food. Fungi must get food from other organisms.

Most fungi get energy by breaking down dead or decaying plants and animals. For example, you might find mushrooms attached to a rotting log in a forest. These fungi break down the log into nutrients they can absorb for energy. The nutrients can also be reused by other organisms.

Fungi can live in almost any dark and wet place, such as a basement. You might also find mold growing on bread, fruit, or other food. Fungi can even grow on the human body, causing itchy ailments such as athlete's foot.

However, many fungi are used by humans. Yeast is a fungus that makes bread rise. Fungi are the original source of the medicines called antibiotics. Antibiotics save lives every day when they destroy organisms that infect humans and animals.

≡ Quick Lab

Bread Mold Activity

1. Trace a slice of bread onto graph paper.

2. Put a drop of water on one corner of the bread. Place the bread in a bag. Place the bag in a warm, dark corner.

3. **Observe** When you first see mold, sketch the shape of the moldy area on the graph paper.

4. For 3 days, use different colors to sketch any new mold growth.

5. **Interpret Data** Each day count the number of squares that are covered with mold.

6. Create a bar graph to show the growth of the mold.

✔ Quick Check

Classify How are fungi different from plants?

Critical Thinking What would happen to a forest without fungi?

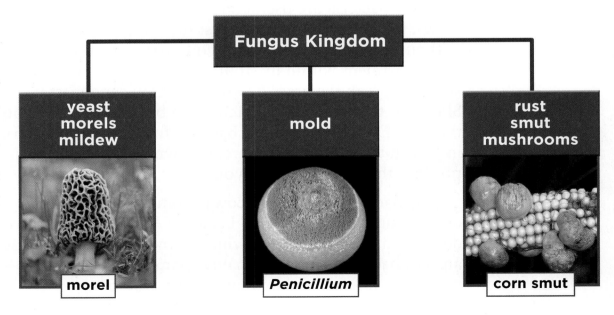

Fungus Kingdom

yeast
morels
mildew

morel

mold

Penicillium

rust
smut
mushrooms

corn smut

Bacteria Kingdoms

Ancient Bacteria

hot springs
bacteria

True Bacteria

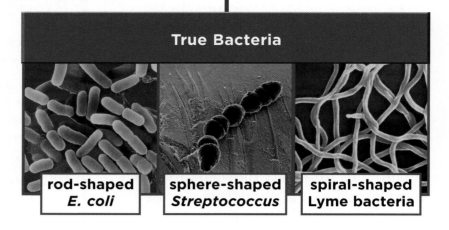

rod-shaped
E. coli

sphere-shaped
Streptococcus

spiral-shaped
Lyme bacteria

What are bacteria and protists?

Bacteria are tiny unicellular organisms. Each *bacterium* (BAK•teer•ee•uhm) is a single cell with no nucleus. Bacteria also lack other membrane-covered organelles, such as mitochondria. Bacteria are classified into two kingdoms called "true" bacteria and "ancient" bacteria.

Organisms in the "true" bacteria kingdom are found just about everywhere—on your toothbrush, in your food, on your skin, even inside your body. Are all of these organisms harmful? Some bacteria can cause disease and infections, such as food poisoning and strep throat. However, many bacteria just live with you and produce no harmful effects.

Some bacteria are useful and helpful to other organisms. Cows, for example, cannot digest the grass they eat without the help of bacteria. A cow stores chewed grass in a compartment within its stomach. Bacteria in the cow's stomach break the grass down so the cow can digest it. Similarly, termites rely on bacteria to break down the wood they eat. Even humans need bacteria for digestion. Bacteria in your intestines help you break down food. They also produce the vitamin K that your body needs.

The bacteria that live in cows' stomachs are members of the kingdom of "ancient" bacteria. Ancient bacteria are descended from the oldest living organisms on Earth. Most ancient bacteria live in harsh environments, such as deep-sea vents, hot springs, and extremely salty water.

spiral-shaped
bacteria

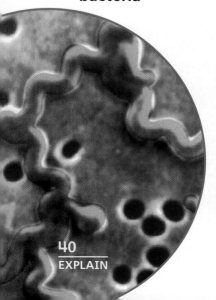

Protist Kingdom

Like the bacteria kingdoms, the protist kingdom contains unicellular organisms. Protists are unicellular or multicellular organisms that either make their own food or eat other organisms. How do you tell these tiny organisms apart? First, protists are much larger than bacteria. Hundreds or even thousands of bacteria cells might fit into a typical protist. Even so, most protists are microscopic. A very large amoeba, for example, might be about 0.5 millimeters in width, barely visible without a microscope.

Besides being larger than bacteria, the cells of protists have a central nucleus and other membrane-bound organelles. Unlike animals, plants, and fungi, protists have simple body structures and lack specialized tissues.

Typical protists include algae, amoebas, and slime molds. There are plantlike, animal-like, and fungilike protists. For example, some protists, like algae, are similar to plants. Algae

diatoms

have chlorophyll and make their own food. Other protists, such as amoebas, share some characteristics with animals. They can move from place to place. Animal-like protists also obtain food by eating other organisms. Fungi-like protists, such as slime molds, break down dead organisms to obtain food.

 Quick Check

Classify A type of bacteria lives in the soil and breaks down waste material. Is it "ancient" or "true" bacteria?

Critical Thinking What stops scientists from classifying plantlike protists as plants?

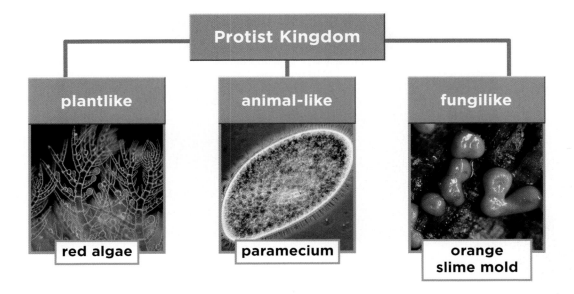

Protist Kingdom

plantlike	animal-like	fungilike
red algae	paramecium	orange slime mold

What are viruses?

They sound like something out of a science fiction movie. *Viruses* bridge the gap between the living and the nonliving. Though they may seem to be alive, many scientists believe that viruses are not living organisms. Viruses cannot be classified in any of the six kingdoms because they do not carry out all of the basic life processes. Viruses carry out only one of these basic processes: reproduction.

Viruses enter the body of a living thing, take over some of its cells, and cause the organism to get sick. The common cold is caused by a virus, as are more serious diseases, including chicken pox, polio, and HIV-AIDS.

Sneezing and coughing send cold viruses from one person to another.

Once inside the body, a virus attaches itself to a cell. When a virus enters a cell it takes control of the cell activities. It "orders" the cell to produce more viruses. Over time the cell becomes filled with virus particles and it bursts open. Now the released viruses can invade other cells, causing an infection and disease.

 Quick Check

Classify How are viruses classified? Explain.

Critical Thinking When could an infected person cause someone else to get sick?

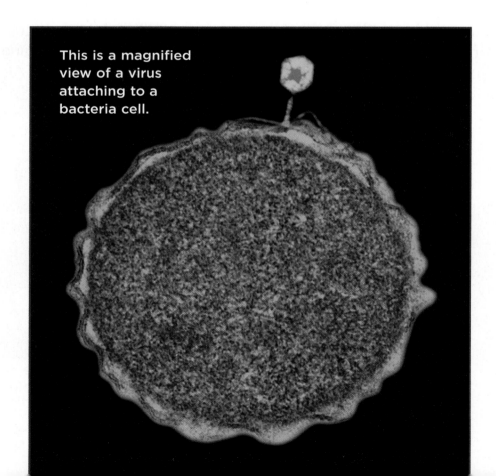

This is a magnified view of a virus attaching to a bacteria cell.

Lesson Review

Visual Summary

All living things are classified into six kingdoms. The **animal kingdom** is divided into vertebrates and invertebrates.

Organisms in the **plant kingdom** and the **fungus kingdom** have cell walls in their cells.

The **bacteria** and **protists** are unicellular organisms. **Viruses** are neither living nor nonliving things.

Make a FOLDABLES™ Study Guide

Make a Trifold Book. Use the titles shown. Tell about each topic on the folds.

Animal Kingdom	Plant Kingdom Fungus Kingdom	Bacteria, protists, viruses

Think, Talk, and Write

1. **Main Idea** Which three kingdoms are mainly multicellular? Mainly unicellular?

2. **Vocabulary** Animals with backbones are _____.

3. **Classify** How would you classify a multicellular organism that has cell walls but no chlorophyll?

4. **Critical Thinking** A computer virus is a program that takes over your computer. How is a computer virus similar to a virus that infects cells?

5. **Test Prep** Which of these kingdoms has organisms with plantlike or animal-like characteristics?
   - **A** plants
   - **B** protists
   - **C** fungi
   - **D** true bacteria

6. **Test Prep** Which one of the following should be classified as a vertebrate?
   - **A** a bird
   - **B** a snail
   - **C** an insect
   - **D** a worm

Writing Link

Explanatory Writing
Write an article that explains how bacteria can be helpful to humans.

Social Studies Link

Alexander Fleming
Write a report on the discovery of antibiotics by Alexander Fleming. How was the discovery an accident? Use books, articles, or the Internet as sources.

Meet Angelique Corthals

How can you fit thousands of organisms into one small room? Angelique Corthals knows. She's a scientist at the American Museum of Natural History, and she's been busy preserving tissue samples of many different organisms from around the globe, including samples from species that have become extinct, or died out.

Angelique works in the museum's frozen tissues lab. She specializes in the preservation of the information in cells. From bacteria to insects to mammals, she collects, preserves, and organizes the cells of all sorts of living things. Angelique stores the organisms' cells and freezes them in small plastic tubes the size of your finger. Just as food stays fresh in the freezer, freezing the cells prevents them from spoiling or decomposing. The tubes are stored in large tanks containing liquid nitrogen. At -150°C (-238°F), this liquid is so cold that all of the cells' biological processes stop.

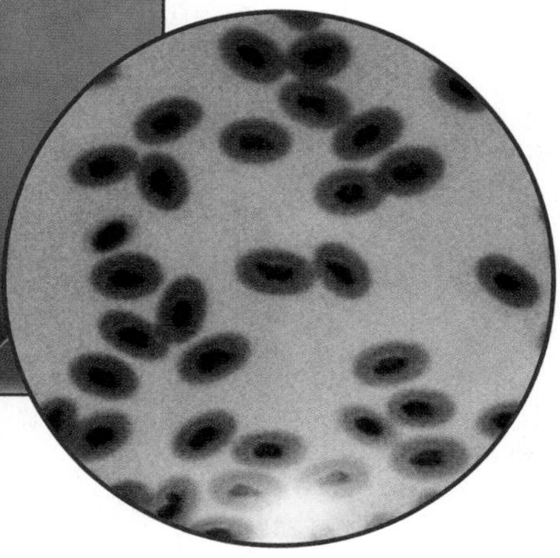

Here is Angelique working in the frozen tissues lab.

reptile cells

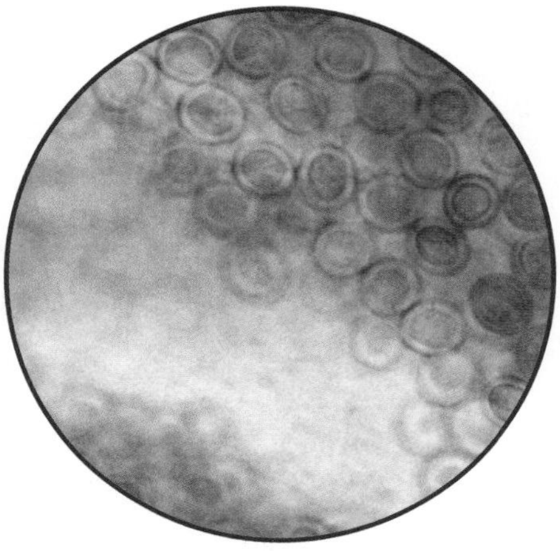

fish cells

Angelique learns about organisms, such as humpback whales and flies, by studying their cells.

By using this freezing process, the cells can be preserved for many years. When a scientist needs to study an organism, she can request a cell sample from the lab. Whether it's from a small fly or a large humpback whale, each cell contains information about the whole organism. Scientists can use this information to learn how different organisms are related. They can also use this information to learn about living things that have already become extinct and to understand why they died out.

Soon, the collection will contain one million frozen tissue samples. One day, the museum expects to have a record of most of the organisms on the planet.

 Write About It

Summarize

1. Make a chart that tells the steps for preserving cells.
2. Use your chart to write a summary of the process Angelique uses to freeze cells from organisms.

LOG ON **e-Journal** Research and write about it online at **www.macmillanmh.com**

Summarize

▶ To summarize a passage, briefly retell it in your own words.

▶ Focus on the most important events or pieces of information.

AMERICAN MUSEUM ö NATURAL HISTORY

Plants

Arizona wildflowers and cacti

Look and Wonder

Some cactus plants can survive for a year on the water they store in their roots and stems. What do cactus plants have in common with other vascular plants?

How is water transported in vascular plants?

Form a Hypothesis

All vascular plants have vessels that transport food and water in the plant. How does the amount of leaves on a plant affect transport through a plant stem? Write your answer as a hypothesis in the form "If the number of the leaves on a plant decreases, then . . .

- 3 plastic cups
- water
- blue food coloring
- 3 celery stalks with leaves
- ruler

Test Your Hypothesis

1. Fill 3 plastic cups with water. Be sure that each cup has the same amount. Put 3 drops of food coloring in each cup of water.

2. Break all the leaves off one celery stalk. Remove all but one leaf on another stalk. Leave the third stalk intact. Place a celery stalk in each cup.

3. **Observe** On the following day, examine each cup. What happened to the water? Note any changes.

4. **Measure** Use a ruler to measure how far up the water traveled in each celery stalk.

Step **1**

Draw Conclusions

5. What are the independent and dependent variables in this experiment?

6. **Interpret Data** Did the amount of leaves affect the transport of water?

7. Did your results support your hypothesis?

Explore More

Step **2**

What other variables can affect the movement of water through a plant? How will adding sugar or salt affect water transport in a plant? Form a hypothesis and test it. Then analyze and write a report of your results.

Read and Learn

▶ **Main Idea**

Plants perform photosynthesis, which provides food and energy for most organisms.

▶ **Vocabulary**

gymnosperm, p. 49
angiosperm, p. 49
xylem, p. 53
phloem, p. 53
cambium, p. 53
photosynthesis, p. 54
transpiration, p. 54
cellular respiration, p. 56

ⓔ-Glossary
at **www.macmillanmh.com**

▶ **Reading Skill** ✓

Draw Conclusions

Text Clues	Conclusions

▶ **Technology** SCIENCE QUEST

Explore photosynthesis and respiration with Team Earth.

How are plants classified?

All plants need space, air, water, and sunlight. In most cases, plants can obtain air and sunlight directly from their environments. Transporting water and other nutrients can be more difficult.

Nonvascular plants are small and survive without a transport system. Mosses, for example, reach heights of a centimeter or less. Their parts are very close to the ground to absorb water directly.

Vascular plants do not have the same size limitations. Trees, for example, can grow to heights of more than 66 meters (200 feet). How do trees get water up to their higher branches and leaves? Inside a tree trunk there is a *vascular system*, which is a series of hollow tubes. These tubes can transport water and nutrients to the top of the tallest redwood tree where they are used by the plant.

Vascular plants are divided into seed plants and seedless plants. Seed plants, like pine trees and flowering plants, produce seeds. A *seed* contains an undeveloped plant, stored food, and a protective covering. The protective covering prevents the seed from drying out or getting damaged. The

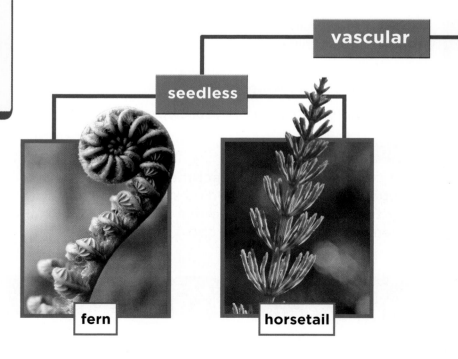

vascular

seedless

fern

horsetail

undeveloped plant uses the stored food to grow and develop.

Seedless plants, like ferns, produce spores. A *spore* is a single cell that can develop into a new plant exactly like the plant that produced it. Spores have a tough outer covering. It protects them from drying out until they find the right conditions for growth.

There are two main types of seed plants: gymnosperms (JIM•nuh•spurmz) and angiosperms (AN•jee•uh•spurmz). A **gymnosperm** is a seed plant that does not produce a flower. They include pines, firs, and other cone-bearing trees. Gymnosperms have hard seeds that are uncovered.

An **angiosperm** is a seed plant that produces flowers. All angiosperms have seeds that are covered by some kind of fruit. Some angiosperms have

familiar fruits, like apples and plums. Other angiosperms, like grasses, have smaller and less-colorful fruits.

Angiosperms are the most plentiful of all plant types. There are about 250,000 different kinds of angiosperms. Some familiar angiosperm plants include tulips, maple trees, rose bushes, and corn.

 Quick Check

Draw Conclusions A plant is 20 meters (65 feet) tall and it does not produce flowers. What conclusions can you draw about this plant?

Critical Thinking How is height an advantage for some vascular plants?

seed

no flowers

Douglas fir

gingko

flowers

hydrangea

gerbera daisy

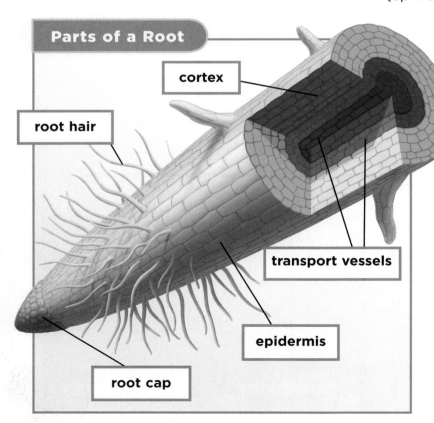

Quick Lab

Observe a Root

1. **Observe** Look at a carrot cut lengthwise. What structures do you see?

2. Look at a cross section of a carrot. Can you identify the epidermis, cortex, and inner transport layers?

3. Draw a diagram of the carrot in cross section. Label the parts.

4. **Infer** Is the carrot a fibrous root or a taproot?

5. Would it be easier to pull a plant with a taproot from the ground or a plant with a fibrous root system? Explain your answer.

What are roots?

Have you ever tried to dig up a plant? If so, then you might have hit a maze of stringy, cordlike roots. A *root* is the part of the plant that absorbs water and minerals, stores food, and anchors the plant.

Roots absorb water using fuzzy root hairs. A *root hair* is a threadlike projection from a plant root. Each root hair is less than 1 millimeter (0.04 inches) in length, but together they soak up moisture like a sponge.

A typical root of a vascular plant is made of three different layers and a root cap. The *root cap* covers the tip of the root. It protects the root tip while it pushes into the ground.

The outer layer of a root and the whole plant is the *epidermis* (ep•i•DUR•mis). The epidermis of a root has root hairs and absorbs water. A *cortex* layer is located just under the epidermis. It is used to store food and nutrients. The vascular system is located in the center of the root. The vascular system transports water and minerals absorbed by the root hairs.

Different plants have different kinds of roots. Some plants have specialized roots for their environment.

Parts of a Root

cortex

root hair

transport vessels

epidermis

root cap

Aerial (AYR•ee•uhl) *roots* are roots that never touch the ground. They anchor the plant to trees, rocks, or other surfaces. Aerial roots absorb water from the air and rain, rather than soil. Many orchids have aerial roots. Some live high up in the rain forest attached to tree bark.

Fibrous (FYE•bruhs) *roots* are thin, branching roots. They do not grow deep into the ground, but they often cover a very wide area. A single clump of grass was found to have some 600 kilometers (390 miles) of fibrous roots.

Plants with *taproots* have a single, main stalk-like root that plunges deep into the ground. Smaller side roots often branch off of a main taproot. Pine trees and plants that live in dry areas often have taproots.

Prop roots usually grow at the bottom of a plant's stem. They prop up and support the plant so it cannot be knocked over. Corn plants and mangrove trees have prop roots.

prop root

 Quick Check

Draw Conclusions An area has many plants with taproots. Where would you expect to find underground water in this area?

Critical Thinking Would you expect a desert plant or a swamp plant to have more root hairs?

taproot

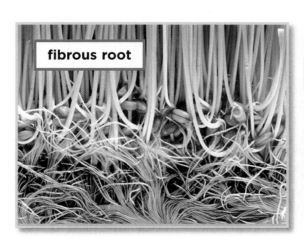

fibrous root

aerial root

What are stems?

Each part of a plant has a special function. The plant's stem has two functions. First, it is a support structure. The stem of a tree, for example, must support the weight of the entire tree. Smaller plant stems support less weight, but most stems must be sturdy enough to support leaves, flowers, and branches.

Stems come in two basic forms—soft stems and woody stems. Soft stems are not as strong as woody stems. They are soft, green, and can bend. Their green color shows that their cells have chlorophyll and produce food. Shrubs and trees have woody stems. Woody stems are often covered with *bark*, a tough outer covering that serves as a protective layer. Woody stems do not contain chlorophyll.

Soft and Woody Stems

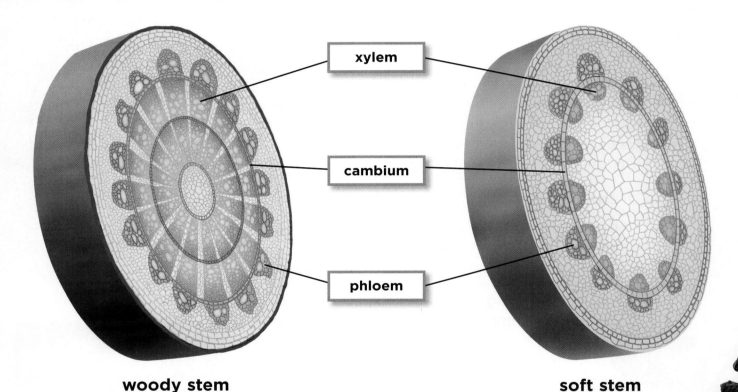

xylem

cambium

phloem

woody stem

soft stem

Read a Diagram

How are the xylem, phloem, and cambium arranged differently in the woody stem and the soft stem?

Clue: Compare the location and shape of the xylem, phloem, and cambium in each stem.

The stem's second function is to serve as a transport system for the plant. The transport system actually begins in the roots of the plant. Two kinds of tissue make up the system. **Xylem** (ZYE•luhm) is a series of tubes that moves water and minerals up the stem. Xylem tissue conducts, or transports, in one direction only—up from the plant roots to the leaves.

Phloem (FLOH•em) moves sugars that are made in the plant's leaves to other parts of the plant. Phloem tissue is a two-way transport route. It flows both up and down in a plant. In a carrot, for example, sugars are brought down from the leaves of the plant to the taproot through the phloem. Phloem also transports sugars up from one part of a plant to another.

The xylem and phloem layers in a plant stem are separated by a layer

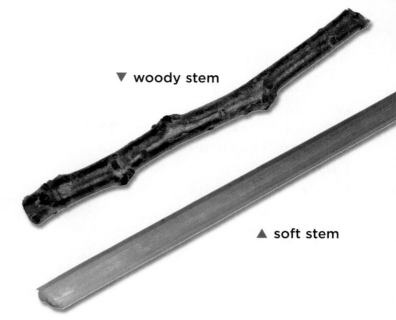

▼ woody stem

▲ soft stem

called the **cambium** (KAM•bee•uhm). Xylem and phloem cells are produced in the cambium, then move inward. When they are alive in the cambium layer, xylem cells are not able to transport water. It is only after the cells die and become hollow that they are able to function as transport vessels.

✔ **Quick Check**

Draw Conclusions In how many directions does xylem move water and minerals?

Critical Thinking Why do most trees have woody stems?

Tree rings are formed by layers of xylem and phloem.

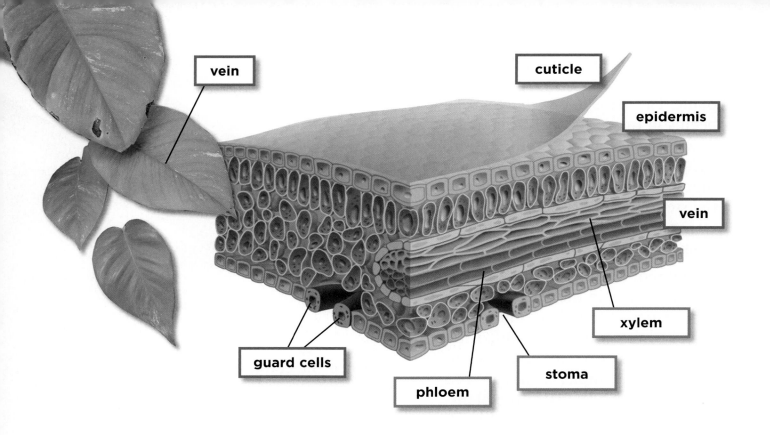

vein · **cuticle** · **epidermis** · **vein** · **xylem** · **stoma** · **phloem** · **guard cells**

What are leaves?

The leaves of a plant have the important function of carrying out **photosynthesis** (foh•tuh•SIN•thuh•sis), or the process of making food. Cells within the leaf's epidermis make up the plant's main food factory. To perform photosynthesis, leaves need three raw materials: sunlight, water, and carbon dioxide from air. Each leaf is designed to obtain these materials.

Many leaves are flat and broad. This allows the leaf to collect the most sunlight possible. Chlorophyll in the chloroplasts of the leaves traps the energy of sunlight.

Water enters the plant through its roots. It is transported up through the xylem tissue in the leaf veins. The top leaf surface also has a waxy *cuticle*, a waterproof layer that prevents moisture from evaporating.

Leaves get carbon dioxide from the air. Air enters and moves out of the plant through tiny pores on the underside of the leaves called *stomata* (STOH•muh•tuh). A single pore is called a *stoma*. Each stoma is controlled by guard cells. When the leaf has plenty of water, the guard cells swell and pull the stomata open. This allows water and air to leave the plant. The loss of water through a plant's leaves is called **transpiration** (tran•spuh•RAY•shuhn). When the plant is low on water, the guard cells shrink. This causes the stomata to close and prevents water from escaping.

As water evaporates from the leaves, more water is carried from the bottom of the plant to the top. Finally, water moves into the leaf, replacing the water that has evaporated.

transpiration

5 Some water evaporates through open stomata.

4 Sugar is then transported in the phloem tissue.

3 Water in the leaves is used to make sugar.

sugar

Now the plant has all the raw materials for performing photosynthesis. Carbon dioxide and water enter the chloroplasts in the plant's cells. They combine in the presence of the trapped energy from sunlight. This reaction produces sugar and oxygen. The sugars are transported to all the plant's cells by the phloem tissue. Excess sugar is stored as starch, which the plant can break down for food. Most of the oxygen leaves the plant through the stomata as a waste product.

Scientists express what happens during photosynthesis by a chemical equation. This equation shows that the materials of photosynthesis react together and what they produce.

2 Water moves through the xylem tissue up to the leaves.

$$6CO_2 + 6H_2O + energy \rightarrow C_6H_{12}O_6 + 6O_2$$

carbon dioxide	water		sugar	oxygen

✓ Quick Check

Draw Conclusions Suppose you did not water a plant for two weeks. What position would you expect the stomata of the plant to be in?

Critical Thinking Would you expect a rain forest tree or a desert cactus to have a thicker cuticle?

Read a Diagram

How does sugar produced in the leaves get to the roots?

Clue: Read the labels to find the answer.

1 Water enters the plant's roots.

sugar

water

How are photosynthesis and respiration related?

The most important source of energy for Earth is the Sun. Plants can use the Sun's energy to make fuel. The sugars produced during photosynthesis are used by most organisms for energy. The energy is released when the cells of organisms use oxygen to break down the sugars stored as starch in the process of **cellular respiration** (SEL•yuh•luhr res•puh•RAY•shuhn).

Cellular respiration occurs in the mitochondria of your cells. Here oxygen combines with stored sugars to release energy which your body uses to do work. Plants also perform cellular respiration. They take in oxygen from the air through their stomata and break down sugars for energy.

During cellular respiration, plant and animal cells produce carbon dioxide and water as waste products, which are then released back into the air. Plants use the released carbon dioxide and water to produce sugars during photosynthesis.

$$6O_2 + C_6H_{12}O_6 \rightarrow 6CO_2 + 6H_2O + energy$$

 Quick Check

Draw Conclusions Do plants produce carbon dioxide? Explain.

Critical Thinking Is it easier for your body to store food or oxygen?

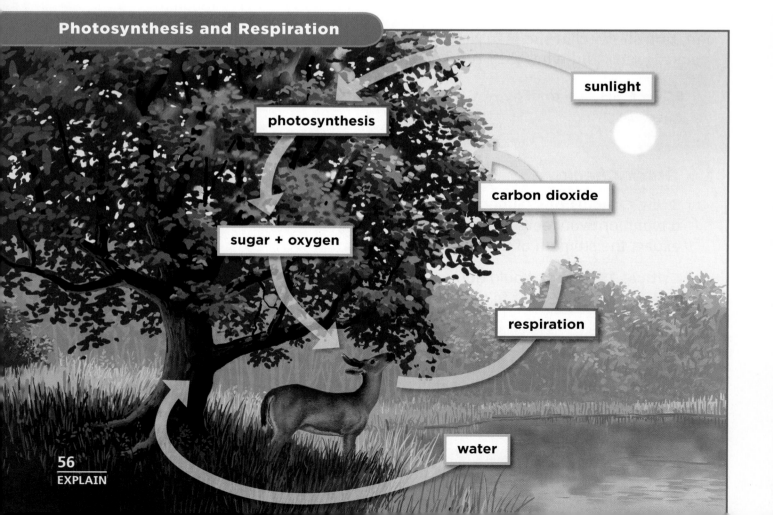

Photosynthesis and Respiration

sunlight

photosynthesis

carbon dioxide

sugar + oxygen

respiration

water

Lesson Review

Visual Summary

Plants are divided into **vascular** and **nonvascular** plants. Vascular plants are divided into seed plants and seedless plants.

Roots anchor a plant and supply it with water and minerals. **Stems** support a plant and transport materials.

Leaves carry out the process of **photosynthesis**. Organisms burn the sugars produced in photosynthesis during **cellular respiration**.

Make a FOLDABLES™ Study Guide

Make a Trifold Book. Use the titles shown. Tell what you learned about each title.

Main Idea	What I learned...	Sketches
Vascular and nonvascular plants		
Roots and stems		
Photosynthesis and cellular respiration		

Think, Talk, and Write

1. **Main Idea** Why do all plants need air, water, and sunlight?

2. **Vocabulary** Flowering plants are called _____.

3. **Draw Conclusions** An insect cannot survive in a covered jar, even though the jar contains food and water. When a plant is added to the jar, the insect can now survive. Explain.

Text Clues	Conclusions

4. **Critical Thinking** Animals depend on plants for food. Could plants make food without animals?

5. **Test Prep** Which kind of plant produces fruit?
   - **A** angiosperm
   - **C** seedless
   - **B** nonvascular
   - **D** gymnosperm

6. **Test Prep** Which of the following is found inside the stem of a plant?
   - **A** epidermis
   - **C** root hairs
   - **B** xylem
   - **D** leaves

Writing Link

Explanatory Writing
Aliens from another planet want to know how organisms on Earth obtain energy. Write a letter to the aliens explaining how organisms on Earth obtain energy.

Math Link

Energy Fractions
A plant uses $\frac{1}{10}$ of the Sun's energy it receives to make sugars. An animal that eats the plant uses about $\frac{1}{10}$ of the plant's stored energy. What fraction of the Sun's energy does the animal use?

Saving Water the Yucca Plant Way

Yucca plants grow in the deserts of California and the southwest parts of North America. They have long, narrow leaves that are adapted to save water. Yuccas use a special kind of photosynthesis called CAM photosynthesis.

Most plants open their stomata during the day. They need carbon dioxide for photosynthesis. Yucca plants only open their stomata at night. This keeps the yucca from losing water through evaporation in the hot desert sun. During the day, the yucca plant uses its stored carbon dioxide to perform photosynthesis. Desert plants that use CAM photosynthesis, like the yucca, lose much less water than other plants.

Explanatory Writing

A good explanation

▶ develops the main idea with facts and supporting details

▶ lists what happens in an organized and logical way

▶ uses time-order words to make the description clear

The yucca plant has long, thin leaves.

Write About It

Explanatory Writing Write an article for young gardeners. Explain the process of CAM photosynthesis. Add a diagram to help explain. Research facts and details for your article.

 LOG ON **e-Journal** Research and write about it online at **www.macmillanmh.com**

Math in Science

Leave It Be

Some leaves, like a tiny pine needle, have a very small surface area. Others, like a maple leaf, have a large surface area. The surface area of leaves is directly connected to the amount of sugar and oxygen they produce. Therefore, one can assume that a single pine needle does not produce as much sugar and oxygen as a maple leaf. How can you find the surface area of a leaf?

Calculating Area

To find the area of an irregular figure

▶ trace the figure on graph paper

▶ count the number of whole square units

▶ count the number of partial square units and divide this number by 2

▶ add the two numbers

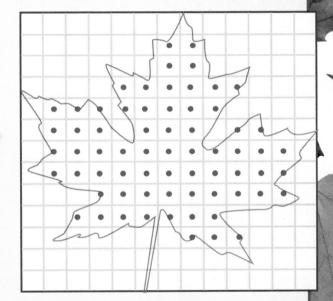

 Solve It

1. Find a leaf. Calculate the area of your leaf.
2. Compare the area of your leaf to the area of the maple leaf in the example above.
3. Which produces more sugar and oxygen?

whole squares	+	partial squares	÷ 2	= area
43	+	24	÷ 2	= area
43	+	12		= 55

Classifying Animals

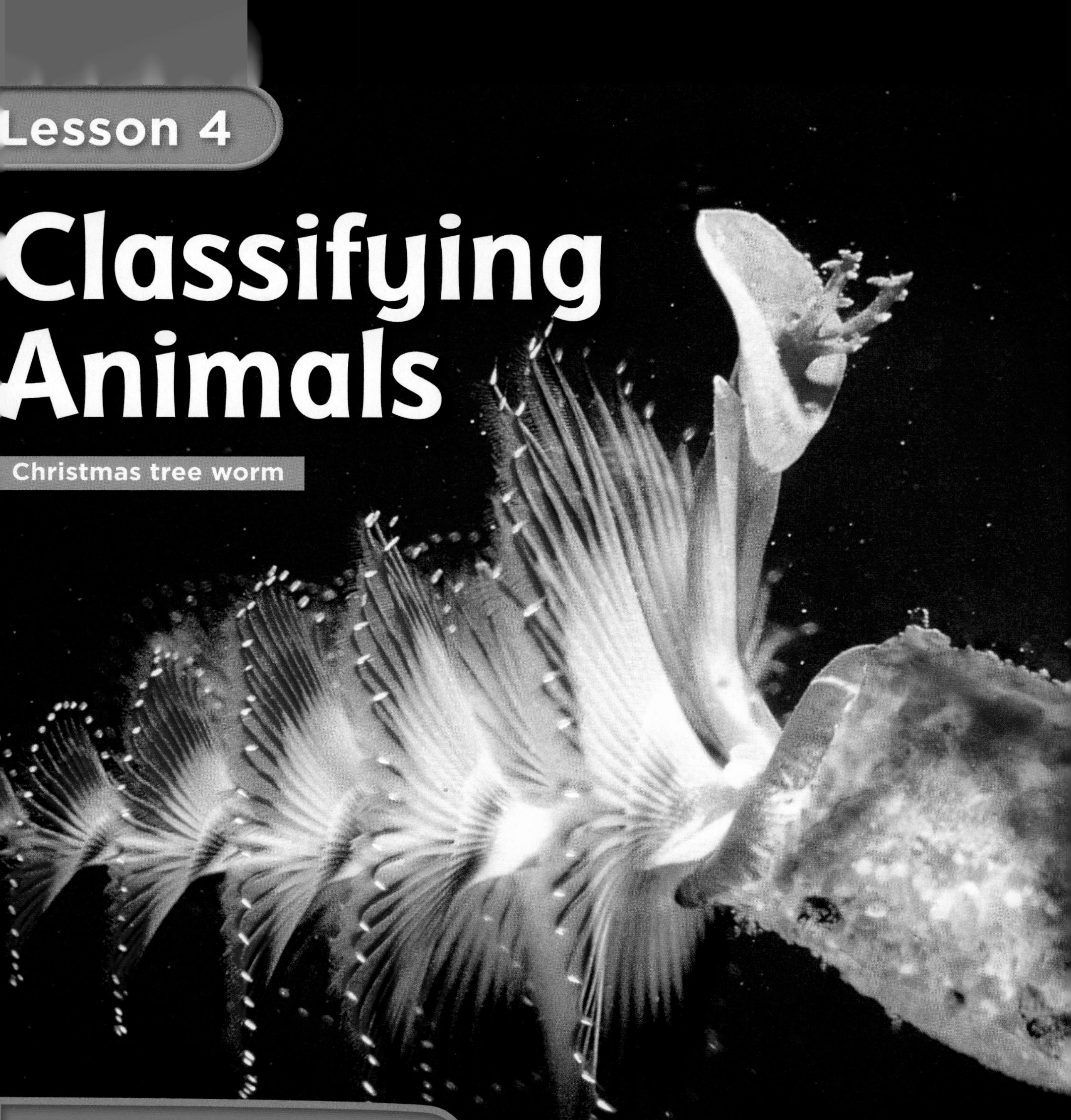

Christmas tree worm

Look and Wonder

These strange-looking organisms are actually animals. What characteristics do all animals have? How do you determine whether an organism is an animal?

How do you classify animals?

Purpose

A dichotomous key lists traits of organisms. It gives directions that lead you to an organism's identity.
Create a dichotomous key to identify the animals shown.

sponge

sea star

insect

Procedure

❶ **Observe** Look at the animals shown. What features do they have? How can you use these features to classify them?

❷ Make a key. The key should include a series of yes or no questions that can help you identify the animals.

❸ Start your key with a general question. Make each additional question eliminate at least one animal. For example, you might start by asking "Is this animal able to move by itself from place to place?" If the animal is a worm, you would answer "Yes," and move on. If the animal is a sponge, you would answer "No."

❹ Keep writing questions until you can single out one animal in your key.

❺ **Communicate** Exchange keys with a partner. Use their key to identify an animal.

worm

spider

Draw Conclusions

❻ **Infer** Could you use your key to identify other animals? Explain.

Dichotomous Key

1. Moves from place to place......Go to 2
   Does not move...............Sponge

2.

3.

4.

Explore More

How would you change your key to make it more useful? Which questions would you change? Which questions would you keep the same?

Read and Learn

▶ **Main Idea**
Form, structure, and behavior can all be used to divide animals into classification groups.

▶ **Vocabulary**
asymmetrical, p.62
radial symmetry, p.62
bilateral symmetry, p.63
monotreme, p.68
marsupial, p.68
placental mammal, p.68

LOG ON **e-Glossary**
at **www.macmillanmh.com**

▶ **Reading Skill** ✔
Main Idea and Details

Main Idea	Details

What are simple invertebrates?

You learned that invertebrates are animals that do not have a backbone. They can live on land or in the water. Most lower invertebrates live in aquatic environments, which are filled with water or are moist. The lower invertebrates include sponges, *cnidarians* (nye•DAYR•ee•uhnz), and worms.

Sponges

Sponges are animals that have no true organization. Sponges are the only animals without real tissues or organs. They are also asymmetrical (ay•si•MET•ri cuhl). An **asymmetrical** body plan cannot be divided into mirror images.

The sponge body structure is arranged around a single tunnel-like canal. The tissue surrounding the canal is filled with lots of tiny pores, or holes. The word *pore* gives sponges their phylum name—*Porifera* (paw•RIF•uhr•uh). All members of the phylum Porifera live in water.

Cnidarians

Jellyfish, sea anemones, corals, and hydras are all cnidarians. Cnidarians are soft-bodied, aquatic creatures. Unlike sponges, they have radial symmetry (RAY•dee•uhl SIM•uh•tree). **Radial symmetry** is a body plan in which all body parts of an organism are arranged around a central point. An organism with radial symmetry has more than one line that divides the organism into two mirror images.

sponge

sea anemone

segmented worm

flatworm

roundworm

Cnidarians have a mouth, tentacles, muscle tissues, and stinger cells. When they hunt, their stingers shoot out like tiny harpoons. The poison inside these cells helps them capture other animals. Once stung, the dart-like stinger cells hold the victim while tentacles move it toward the animal's mouth.

Worms

There are three main worm groups: flatworms, roundworms, and segmented worms. All worms have bilateral symmetry (bye•LAT•uhr•uhl). **Bilateral symmetry** is a body plan in which an organism can be divided along only one plane of their body to produce two mirror images.

Flatworms, or *platyhelminthes* (plat•uh•HEL•minths), have a flat body and a head with simple eyes and a mouth. With only one body opening, undigested food must leave the flatworm's body through its mouth.

Roundworms, or *nematodes* (NEM•uh•tohdz), have simple digestive and nervous systems. Roundworms are some of the most abundant animals on Earth. Like flatworms, they often live inside the bodies of other animals.

Segmented worms, or *annelids* (AN•uh•lidz), have a body plan that is divided into sections, or segments. They have a two-way digestive system and organs, including a stomach, heart, and brain.

 Quick Check

Main Idea and Details What are some lower invertebrates?

Critical Thinking Which characteristics do scientists use to classify cnidarians?

What are complex invertebrates?

Invertebrates include a very diverse group of animals that live in many different environments. Some of these animals have specialized organs and complex body structures. Mollusks, echinoderms, and arthropods are phyla with very different characteristics.

Mollusks

All mollusks share the same body plan. They have a muscular foot or tentacles, a fold of tissue called the *mantle*, and a mass of internal organs. They all have bilateral symmetry. Mollusks include snails, clams, and squids. Almost all mollusks have a shell, which is secreted by the mantle.

Mollusks have several specialized organs, including a heart, gills for breathing, and a well-developed nervous system. The squid and octopus have extremely good eyesight and a very large brain.

Echinoderms

Echinoderms include sea stars, sea urchins, and sea cucumbers. Echinoderms have a hardened

blue sea star

skeleton located inside the body called an *endoskeleton*. Spines from the endoskeleton often poke through their thin, bumpy skin. Echinoderms usually have radial symmetry.

Echinoderms use a water pressure system that helps them feed, breathe, and move. Seawater enters the system and moves to different parts of the animal's body under pressure. The system ends in the echinoderm's many tube feet, which cling to surfaces like small suction cups.

garden snail

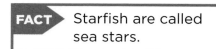

FACT ▶ Starfish are called sea stars.

Arthropods

Arthropods are the most numerous animal group on Earth. More than half of the world's animal species are arthropods, including spiders, crabs, and insects. Arthropods have a very efficient body plan. They are small and light with a hard skeleton on the outside of the body called an *exoskeleton*. It provides both strength and protection. All arthropods have bilateral symmetry.

Arthropods have a segmented body, with paired limbs on either side of their body. In some cases, these limbs are used as wings or claws. In other cases, their many legs help arthropods move quickly. Arthropods have simple but very efficient nervous systems and good sensory systems. If you have ever tried to swat a fly, you know just how quickly an arthropod can move!

✔ Quick Check

Main Idea and Details What body part do arthropods have that echinoderms do not have?

Critical Thinking Is the statement "the smaller the animal, the simpler the animal" generally true? Explain.

Read a Photo

What do arthropods have in common?

Clue: Look at the basic body plan of both arthropods. What features do they share?

≡Quick Lab

Observe Insects

1. **Observe** Look at each insect with a hand lens. Draw what you see.

2. Compare the structures of each insect. Create a table to record your observations.

3. **Infer** Choose any two structures that you observed. What is special about these structures? How does the design of these structures help the insect survive?

What are vertebrates?

Recall that vertebrates are animals that have a backbone. They all have bilateral symmetry and endoskeletons.

Fish

There are three classes of fish: jawless fish, cartilaginous fish, and bony fish. Lamprey and hagfish are jawless fish. Instead of a true backbone, they have a flexible nerve cord. Without jaws, they are forced to suck in their food.

Sharks, skates, and rays are cartilaginous fish. They have skeletons made of cartilage rather than bone. *Cartilage* is the material that gives your ear its stiffness. They have paired fins and jaws.

Bony fish have a nerve cord covered by bone, not cartilage. Like sharks, they have jaws and paired fins. They have balloonlike swim bladders that allow them to easily go up or down in water. Bony fish also have moving flaps that push water into their gills. This allows them to get oxygen while not moving in the water.

tree frog

Amphibians

Frogs, salamanders, and other amphibians bridge the gap between land and water vertebrates. An amphibian is an animal that spends part of its life in the water and part of its life on land.

A frog begins its life in the water as a tadpole with gills. As it matures, the frog develops four legs and lungs for breathing air. Most adult amphibians do not leave the water for too long. Many frogs, for example, breathe through their skin as well as their

white-tip reef shark

lungs, so they need to stay moist at all times. They also return to the water to lay their eggs.

Reptiles

There are about 7,000 different kinds of reptiles. A reptile is a true land animal with one lung or two. They have thick, scaly, waterproof skin. Reptiles include lizards, snakes, turtles, alligators, and crocodiles.

Reptiles do not generate much body heat. Instead, reptiles stay warm by sunning themselves and taking advantage of heat in their environments. They are cold-blooded animals. This means they cannot automatically keep their body temperature steady.

Birds

Birds have several features that make them different from other vertebrates. Rather than four legs, birds have two legs and two wings. Birds have hollow bones to reduce their weight.

Feathers are a unique characteristic of birds. They are strong and incredibly light. Feathers help keep heat inside the bodies of birds. Birds are warm-blood animals that can keep their body temperatures constant. Their feathers are sturdy enough to hold up to the stresses and strains of flight. The world's fastest bird, the peregrine falcon, can fly at speeds of more than 322 kilometers per hour (200 miles per hour).

 Quick Check

Main Idea and Details What are characteristics of vertebrates?

Critical Thinking Why do many reptiles hide at night and stay completely inactive?

Birds and Reptiles

hummingbird

gila monster

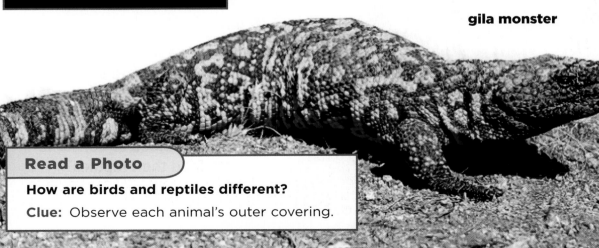

Read a Photo

How are birds and reptiles different?

Clue: Observe each animal's outer covering.

What are mammals?

Milk, hair, and large brains are the key characteristics of mammals. Mammals are unique because they produce milk to feed their young. Most mammals have hair or fur. Some mammals, like humans and whales, have little body hair. Like birds, mammals are warm-blooded. They generate their own body heat by burning food.

Mammals are divided into three subclasses: monotremes (MON•uh•treemz), marsupials (mahr•SEW•pee•uhlz), and placental mammals (pluh•SEN•tuhl). A **monotreme** is a mammal that lays eggs. After the young hatch, they are fed milk from their mothers. The duck-billed platypus and the spiny anteater are members of this group.

A **marsupial** is a pouched mammal. These mammals give birth to partially developed offspring. They carry their developing young in a pouch on the front of their bodies. Kangaroos and koala bears are marsupials. Most monotremes and marsupials are found in Australia.

Koala bears are marsupials.

A spiny anteater is a monotreme.

What do humans have in common with dogs, tigers, elephants, and whales? All are placental mammals. The young of a **placental mammal** develops within its mother. Placental mammals are born more mature than the offspring of marsupials.

A tiger is a placental mammal.

✓ Quick Check

Main Idea and Details What characteristics are shared by all mammals?

Critical Thinking Why do mammals need to consume more food than reptiles?

Lesson Review

Visual Summary

Invertebrates do not have a backbone. Porifera, cnidaria, and worms are **simple invertebrates**.

Mollusks, echinoderms, and arthropods are **complex invertebrates**.

All **vertebrates** have backbones. Fish, amphibians, reptiles, birds, and mammals are vertebrates.

Make a FOLDABLES™ Study Guide

Make a Three-Tab Book. Use the titles shown. Tell what you learned about each topic on the tabs.

Simple invertebrates

Complex invertebrates

Vertebrates

Think, Talk, and Write

1 **Main Idea** What is the main difference between vertebrates and invertebrates?

2 **Vocabulary** Snails, squid, and clams are all _____.

3 **Main Idea and Details** What characteristics identify birds?

Main Idea	Details

4 **Critical Thinking** Some scientists think dinosaurs were warm-blooded. How would this make dinosaurs more likely to be related to birds?

5 **Test Prep** Which animal group has the least organized body plan?
  A flatworms
  B mollusks
  C porifera
  D reptiles

6 **Test Prep** Which of the following does not have bilateral symmetry?
  A worms
  B fish
  C mammals
  D sponges

Math Link

Number of Beetles
Suppose about 350,000 different beetles were identified. If 1,900,000 animals were identified, what percent are beetles?

Social Studies Link

Australian Mammals
Research marsupials and monotremes. Find out why these animals are found mostly in Australia.

The Underground Life of Mole Rats

Some animals are cute, like pandas or koala bears. But this animal—the naked mole rat—is considered by many people to be quite ugly. It reminds most people of a pink, wrinkled, sausage with feet!

Mole rats spend most of their lives digging a maze of underground tunnels. The tunnels bring the mole rats closer to the plant roots they eat. The tunnels also offer protection from heat and other animals. A colony of mole rats can dig tunnels that would be several kilometers long if they were laid out in a straight line.

Scientists call this animal *Heterocephalus glaber*. Its scientific name is important because this animal is neither a mole nor a rat. It is actually related to porcupines, chinchillas, and guinea pigs. You would hardly recognize any family likenesses if you looked at these animals. That is why scientific names are important. You can learn a lot about an organism when you know its scientific name.

chinchilla

Descriptive Writing

A good description

▶ uses sensory words to describe how something looks, sounds, smells, tastes, and/or feels

▶ includes vivid details to help the reader experience what is being described

Write About It

Descriptive Writing Find out the scientific name of an animal you think is cute or ugly. Write a description of the animal. Use words and details that appeal to the senses in your description.

 LOG ON **e-Journal** Research and write about it online at **www.macmillanmh.com**

Math in Science

Animal Symmetry

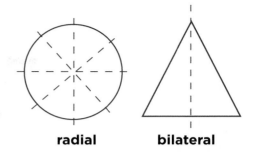

radial   bilateral

As you learned, most animals have symmetry. This means they have at least two sides that are mirror images. Animals can have bilateral symmetry or radial symmetry. A dog has bilateral symmetry because there is only one line of symmetry that divides it into mirror images. A sand dollar has radial symmetry because there is more than one line of symmetry that divides it into mirror images. You can use graph paper to determine which kind of symmetry a figure or an object has.

Line Symmetry

To find the symmetry of a figure

▶ trace the figure onto graph paper

▶ fold the graph paper in half to find the line of symmetry

▶ if the two halves of the figure match exactly, draw the line of symmetry

 Solve It

1. Look at the picture of the butterfly. What kind of symmetry does it have?

2. Draw a figure with at least two lines of symmetry. Show all lines of symmetry.

71
EXTEND

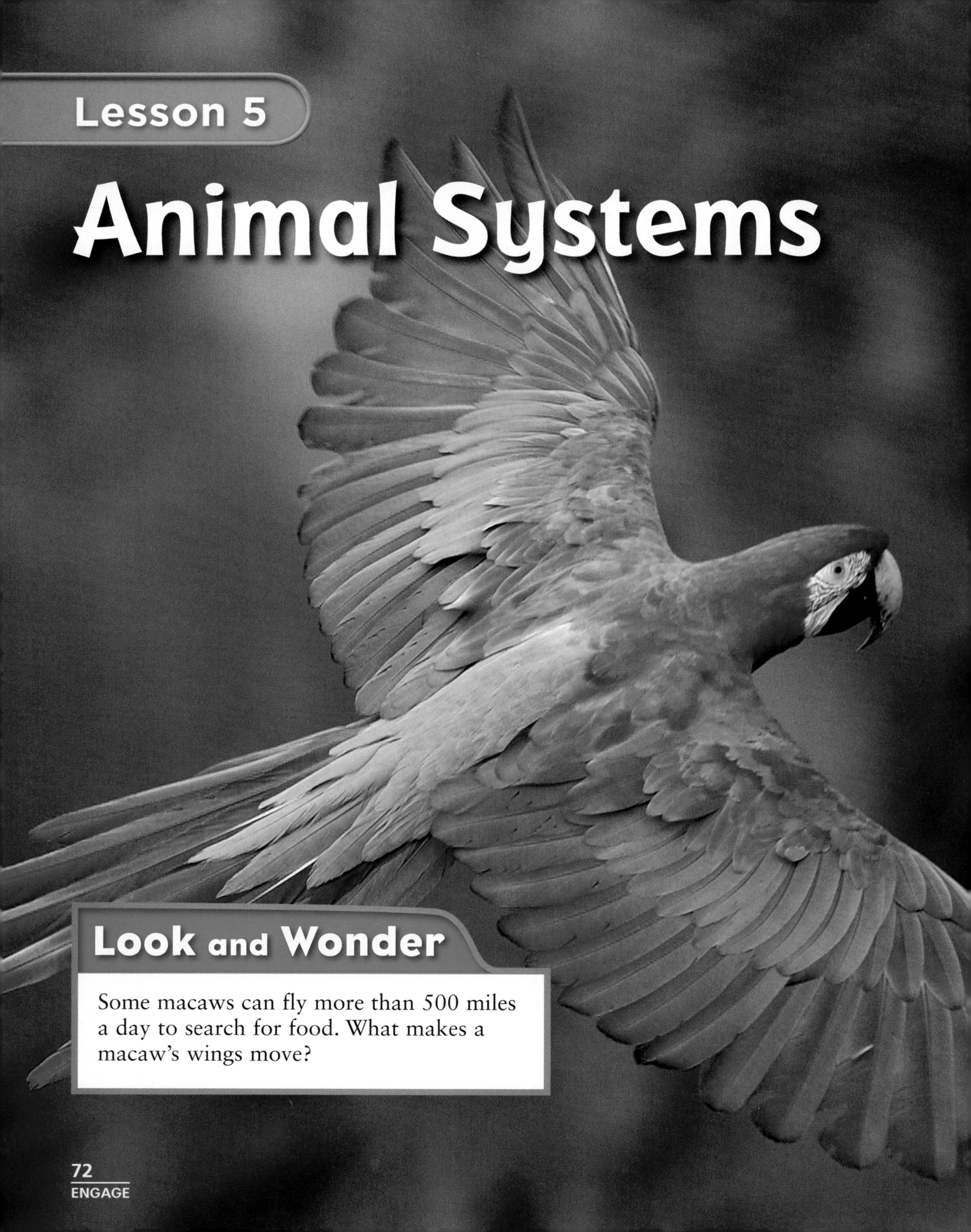

Lesson 5

Animal Systems

Look and Wonder

Some macaws can fly more than 500 miles a day to search for food. What makes a macaw's wings move?

How does a muscle work?

Make a Prediction

How do your muscles make you move? What happens if you shorten a muscle that pulls on a bone? Make a prediction.

Test Your Prediction

1 **Make a Model** Cut a small slit in the straw so it will bend in one direction.

2 Attach modeling clay to both ends of the straw. Use a larger piece of clay on one end.

3 Push paper clips into the clay so they stick up as shown. Tie the string to the paper clip in the smaller piece of clay.

4 Pull the string through the paper clip on the larger piece of clay.

5 **Experiment** Pull on the string to model how a muscle works. What happens when the muscle shortens? What happens when the muscle returns to its original length?

Draw Conclusions

6 Which parts of your model represent bones? Which part represents muscle?

7 **Infer** Which muscle in your body is similar to this model? Explain.

8 How do muscles work? What happens when muscles shorten and lengthen? Explain.

Explore More

What would happen if you did not cut a slit in the straw? Make a prediction and plan an experiment to test it.

Materials

- **straw**
- **scissors**
- **modeling clay**
- **paper clips**
- **string**

Step 3

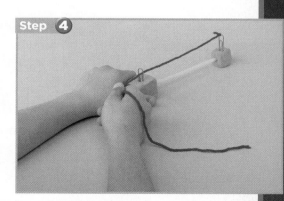
Step 4

► Main Idea

Body systems work together to allow the body to get energy, move, and respond to the outside world.

► Vocabulary

skeletal system, p. 74
muscular system, p. 74
digestive system, p. 76
excretory system, p. 77
respiratory system, p. 78
circulatory system, p. 78
nervous system, p. 80
endocrine system, p. 80

LOG ON **ⓔ-Glossary**
at **www.macmillanmh.com**

► Reading Skill ✓

Summarize

► Technology

Explore the body systems with Team Earth.

SCIENCE QUEST

What are the skeletal and muscular systems?

The **skeletal system** (SKEL•uh•tuhl SIS•tuhm) is made up of bones, tendons, and ligaments. The bones in the skeletal system have two main functions. First, bones provide protection for some soft body organs. For example, the rib cage protects the heart and lungs. Bones are living tissues that are both light and strong. The skull, for example, is hard enough to shield the delicate brain from injury, but light enough to allow an organism to hold its head up.

The second function of the skeletal system is to provide a solid framework for the body. Bones give the body structure and allow it to move. Bones move freely, but they do not have any power to move on their own. The power to actually produce movement is provided by the **muscular system** (MUS•kyuh•luhr).

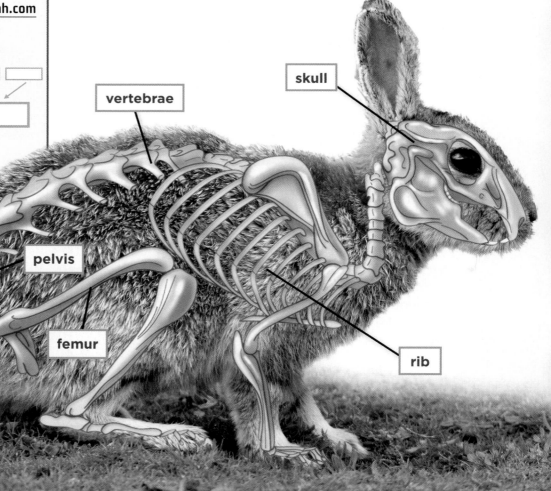

Skeletal muscles are attached to bone by tough, rubbery tendons. When the muscle shortens, the bone moves. Muscles that cause movement work in pairs or opposing groups. When a rabbit runs, one muscle group pulls its leg up. The opposing muscle pulls the leg down.

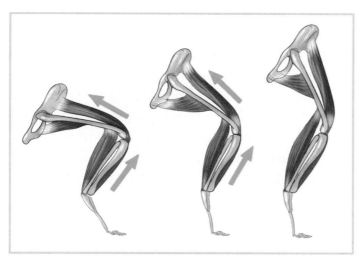

When a muscle contracts it pulls on the bone it's attached to. At the same time another muscle in the leg relaxes, allowing the bone to move.

As a rabbit runs, an electrical nerve signal is sent from the brain to the muscles in its legs. This causes the muscles to shorten, or contract. The contracted muscle pulls on the tendon, which moves the leg bone up. Keep in mind that muscles can only pull; they never push. As it runs, the pair of muscles in the rabbit's legs contract and relax. When the muscle that pulls up contracts, the muscle that pulls down relaxes. This continues over and over again as long as the rabbit moves.

cicada shedding its exoskeleton

Exoskeleton

The skeletal system of arthropods is found on the outside of their bodies. The exoskeleton is a tough, rigid structure connected by flexible joints. Like the vertebrate skeletal system, the exoskeleton provides protection and support and helps with movement. For an arthropod, such as a beetle, to grow it needs to shed its exoskeleton and create a larger one.

 Quick Check

Summarize Tell what happens to the rabbit's muscles as it runs.

Critical Thinking Muscles that move your fingers are in your arm. How are your fingers able to move?

What are the digestive and excretory systems?

The **digestive system** (di•JES•tiv) is a long tube in which food is broken down into nutrients an organism can use. When a rabbit eats a carrot, its teeth grind the carrot. At the same time, the rabbit's saliva starts the chemical breakdown of the carrot.

From the mouth, the chewed pieces of carrot travel through a muscular tube called the *esophagus* (ee•SAH•fuh•guhs). The smooth muscles in the esophagus contract and expand to squeeze food down to the stomach. The stomach holds partially digested food. Strong acids in the stomach help break down food, like the carrot.

The stomach empties its partially digested contents into the rabbit's coiled small intestine. Digestive juices produced by the liver and pancreas combine with the food in the small intestine. These chemicals break down sugars, proteins, and fats into nutrients. Nutrients are then absorbed in the small intestine.

Tiny folds in the small intestine touch tiny blood vessels that deliver the nutrients to the blood. During digestion, there is always some undigested food. The large intestine reabsorbs water from the undigested food. The remaining solid waste is processed in the large intestine and leaves through the second opening in the digestive system.

Digestive and Excretory Systems

Read a Diagram

Near which digestive organ would you expect to find a lot of blood vessels?

Clue: Where does absorption take place?

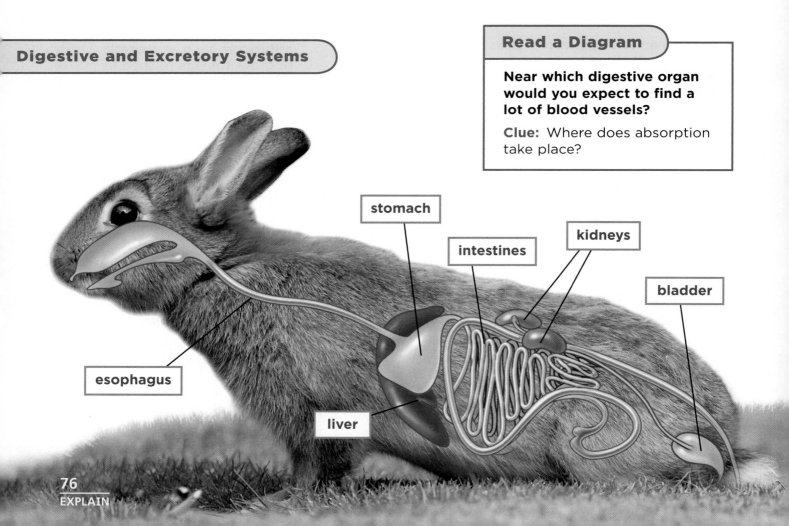

stomach

intestines

kidneys

bladder

esophagus

liver

Undigested food is not the only waste that the rabbit's body needs to remove. The **excretory system** (EKS•kruh•tor•ee) removes waste products from the body. The rabbit removes excess water, salt, and other waste through its skin in the form of sweat. Its cells also create wastes, including carbon dioxide and nitrogen-containing wastes. Carbon dioxide is removed through the lungs. Nitrogen wastes are sent to the blood. The waste is carried in the blood from the liver to the kidneys. The liver breaks down toxins in the blood. A *toxin* is a poisonous substance. The kidneys then filter wastes from the blood.

The rabbit's kidneys remove waste from the blood using millions of tiny filters called *nephrons* (NEPH•ruhnz). Nephrons separate waste from useful materials in the blood. Nephrons have membranes with very small openings that allow some substances to pass through but not others. They send any useful substances back into the blood and collect the wastes.

The kidneys produce urine from excess water and the collected wastes from the nephrons. Urine flows from the kidneys to the bladder. Once the bladder is full, the *urethra* (yew•REE•thruh) carries urine from the bladder to the outside of the body.

 Quick Check

Summarize What key steps take place in the process of digestion?

Critical Thinking Why are nephrons surrounded by many tiny blood vessels?

Invertebrate Digestion

Some animals, such as flatworms and jellyfish, have a digestive system with only a single opening. Food must enter and leave the system through the same opening. A jellyfish captures prey with its stinging cells. Then it brings the food to its mouth with its tentacles. Once the food is eaten, digestion continues. All undigested waste leaves the jellyfish's body through its mouth.

What are the respiratory and circulatory systems?

The cells of all animals are living things that require oxygen and food. The respiratory (RES•pur•uh•tor•ee) and circulatory (SIRK•kyuh•luh•tor•ee) systems are two organ systems that work together to provide oxygen and food to cells. The **respiratory system** of a rabbit is made up of the lungs and the passageways that lead to them. Many other structures are used for gas exchange in animals, such as the gills of fish. The **circulatory system** consists of the heart and blood vessels.

When the rabbit breathes in, air enters its body through its nose and mouth. The air travels through a series of tubes. The tubes inside the lungs branch out into smaller tubes called *bronchi* (BRON•kye). Bronchi have branches like a tree. These branches empty into very thin-walled air sacs called *alveoli* (al•vee•OH•lye). The walls of the alveoli are so thin that gasses like oxygen and carbon dioxide can pass through them.

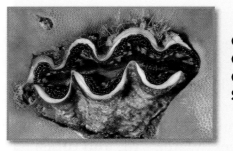

Open and Closed Circulatory Systems

All vertebrates and some other animals have closed circulatory systems. In a closed circulatory system, the animal's blood is carried in blood vessels. The blood vessels have thin walls that allow gas exchange to occur.
Arthropods, most mollusks, and many other invertebrates have open circulatory systems. In an open circulatory system the animal's blood moves around the body cavity. It is not closed in tubes. The blood bathes the cells directly and substances can be exchanged.

right atrium

left atrium

right ventricle

left ventricle

Heart

alveoli

bronchi

Lungs

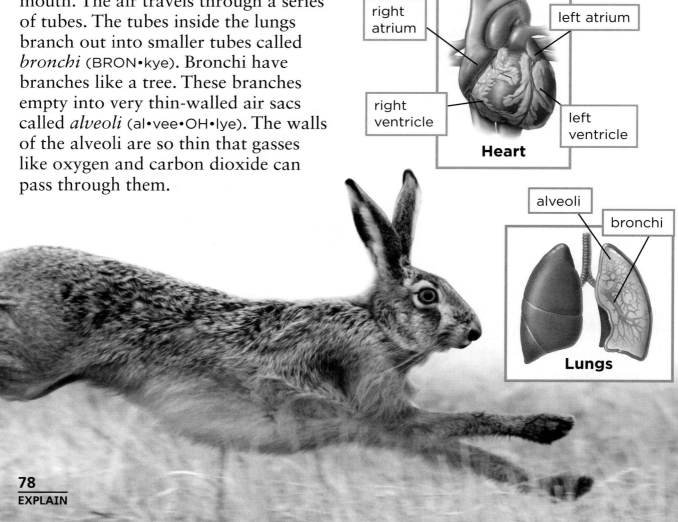

Circulation and Respiration

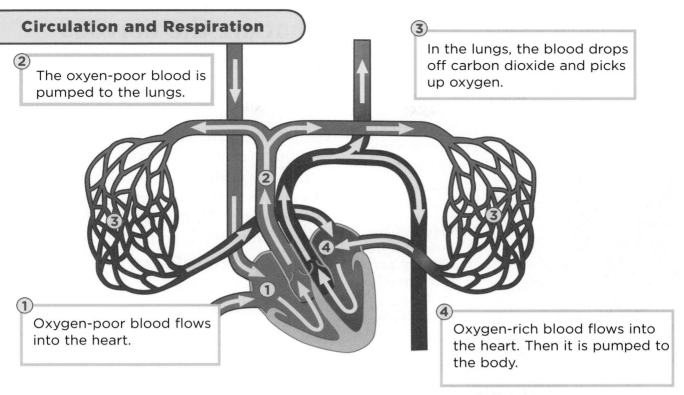

② The oxyen-poor blood is pumped to the lungs.

③ In the lungs, the blood drops off carbon dioxide and picks up oxygen.

① Oxygen-poor blood flows into the heart.

④ Oxygen-rich blood flows into the heart. Then it is pumped to the body.

Circulation begins as oxygen-poor blood gets pumped by the rabbit's heart to its lungs. In the lungs, blood carrying carbon dioxide picks up oxygen from the air in the alveoli. At the same time the blood releases carbon dioxide. Carbon dioxide is a waste product of cellular respiration. Carbon dioxide leaves the rabbit's body through its respiratory system when the rabbit breathes out.

The oxygen-rich blood flows back to the heart. The heart pumps the blood around the body. When it reaches the small intestine it picks up nutrients. Now this nutrient-rich and oxygen-rich blood can be delivered to the rabbit's body cells. The blood moves through the body, traveling through blood vessels. Finally, the blood reaches tiny vessels called *capillaries* (KAP•uh•layr•eez). Dissolved nutrients and oxygen can pass through their thin walls and enter the cells.

Read a Diagram

Where is the oxygen-poor blood pumped?

Clue: Red represents oxygen-rich blood. Blue shows oxygen-poor blood.

 Science in Motion Watch how the circulatory and respiratory systems work together at **www.macmillanmh.com**

Cells pass their waste materials back through the capillaries and into the blood. This oxygen-poor blood gets pumped back to the heart and the process continues.

✓ Quick Check

Summarize What steps does the circulatory system take to deliver blood to the body?

Critical Thinking Is the respiratory system part of the excretory system? Explain.

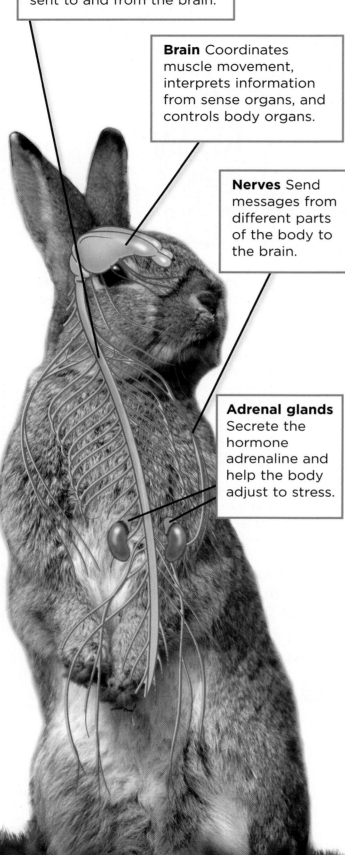

Spinal cord Passes signals sent to and from the brain.

Brain Coordinates muscle movement, interprets information from sense organs, and controls body organs.

Nerves Send messages from different parts of the body to the brain.

Adrenal glands Secrete the hormone adrenaline and help the body adjust to stress.

What are the nervous and endocrine systems?

The vertebrate **nervous system** (NUR•vuhs) includes the brain, nerve cord, nerves, and sense organs. The nervous system works with the **endocrine system** (EN•duh•krin), which has glands that produce hormones (HOR•mohnz). *Hormones* are chemicals released into the bloodstream that change body activity.

Suppose a rabbit sees a fox running toward it. Its response begins when the light reflecting off the fox hits the rabbit's eyes. Nerve cells in the rabbit's eyes send an electrical nerve message to the rabbit's brain. Its brain then sends a message back in response. The message travels through nerves to the *spinal cord*. At the spinal cord, nerve cells send a message to the rabbit's leg muscles telling them to start running.

At the same time, the rabbit's endocrine system sends out *adrenaline* (uh•DREN•uh•lin), a hormone that increases heart rate and sends extra blood to the muscles. As the rabbit's heart pounds, it gets ready to fight or run. Adrenaline has sped up the rabbit's body so it's ready for anything.

 Quick Check

Summarize What steps take place in a rabbit's nervous system when a rabbit sees a fox?

Critical Thinking What would happen if it took a minute for a message to travel from a rabbit's brain to its leg muscles?

Lesson Review

Visual Summary

The **skeletal** and **muscular systems** work together to enable the body to move.

The **digestive system** breaks down food so it can be absorbed into the bloodstream. The **excretory system** removes waste from the blood.

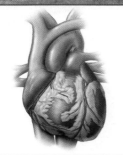

The **circulatory** and **respiratory systems** transport food and oxygen. The **nervous system** and **endocrine system** help an organism to respond to the outside world.

Make a FOLDABLES™ Study Guide

Make a Trifold Book. Use the titles shown. Summarize what you learned about each topic.

| Skeletal and muscular systems | Digestive and excretory systems | Circulatory, respiratory, nervous, and endocrine systems |

Think, Talk, and Write

1. **Main Idea** How are the circulatory, respiratory, skeletal, muscular, and nervous systems put into use when a rabbit runs from danger?

2. **Vocabulary** Hormones to change body activity are sent out by the _____.

3. **Summarize** What steps take place in the rabbit's body to bring blood to the body cells?

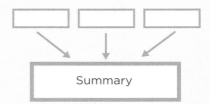

Summary

4. **Critical Thinking** Carbon dioxide is considered a waste product in animals. How do plants use carbon dioxide?

5. **Test Prep** Which organ systems get rid of wastes?
   - **A** circulatory and muscular systems
   - **B** digestive and excretory systems
   - **C** excretory and nervous systems
   - **D** endocrine and nervous systems

6. **Test Prep** Which of the following has an exoskeleton?
   - **A** rabbit
   - **B** dog
   - **C** grasshopper
   - **D** jellyfish

Writing Link

Descriptive Poem
Write a poem about one of the systems you studied in this lesson. Your poem may rhyme or not—it's up to you!

Math Link

Number of Beats
Suppose your heart pumps an average of 80 times per minute. On average, how many times will it pump in a day?

Be a Scientist

Materials

stopwatch

graph paper

When does your heart work the hardest?

Form a Hypothesis

Driven by the heart, your blood cells travel around your body carrying oxygen to your body cells. When you exercise, your body requires more oxygen. What happens to your heart rate when you exercise? Write your answer as a hypothesis in the form "If the body requires more oxygen, then . . ."

Test Your Hypothesis

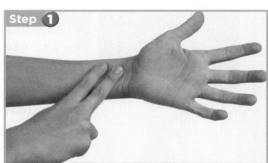

Step **1**

1 **Experiment** Take your pulse when you are resting. Press lightly on the skin of your wrist until you feel a beat. Then count how many beats you feel in 30 seconds. Record your pulse in a chart.

2 Walk in place for 1 minute. Without taking time to rest, take your pulse for 30 seconds. Record your data.

3 Jog in place for 1 minute. Without taking time to rest, take your pulse for 30 seconds. Record your data.

4 Rest for 5 minutes then repeat steps 1–3 four more times.

5 **Use Numbers** Use the data you collected to make a graph of your heart rate when resting, walking, and jogging.

Step **2**

Draw Conclusions

6 **Interpret Data** Did your pulse change as your activity increased?

7 Did the experiment support your hypothesis? Explain.

8 What were the independent and dependent variables?

Guided Inquiry

When do your lungs work the hardest?

Form a Hypothesis

You have already tested the effects of exercise on your heart. Do you think exercise affects your lungs? What happens to your breathing rate when you exercise? Write your answer as a hypothesis in the form "If the body requires more oxygen, then. . ."

Test Your Hypothesis

Design a plan to test your hypothesis. Then write out the materials, resources, and steps you need. Record your results and observations as you follow your plan and conduct your experiment.

Draw Conclusions

Did your experiment support your hypothesis? Why or why not? Present your results to your classmates.

Open Inquiry

How are your heart rate and your breathing rate affected by other activities? Would your heart or breathing rates increase if you read a book, play a sport, or try to solve a math problem? Determine the steps you would follow to answer your question. Record and document the resources you would use during your investigation.

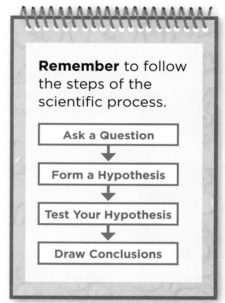

Remember to follow the steps of the scientific process.

Ask a Question
↓
Form a Hypothesis
↓
Test Your Hypothesis
↓
Draw Conclusions

Visual Summary

Lesson 1 Living things are all made of the same basic building blocks—cells.

Lesson 2 All organisms can be classified into six separate kingdoms.

Lesson 3 Plants perform photosynthesis, which provides food and energy for most organisms.

Lesson 4 Form, structure, and behavior can all be used to divide animals into classification groups.

Lesson 5 Body systems work together to allow the body to get energy, move, and respond to the outside world.

Make a FOLDABLES™ Study Guide

Assemble your lesson study guide as shown. Use your Lesson 5 study guide to review what you have learned in this chapter.

Fill in each blank with the best term from the list.

bilateral symmetry, p. 63

cell, p. 22

classification, p. 34

kingdom, p. 34

muscular system, p. 74

organism, p. 22

photosynthesis, p. 54

respiratory system, p. 78

1. Any living thing is a(n) _____.

2. The science of finding patterns is called _____.

3. The body gets the power to produce movement from the _____.

4. An animal's lungs and the passageways that lead to them make up its _____.

5. The smallest unit of living things that can carry out the basic processes of life is a(n) _____.

6. The leaves of a plant carry out _____ to make food for the plant.

7. An animal that can be divided into two halves that are alike has _____.

8. The broadest group into which organisms are classified is a(n) _____.

LOG ON **e-Review** Summaries and quizzes online at **www.macmillanmh.com**

Answer each of the following in complete sentences.

9. Draw Conclusions You are observing a cell under a high-power microscope. The cell has a stiff cell wall and a large, central vacuole. What can you conclude about this cell? Explain your answer.

10. Classify To which kingdom and phylum does the organism below belong?

11. Experiment You want to find out which of two types of fungi grows fastest. Describe a simple experiment you could conduct to find the answer.

12. Critical Thinking Do you think a lizard could survive in Antarctica? Explain why or why not.

13. Descriptive Writing Describe the two types of plant stems.

14. How are living things similar?

Name That Animal!

What to Do

1. Visit a public space where lots of different types of animals might be observed, such as a park or zoo.

2. Make a list of the different animals you see on your trip. Include at least five different animals.

3. When you return, use reference materials to look up the scientific name of each animal, which also would explain its genus and species. Add these names to your list.

Example: dog = *Canis familiaris*
(genus) (species)

Analyze Your Results

▶ Were any of the animals you saw the same genus? Why do you think this was or was not so?

1. Which body system are the organs in the diagram below a part of?

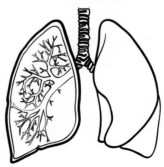

A circulatory system

B digestive system

C nervous system

D respiratory system

CHAPTER 2

Parents and Offspring

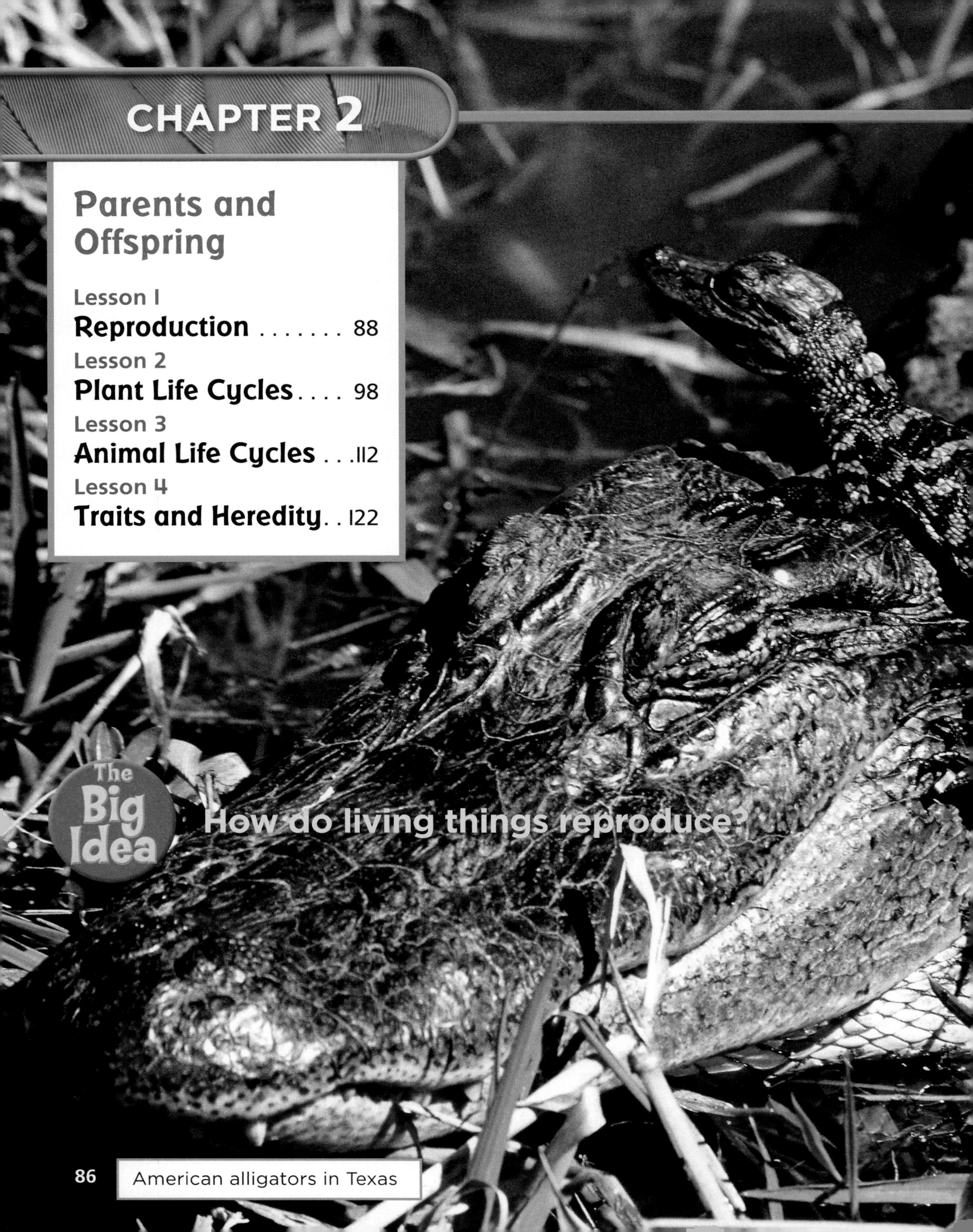

The Big Idea
How do living things reproduce?

American alligators in Texas

Key Vocabulary

sexual reproduction
the production of a new organism from a female sex cell and a male sex cell (p. 90)

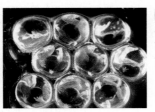

asexual reproduction
the production of new organisms from only one cell (p. 91)

pollination
the transfer of a pollen grain to the egg-producing part of a plant (p. 104)

embryo
a developing organism that results from fertilization (p. 106)

metamorphosis
a series of distinct growth stages that are different from one another (p. 114)

heredity
the passing down of inherited traits from one generation to the next (p. 124)

Lesson 1

Reproduction

Look and Wonder

Each little plant is an identical copy of the kalanchoe plant it is growing on. How did these plants grow without seeds or spores?

Explore

Can some flowering plants grow without seeds?

Make a Prediction

You have learned that flowering plants use seeds to reproduce. Can some flowering plants reproduce without seeds? Can you use part of a plant to create a new plant? Make a prediction.

- philodendron plant
- safety scissors
- hand lens
- plastic cup
- water
- 2-week-old cutting in a plastic cup (optional)

Test Your Prediction

1 Cut a piece of stem from the philodendron plant that measures 15 cm (6 in.) in length. Cut off the leaves that are closest to the plant. Leave 2 leaves at the very tip of the cutting.

2 **Observe** Look at your cutting with the hand lens. Record your observations.

3 Fill the plastic cup $\frac{2}{3}$ of the way with water. Place the cutting into the plastic cup.

4 **Interpret Data** Examine your cutting each day with the hand lens. Record your observations and any changes.

Draw Conclusions

5 **Infer** What happened to the cutting in the plastic cup with water?

6 Is it possible to grow a new plant without planting a seed? Explain.

Explore More

Could other plants grow in a way that is similar to the philodendron plant? Plan an investigation to answer the question. Write a report of your results and present it to the class.

Step **1**

Step **3**

▶ Main Idea

All living things come from other living things.

▶ Vocabulary

sexual reproduction, p.90

fertilization, p.90

asexual reproduction, p.91

vegetative propagation, p.93

runners, p.93

LOG ON ℮-Glossary

at **www.macmillanmh.com**

▶ Reading Skill ✓

Sequence

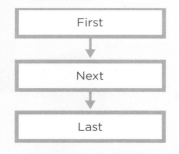

| First |
| Next |
| Last |

What are sexual and asexual reproduction?

Where do living things come from? Living things come from other living things. The survival of a species depends on its ability to produce new members. Every organism comes from a parent organism through the process of reproduction.

Reproduction (ree•pruh•DUK•shuhn) involves the transfer of genetic material from parent to offspring. The genetic material contains information that controls how the new individual will look and function. There are two main types of reproduction: sexual (SEK•shew•uhl) and asexual (ay•SEK•shew•uhl).

Sexual Reproduction

Sexual reproduction is the production of a new organism from two parents. A sperm cell from a male and an egg cell from a female join into a single unit in a process called **fertilization** (fur•tuh•luh•ZAY•shuhn). The fertilized egg cell contains genetic material from both parents. The fertilized egg then goes on to develop into a new individual. The new individual will receive some characteristics, or traits, from each parent. A trait is any characteristic of a living thing. For example, a plant may be tall and have pink flowers. Height and flower color are traits that may be passed to offspring from parents.

All mammals reproduce sexually. A lion cub receives traits from both of its parents.

Types of reproduction	Number of parents	Sex cells	Offspring	Mixing of traits
Asexual reproduction	1	not needed	identical to parent	no
Sexual reproduction	2	needed	different from parent	yes

Spider plants reproduce asexually. Each tiny plant is a new offspring.

Asexual Reproduction

Asexual reproduction is the production of a new organism from a single parent. It produces a new offspring that has the same genetic information as the parent. No male and female sex cells combine during asexual reproduction. Since there is only one parent in asexual reproduction, genetic information is not mixed. The offspring are identical to the original parent.

You can find organisms that reproduce asexually in all six kingdoms. All members of the bacteria kingdoms and most unicellular protists reproduce asexually. Most fungi and many plants can reproduce asexually during a part of their lives and sexually during another.

Animals such as jellyfish, corals, worms, and some echinoderms can form new offspring asexually. Some kinds of lizards, frogs, fish, and insects can also reproduce asexually.

 Quick Check

Sequence What is the first step of sexual reproduction?

Critical Thinking Asexual reproduction produces an exact copy of the parent organism. When could this be a disadvantage?

FACT Multicellular organisms can reproduce asexually.

Asexual Reproduction Poster

1. Research three types of asexual reproduction. Use the Internet, books, and magazines as sources.

2. Find out what organisms use these types of asexual reproduction.

3. Make a poster that compares the types of asexual reproduction you researched. Your poster can be a chart, graph, diagram, or table.

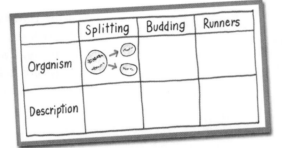

4. **Communicate** Make illustrations or cut out pictures of organisms that use these types of asexual reproduction. Place them in your poster and describe them.

5. How are the types of asexual reproduction similar and different?

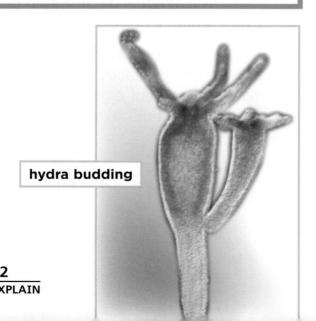

hydra budding

How do organisms reproduce asexually?

There are several methods of asexual reproduction shared by a wide variety of organisms. All forms of asexual reproduction eliminate the need for an organism to find a mate. They all produce genetic copies of the parent organism.

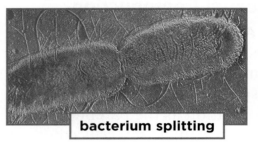

bacterium splitting

Splitting

Most unicellular protists and bacteria reproduce simply by splitting into two cells. Before splitting, the organism first makes a copy of its own genetic material. The two new offspring organisms will have a copy of the genetic material they need to carry out life processes. Some bacteria can divide into 2 new organisms every 10 to 20 minutes.

Budding

Organisms such as cnidarians, sponges, and some fungi can reproduce through *budding*. During budding, a small part of the parent's body grows into a tiny and complete version of the parent. In some cases, the bud breaks off from the parent organism and continues to grow. In other organisms, such as coral, the newly formed bud remains attached to the parent.

worker   queen   drone

Other Forms of Asexual Reproduction in Animals

Some species of fish, insects, frogs, and lizards go through asexual reproduction in a different way. The females of these animals produce eggs. Normally these eggs would be fertilized by male sex cells, but in some cases, fertilization never takes place. The eggs just develop into a new animal without fertilization. For example, when queen honeybees lay eggs, some are fertilized and others are not. The fertilized eggs develop into females, or worker bees. The unfertilized eggs become males, or drone bees.

Vegetative Propagation

Plants can undergo a form of asexual reproduction. **Vegetative propagation** (vej•i•TAY•tiv prop•uh•GAY•shuhn) is asexual reproduction in plants that produces new plants from leaves, roots, or stems. Many plants commonly reproduce this way by producing runners. **Runners** are plant stems that lie on or under the ground and sprout up as new plants. Runners can also grow downward from hanging plants. Strawberry plants, most grasses, aspen trees, and ferns can reproduce using runners.

✔ Quick Check

Sequence Describe the steps in bacteria reproduction.

Critical Thinking What is the difference between a drone and a worker bee?

Strawberry Reproduction

Read a Photo

Which part of the strawberry plant can produce new plants without seeds?

Clue Look at the photo. Which part of the original plant is attached to the new strawberry plant?

How do sexual and asexual reproduction compare?

You might wonder why some organisms reproduce asexually while others reproduce sexually. Asexual reproduction is convenient. An organism that reproduces asexually does not have to depend on another organism. It can live in isolation and still reproduce. Organisms that reproduce asexually tend to be well suited to their environment and produce equally well-suited offspring.

So why do organisms bother with sexual reproduction? One major advantage of sexual reproduction is that it promotes variety in a species. Sexual reproduction gives rise to offspring that may be better suited to environmental changes than either parent. The offspring produced are not identical to either parent. Some would be smaller, larger, or faster than others.

For example, the ability to run fast is an advantage for some organisms, like mice. Slower mice are more likely to be captured and eaten by other animals, such as owls or snakes. Faster mice might survive more frequently than slower mice. Over time, fast mice will reproduce and pass on this trait to their offspring.

 Quick Check

Sequence Describe the sequence of events that could happen to a mouse population if a new enemy were to appear in its habitat.

Critical Thinking What is an advantage of asexual reproduction?

Variation

Read a Photo

How can you tell that these puppies are not the result of asexual reproduction?

Clue Look at the puppies in the photo. Do they look the same?

Lesson Review

Visual Summary

Living things come from other living things through **reproduction**.

Splitting, budding, and vegetative propagation are methods of **asexual reproduction** used by a variety of organisms.

Sexual reproduction promotes variety in a species.

Make a FOLDABLES™ Study Guide

Make a Three-Tab Book. Use the titles shown. Tell about each topic on the tabs.

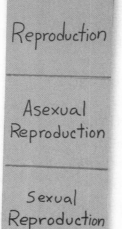

Reproduction

Asexual Reproduction

Sexual Reproduction

Think, Talk, and Write

1 **Main Idea** How do sexual and asexual reproduction differ?

2 **Vocabulary** A runner is a form of asexual reproduction called _____.

3 **Sequence** What happens after a bud forms on an organism?

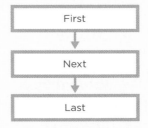

First

Next

Last

4 **Critical Thinking** What are the advantages of sexual reproduction?

5 **Test Prep** Which of the following best describes budding?

  A offspring develop from fertilized egg
  B offspring develop on parent
  C offspring develop from stem
  D offspring develop from two parents

6 **Test Prep** Which organisms reproduce by splitting?

  A bacteria
  B jellyfish
  C strawberries
  D honeybees

 Writing Link

Sheep Clone

Cloning is an artificial way to produce an organism. Scientists were able to create a cloned sheep named Dolly. Write a report explaining Dolly and cloning.

 Math Link

Bee Math

Drone bees make up 25% of a beehive population. If there are 30,000 bees in a hive, how many of them are drones?

How Do Sea Stars Regenerate?

A new sea star regenerating from a ray.

What's so amazing about sea stars?

They have the ability to asexually reproduce and heal themselves through regeneration! How does this process work? The sea star's body provides an important clue. It is made up of a central disk and five or more arms, or rays. The rays are extensions of the main body and contain part of the central disk.

What happens when the sea star's body is split in two? First, skin cells cover the wound. Then stem cells work to fix the problem. These cells divide and become the types of specialized cells needed to create new body parts. Finally, the two halves regenerate and become two new sea stars. At times a sea star's body may be split into four. Then each part repairs itself and becomes four new sea stars.

Sometimes, an entire arm or ray is broken off. Then, if the ray contains at least one fifth of the central disk, it regenerates and becomes a copy of the original sea star. What happens to the sea star with the cut-off arm? It regenerates and grows a new one. Regeneration is a slow process. It can take over a year.

Explanatory Writing

A good explanation

▶ gives clear details about a main idea

▶ lists what happens in an organized and logical way

▶ uses time-order words to make the description clear

Write About It

Explanatory Writing Explain how sea stars produce offspring using regeneration. Choose another animal that reproduces asexually. Write an explanation of how this process takes place.

 LOG ON **e-Journal** Research and write about it online at www.macmillanmh.com

Math in Science

Some organisms, like bacteria, increase in number very quickly. By splitting, one bacterium can generate more than one billion offspring in one day! This chart represents the reproduction rate of one type of bacteria in minutes. Using the data in the chart, you can create a graph. Each point on the graph represents an ordered pair. An ordered pair gives the location of a point on a graph.

Making a Line Graph

To make a line graph

▶ decide on a scale for the x-axis and y-axis

▶ label each axis and give the graph a title

▶ use the data from your chart to plot the ordered pairs

▶ connect the points

Bacteria Growth

time in minutes	0	20	40	60	80	100
number of bacteria	20	40	80	160	320	640

 Solve It

1. Assume that 1 strawberry plant can produce 3 new plants a year using runners. Fill in the chart below with the data. How many strawberry plants will be produced after 6 years?

2. Use the data in the chart below to graph the reproduction rate for strawberry plants. Give your graph a title and label each axis.

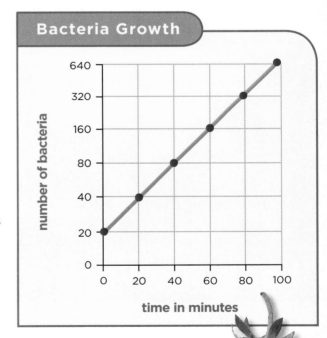

Strawberry Reproduction Data

year	1	2	3	4	5	6
number of plants	9	27				

Plant Life Cycles

Look and Wonder

One sunflower can produce more than 1,000 seeds! What conditions do these seeds need to grow into new sunflower plants?

How do flowering plants reproduce?

Form a Hypothesis

Do seeds require moist or dry conditions to grow? Write your answer as a hypothesis in the form "If seeds are placed in a moist environment, then . . ."

Test Your Hypothesis

1. Moisten a paper towel and place it at the bottom of a plastic cup.

2. Place 1 lima bean seed inside the cup with the paper towel. Fold the paper towel over the seed.

3. Repeat steps 1 and 2 with a dry paper towel.

4. **Experiment** Place the cups in a sunny spot and observe them daily for 5 days. Record your observations in a data table.

5. After 5 days, fill 2 cups with potting soil.

6. Take each seed and place it into a cup with soil. Gently cover each seed with soil and sprinkle some water on top.

7. **Observe** Place the cups in a sunny spot. Water the seeds daily and look for any changes. Record your observations in your data table.

Draw Conclusions

8. What were the independent and dependent variables in this experiment?

9. **Infer** What conditions were needed for your seeds to grow?

10. Did your results support your hypothesis?

Explore More

Keep observing your plant over time. What does your plant need to produce seeds? Make a prediction. Test your prediction and present your results to the class.

Materials

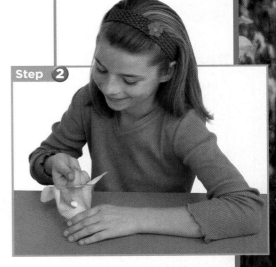

- paper towels
- 4 plastic cups
- water
- lima bean seeds
- potting soil

Step 2

Step 6

▶ **Main Idea**

The life cycles of all plants involve different stages of development.

▶ **Vocabulary**

pollination, p.104

pollen, p.104

embryo, p.106

seed coat, p.106

germination, p.106

monocot, p.107

dicot, p.107

conifer, p.108

LOG ON **e-Glossary**
at **www.macmillanmh.com**

▶ **Reading Skill** ✔

Cause and Effect

Cause → Effect
→
→
→
→

What are seedless plant life cycles?

All living things have life cycles. A *life cycle* is a series of differing stages of development. Plant life cycles include both an asexual and a sexual stage. This process of alternating between asexual and sexual reproduction is called an *alternation of generations*.

Moss Life Cycle

The life cycle of moss begins with asexual reproduction. Moss plants grow thin brown stalks with capsules at the top. The capsules contain tiny spores. Spores are cells that can develop into a new plant without fertilization. When the capsule opens, spores are released and carried by the wind. Spores that land in shady, moist soil are likely to grow.

In the sexual stage, the spore develops into the green, carpetlike adult plant. The adult plant has male structures that produce sperm and female structures that produce eggs. Sperm gets carried by water to the female structure where fertilization takes place. The fertilized egg grows on the female structure. It develops into a brown stalk with a spore capsule, and the cycle continues.

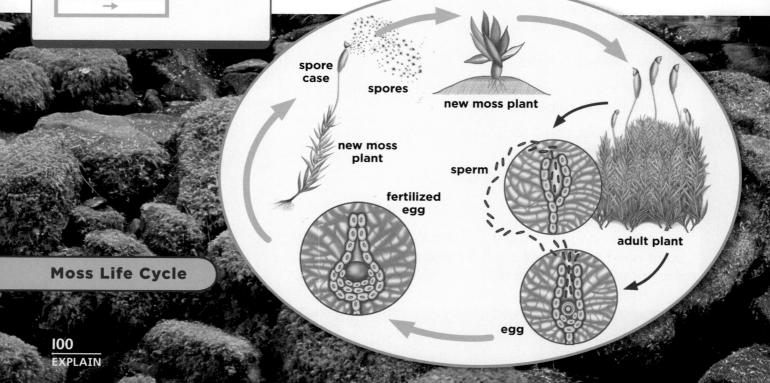

Moss Life Cycle

spore case

spores

new moss plant

new moss plant

sperm

fertilized egg

adult plant

egg

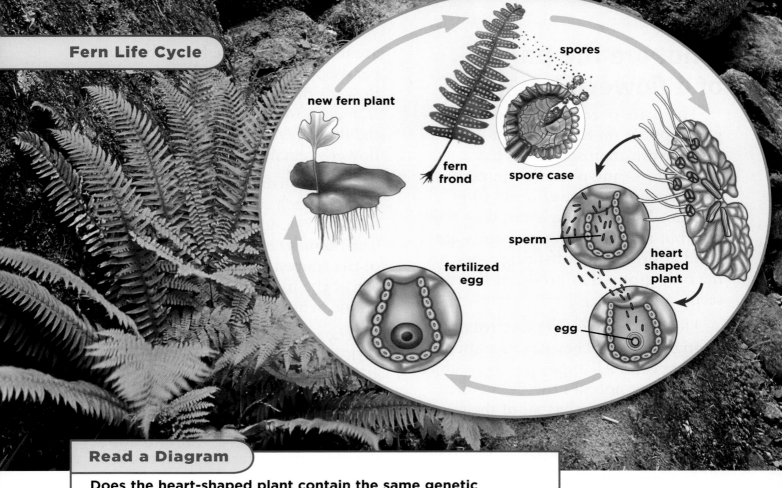

Fern Life Cycle

spores

new fern plant

fern frond

spore case

sperm

heart shaped plant

fertilized egg

egg

Read a Diagram

Does the heart-shaped plant contain the same genetic information as the leafy fern plant? Explain.

Clue: Look at where the spores are produced on the diagram.

 Science in Motion Watch fern life cycle at **www.macmillanmh.com**

Fern Life Cycle

Like mosses, ferns begin their life cycle with asexual reproduction. Ferns produce spores on the undersides of their fronds, or leaves. Spores are usually in clusters inside a spore case. When the spore case opens, the spores are released.

Fern spores that find the right conditions grow into a small heart-shaped plant with male and female structures. Now the fern undergoes sexual reproduction. The heart-shaped plant produces male and female sex cells. If a male sex cell fertilizes a female sex cell, the fertilized egg forms a new plant. The new plant develops into a leafy fern plant. Spore cases on the fern's fronds produce spores, and the cycle continues.

 Quick Check

Cause and Effect What causes the stalk and spore capsule to form in mosses?

Critical Thinking Spore cases can spray spores a meter (3 feet). Why would the plant spray spores so far?

What are the parts of a flower?

There are over 300,000 identified plant species on Earth. About 250,000 of these are angiosperms. What makes flowering plants so numerous? Flowering plants are efficient food-makers. They are tough and they grow fast, but mostly, flowering plants are good at producing offspring. They are the only group that produces flowers, seeds, and fruits.

Flowers are the reproductive organs of angiosperms. They produce sperm and egg cells. All angiosperms produce flowers, but not all flowers are alike. A *complete flower* has all of the four main parts: petals, sepals, stamens, and pistils. *Petals* are the brightly colored outer parts of a flower. Sepals, which are usually green, are found below petals. *Sepals* cover and protect the flower's parts when it is just a bud. A *stamen* (STAY•muhn) is the male part of the flower, while the central *pistil* is the flower's female organ.

Flowers usually have more than one stamen. Each stamen is made up of a filament (FIL•uh•muhnt) and an anther. The *filament* is the thin stalk portion of the stamen. The *anther* is at the top of the filament. It produces pollen grains, which contain sperm cells.

The pistil is made up of a stigma (STIG•muh), a style, and an ovary. The *stigma* is the opening at the top of the pistil. The *style* is the long, necklike structure that leads down to the ovary. The *ovary* houses the egg cells and it is the place where fertilization occurs.

Structure of Flowers

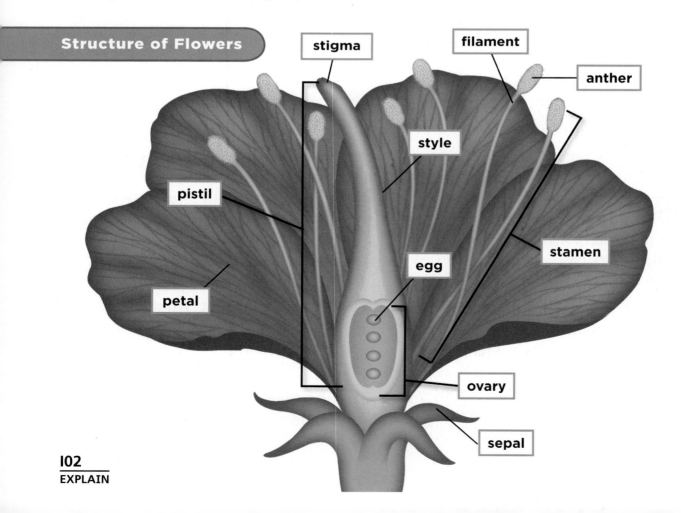

stigma · filament · anther · style · pistil · stamen · egg · petal · ovary · sepal

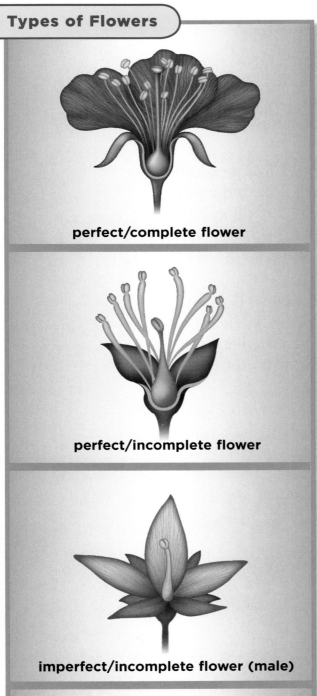

perfect/complete flower

perfect/incomplete flower

imperfect/incomplete flower (male)

imperfect/incomplete flower (female)

An *incomplete flower* is missing one or more of the flower parts of a complete flower. A *perfect flower* has both stamens and a pistil—the male and female parts. Typical perfect flowers include lilies, gladioli, tulips, and most fruit blossoms.

A flower can be incomplete and perfect. A windflower, for example, does not have petals, but it has both male and female structures, and sepals. The windflower's sepals have a petal-like appearance.

Imperfect flowers lack either a stamen or pistil. In other words, these flowers are either male or female. Some plants, such as willow trees, produce only male or female flowers. Other plants produce separate male and female flowers on the same plant. For example, a single corn plant will have both male flowers and female flowers.

 Quick Check

Cause and Effect What is the main effect of being an imperfect flower?

Critical Thinking Can a flower be imperfect and complete? Explain.

Read a Chart

How do complete and incomplete flowers differ?

Clue Look at the flower features that make a flower perfect or imperfect. Then look at the features that make it complete or incomplete.

What is the angiosperm life cycle?

Before fertilization can occur, **pollination** (pol•uh•NAY•shuhn) must take place. Pollination is the transfer of pollen (POL•uhn) from the stamen to the pistil. **Pollen** is a yellow powder that contains sperm cells. The trouble is that pollen cannot move on its own. How do plants get pollinated?

One way is through pollinators, such as bees, birds, and other animals. Why should these animals want to help out in the job of pollinating a flower? Because pollinators get something out of it: nectar. *Nectar* is a sweet liquid produced by flowers to attract pollinators. Flowers also have colorful petals, interesting shapes, and scents that appeal to pollinators.

As flowers open, bees and other pollinators arrive. Bees and other pollinators are attracted to the sugary nectar. As each bee drinks, grains of pollen rub off on its body. When the bee goes on to the next flower, some of that pollen rubs off on the flower's pistil and pollination occurs.

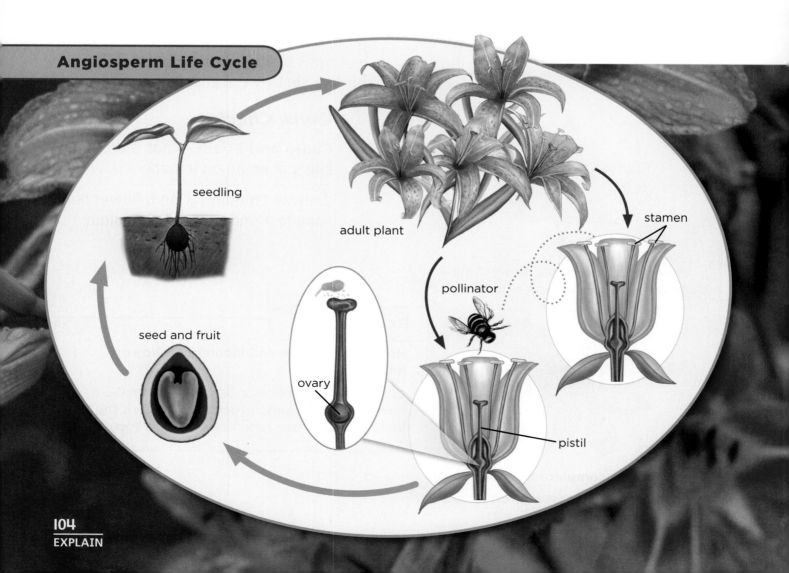

Angiosperm Life Cycle

seedling

adult plant

stamen

pollinator

seed and fruit

ovary

pistil

Animals are not the only means of flower pollination. Some flowering plants rely on the wind to blow pollen from stamen to pistil. Wind-pollinated plants include grasses and some trees.

You can usually tell how a flower gets pollinated by looking at it. Large, scented blossoms are usually pollinated by animals. Their colors and scents attract pollinators. Wind-pollinated plants do not attract animals, so their flowers tend to be small and dull.

Pollination can be carried out in a number of ways. *Self-pollination* occurs when a perfect flower with both male and female parts pollinates itself. *Cross-pollination* occurs when the pollen from one plant pollinates a flower on a different plant. Note that both self-pollination and cross-pollination can be carried out by the wind or an animal pollinator. Once pollinated, sperm cells from the pollen move down the style of the pistil to the ovary. Fertilization takes place when the sperm and egg cell combine to form a seed.

A moth drinking nectar.

✓ *Quick Check*

Cause and Effect What is the effect of an animal pollinator moving from one flower to the next?

Critical Thinking Can pollination take place without fertilization? Explain your answer.

FACT ► Plants use sexual reproduction.

What is in a seed?

Inside the ovary of a flower, the fused sperm cell and egg cell develop into an embryo (EM•bree•oh). An **embryo** is the beginning of a new offspring. As it grows, the embryo is packaged inside a seed. As the seed develops, the ovary enlarges until it becomes a fruit. The fruit protects the seeds inside it.

The seed itself has three main parts. The tiny embryo is the offspring that can grow into a new plant. Surrounding the embryo is its food supply, or *cotyledon* (kot•uh•LEE•duhn). The new, growing embryo lives off this food supply until it gets big enough to make food on its own. Finally, the entire seed is surrounded by a tough outer covering called a **seed coat**.

Once formed, seeds must be *dispersed*, or spread, to a favorable location in order to germinate. **Germination** (jur•muh•NAY•shuhn) is the development of a seed into a new plant. Seeds can wait to sprout—sometimes for years—until conditions favor growth. These conditions include water, sunlight, and space to grow.

Seeds are dispersed in many ways. Some seeds have fuzzy, parachute-like structures that help them float or get carried by the wind to a new location. Certain seeds, like a coconut, can float and are dispersed by water. Other seeds use animals for dispersal. Burrs have sticky hooks that cling to an animal's fur or feathers and get carried to new locations. The most common seed-dispersal method is for the plant to surround the seed with a sweet, fleshy fruit. When eaten, fruits pass through animals' digestive systems and get deposited in a new location.

Germination

Monocots	Dicots

flower parts in multiples of 3	**flower parts in multiples of 4 or 5**

parallel veins	**branched veins**

one cotyledon	**2 cotyledons**

Quick Lab

Comparing Seeds

1 **Observe** Take a look at each seed type.

2 Record the characteristics of each seed in a table. Use the following headings: size, shape, weight, toughness.

3 **Predict** How do you think each of the seeds you observed gets dispersed? Explain your answer.

Flowering plants are divided into two groups based on the types of seeds they produce. A **monocot** (MON•uh•kot) produces seeds with a single cotyledon. Leaves of monocot plants show a parallel vein pattern. Monocot flower petals come in groups of three. Typical monocots include corn plants, orchids, and grasses.

A **dicot** (DYE•kot) produces seeds with two cotyledons. Leaves of dicot plants show a branched vein pattern. Dicot flower petals come in groups of four or five. Typical dicots include bean plants and roses.

 Quick Check

Cause and Effect Why are seeds able to wait for the proper conditions to germinate?

Critical Thinking A volcano island in the ocean has no plants growing on it. Soon seed plants appear. How might these seeds have reached the island?

What is the conifer life cycle?

A **conifer** (KON•uh•fuhr) is a gymnosperm, a plant that has seeds but not flowers. Conifers include evergreens such as pines, firs, and other cone-bearing trees. Gymnosperms differ from angiosperms in two key ways. First, gymnosperms produce cones for reproduction rather than flowers. Second, gymnosperms have "naked" seeds. Their seeds are not packaged inside a fruit like the seeds of flowering plants.

A single conifer tree will produce both male and female cones. The smaller male cone releases clouds of powdery pollen that blow in the wind. The larger female cone produces a sticky fluid. Pollination occurs when the pollen grains land on the sticky fluid. After fertilization occurs, the developing egg remains attached to the female cone, where it matures into a seed.

How do conifer seeds get dispersed? The seeds have papery, winglike structures that help them whirl their way to the ground. In a strong wind, these winged seeds can end up far from the tree they were launched from. Under the right conditions, these seeds can then grow into new trees.

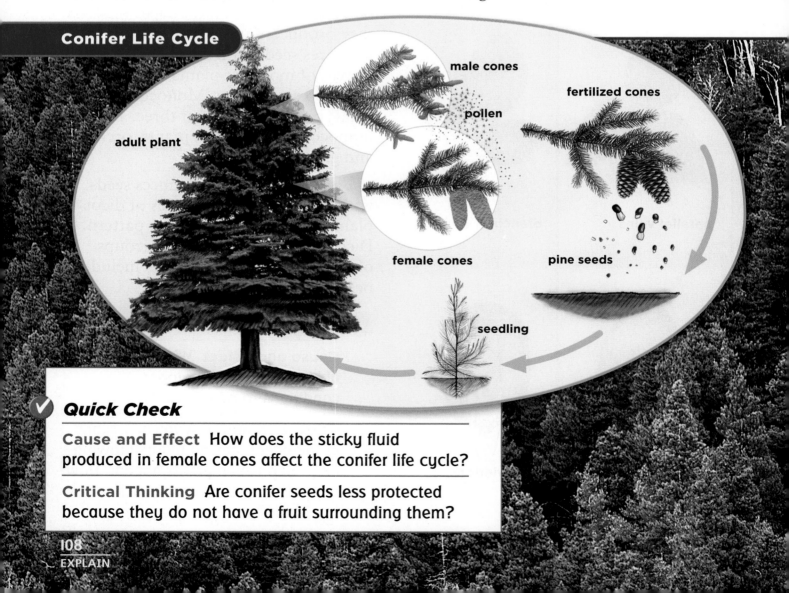

Conifer Life Cycle

male cones

pollen

fertilized cones

adult plant

female cones

pine seeds

seedling

Quick Check

Cause and Effect How does the sticky fluid produced in female cones affect the conifer life cycle?

Critical Thinking Are conifer seeds less protected because they do not have a fruit surrounding them?

Visual Summary

All plants have life cycles that include an **alternation of generations**.

Flowers are the reproductive organs of angiosperms. Angiosperm embryos are packaged in **seeds**.

Conifers are seed plants that produce cones rather than flowers.

Make a FOLDABLES™ Study Guide

Make a Three-Tab Book. Use the titles shown. Tell what you learned about each title.

Main Ideas	What I learned	Sketches
Alternation of generations		
Flowers and seeds		
Conifers		

Think, Talk, and Write

1 Main Idea What three features help make angiosperms the most plentiful plant group?

2 Vocabulary Flowers with petals that appear in groups of three are _____.

3 Cause and Effect What effect does water have in the moss life cycle?

Cause → Effect
→
→
→
→

4 Critical Thinking Can fertilization take place without pollination occurring in flowering plants?

5 Test Prep Flowers that are wind-pollinated are generally

A small and dull.
B colorful and small.
C dull and scented.
D large and showy.

6 Test Prep Which is not part of a seed?

A embryo
B seed coat
C cotyledon
D stamen

 Writing Link

Flower Ad

Write an advertisement for a flower that is trying to attract new pollinators. Your ad can be in the form of a cartoon, a radio skit script, a poster, or any other form you would like to use.

 Social Studies Link

Pollination and Food

How important are insects in the pollination of crops? Write a report that focuses on the importance of pollination in the food industry. Use the Internet, magazines, or books for source material.

Focus on Skills

Inquiry Skill: Observe

You just learned about plant life cycles and plant structures. For example, flowering plants reproduce sexually by forming seeds when sperm from pollen fertilizes an egg cell inside the pistil.

Perfect flowers have both a pistil (female part) and stamen (male part). Imperfect flowers have either a pistil or stamen, but not both. How do scientists know this? They **observe** real flowers!

▶ Learn It

When you **observe**, you use one or more of your senses to learn about an object. It is important to record what you observe. One way is to draw a diagram with labels to identify exactly what you saw. You can record other observations, such as odors and sounds, under the diagram. Then you can use the information to help identify other plants and their parts.

This diagram is a record of someone's observations. Each flower part is labeled. Note the observations under the diagram.

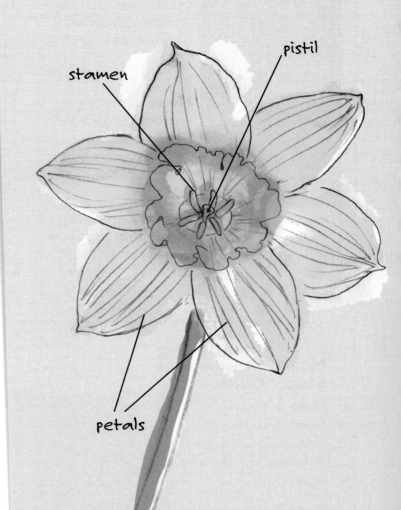

stamen

pistil

petals

Petals feel smooth and velvety to the touch. Pollen feels like soft powder. Petals have no smell. Plant parts make a crisp "SNAP" when broken.

▶ Try It

Materials **flower, paper, pencil, colored pencils or markers, hand lens**

1 **Observe** Look at a flower.

2 Make a diagram like the one shown. Be sure to include labels and color your flower and its parts.

3 Write any other observations about your flower under the diagram.

▶ Apply It

1 Now it is time to use your diagram and other observations to answer questions. Which senses were used to observe this flower? Is this a perfect or imperfect flower? How can you tell?

2 Continue to use your observation skills. Choose an object in your classroom, such as a stapler, pencil sharpener, TV, or the intercom system.

3 **Observe** Look at the object you chose. Then make a diagram of the object. Include labels to identify any parts and how they are used. Write any other observations, such as how it feels to the touch or the sound it makes, under the diagram.

4 Share your diagram and observations with your classmates.

Animal Life Cycles

Look and Wonder

After a duck lays eggs, it takes about 30 days for the eggs to hatch. How do ducklings develop into adults?

What are the stages in an animal's life cycle?

Purpose

You are part of an expedition that is studying the life cycle of frogs. You have collected some data about the frogs you observed. Interpret your data and photos to determine how long each stage of a frog's life cycle lasts.

Stage 1: Fertilized eggs
Date: April 1

Procedure

1 **Observe** Take a look at the stages involved in frog development.

2 Create a table to record changes in the frog's body structure during each stage of development.

3 **Interpret Data** Use the photos to determine how long each stage lasts. Record the information in your chart.

Stage 2: Tadpole
Date: April 5

Stage 3: Tadpole
Date: June 23

Draw Conclusions

4 What was the shortest stage in frog development? What was the longest stage?

5 **Infer** When did the organism seem to change the most?

6 How is the organism in stage 2 different from the organism in stage 4?

Stage 4: Froglet
Date: July 7

Explore More

How does the fertilized frog egg develop into a tadpole? Use the Internet or other sources to find photographs of the first four days of a tadpole's life. Describe the changes you see.

Stage 5: Mature Frog
Date: July 21

What are animal life cycles?

Like plants, animals have life cycles. When most animals begin life, they resemble the adults they will become. For example, after a baby chameleon hatches, its body gradually increases in size until it is an adult. Other animals, such as amphibians and insects, go through metamorphosis (met•uh•MAWR•fuh•sis). **Metamorphosis** is a series of distinct growth stages that are different from one another.

Complete Metamorphosis

Some animals, including butterflies, moths, flies, and beetles, go through complete metamorphosis. In **complete metamorphosis** the animal goes through four distinct stages. The adult body form looks very different from the newly hatched animal. A butterfly, for example, emerges from the egg as a plump larva (LAHR•vuh). A **larva** is an immature stage that does not resemble the adult. The wormlike butterfly larva, or caterpillar, has no wings. It often eats different food than the adult.

After hatching, caterpillars begin a period of nonstop feeding. As the caterpillar grows, its outer skin stretches. This stretching stimulates a release of hormones that brings on the next stage of the life cycle—the pupa (PYEW•puh). The **pupa** is a nonfeeding stage during which a hard, caselike cocoon surrounds the organism.

The pupa is often thought to be a resting stage. Although it seems quiet, the organism inside the cocoon is really very active. The entire body is changing. Wings, different mouth parts, new muscles, and new legs appear. When the cocoon opens, an adult butterfly with a completely restructured body emerges.

adult chameleon

hatchling chameleon

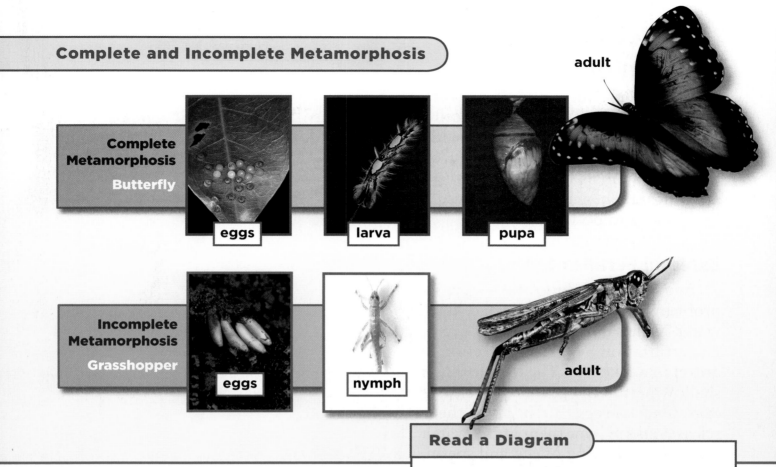

Complete and Incomplete Metamorphosis

Complete Metamorphosis
Butterfly

eggs

larva

pupa

adult

Incomplete Metamorphosis
Grasshopper

eggs

nymph

adult

Read a Diagram

Which metamorphosis stage is skipped in incomplete metamorphosis?

Clue: Compare the diagrams. How are they different?

Incomplete Metamorphosis

Some insect species, including grasshoppers, termites, and bedbugs, go through incomplete metamorphosis. During **incomplete metamorphosis** the animal goes through three stages that occur gradually. Young grasshoppers, for example, take on the nymph body form after hatching from eggs. A **nymph** (NIMF) is similar to an adult form, but it is smaller and lacks wings and reproductive structures. There may be several different nymph stages before the animal becomes an adult.

Because it has a rigid exoskeleton, an insect cannot grow gradually as a mammal or bird does. Instead, an insect must shed its hard skeleton all at once to make room for a larger body size. Grasshoppers go through five separate shedding stages before they reach adulthood. During each stage, the wings emerge a little bit more. By the time the grasshopper reaches its final stage, the adult body form is complete.

 Quick Check

Compare and Contrast How are larvae and adult butterflies different?

Critical Thinking Why can't grasshoppers grow gradually like mammals, reptiles, and birds do?

How does fertilization occur in animals?

Sexual reproduction in animals starts with fertilization. When a sperm cell combines with an egg cell, the resulting fertilized egg starts growing. The problem that most animals face is that sex cells are delicate. They must be protected for fertilization to occur.

External Fertilization

Amphibians and most fish solve this problem by releasing their sex cells into water. Male salmon, for example, prepare for fertilization by finding a section of gravel in a lake bed. The female digs a shallow nest in the gravel and releases her eggs. Once the eggs are in place, the male releases sperm over them in the water. This out-of-body joining of egg and sperm is called **external fertilization**.

External fertilization is a high-risk process. Ponds, lakes, rivers, and oceans contain vast amounts of water. The chances of sperm cells finding and fertilizing the egg cells are decreased in large amounts of water. Many sex cells are lost and some are eaten by other animals. The sex cells can also be exposed to extreme temperatures and pollution in the water.

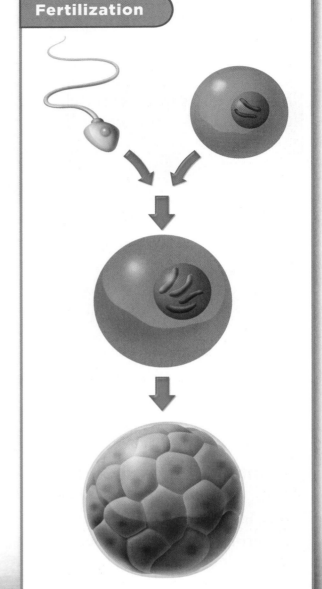

Fertilization

This male salmon is spreading sperm cells onto egg cells that have been released into the water by the female.

How do these animals manage to produce any offspring at all in these conditions? These animals ensure the production of offspring by releasing a great many sex cells at one time. The large numbers increase the chances that fertilization will occur. Typically, only one or two out of every thousand eggs laid survive to adulthood. For this reason, many fish and amphibians tend to produce a lot of eggs.

Internal Fertilization

How do the sex cells of land animals survive in the dry conditions on land? Reptiles, birds, and mammals solved this problem with internal fertilization. **Internal fertilization** is the joining of sperm and egg cells inside a female's body.

Internal fertilization increases the chances of fertilization and the offspring's survival. It protects sex cells and fertilized eggs from drying out. It also protects them from the dangers of harsh environments and other organisms. Since the chances of fertilization occurring are much greater with this method than through external fertilization, fewer eggs need to be produced.

 Quick Check

Compare and Contrast How are external and internal fertilization similar? How are they different?

Critical Thinking Suppose a fish lays eggs on a day in which strong currents disturb the water. How might this affect reproduction?

Quick Lab

Model External Fertilization

1. **Make a Model** Cover the bottom of a tank with about 1 cm ($\frac{1}{2}$ in.) of fish tank gravel. Fill the tank about $\frac{2}{3}$ full with water.

2. Scatter 15 blue marbles in the water. The blue marbles represent unfertilized eggs.

3. After the blue marbles settle, scatter 15 red marbles in the tank.

4. How many blue marbles touched, or "fertilized," red marbles?

5. **Infer** What does this model tell you about the accuracy of external fertilization?

What happens to a fertilized egg?

Whether fertilized externally or internally, a successful fertilization produces an egg with a developing embryo inside of it. Animals have different eggs depending on their structures and the environments in which they live.

Fish, frogs, reptiles, birds, and some mammals lay eggs. Fish and frogs lay their eggs in the open water. The eggs are exposed to any hungry creature that finds them. The embryos are somewhat protected by a jellylike layer around the eggs. These embryos get food from the yolks of the eggs.

Reptile and bird eggs have tough shells that are filled with watery liquid. This liquid gives the embryo the wet environment it needs to develop and protects it from drying out.

The tough outer shell also protects the growing embryo from being eaten by other animals. The yolk inside the egg provides the embryo with food.

Most mammals take the safety of the growing embryo a step further. Instead of hatching outside of the mother's body, the eggs of most mammals develop inside the mother. Embryos are fed by their mothers' bodies as they develop. Monotremes are the only mammals that lay eggs. All other known mammals give birth to live young.

✔ Quick Check

Compare and Contrast How are the eggs of animals alike? How are they different?

Critical Thinking Animals that lay more eggs, such as reptiles, tend to give less care to their offspring. Why might this be true?

Comparing Eggs

crocodile egg

frog egg

chicken eggs

Read a Photo

Which egg provides the least protection for the developing embryo?

Clue: Compare the outside layers of the eggs.

Lesson Review

Visual Summary

Insects and amphibians develop in distinct growth stages going through the process of **metamorphosis**.

Eggs get fertilized outside of the body in the process of **external fertilization**. Land animals use **internal fertilization** to protect their eggs and developing offspring.

Animals have different types of **eggs** depending on their environments and life cycles.

Make a FOLDABLES™ Study Guide

Make a Three-Tab Book. Use the titles shown. Write what you know about the topic on each tab.

Metamorphosis

External and Internal Fertilization

Eggs

Think, Talk, and Write

1. **Main Idea** Why do most aquatic animals use external fertilization and most land animals use internal?

2. **Vocabulary** A hard, caselike cocoon forms during the _____ stage.

3. **Compare and Contrast** How do complete metamorphosis and incomplete metamorphosis compare?

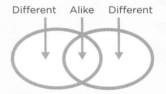

Different   Alike   Different

4. **Critical Thinking** Chicken eggs have a large food supply for the embryo. Why don't mammal eggs have this? Explain.

5. **Test Prep** Animals that carry out internal fertilization generally
   A produce a great many eggs.
   B produce only one egg their entire life.
   C produce thousands of offspring.
   D produce a small number of eggs.

6. **Test Prep** Which stage is a caterpillar?
   A egg
   B larva
   C pupa
   D adult

Math Link

Fish Eggs

On average, 4 out of every 1,000 fish hatch and survive to adulthood. How many eggs are needed to expect 100 offspring to survive to adulthood?

Health Link

Parts of Chicken Eggs

The chicken eggs we eat are unfertilized. Research egg development. Which part of the egg keeps the embryo from drying out? Which part is a food supply?

Be a Scientist

Materials

2 plastic cups with lids

wax-worm food

forceps

wax worms

black construction paper

lamp

petri dish

metric ruler

How does light affect the life cycle of the wax moth?

Form a Hypothesis

Some insects have different stages in their life cycle in which the body looks nothing like the adult stage. This is called complete metamorphosis. For example, a butterfly's larval stage is a caterpillar. A caterpillar does not have wings and looks very little like the adult. Wax moths have a similar life cycle.

How does the amount of light affect the wax moth's life cycle? Write your answer as a hypothesis in the form "If the amount of light is decreased, then wax-moth larvae will . . ."

Test Your Hypothesis

1. Create wax-worm habitats by filling 2 plastic cups halfway with wax-worm food. Use forceps to place 5 wax worms in each cup. Then put the lids on each cup.

2. Tape black construction paper around one of the cups. Be sure to cover the cup completely.

3. **Use Variables** Place the uncovered cup under a lamp or a bright window. Place the covered cup away from the light source.

4. **Measure** Remove a wax worm from one of the cups and place it on a petri dish. Use a metric ruler to measure the length and width of the wax worm. Be sure to place the larva back in the correct cup.

5. Repeat step 4 with each of the wax worms. Record the measurements in a data table.

6. **Observe** Measure the wax-moth larvae every two days until they pupate. Record the measurements in your data table. Record any other changes you see.

Step **2**

Step **3**

Draw Conclusions

7 Did the growth data of wax-moth larvae under the light source differ from the larvae in the covered cup?

8 **Infer** Did the amount of light in the wax worms' environment have any effect on their life cycle?

How does temperature affect the life cycle of the wax moth?

Form a Hypothesis

Does temperature affect the rate of the wax worm's life cycle? Write your answer as a hypothesis in the form "If the temperature of the environment of wax-moth larvae is increased, then wax-moth larvae will . . ."

Test Your Hypothesis

Design an experiment to test your hypothesis. Write out the materials you need and the steps you will follow. Record your results and observations.

Draw Conclusions

Did your results support your hypothesis? Why or why not? Present your results to your classmates.

What else can you learn about insect life cycles? For example, how does light or temperature affect the life cycle of a butterfly? Design an experiment to answer your question. Your experiment must be organized to test only one variable. It must be written so that other students can complete the experiment by following your instructions.

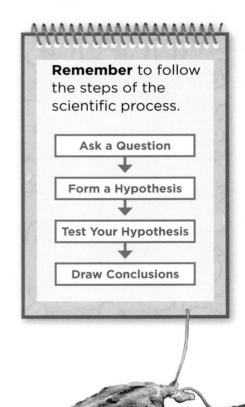

Remember to follow the steps of the scientific process.

Ask a Question

↓

Form a Hypothesis

↓

Test Your Hypothesis

↓

Draw Conclusions

Traits and Heredity

Look and Wonder

These brown bear cubs resemble their mother. Is it just a coincidence, or do parents pass traits on to their offspring?

What are some inherited human traits?

Purpose

Everyone has different physical characteristics, but some traits are similar in different people. Do you have any of the same traits as your classmates? Perform a survey to find out which traits your classmates possess. Use your data to determine which traits appear most frequently.

Materials

- paper
- pencils

Procedure

1. Have a partner check you for each of the traits shown. Record which form of the trait you have in a chart.

2. Reverse roles and repeat.

3. **Communicate** Tally your results and combine them with your classmates' results in a classroom chart.

4. **Interpret Data** Use the data from the classroom chart to create a bar graph of your results.

Draw Conclusions

5. **Use Numbers** Find the percent of each trait in the class.

6. Which form of each trait appears more frequently?

7. **Infer** Are some forms of traits more common than others? If so, why?

Explore More

How do your results compare with a larger group? Plan an experiment that would answer this question.

regular thumb

hitchhiker's thumb

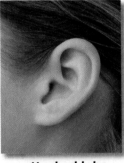

attached lobe

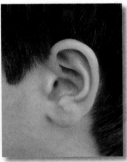

unattached lobe

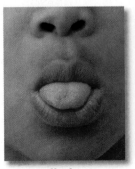

non-rolled tongue

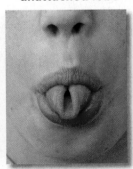

rolled tongue

► Main Idea

Traits that are passed from parents to offspring control how organisms look and what they do.

► Vocabulary

heredity, p.124

inherited trait, p.124

instinct, p.124

gene, p.126

dominant trait, p.126

recessive trait, p.126

pedigree, p.128

carrier, p.128

LOG ON e-Glossary
at www.macmillanmh.com

► Reading Skill ✔

Fact and Opinion

Fact	Opinion

What is heredity?

Have you ever stopped to look at a beautiful flower? What causes that flower to be pink, red, or some other color? It is the same thing that causes you to have blue, brown, or green eyes. It is **heredity** (huh•RED•i•tee), which is the passing down of traits from parents to offspring.

Heredity applies to all organisms. In plants, for example, flower color and plant height are inherited traits. An **inherited trait** is a trait that an offspring receives from its parents. Some inherited human traits include dimples, hair and eye color, facial features, and even the way you laugh.

Can heredity affect an organism's behavior? Some behaviors, such as instincts (IN•stingkts), are inherited.

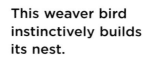

This weaver bird instinctively builds its nest.

An **instinct** is a way of acting or behaving that an animal is born with and does not have to learn. Take, for example, a spider spinning its web. Does the spider learn to build this complicated structure? Or is web building an inherited instinct? It turns out that web building is an instinct. Just as human babies do not learn how to breathe, spiders do not need to learn how to build a web—they can just do it.

Web building is an instinct in spiders.

Similarly, birds are born with the instinct to build a nest. Different types of birds build different types of nests. For example, weaver birds always build elaborate hanging nests out of small plant pieces. Other birds, such as some penguins, build nests out of pebbles. Young birds instinctively build nests that are like their parents'.

Other behaviors are learned rather than inherited. A learned behavior is developed during the course of an animal's lifetime. Learning results from practice and experience. For example, your pet dog might be able to sit, bark, or catch a ball on your command. Your pet did not inherit these abilities from its parents—it had to learn how to do them.

The ability to learn helps animals survive. Learning enables animals to respond better to changes in their environment. For example, one learned behavior, called *imprinting*, helps young birds survive. Imprinting occurs when an animal forms a social bond with another organism shortly after birth or hatching.

When ducks hatch they learn to recognize and follow their mother. The mother duck is much more experienced at finding food and avoiding danger. By learning to recognize and follow the mother duck, young ducklings increase their chances of survival.

✔ Quick Check

Fact and Opinion Breathing and blinking are inherited behaviors. Is this statement a fact or an opinion?

Critical Thinking A weaver bird hatches in a zoo rather than the wild. It is kept in a cage with other birds, such as robins. What type of nest will this bird build? Why?

Catching a flying disk is a learned behavior for this dog.

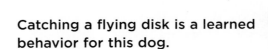

FACT ▶ Traits come from both parents.

How are traits inherited?

What controls the traits you inherit? Why do some people look more like one parent than another? To answer these questions, you need to learn about the work of Gregor Mendel.

Gregor Mendel was an Austrian monk who studied and discovered the basic principles of heredity. Mendel grew up on a farm where he became curious about how traits were passed on in plants and animals. In 1856 Mendel began experimenting with garden pea plants at his monastery. He crossed plants that had different traits and observed how those traits were passed on. He studied pea plants because they produce seeds quickly and their traits are easy to trace from one generation to another.

Mendel spent over seven years carefully conducting his experiments with pea plants. Based on his results, Mendel determined that inherited traits are passed from parents to offspring during reproduction. He believed that each inherited trait is controlled by two factors. The offspring receive one of these factors from each parent. Today scientists refer to these factors as genes (JEENZ). A **gene** contains chemical instructions for inherited traits. They are stored on cell structures called *chromosomes* (KROH•muh•sohmz), which are found in the nucleus of the cell.

Mendel found that for each trait he studied, one form of the trait could mask the other. For example, pea plants can have purple or white flowers. When Mendel crossed a purple pea with a white pea, all the offspring had purple flowers. What happened to the trait for white flowers? When Mendel crossed two of the purple offspring, the trait for white flowers reappeared in the next generation. The trait for white flowers had not disappeared. It had been hidden by the trait for purple flowers.

Mendel concluded that for every trait there is a dominant (DOM•uh•nuhnt) form and a recessive (ri•SES•iv) form. A **dominant trait** is one that dominates, or masks, another form of that trait. A **recessive trait** is one that is hidden, or masked, by another form of the trait. Each form of the trait can be represented by letters. A capital letter is used for the dominant form of the trait. A lowercase letter is used for the recessive

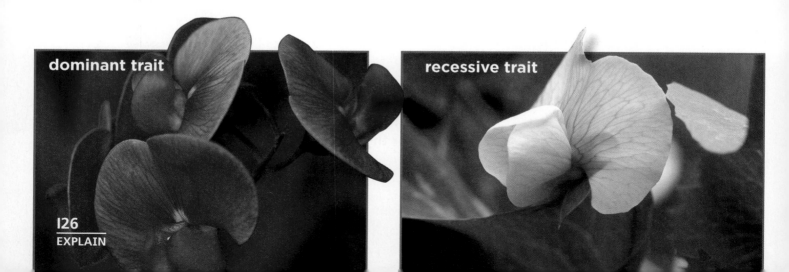

dominant trait

recessive trait

form of the trait. For example, in pea plants purple flowers (P) are dominant to white flowers (p).

Mendel's findings are important because they apply to all organisms, not just pea plants. For example, in humans the genes that determine the shapes of your earlobes, hairline, and thumb have dominant and recessive forms. Dominant traits tend to be expressed more frequently than recessive traits. The recessive traits are masked by the dominant forms.

 Quick Check

Fact and Opinion A cross between white and purple pea plants creates purple offspring. White flowers are prettier than purple flowers. Are these statements facts or opinions?

Critical Thinking You have a red flower. Can you tell the color of its offspring? Explain.

≡Quick Lab

Inherited Traits in Corn

Each corn kernel is a separate seed that inherits traits, such as kernel color, from a parent plant.

1. **Observe** Look at an ear of corn. What do you notice?

2. Count the number of purple kernels on your ear of corn. Record the number.

3. Count the number of yellow kernels on your ear of corn. Record the number.

4. **Interpret Data** Which color kernel occurs more frequently?

5. Is the trait for purple kernels dominant or recessive? Explain.

Pea Crossing

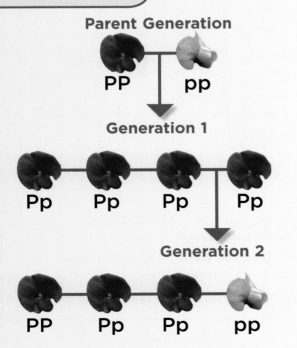

Parent Generation

PP   pp

Generation 1

Pp   Pp   Pp   Pp

Generation 2

PP   Pp   Pp   pp

| ■ Purple flower (dominant trait) | □ White flower (recessive trait) |

Read a Diagram

Why are the purple flowers in the first generation represented by Pp?

Clue: Which forms of the trait do the parents possess?

How do we trace inherited traits?

Some of the traits controlled by genes are easy to see, such as hair color. Genes also control many things you cannot see. Some individuals "carry" a trait without showing evidence of that trait themselves. For example, how can two parents with dimples have a child without dimples?

You can find out by using a pedigree (PED•i•gree). A **pedigree** is a chart used to trace the history of traits in a family. They are used to study heredity patterns. Parents and offspring are shown in a pedigree. Horizontal lines connect parents, and vertical lines connect parents to offspring. Males are represented as boxes; females are shown as circles. Individuals with a dominant trait have a shaded shape. Unshaded shapes represent recessive individuals.

In the pedigree below, you can see that both parents have dimples, but they carry the recessive gene for no dimples. A **carrier** (KAR•ee•uhr) is any individual who has inherited the gene for a trait, but does not show that trait physically. This couple can have three different types of offspring—one type does not have dimples (dd). They can also have children who are carriers of the recessive gene (Dd) or dominant for the gene (DD).

 Quick Check

Fact and Opinion Give a fact and an opinion about pedigrees.

Critical Thinking In the pedigree shown, can the child without dimples have children with dimples?

Pedigree Chart

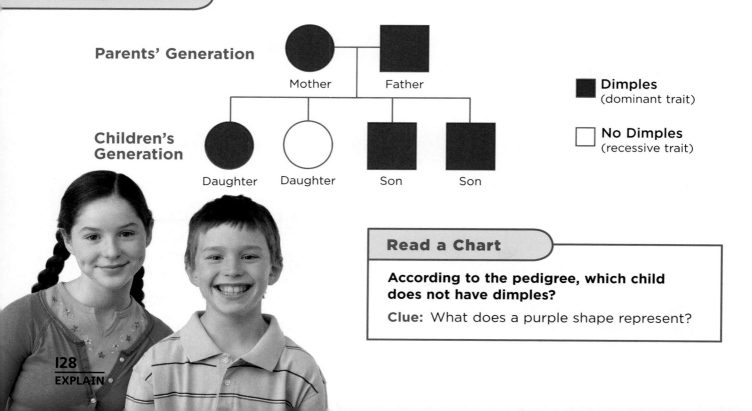

Parents' Generation

Mother   Father

Children's Generation

Daughter   Daughter   Son   Son

■ **Dimples** (dominant trait)

☐ **No Dimples** (recessive trait)

Read a Chart

According to the pedigree, which child does not have dimples?

Clue: What does a purple shape represent?

Lesson Review

Visual Summary

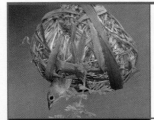

Heredity is the passing down of traits from parents to offspring.

Gregor Mendel found that **dominant** traits mask **recessive** traits.

Carriers can pass on genes for a trait to offspring without showing evidence of the trait themselves. **Pedigree** charts help us study heredity patterns.

Make a FOLDABLES™ Study Guide

Make a Trifold Book. Use the titles shown. Tell facts about each topic.

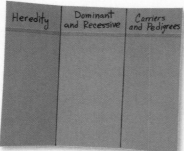

Heredity | Dominant and Recessive | Carriers and Pedigrees

Think, Talk, and Write

1. **Main Idea** Why are offspring similar to parents, but not exactly like either one of them?

2. **Vocabulary** Traits are controlled by cell structures called _____.

3. **Fact and Opinion** A friend claims that with practice anyone can roll their tongue. Is this fact or opinion? Explain.

Fact	Opinion

4. **Critical Thinking** Why is it a good idea for carriers of disease genes to get tested before having children?

5. **Test Prep** Which of the following controls traits in organisms?
   - **A** genes
   - **B** pedigrees
   - **C** instincts
   - **D** cell membranes

6. **Test Prep** Which of the following is a learned behavior?
   - **A** a bird building its nest
   - **B** a duck imprinting
   - **C** a spider building its web
   - **D** a baby breathing

 Writing Link

Fictional Narrative
Write a story about Gregor Mendel. Tell the story from his point of view. Describe his experiments with pea plants and his conclusions.

 Math Link

Human Genes
A human sperm or egg has about 20,000 genes on 23 different chromosomes. On the average, how many genes does each chromosome carry?

Real World

Genetically Modified Corn

Did you know that 560 million tons of corn are produced worldwide each year? You might think "Wow! That's a lot of popcorn!" But corn is much more than a tasty snack. It is a major source of animal feed for hogs, poultry, and beef and dairy cattle. Corn is also refined and used as a sweetener for processed foods, such as cereals, soft drinks, salad dressings, and candies. Ethanol is an alcohol refined from corn that is added to gasoline. And, of course, you can eat corn at the dinner table.

While millions of tons of corn are produced each year, about 39 million tons never make it to the market. The reason is a small insect, the European corn borer. These insects eat so much of the corn that the plant topples over or stops growing.

European corn borer

monarch butterfly

To fight these pests, farmers spray their crops with a special powder. It is made of a naturally occurring type of bacteria, *Bacillus thuringiensis*, also known as Bt. Bt has a gene that produces a protein that is a toxin. It is poisonous to many corn borers.

The Bt powder cannot reach the whole crop, so it does not kill all the corn-borer larvae. Scientists had to find another solution. Starting in 1993, they used genetic engineering to insert the toxin gene from the Bt directly into the corn's genetic material. The genetically modified plant, known as Bt corn, produces the same toxin as the bacteria. Now the plant can protect itself when the corn borer attacks. Bt corn has proven to be 99 percent effective in killing corn-borer larvae.

While this is good news for many farmers, genetic engineering may affect the health of ecosystems and humans. Scientists are examining how genetically modified organisms might affect other plants and animals. Environmentalists are concerned that pollen from Bt corn might be killing harmless insects, like the monarch butterfly. Some scientists also wonder about the effect of genetically modified food on human health. Scientists continue to look for answers to questions about genetically modified foods.

Bt bacteria

Cause and Effect

► Look for the reason why something happens to find a cause.

► An effect is what happens as a result of a cause.

Write About It
Cause and Effect

1. Explain how the bacterium Bt affects corn borers.
2. Tell how genetically modified corn might cause problems for other insects and the environment.

LOG ON **e-Journal** Research and write about it online at **www.macmillanmh.com**

AMERICAN MUSEUM ŏ NATURAL HISTORY

Visual Summary

Lesson 1 All living things come from other living things.

Lesson 2 The life cycles of all plants involve different stages of development.

Lesson 3 Animals use different strategies to reproduce and ensure the survival of their offspring.

Lesson 4 Traits that are passed on from parents to offspring control how organisms look and what they do.

Make a FOLDABLES™ Study Guide

Assemble your lesson study guide as shown. Use your study guide to review what you have learned in this chapter.

Fill in the blank with the best term from the list.

embryo, p. 106

fertilization, p. 90

gene, p. 126

heredity, p. 124

inherited trait, p. 124

metamorphosis, p. 114

pollination, p. 104

pupa, p. 114

1. The beginning of a new life form is called a(n) _____.

2. The stage in which a caselike cocoon forms around the organism is called the _____.

3. The passing down of traits from one generation to the next is called _____.

4. Pollen is transferred from the stamen to the pistil of a flower during _____.

5. A characteristic passed down from parent to offspring is a(n) _____.

6. The chemical instructions for inherited traits are carried in a _____.

7. A sperm cell and an egg cell join into a single new unit during _____.

8. During its life cycle, a butterfly goes through a complete _____.

Answer each of the following in complete sentences.

9. **Sequence** Describe in order the steps that take place during budding.

10. **Compare and Contrast** Compare wind-pollinated flowers and animal-pollinated flowers. Do you think this flower is pollinated by animals or the wind?

11. **Observe** Look at a flower in your area. Based on your observations, make a drawing of the flower. On your drawing, label any traits you notice. These may include the color of the flower's center and petals, the number of petals, and the length of the stem.

12. **Critical Thinking** If a child has two parents who both have a dominant gene for brown eyes, will the child have brown eyes as well? Explain.

13. **Explanatory Writing** Explain the disadvantages of external fertilization.

14. How do living things reproduce?

Family Traits

Identify inherited traits in either your own family or the family of a well-known person.

What to Do

1. Gather pictures that show at least three generations of people in the family. Try to find pictures of more than one person in each generation. If possible, select pictures that show the people at similar ages.

2. Look at the photographs to identify physical traits that each person has.

3. List traits that people in the family share, and who shares them.

Analyze Results

▶ Review the traits of the people in the youngest generation. From which people might they have inherited each of these traits?

Test Prep

1. **What part of the flower below is the arrow pointing to?**

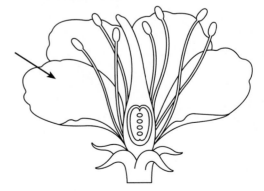

A a sepal   **C** a pistil

B a stamen   **D** a petal

Careers in Science

Health Care Technician

Doctors give patients checkups and medical treatment, but they do not work alone. They need the help of health care technicians. If you like working with people, then you might like this career. You would train while working in a doctor's office, clinic, hospital, or nursing home. You might schedule appointments and monitor the flow of patients. You might take X rays of patients' muscles and bones and then develop the films. You might take blood samples for tests or treatment. You might collect lab specimens and perform tests. As you gained experience, you would take on more and more responsibility. Both doctors and their patients would depend on you doing a good job.

▲ Health care technicians prepare samples to be tested.

▼ Botanists expertly identify plants.

Botanist

If you are a "plant person" then you may have thought about making plants your life's work. A college degree in botany, the scientific study of plants, can lead to a variety of career paths. Botanists may work in laboratories or out in the field. Some botanists specialize in identifying and classifying plants. Others research plant diseases or work on the uses of plants in foods, medicines, building materials, and fibers. Some botanists work in museums or botanical gardens and some teach at universities. Horticulture, which includes ornamental plants, is another special field of botany. Which field do you think you might choose?

Ecosystems

Desert adder in Namibia, Africa

Watch out! This snake is hiding in the sand and waiting for its next meal.

dragonfly

dragonfly nymph

midge

midge larva

THE CASE FOR CLEAN WATER

People, plants, and animals all need clean water. How do we know if a body of water is clean and healthy?

Clue 1. How does it look?

Begin your investigation with obvious evidence. What does the water look like? How does it smell? Even if water looks clear and clean, it might not be.

Clue 2. What is in it?

Use equipment to do some tests. Testing the pH and chemistry of the water are ways to check its health.

Clue 3. Who lives there?

Look for insects that spend part or all of their life living in the water you are investigating. Dragonfly nymphs like to live in clean water, but if necessary they can live in polluted water. Therefore you cannot know whether or not the water is clean if you see dragonfly nymphs. Midge larvae can live in polluted water until they hatch into adult midges. If you notice many midge larvae and not much else, it is a sign that the water is in trouble. Some insects, including caddisfly larvae, mayfly nymphs, and stonefly nymphs, cannot live in polluted water. If you find these insects, you know the water must be clean.

Write About It

Response to Literature This article tells how to find out if a body of water is clean. Research additional information about the insect larvae mentioned in the article. Write a report about the effects of pollution on these insects. Include facts and details from the article and from your research.

LOG ON **e-Journal** Write about it online at **www.macmillanmh.com**

mayfly nymph

mayfly

Interactions in Ecosystems

The Big Idea How do organisms interact?

Katmai National Park, Alaska

Key Vocabulary

ecosystem
all the living and nonliving things in an environment, including their interactions with each other (p. 142)

food web
the overlapping food chains in an ecosystem (p. 146)

energy pyramid
a diagram that shows the amount of energy available at each level of an ecosystem (p. 148)

carrying capacity
the maximum population size that an area can support (p. 157)

commensalism
relationship between two kinds of organisms that benefits one without harming the other (p. 161)

camouflage
an adaptation in which an animal protects itself against predators by blending in with the environment (p. 173)

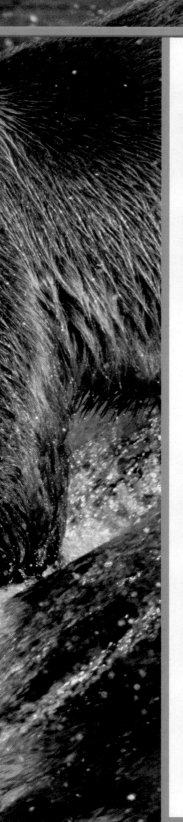

Energy Flow in Ecosystems

Look and Wonder

A cheetah chasing prey can run 112 km/h (70 mph). That takes a lot of energy! How do organisms depend on one another for energy?

How do organisms in a food chain interact?

Purpose

A food chain models how food energy is transferred from one organism to another. Producers make their own food. Herbivores consume producers. Carnivores consume herbivores. Create food chains using the list below.

Materials

- blank note cards
- construction paper
- glue stick
- magazines
- markers
- scissors

PRODUCERS	HERBIVORES	CARNIVORES
algae	grasshopper	wolf
berries, flowering plants	deer	otter
shrubs	chipmunk	hawk
seeds, grass	squirrel	robin
acorn, oak tree	fish	owl

Procedure

1. Make cards for the organisms listed in the chart above. Draw or glue a picture of an organism on each card.

2. Create a three-column chart on the paper. Label the columns as shown.

3. Use your organism cards to make five food chains. Place the organism cards on your chart under the correct columns.

Step 2

First Level	Second Level	Third Level
1. Acorn 2.	Squirrel	Owl

Draw Conclusions

4. **Communicate** Compare your food chains with a classmate's food chains. Explain how they compared.

5. **Infer** Can food chains overlap? How does this affect the ecosystem?

Explore **More**

Research one of your food chains. What ecosystem is it part of? What other organisms are part of this ecosystem? How are these organisms connected to your food chain?

Read and Learn

Main Idea

Food chains, food webs, and energy pyramids show the energy flow between organisms in an ecosystem.

Vocabulary

ecosystem, p.142

population, p.143

community, p.143

food chain, p.144

food web, p.146

predator, p.147

prey, p.147

energy pyramid, p.148

e-Glossary
at www.macmillanmh.com

Reading Skill ✓

Sequence

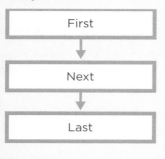

What is in an ecosystem?

You are on a hike in a beautiful forest. What do you see? Plants, including spruce trees, wildflowers, and grasses, grow along your path. Chipmunks scurry across the trail and birds fly overhead. These are some of the living things, or *biotic factors* (bye•OT•ik FAK•tuhrz), of the environment.

Plenty of nonliving things, or *abiotic factors* (ay•bye•OT•ik), are also in view. Fresh air fills your lungs. Rocks lie on the trail. Below you hear the gurgle of a nearby stream filled with water. Together, these biotic and abiotic factors make up the forest ecosystem (EK•oh•sis•tuhm). An **ecosystem** includes all living and nonliving things in an environment.

Biotic and abiotic factors in an ecosystem interact and supply the needs of living things. Recall that plants need abiotic factors to survive, including soil, sunlight, air, and water. Plants, in turn, provide food for most of the animals in an ecosystem.

Forest Ecosystem

The organisms in an ecosystem can be sorted into different populations (pop•yuh•LAY•shuhnz). A **population** includes all members of a single species in an area at a given time. For example, all the blue spruce trees in a forest form a population. Each species forms its own population. The monarch, painted lady, and buckeye butterflies all form separate butterfly populations in an ecosystem.

Together, the many different populations make up a community (kuh•MYEW•ni•tee). A **community** includes all the living things in an ecosystem. In addition to plants and animals, a community also has bacteria, protists, and fungi. For example, the soil of a forest community has huge populations of molds and bacteria living in it. The living community of most ecosystems might include thousands of populations.

This fallen log is part of a tiny ecosystem that includes fungi, moss, and bacteria.

An ecosystem can be local or widespread. An entire forest that covers a huge area can be an ecosystem. But one fallen log in the middle of that forest can also make up an ecosystem.

 Quick Check

Sequence List the parts that make up an ecosystem from the smallest to the largest.

Critical Thinking Soil is usually called an abiotic factor in an ecosystem. Why can soil also be considered a biotic factor?

Read a Photo

What biotic and abiotic factors can you see in this photo?

Clue: Make a list of the living and nonliving things.

How are food chains alike?

The path that energy and nutrients follow in an ecosystem is called a **food chain**. Food chains model the feeding relationships between organisms in an ecosystem. Energy flows in one direction in food chains. Once energy is used by an organism, it leaves the organism's body as heat. It becomes unavailable for other organisms in the ecosystem.

The energy in a food chain starts with the Sun. It is the energy source for almost all organisms on Earth. *Producers* are organisms that use the Sun's energy to make sugar and oxygen. Producers are at the base of every food chain.

During photosynthesis, producers, such as plants and algae, build sugar molecules out of carbon dioxide and water. The sugar molecules are the original source of food for consumers. A *consumer* is any animal that eats plants or other animals.

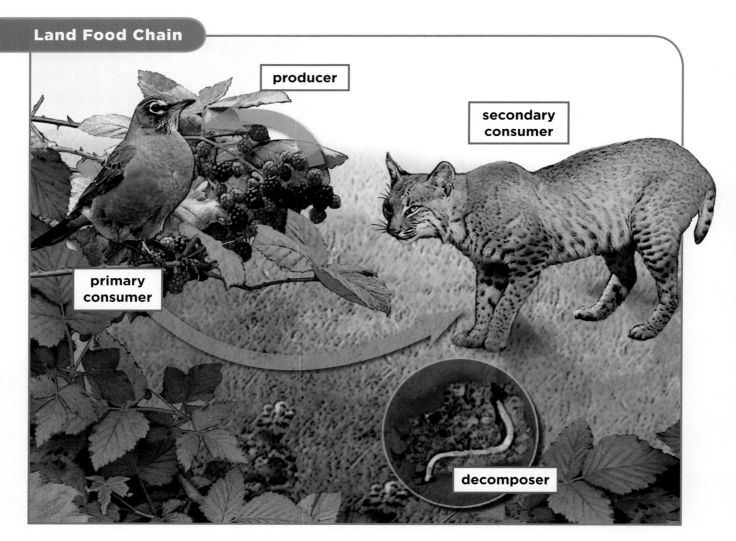

Land Food Chain

producer

secondary consumer

primary consumer

decomposer

Animals that eat producers directly are called herbivores. These consumers include squirrels, some birds, some insects, and grazing animals. Animals that eat other animals rather than producers are called *carnivores* (KAHR•nuh•vawrz). Bobcats and hawks are carnivores. Animals that eat both plants and other animals are called *omnivores* (OM•nuh•vawrz). Raccoons, woodpeckers, mice, and some crabs are omnivores.

Finally, there are decomposers in a food chain. *Decomposers* break down dead or decaying plant and animal material. Decomposers include fungi,

bacteria, termites, and many worm species. *Scavengers* are not included in these food chains. Scavengers are consumers that eat leftover bodies after they have started to rot. Common scavengers include vultures, raccoons, and some crabs.

 Quick Check

Sequence What general pattern do all food chains follow?

Critical Thinking What is the fewest number of links a food chain could have? The greatest number?

Water Food Chain

producer

primary consumer

secondary consumer

decomposer

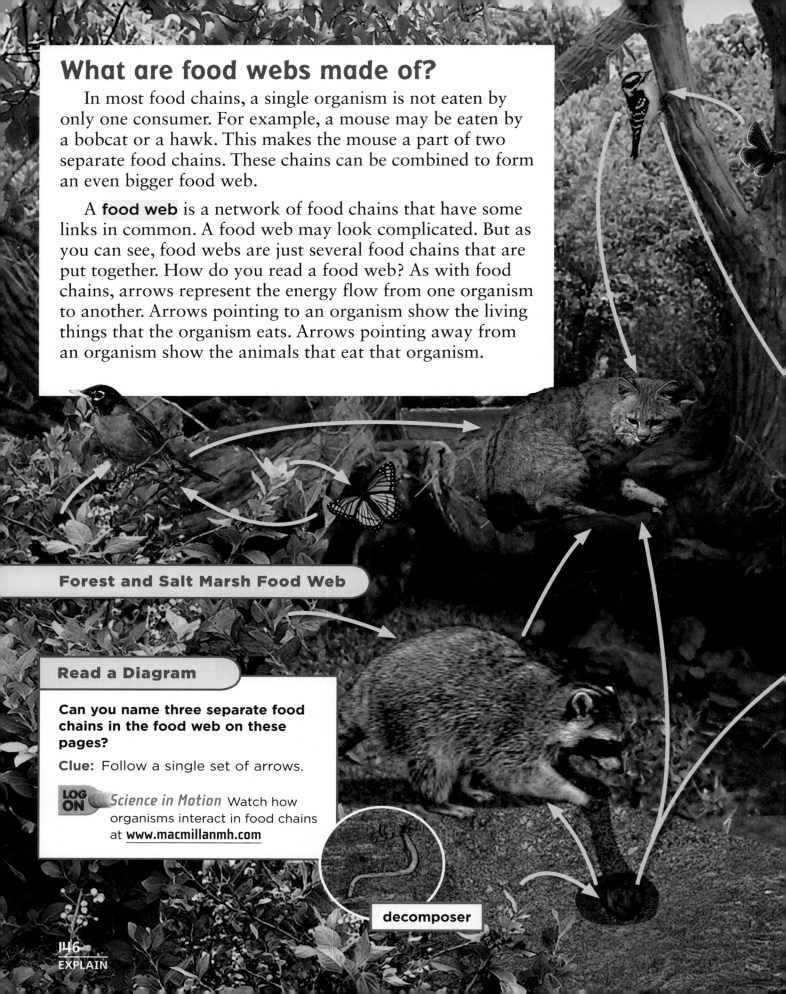

What are food webs made of?

In most food chains, a single organism is not eaten by only one consumer. For example, a mouse may be eaten by a bobcat or a hawk. This makes the mouse a part of two separate food chains. These chains can be combined to form an even bigger food web.

A **food web** is a network of food chains that have some links in common. A food web may look complicated. But as you can see, food webs are just several food chains that are put together. How do you read a food web? As with food chains, arrows represent the energy flow from one organism to another. Arrows pointing to an organism show the living things that the organism eats. Arrows pointing away from an organism show the animals that eat that organism.

Forest and Salt Marsh Food Web

Read a Diagram

Can you name three separate food chains in the food web on these pages?

Clue: Follow a single set of arrows.

LOG ON *Science in Motion* Watch how organisms interact in food chains at **www.macmillanmh.com**

decomposer

For example, arrows pointing to the hawk show that it hunts fish, frogs, mice, and small birds. A **predator** (PRED•uh•tuhr) is an animal that hunts other animals for food. Top carnivores are the highest-level predators in a food web. Arrows pointing away from the mouse show that it is hunted by hawks, raccoons, and bobcats. **Prey** (pray) are organisms that are eaten by other animals.

Predators are important in food webs and food chains. They limit the size of prey populations. When the number of prey animals is reduced, producers and other resources in an ecosystem are less likely to run out.

✔️ **Quick Check**

Sequence Describe the steps in constructing a food web.

Critical Thinking Can one organism be a consumer, omnivore, predator, and prey? Give an example that explains your answer.

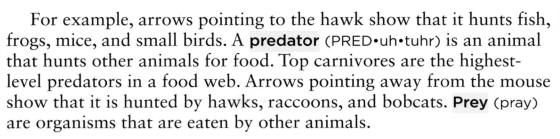
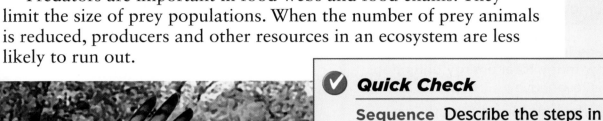

decomposer

Quick Lab

Energy Transfer

1. Make 100 energy cards. Each card represents 100 energy units.

2. Make an energy-level poster with four levels: Producers, Herbivores, Carnivores, and Top Carnivore.

3. Place 100 cards on the Producers. How many total energy units does this level have?

4. Continue to take 10% of the energy units from one level to the next level up to the Top Carnivore level. Use scissors if necessary.

5. **Draw Conclusions** How many energy units does the Top Carnivore level have?

How do energy pyramids compare?

An **energy pyramid** (EN•uhr•jee) is a diagram that shows the amount of energy available at each level of an ecosystem. How much of the Sun's original energy actually gets used during photosynthesis? In fact, only about 10 percent of the Sun's energy gets turned into food energy by a producer.

When a producer is eaten, only about 10 percent of the food energy it contains gets turned into herbivore or omnivore tissue. The rest is utilized or transferred into heat energy. For example, a butterfly drinks nectar from flowers to obtain energy. The butterfly's body then uses the energy from the nectar to support its life processes, such

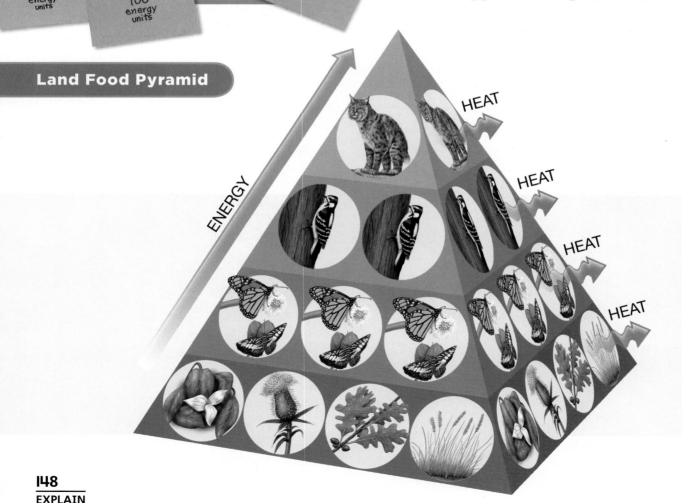

Land Food Pyramid

as respiration and digestion. If only 10 percent of plant tissue gets turned into butterfly tissue, 90 percent of the plant's energy is not used by the butterfly! This pattern continues with each level of an energy pyramid. When a bird eats the butterfly, it obtains even less energy. At each stage, about 90 percent of the available energy is not utilized. What does this mean? It means that most feeding patterns are not very efficient.

Energy pyramids illustrate that it takes a huge number of organisms to support an ecosystem. The bottom of the pyramid represents the producers. It is the largest level because it contains the most organisms, and therefore the most energy. There are fewer numbers of organisms and less available energy at each level of the pyramid.

In any ecosystem the number of producers is greater than the number of herbivores. Similarly, there are many more herbivores than carnivores. In a forest, for example, there are more flowers than butterflies. There are many more butterflies and other insects than birds. There are many more birds than bobcats, the top carnivores.

 Quick Check

Sequence Explain how energy is utilized in an energy pyramid.

Critical Thinking In a prairie ecosystem, would you expect a rabbit or a hawk population to be larger? Explain.

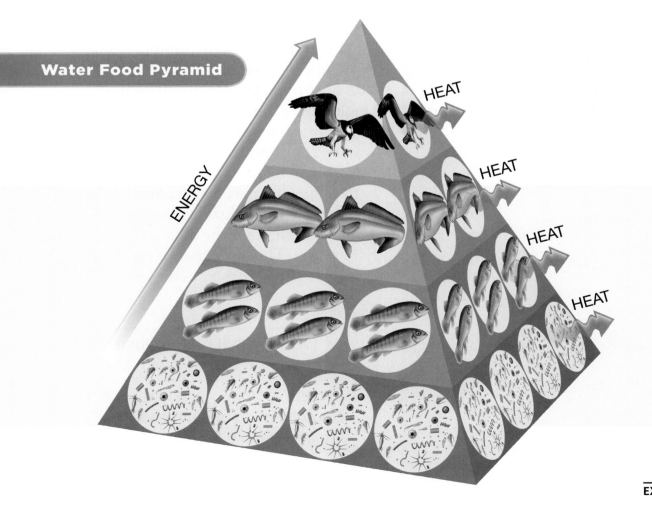

Water Food Pyramid

ENERGY

HEAT

HEAT

HEAT

HEAT

red tide algae

How does change affect a food web?

Most ecosystems stay in balance most of the time. What happens when a top carnivore is removed from a food web? What happens when a population in a food web increases in number? These changes set off a chain of events that affect all the organisms in a food web.

When top carnivores are removed from a food chain, prey populations are no longer controlled. Now prey organisms can reproduce without limits. When prey populations increase in number, more producers are required to supply them with energy. For example, if you removed the bobcat from the forest food web, the populations of birds, mice, and raccoons would increase. Soon there would be less grass, trees, and other producers to support these organisms.

Sometimes, a single population can grow out of control. For example, a *red tide* is a sudden explosive growth of single-celled algae in coastal areas. Red tides can occur when nutrient-rich deep water gets brought to the surface after a storm. With so many nutrients in the water, the algae keep reproducing. Toxins produced by the algae can cause the organisms that eat the algae, such as small fish, to die. This reduces the food energy available for predators who eat the fish.

 Quick Check

Sequence What occurs when a top carnivore is removed from a food web?

Critical Thinking What might happen if a population of producers was removed from a food web?

Lesson Review

Visual Summary

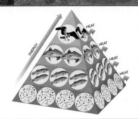

Producers, consumers, and decomposers all play important roles in **food chains**.

Food webs are networks of food chains that share common links.

Energy pyramids show the amount of energy available at each level of an ecosystem.

Make a FOLDABLES™ Study Guide

Make a Trifold Book. Use the titles shown. Tell about the topic on the inside of each fold.

Food chains Food webs Energy pyramids

Think, Talk, and Write

1. **Main Idea** How do producers and consumers obtain energy?

2. **Vocabulary** A consumer that hunts for its food is called a(n) _____.

3. **Sequence** Describe the events that take place as energy from the Sun travels through an energy pyramid.

First

↓

Next

↓

Last

4. **Critical Thinking** Where would you place decomposers on an energy pyramid? Explain.

5. **Test Prep** A food web can be broken down into separate

   A producers.   **C** food chains.
   B decomposers.   **D** food pyramids.

6. **Test Prep** Which of the following is the largest group in an energy pyramid?

   A consumers   **C** carnivores
   B producers   **D** herbivores

 ## Writing Link

Your Food Chain

What food chains are you a part of? Describe what you ate for lunch today. Tell how the foods you ate link back to a food chain or food web.

 ## Math Link

Food Pyramid

Suppose all the faces, or sides, of a food pyramid are triangles. How many sides, edges (where two sides meet), and vertices (where two or more edges meet) does it have?

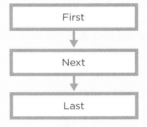

Writing in Science

Two Desert Creatures

Fictional Narrative

A good fictional story

▶ has an interesting beginning, middle, and end

▶ describes a setting that tells when and where the story takes place

▶ has a plot that centers around a problem or conflict

▶ has characters whose actions move the plot along

It was a cool night in the desert. The kangaroo rat crawled out of his underground burrow. He hopped on his long back legs to some nearby bushes. There he found some seeds on the ground. He was so busy stuffing seeds into his cheek pockets that he did not hear the soft rattling noise coming from behind him.

"Hello, furry friend," said the rattler. The moonlight shone on the brown diamond shapes along his back. "I'm very hungry. Are those seeds-s-s-s any good?"

"Stay back," screeched the kangaroo rat when he saw the snake slithering closer. "Don't be silly. I won't eat you, I just want some of your seeds-s-s-s," hissed the snake. But he quickly moved to within striking distance. The kangaroo rat tried to hop away, but it was too late.

Write About It

Fictional Narrative Choose two other organisms that share a predator/prey relationship. Write a fictional narrative in which these two organisms are in conflict.

LOG ON e-Journal Research and write about it online at **www.macmillanmh.com**

Calculate How Much Energy Is Used

In the energy pyramid, only about ten percent of the available energy gets used as food energy at each level. This is an inefficient use of energy. You can calculate how much energy is used for food and how much energy is transferred to heat at each level to see exactly how inefficient it is.

Suppose that the Sun gave off 95,000 energy units. To calculate how many units of energy are available at the producer level, you must start by finding 10 percent of 95,000. You can see that 10 percent of 95,000 is 9,500. This is how much energy would be available at the producer level.

Calculate Percents

To calculate percents

▶ **you write a percent as a decimal**

10% = 0.10

▶ **multiply the decimal by the total number**

95,000 x 0.1 = 9,500

 Solve It

1. Suppose that there are 9,500 energy units at the producer level. Calculate how many units of available energy each level would receive. How many units of available energy are transferred to heat energy at each level? **Hint**: To find out how much energy is transferred you must find 90% of available energy.

2. Suppose that a plant used 6,400 energy units from the Sun. How many units of available energy will a carnivore receive when it eats a herbivore? How many energy units did the plant transfer as heat?

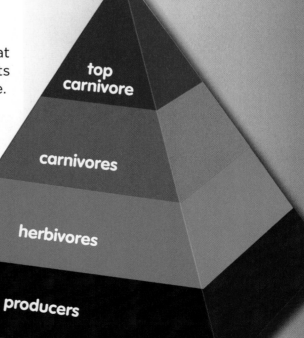

top carnivore

carnivores

herbivores

producers

Relationships in Ecosystems

Look and Wonder

African jacana birds spend hours picking small insects and ticks off the backs of hippos. How does this relationship help both organisms survive?

What do organisms need to survive?

Make a Prediction

What do organisms need to survive? Do organisms in an aquatic environment need different things than organisms in land environments? Make a prediction.

Test Your Prediction

1 Make a water environment. Place gravel in one container. Fill the container with pond water. Add water plants and snails.

2 Make a land environment. Place gravel in the other container and cover it with a layer of soil. Add grass seeds and earthworms and cover them with additional soil. Water the seeds.

3 Cover each container with a lid. Place the containers in a well-lit place out of direct sunlight.

4 **Observe** Examine your containers for changes every day for a week. Do the organisms in each environment interact? Record your observations.

Draw Conclusions

5 What are the abiotic and biotic parts of water and land environments?

6 **Infer** How do the plants help the animals survive in the water environment? The land environment?

7 What would happen to each environment if the plants or animals were removed?

Explore More

What other factors affect an organism's survival? Try adding more plants or animals to your environments. Try placing your environments in the dark for a few days. How do the environments change?

Materials

- gravel
- 2 containers with lids
- pond water
- water plants
- water snails
- soil
- grass seed
- earthworms

Step **2**

Step **3**

▶ Main Idea

Abiotic factors and interactions between organisms control the size of populations in a community.

▶ Vocabulary

limiting factor, p.156
carrying capacity, p.157
habitat, p.158
niche, p.158
symbiosis, p.160
mutualism, p.160
commensalism, p.161
parasitism, p.162

-Glossary
at **www.macmillanmh.com**

▶ Reading Skill ✔

Infer

Clues	What I Know	What I Infer

Why do organisms compete?

Life in an ecosystem is a constant struggle. Food, water, space, and other resources are restricted. Organisms struggle to get their share of each resource. This fight for limited resources is called *competition*.

Who competes in an ecosystem? Organisms within a population compete with one another. A fox must compete with other foxes to catch rabbits. Populations also compete. Foxes and hawks, for example, both eat rabbits. Since there is a limited number of rabbits, the two predator populations compete for food. The rabbits must also compete with other herbivore populations for their food.

Ultimately, the survival of populations comes down to resources. A **limiting factor** (LIM•i•ting FAK•tuhr) is any resource that restricts the growth of populations. A forest, for example, gets more rainfall and is much warmer in summer than winter. In summer, the forest can support many more populations than in winter. In this case, rainfall and temperature are limiting factors. Common abiotic limiting factors include water, temperature, weather, soil type, space to grow, shelter, and sunlight.

In winter, bison must search for food.

This pond is overcrowded with algae.

≡ **Quick Lab**

Limiting Factors

1. ⚠ **Be Careful.** Use scissors to cut out twenty 2.5 cm (1 in.) circles. Each circle represents the range that the roots of the plant extend.

2. **Measure** Create an environment for these plants by making a 20 cm (8 in.) square box on your desk.

3. Toss 8 plants into the environment. If a plant does not touch another plant, it "survives." If the plant touches another plant, remove the plant and any plant that it touches. Record your results in a data table.

4. Increase the number of plants that you toss to 10, 12, 14, and so on. Record your results. Which number of plants tossed allows the most plants to survive?

5. **Infer** How can crowding be a limiting factor for a population?

Biotic factors can also limit ecosystems. A prairie ecosystem has more producers than a desert ecosystem. As a result, the prairie can support more herbivores, which support more carnivores. In this case, the amount of available food is the biotic limiting factor for the desert ecosystem. With more available food, the prairie ecosystem can support more populations.

Together, biotic and abiotic factors determine the carrying capacity (KAR•ee•ing kuh•PAS•i•tee) for each population. The **carrying capacity** is the greatest number of individuals within a population that an ecosystem can support. For example, a rain forest can support a certain number of jaguars. If the jaguar population starts to rise, food becomes harder to find. Soon, some of the jaguars die and the population returns to its former level.

Overcrowding also limits growth. An algae population in a nutrient-rich pond may seem like it can grow indefinitely. But the algae will eventually get so thick that they start to use up the oxygen in the pond. Without enough oxygen for respiration, the algae and other organisms begin to die off.

✔ Quick Check

Infer Compared to the surface, the bottom of the ocean is dark and has very few organisms. What might be a limiting factor in this ecosystem?

Critical Thinking Why is a sudden increase in a predator population usually temporary?

FACT Populations cannot grow indefinitely.

How do organisms avoid competition?

An organism avoids competition by having a specific territory and a unique role within its ecosystem. A **habitat** (HAB•i•tat) is the physical place where an organism lives and hunts for food. Some creatures have very small habitats. Pill bugs, for example, spend most of their time under and around a stump or rock. A bee's habitat is larger. It is not only the hive where the bee lives. It also includes the fields and forests where the bee searches for flowers.

A **niche** (nich) is the special role that an organism plays in a community. For example, two birds might live in the same location and eat the same food. But one bird is active at night while the other is active during the day. Therefore, the two birds occupy different niches.

In a similar way, two birds might share the same rain-forest habitat but eat different foods. One bird eats plants while the other eats insects. The two birds occupy two different niches in the community. For example, honeycreepers are a group of related birds found on the islands of Hawaii. These birds all share the same habitat, but are able to avoid competing with each other by eating different foods.

The **akiapolaau** removes insects from beneath tree bark.

The **iiwi** sips nectar from long, tube-shaped flowers.

The **Maui parrotbill** finds insects and grubs by crushing twigs.

The **Maui creeper** eats insects and grubs it finds on the leaves, branches, and bark of trees.

The **apapane** sips nectar from flowers high in the tree tops of the rain forest.

Read a Photo

Why does each honeycreeper have a uniquely shaped beak?

Clue: Compare the beak shapes with the methods for finding food. How would certain beaks help honeycreepers to obtain different foods?

✓ Quick Check

Infer Two populations share the same food and habitat. W key difference could cause them to occupy different niches

Critical Thinking What might happen to organisms when habitats are destroyed?

How do organisms benefit from interactions?

Living things in an ecosystem depend on one another. For example, all animals in an ecosystem depend on plants and other producers for food. Plants depend on animals for carbon dioxide. These interlocking relationships are examples of interdependence. *Interdependence* is the reliance of organisms on other organisms for their survival. Some forms of interdependence are linked more closely than others. **Symbiosis** (sim•bye•OH•sis) is a relationship between two or more kinds of organisms that lasts over time.

Mutualism

A symbiotic relationship that benefits both organisms is called **mutualism** (MYEW•chew•uh•liz•uhm). A pollinator and a flowering plant provide an example of mutualism. The pollinator, usually an insect or bird, gets sweet nectar from the flower. The plant gets its pollen transported to the pistil of another flower. Both organisms gain from the relationship.

A fascinating example of mutualism is seen in the relationship between ants and acacia trees. The tree provides a home and food for the ants. The ants in turn defend the tree against other insect pests. How successful is this relationship? Scientists used chemicals to get rid of the ants on an acacia. Without its ants, the tree soon died!

Another example of mutualism can be seen in lichens. A *lichen* is actually two different organisms— a fungus and an alga—that live together. The fungus provides the alga with a home and nutrients. As a result, the alga does not dry out. The alga, in turn, provides the fungus with food and oxygen.

These ants are defending an acacia tree from other insects.

British soldier lichen

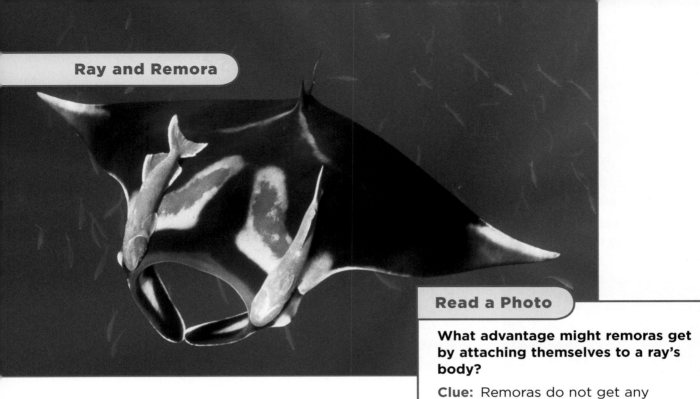

Ray and Remora

Read a Photo

What advantage might remoras get by attaching themselves to a ray's body?

Clue: Remoras do not get any nutrition from the ray itself.

Commensalism

Remora are fish that attach themselves to the bodies of rays and sharks. The remora gets food scraps, transportation, and protection from the ray. What does the ray get from the remora? While the remora does not hurt the ray in any way, it does not help the ray either. A symbiotic relationship that benefits one organism without harming the other is called **commensalism** (kuh•MEN•suh•liz•uhm).

Other examples of commensalism include the growth of orchids on trees in the rain forest. Rather than root in the ground, orchids anchor themselves high in a tree. This situation helps the orchid. It does not hurt the tree, so it is an example of commensalism. Barnacles growing on the backs of whales are also commensal. The barnacles gain a home. The whales are not hurt by the barnacles.

Sometimes it is difficult to tell whether a relationship between organisms really is an example of commensalism. The clownfish, for example, lives among the tentacles of the sea anemone. It uses the anemone for protection. When chased by predators, the clownfish retreats to the anemone's tentacles. In this relationship, the clownfish is clearly helped by the anemone. However, it is hard to tell whether the anemone gains from the clownfish. Most scientists think that the relationship is an example of commensalism.

 Quick Check

Infer How do algae and fungi benefit from living as a lichen?

Critical Thinking Oxpecker birds eat pests that bother rhinoceroses. Is this an example of mutualism or commensalism? Why?

▲ A magnified view of a wood tick on human skin.

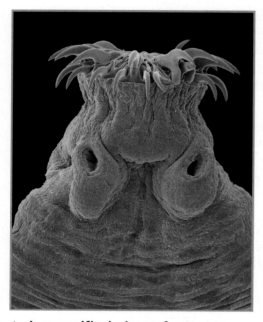

▲ A magnified view of a tapeworm.

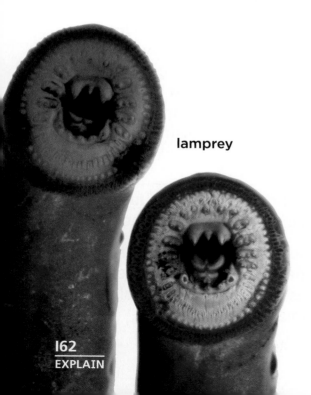

lamprey

What are parasites?

Some partnerships are harmful for individuals in the relationship. **Parasitism** (PAR•uh•sye•tiz•uhm) is a symbiotic relationship where one organism benefits and the other is harmed. A *parasite* lives in or on a host organism and benefits from the relationship. For example, ticks are parasites on dogs and other animals. Ticks use their host's body for a home and a food source. A tick attaches itself to a host and harms the host by taking in the host's blood. The tick's host gets no benefit from the relationship.

Some parasites are very harmful for the host organism. Millions of people around the world have parasites called tapeworms. These worms live inside a person's intestinal tract. Tapeworms more than 70 centimeters (2 feet) in length have been found in humans. Tapeworms can harm their hosts by causing fevers and digestive problems. Another dangerous parasite is a lamprey. Lampreys are parasitic fish. They use their suckerlike mouth to attach themselves to other fish. They harm their host by sucking out its blood and other body fluids.

Some parasites are protists, including the species of amoeba that causes a disease called *dysentery* (DIS•uhn•ter•ee). Dysentery amoebas enter the host's body through contaminated food or water. The protist that causes sleeping sickness in Africa lives in the bodies of cows and other large animals. When these animals are bitten by flies, the flies transfer the protists to humans, causing the disease.

✔ Quick Check

Infer Why do parasites often harm, but not kill, their hosts?

Critical Thinking How is a parasitic relationship like a predator-prey relationship?

Lesson Review

Visual Summary

Competition and **limiting factors** control the size of populations in an ecosystem.

Organisms avoid competition by occupying different **niches** and **habitats**.

Mutualism, commensalism, and parasitism are examples of **symbiosis**.

Make a FOLDABLES™ Study Guide

Make a Three-Tab Book. Use the titles shown. Tell about the topics on the inside of each tab.

Competition and limiting factors

Niche and habitats

Symbiosis

Think, Talk, and Write

1 Main Idea How can biotic and abiotic factors affect the size of a population?

2 Vocabulary The role of an organism in the community is its _____.

3 Infer A predator population suddenly decreases even though the prey stays the same. Besides disease, what could explain this change?

Clues	What I Know	What I Infer

4 Critical Thinking How do humans change the abiotic factors in their habitat? Explain.

5 Test Prep Which of the following determines the carrying capacity of a population in an ecosystem?
   A plants and animals
   B abiotic limiting factors
   C biotic limiting factors
   D abiotic and biotic limiting factors

6 Test Prep Which term represents all living things in an ecosystem?
   A a community   **C** a limiting factor
   B a population   **D** a habitat

Writing Link

Personal Narrative
What niche do you occupy? Write a personal narrative that tells about your unique "niche."

Math Link

Determine Area
Suppose a wolf's habitat is a rectangle that measures 4.5 km on one side and 6.4 km on the other side. What is the area of this habitat?

Focus on Skills

Inquiry Skill: Predict

You just read about how some organisms get food by eating other organisms. Can anyone know in advance what effect this will have on the population size? When scientists have questions like that, they conduct simulations and study the results. Then they can **predict** what might happen in a similar situation.

▶ Learn It

When you **predict**, you state the possible results of an event or experiment. Then you conduct a test and interpret the results to determine if your prediction was correct.

It is important to record your predictions, as well as any measurements or observations you make during the test. Your observations and measurements provide written proof of whether or not your prediction was correct. In this activity, you will predict how population size will change.

▶ Try It

How many deer do you **predict** will survive in a population of wolves? Use what you have learned about predators and prey to write your prediction. Then use the model to see if your prediction was correct.

Materials masking tape, 8 7.5 cm cardboard squares, 100 2.5 cm construction paper squares, graph paper

1. Use tape to mark off a 60 cm by 60 cm square. This square represents a meadow. Distribute 10 of the 2.5 cm paper deer squares in the meadow.

2. Toss the 7.5 cm cardboard wolf square in the meadow. Remove any deer that touch the wolf. In order to survive, the wolf must catch, or touch, 3 deer. If the wolf survives, it produces 1 offspring. If the wolf does not catch any deer, it starves.

Predator-Prey Results

Trial	Deer	Wolf	Deer Caught	Wolves Starved	Wolves surviving	New Baby Wolves	Deer left
1	10	1	2	1			8
2	16	1	3		1	1	13
3	26	2	10		2	2	16
4	32	4	18		4	8 4	14
5	28	8	23	3	5	5	5
6	10	10	9	9	1	1	1
	2	2		2			2

3 Record your results in a data table. What happened to the wolf and deer in this trial?

4 At the start of the next trial, double the deer remaining from the first trial to represent new deer offspring. Disperse these new deer in the meadow.

5 If the entire deer population was caught by the wolf in the previous trial, then add 3 new deer to the meadow.

6 In each additional trial throw a wolf square once for each wolf. This includes any surviving wolves from previous trials and any of the offspring produced in previous trials. Record your results in your data table.

7 Repeat steps 1 through 6 for a total of 14 trials.

▶ **Apply It**

Predict the outcomes for 6 trials. Base your prediction on the pattern you observed during the first 14 trials. Then actually model trials 15 to 20. Were your predictions correct?

Graph the data for your 20 trials. Place the deer and wolf data on the same graph so that the interrelationship can be easily observed. Label the vertical axis "Number of Animals" and the horizontal axis "Trials." Use one color for the deer data and another for wolf data.

Lesson 3

Adaptation and Survival

Look and Wonder

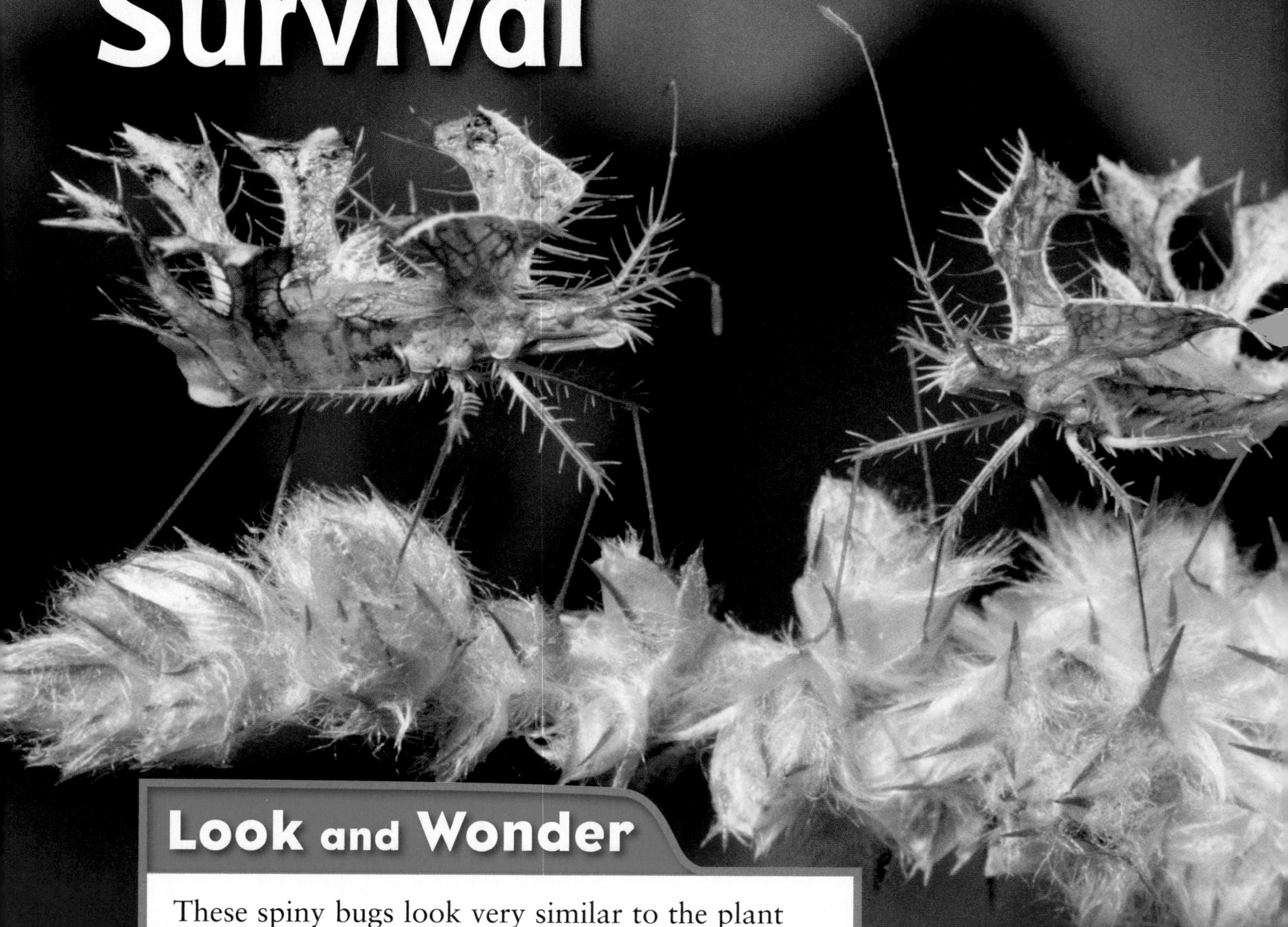

These spiny bugs look very similar to the plant they are standing on. How does blending in with an environment help an organism?

How do adaptations help animals survive in their environment?

Materials

Form a Hypothesis

Sow bugs are animals that live under logs, leaves, and rocks. Are sow bugs adapted to prefer damp or dry environments? Write your answer as a hypothesis in the form "If moisture in the sow bug's environment is increased, then . . ."

Test Your Hypothesis

1 **Observe** Place 15 sow bugs on the tray. Examine the sow bugs with the hand lens. Record your observations.

2 **Experiment** Tear four paper towels in half. Make sure they are the same size. Dampen two of the halves.

3 Move the sow bugs to the center of the tray. Place the moist paper towels in one end of the tray. Place the dry paper towels on the opposite side of the tray.

4 Watch the sow bugs for several minutes. Look for changes in their behavior.

5 After 10 minutes, count the sow bugs on each side of the tray. Record your results. ⚠ **Be Careful.** Wash your hands after handling sow bugs.

Draw Conclusions

6 Based on your observations, what traits help sow bugs survive in their environments?

7 What were the independent variable and dependent variable? What variables remained constant?

8 **Infer** Did your results support your hypothesis? Explain why or why not.

- sow bugs
- tray
- hand lens
- paper towels
- water

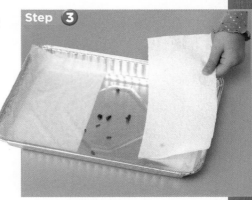

Step **1**

Step **3**

Explore More

Are sow bugs adapted to prefer dark or light environments? Form a hypothesis and test it. Then analyze and write a report of your results.

Read and Learn

▶ **Main Idea**

Organisms have adaptations that help them survive in their environments.

▶ **Vocabulary**

adaptation, p.168
camouflage, p.173
protective coloration, p.173
protective resemblance, p.173
mimicry, p.174

LOG ON ⊜-Glossary
at www.macmillanmh.com

▶ **Reading Skill** ✓

Problem and Solution

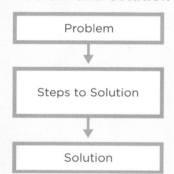

What is adaptation?

Survival in any ecosystem is a constant struggle. An **adaptation** (ad•uhp•TAY•shuhn) is any characteristic that helps an organism survive in its environment. Over time, organisms with successful adaptations survive more frequently than other organisms. Their offspring inherit these adaptations. Adaptations can be structural or behavioral.

Structural Adaptations

Structural adaptations are adjustments to internal or external physical structures. Fur color, long limbs, strong jaws, and the ability to run fast are structural adaptations. Some structural adaptations help organisms survive in certain environments. For example, ducks have webbed feet that help them survive in water. Cactuses have a thick, waxy cuticle that prevents water loss in their dry environment.

Other structural adaptations protect prey from predators or enable predators to hunt more successfully. Turtles have hard shells that protect them from predators. Predators such as sharks have an excellent sense of smell and sharp teeth. Both of these traits help sharks catch their prey.

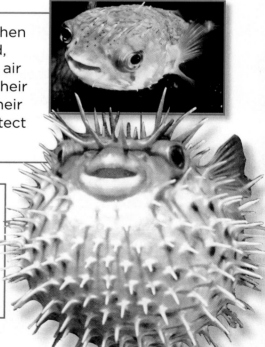

Structural Adaptation When puffer fish are threatened, they fill their bodies with air or water. As they fill up, their spines are pushed out. Their spines and large size protect them from predators.

Structural Adaptation Many plants, such as roses and cactuses, have thorns or spines on their stems. These modified leaves protect the plant from herbivores.

Behavioral Adaptations

An adjustment in an organism's behavior is a *behavioral adaptation*. For example, wolves traveling in packs is a behavioral adaptation. Wolf packs can hunt large prey that one wolf alone could not capture. Many prey animals also travel in groups. Some fish swim in schools which protects them from predators. Symbiotic relationships are also behavioral adaptations.

Some behavioral adaptations help animals survive seasonal changes in the climate. Many animals such as birds, butterflies, and fish migrate. *Migration* (mye•GRAY•shuhn) is a seasonal movement of animals to find food, reproduce in better conditions, or find a less severe climate. Other animals such as bats, snakes, turtles, and frogs hibernate to escape the cold. *Hibernation* (hye•ber•NAY•shuhn) is a period of inactivity during cold weather. The animals remain inactive until warmer temperatures return in spring.

 Quick Check

Problem and Solution How do sea otters eat animals with shells?

Critical Thinking What structural and behavioral adaptations do humans have?

Behavioral Adaptation Sea otters eat shelled animals, such as crabs and clams. They crack open the shells using rocks. An otter will hold a rock on its stomach and smash the crab or clam against the rock.

Behavioral Adaptation Elephants have complex social behaviors. Adult elephants form herds which protect their young from predators and other dangers. Young elephants will often hold on to their mothers' tails to stay close to the herd.

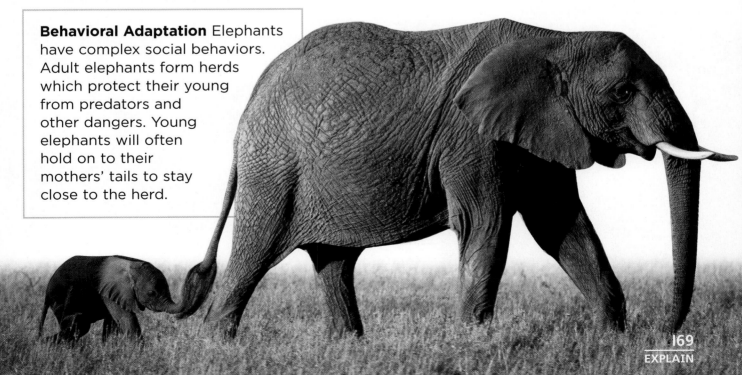

What are some plant adaptations?

Angiosperms have scented flowers that attract certain pollinators. They have leaves that catch sunlight and roots that soak up water. These and other adaptations help plants survive.

Some plants have specific structural adaptations to different environments. Rain-forest plants, like orchids, have adaptations that help them survive wet, hot temperatures. Orchid stems have storage organs called *pseudobulbs* (SEW•doh•bulbz). They store water for the plant. An orchid's aerial roots help secure it to a tree high in the rain forest. These roots also absorb water

Rain-Forest Adaptations

Stems Pseudobulbs store water.

Leaves Drip-tip leaves drain excess water.

Roots Aerial roots absorb water from air.

Read a Diagram

Which part of the orchid is the pseudobulb a part of?

Clue: Look at the diagram and the photo. What is the pseudobulb attached to?

from the moist air. Like many rain-forest plants, orchids have drip-tip leaves. These leaves are adapted to the constant wet conditions in a rain forest. Their tips drain excess water.

Plants, like cactuses, that live in hot and dry environments have thick, waxy stems that prevent water loss. They have very dense, shallow roots that soak up rain quickly. Plants that live in forests, like oak trees, lose their leaves in the winter. This helps them prevent water loss. Cold climate plants, such as moss, are able to complete their life cycle in a shortened growing season. Some aquatic plants, such as water lilies, have stomata on the top surface of the leaf instead of the bottom. This enables the stomata to take in and release carbon dioxide and oxygen.

Many plants have adaptations that defend them from herbivores. For example, some plants produce chemicals that give them a bad taste. When most herbivores eat the leaves, they do not like the taste and stop eating the plant. Other plants, such as milkweeds, produce chemicals that are poisonous to most animals. Both of these adaptations protect the plants from predators.

Quick Check

Problem and Solution How do water plants release oxygen and take in carbon dioxide?

Critical Thinking Why do adaptations always "fit" the environment? For example, why don't cactuses have drip-tip leaves?

Quick Lab

Leaf Adaptations

1 Examine an oak leaf, pine needles, and an elodea leaf. Draw what you see.

2 **Measure** Use a ruler to measure each leaf. Record your data.

3 Break open each leaf. How do the leaves compare?

4 **Infer** Which environments are each of the leaves adapted to? Explain.

Water Adaptation These water lilies have stomata on top of their leaves.

What are some animal adaptations?

Like plants, animals have adaptations that help them survive in specific environments. Animals that live in cold climates have thick fur and extra body fat that keep them warm. Desert animals are often active at night, or *nocturnal*. They stay in shelters or underground burrows during the day and avoid the heat. Nocturnal animals come out at night to search for food.

Animals that live in water also have adaptations. Aquatic animals are usually much more streamlined than land animals. This allows them to swim quickly through the water. Aquatic mammals can hold their breath for long periods of time. Other aquatic animals breathe underwater using gills.

Many animal adaptations develop because of predator and prey relationships. Prey have adaptations that enable them to avoid predators. Predators have adaptations that help them hunt and capture prey. Prey animals, such as gazelles, are able to run at speeds of up to 80 kilometers per hour (49.7 miles per hour). Some animals use chemicals to escape predators. When skunks are threatened, they spray a bad-smelling liquid. These adaptations help prey escape predators.

Predators also have adaptations that make them more efficient hunters. Owls, for example, have several adaptations that make them successful night hunters.

Head Owls have excellent hearing which helps them hunt. One of their ears is higher than the other. This increases their ability to distinguish where sounds are coming from and how far away a sound is.

Eyes Owls have large eyes which help them see tiny prey, such as mice, in the dark. Their eyes are positioned at the front of their head which gives them better vision.

Wings An owl's large, muscular wings help it swiftly hunt for prey. Special tips on the wing feathers muffle the sound of air rushing over the wings as the owl flies. This helps the owl fly silently.

Feet An owl's feet are also adapted for hunting. They have large talons, or claws, for accurately grabbing prey. This adaptation helps them pick up larger prey animals.

Camouflage

Some organisms increase their survival in an environment by blending in. Any coloring, shape, or pattern that allows an organism to blend in with its environment is called **camouflage** (KAM•uh•flahzh). Predators with camouflage can sneak up on prey. Camouflage also helps prey animals hide from predators.

Protective coloration (pruh•TEK•tiv kul•uh•RAY•shuhn) is a type of camouflage in which the color of an animal helps it blend in with its background. In winter, the arctic fox has a white coat that blends in with the snow. In summer, the fox's coat changes color to help it blend in with the plants that grow in the warm weather. Similarly, a tiger's stripes make it difficult to see in the grass. Stripes help a tiger conceal itself from its prey.

Some organisms go beyond protective coloration. Matching the color, shape, and texture of an environment is called **protective resemblance** (ri•ZEM•bluhns). The walking stick insect, for example, resembles a stick or a small branch.

This pipefish resembles the sea grass in its environment.

 Quick Check

Problem and Solution How could you tell whether a rabbit comes from a cold weather or a warm weather environment?

Critical Thinking Many flowering plants have brightly colored flowers that are very noticeable. Why don't these plants use camouflage?

Protective coloration helps arctic hares blend in with their snowy environment.

What is mimicry?

Some animals have adapted to their environment by copying other well-adapted organisms. An adaptation in which an animal is protected against predators by its resemblance to an unpleasant animal is called **mimicry** (MIM•i•kree). The viceroy butterfly, for example, is protected from predators because it looks just like the bad-tasting, poisonous monarch butterfly.

Mimic organisms can look so much like a dangerous or unpleasant animal that their enemies stay away. The harmless robber fly resembles the dangerous bumblebee. The king snake mimics the coloring of the poisonous coral snake.

Predators also use mimicry. Instead of warning their prey, they use mimicry to deceive it. Some snapping turtles, for example, have the ability to wag a fleshy "lure" in their mouth. The lure looks like a worm. When fish come closer to try to eat the "worm," the turtles catch the fish.

 Quick Check

Problem and Solution How do snapping turtles solve the problem of catching fish?

Critical Thinking How does mimicry increase an organism's chance of survival?

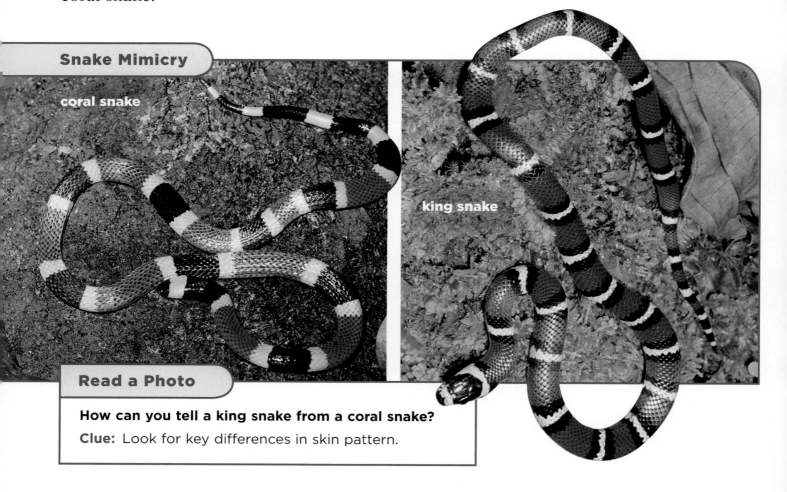

Snake Mimicry

coral snake

king snake

Read a Photo

How can you tell a king snake from a coral snake?

Clue: Look for key differences in skin pattern.

Lesson Review

Visual Summary

Adaptations are traits that help organisms survive in their environments.

Plant adaptations include variations in their leaves, flowers, stems, and roots that help them survive in different environments.

Animal adaptations include **camouflage** and **mimicry**.

Make a FOLDABLES™ Study Guide

Make a Trifold Book. Use the titles shown. Tell what you learned about each topic.

Main Idea | What I learned... | Sketches
Adaptations
Plant adaptations
Camouflage and mimicry

Think, Talk, and Write

1. **Main Idea** What are structural and behavioral adaptations?

2. **Vocabulary** An organism imitating a harmful organism is called _____.

3. **Problem and Solution** How are aquatic animals able to survive in water?

Problem

Steps to Solution

Solution

4. **Critical Thinking** Can adaptations be both behavioral and structural? Explain.

5. **Test Prep** Which of the following are adaptations for cold weather?
   - **A** thick fur, big ears
   - **B** thick fur, body fat
   - **C** body fat, gills
   - **D** sleek shape, gills

6. **Test Prep** Which of the following is a behavioral adaptation?
   - **A** scaly skin
   - **B** sharp teeth
   - **C** hibernation
   - **D** camouflage

Writing Link

Fictional Narrative
Why does the giraffe have a long neck? How does its neck help the giraffe survive in its environment? Write a story about how the giraffe might have acquired this adaptation.

Art Link

Adaptation Art
Make a painting or drawing that illustrates an animal using camouflage, protective coloration, protective resemblance, or mimicry.

Meet Caroline Chaboo

Plants in the tropical forests of the Caribbean face many challenges. They endure pounding rain, drought, and 160-kilometer-per-hour (100-mile-per-hour) hurricane winds. Yet plants like the Sabal palm (*Sabal causarium*) have adapted to meet these challenges.

This tall, regal palm resists the power of the wind very well. Its root system holds the tree in place and prevents it from being knocked over by strong storms. The palm's long, flexible leaves also help it survive high winds.

The tree can live through hurricanes, but faces another obstacle—a plant-eating beetle. Caroline Chaboo is a scientist at the American Museum of Natural History. Caroline studies the relationships between plants and insects. She researches the tiny tortoise beetle (*Hemisphaerota palmarum*), which is found in the Dominican

Female tortoise beetles place their eggs within the tissues of the plant.

These scars were caused by tortoise beetles.

Republic. This beetle and its larvae feed on the leaves of the Sabal palm. They scrape the palm's leaves with their mandibles, or jaws. This produces long scars that cause the leaves to dry out and die. Since palm trees have few leaves, losing even one can harm the growth of the entire tree.

While the tortoise beetle weakens the palm, the tree itself does not die. Scientists have found that many plants produce proteins that serve as a defense against insects. Caroline is studying the Sabal palm to find out whether it too produces a natural pesticide against the beetles.

By studying the tortoise beetle and the Sabal palm, Caroline hopes to learn more about how plants and animals adapt to their habitats.

 Write About It

Infer

1. How might a natural pesticide produced by the Sabal palm help other organisms?
2. Research tortoise beetles. What other plants do they eat? Write a report that tells how such a pesticide could help other plants.

LOG ON ℮-Journal Research and write about it online at **www.macmillanmh.com**

Infer

▶ Review the text to make inferences about information not stated explicitly.

▶ List the details that support the inferences you make.

AMERICAN MUSEUM OF NATURAL HISTORY

Visual Summary

Lesson 1 Food chains, food webs, and energy pyramids show the energy flow between organisms in an ecosystem.

Lesson 2 Abiotic factors and interactions between organisms control the size of populations in a community.

Lesson 3 Organisms have adaptations that help them survive in their environments.

Make a FOLDABLES™ Study Guide

Assemble your lesson study guides as shown. Use your study guides to review what you have learned in this chapter.

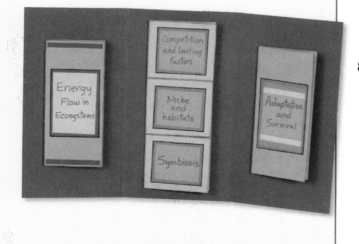

Fill in each blank with the best term from the list.

adaptation, p. 168   **habitat**, p. 158

camouflage, p. 173   **parasitism**, p. 162

ecosystem, p. 142   **prey**, p. 147

food chain, p. 144   **symbiosis**, p. 160

1. All living and nonliving things in an environment make up a(n) _____.

2. A relationship where one organism benefits and the other is harmed is _____.

3. An organism lives and hunts for food in its _____.

4. Animals that are eaten by other animals are called _____.

5. A characteristic that helps an organism survive in its environment is a(n) _____.

6. Some organisms blend in with their environment using _____.

7. The path that energy and nutrients follow in an ecosystem is a(n) _____.

8. Mutualism and commensalism are different types of _____.

LOG ON **e–Review** Summaries and quizzes online at **www.macmillanmh.com**

Answer each of the following in complete sentences.

9. **Problem and Solution** Desert ecosystems are dry and often hot. What structural and behavioral adaptations do organisms living in a desert have to solve this problem?

10. **Infer** How do the abiotic characteristics of this pond environment act as limiting factors in this environment?

11. **Predict** A rabbit with brown fur lives in a snowy environment. What do you think will happen to the rabbit?

12. **Critical Thinking** What would happen if an organism at the bottom of a food chain died off?

13. **Fictional Narrative** Write a short story set in the future. Suppose that some people have settled, with their pets, on a new planet. Create an ecosystem for the planet. What adaptations will the humans and animals develop to live in the new planet's ecosystem?

The Big Idea

14. How do organisms interact?

An Ecosystem in Action!

Create a skit about how animals in an ecosystem interact.

What to Do

1. Working with a group, choose an ecosystem in which to set your skit. What animals, plants, and other organisms live in this ecosystem?

2. Choose several animals from your ecosystem that interact with each other. They may be predator and prey, or they may compete for food. They may also interact through symbiosis.

3. Write a skit showing how the animals interact. Perform your skit for the class.

Test Prep

1. **Which of the following organisms is a producer in this food chain?**

A blueberry bush

B robin

C bobcat

D earthworm

CHAPTER 4

Ecosystems and Biomes

The Big Idea

How are ecosystems different?

Coral reef, Papua New Guinea

Key Vocabulary

water cycle
the continuous movement of water between Earth's surface and the air, changing from liquid to gas to liquid (p. 184)

endangered species
a species that is in danger of becoming extinct (p. 199)

succession
the process of one ecosystem changing into a new and different ecosystem (p. 200)

pioneer species
the first species living in an otherwise lifeless area (p. 200)

climax community
the final stage of succession in an area, unless a major change happens (p. 201)

estuary
the boundary where a river feeds into an ocean (p. 226)

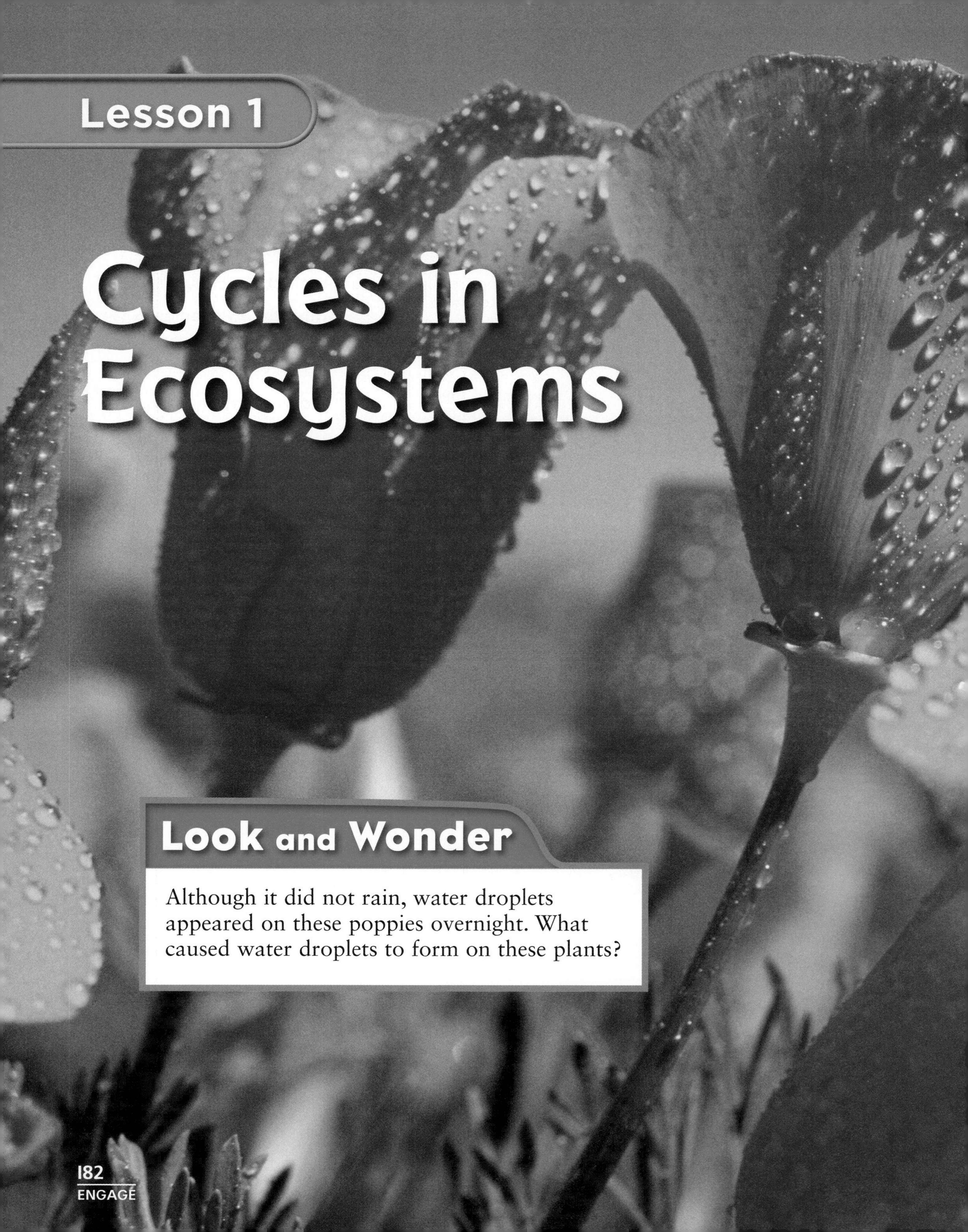

Cycles in Ecosystems

Look and Wonder

Although it did not rain, water droplets appeared on these poppies overnight. What caused water droplets to form on these plants?

Explore

How do water droplets form?

Form a Hypothesis

Water droplets occur when water changes from a gas to a liquid. Does temperature affect water droplet formation on an object? Write your answer as a hypothesis in the form "If the temperature of a glass is decreased, then . . ."

Test Your Hypothesis

1. Fill one glass completely with ice. In a separate glass, add a few drops of food coloring to some cold water and stir. Then pour the water into the glass that is full of ice.

2. Fill an empty glass with room-temperature water. Add a few drops of food coloring to the water and stir. Be sure to use the same amount of food coloring and water in each glass.

3. **Experiment** Sprinkle salt onto each saucer. Then put one glass on each saucer. Allow the glasses to sit for 30 minutes.

4. **Observe** What do you see on the sides of each glass?

Draw Conclusions

5. What does the color of the droplets indicate about where the water droplets came from?

6. **Use Variables** What were the independent and dependent variables in this experiment? Which variables were controlled?

7. **Infer** Why do you think water droplets formed where they did?

Explore More

What happened to the salt under the glass with water droplets? Plan and carry out an experiment that shows where the salt is.

Materials

- 2 glasses
- ice
- food coloring
- water
- spoon
- salt
- 2 saucers

Step 1

Step 3

▶ **Main Idea**

The important chemicals for life—water, carbon, nitrogen, and oxygen—are used and reused as they flow through ecosystems.

▶ **Vocabulary**

water cycle, p.184

evaporation, p.184

condensation, p.184

precipitation, p.184

watershed, p.184

runoff, p.184

groundwater, p.184

carbon cycle, p.186

nitrogen cycle, p.188

compost, p.190

 e-Glossary
at www.macmillanmh.com

▶ **Reading Skill** ✔

Summarize

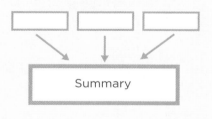

Summary

What is the water cycle?

Water in your environment can seem to change every day. One day it is raining and the next day it is dry as a desert. Where did all the water go? All water on Earth is *recycled*, or reused, constantly. This continuous movement of water between Earth's surface and the air, changing from liquid to gas to liquid, is the **water cycle**.

The water cycle is driven by the Sun's energy. Water in oceans, seas, lakes, ponds, and streams absorbs the Sun's heat. The heat helps speed the evaporation (i•vap•uh•RAY•shuhn) of the water. **Evaporation** is the changing of a liquid into a gas. The evaporated water rises into the *atmosphere* and cools. As it cools, it condenses into water droplets. **Condensation** (kon•den•SAY•shuhn) is the changing of a gas into a liquid. The water droplets gather with dust particles and form clouds. In time, the condensed water may become too heavy and drop out of the clouds as precipitation (pri•sip•i•TAY•shuhn). **Precipitation** is any form of water that falls from the atmosphere and reaches the ground, such as rain, sleet, snow, or hail.

The water cycle continues as precipitation falls back to the surface of Earth. Some of the water that falls as precipitation collects on land and flows downhill. A **watershed** is an area from which water is drained. Precipitation that flows across the land's surface and is not absorbed will flow into rivers, lakes, and streams as **runoff**. Most of the water will flow from rivers to the ocean. Some of the water will settle underground and become **groundwater**. Groundwater is stored in tiny holes, or pores, in soil and rocks.

Plants and animals also play a role in the water cycle. Plant roots soak up groundwater. Excess water evaporates out of the plant's leaves through transpiration. Animals also take in water and return some water to the atmosphere through respiration.

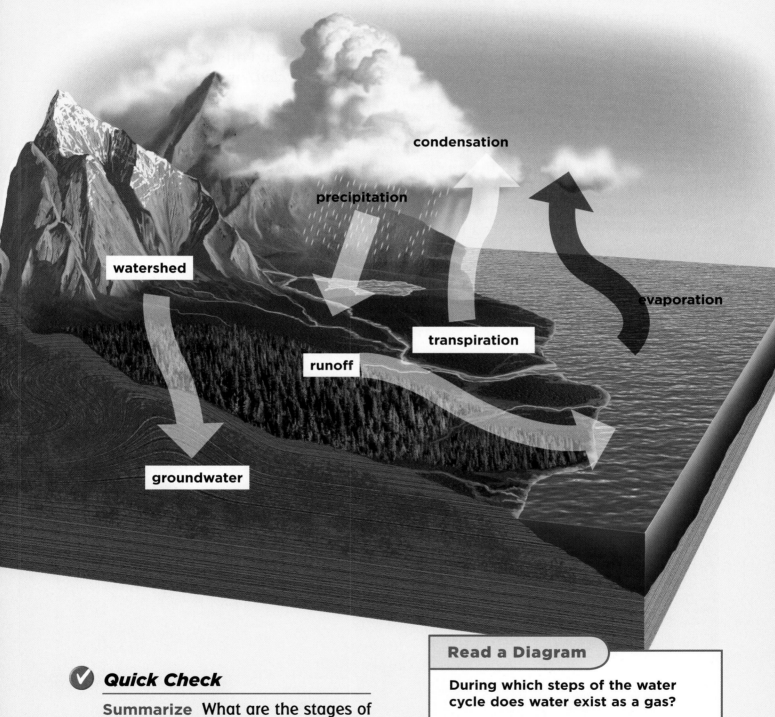

condensation

precipitation

watershed

evaporation

transpiration

runoff

groundwater

✅ *Quick Check*

Summarize What are the stages of the water cycle?

Critical Thinking Which would you expect to have a higher rate of evaporation – hot water or cold water? Why do you think so?

Read a Diagram

During which steps of the water cycle does water exist as a gas?

Clue: If water is not in a liquid or solid state, then it must be a gas.

 Science in Motion Watch how the water cycle works at **www.macmillanmh.com**

What is the carbon cycle?

Carbon is an important element in every living thing. How important is carbon? About 18 percent of your body is carbon. Carbon in the atmosphere is plentiful as CO_2, or carbon dioxide gas. It is also present in rocks, such as limestone. However, your body cannot use these sources of carbon directly.

How do people and other living things get the carbon they need? The continuous exchange of carbon among living things is the **carbon cycle**. Plants and other photosynthetic organisms take in carbon dioxide from the air. They combine it with water to make sugars and other chemicals, such as fats and proteins. These carbon-rich chemicals are then eaten directly by herbivores or omnivores and indirectly by carnivores.

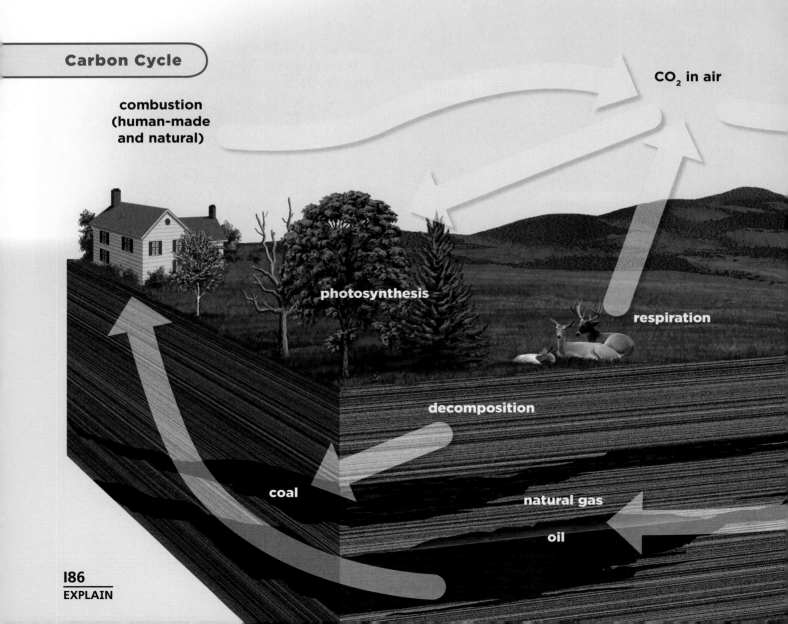

Carbon Cycle

CO_2 in air

combustion
(human-made
and natural)

photosynthesis

respiration

decomposition

coal

natural gas

oil

Both animals and plants burn carbon-rich foods for fuel during cellular respiration. The end product of cellular respiration, carbon dioxide, then returns to the atmosphere. Sometimes the carbon may not be recycled for a long period of time. The wood of a tree, for example, contains a large amount of carbon that will remain stored in a tree for as long as it lives. Carbon stored in plants and other organisms cannot be reused until they are eaten or decompose.

Decomposers, such as bacteria and insects, break down dead or decaying plants and animals. This breakdown releases some additional carbon dioxide to the atmosphere. Other decaying plant and animal materials go deep into the ground. Over a long period of time and under extreme pressure from the layers of Earth above, it may get turned into fossil fuels, such as oil, natural gas, and coal. The carbon in these materials is released back into the atmosphere as carbon dioxide when people burn them for energy.

✔ Quick Check

Summarize Write a brief summary of the carbon cycle.

Critical Thinking Would removing animals from the carbon cycle stop the cycle? Explain.

dissolved CO$_2$ in water

marine plankton remains

rock

Read a Diagram

Where is carbon likely to get trapped and stay out of the atmosphere for the longest period of time?

Clue: Follow each pathway. Where is carbon trapped for a long time?

What is the nitrogen cycle?

Nitrogen (NYE•truh•juhn) is another key element for all organisms. The proteins that make up the muscles, skin, nerves, bones, blood, and enzymes in your body contain nitrogen. Nitrogen is also a part of the genetic material in all cells.

Where do cells get nitrogen? The air is 78 percent nitrogen gas. But few living things can use nitrogen gas. Nitrogen must first be changed, or fixed into a form organisms can use. The continuous trapping of nitrogen gas into compounds in the soil and its return to the air is called the **nitrogen cycle**.

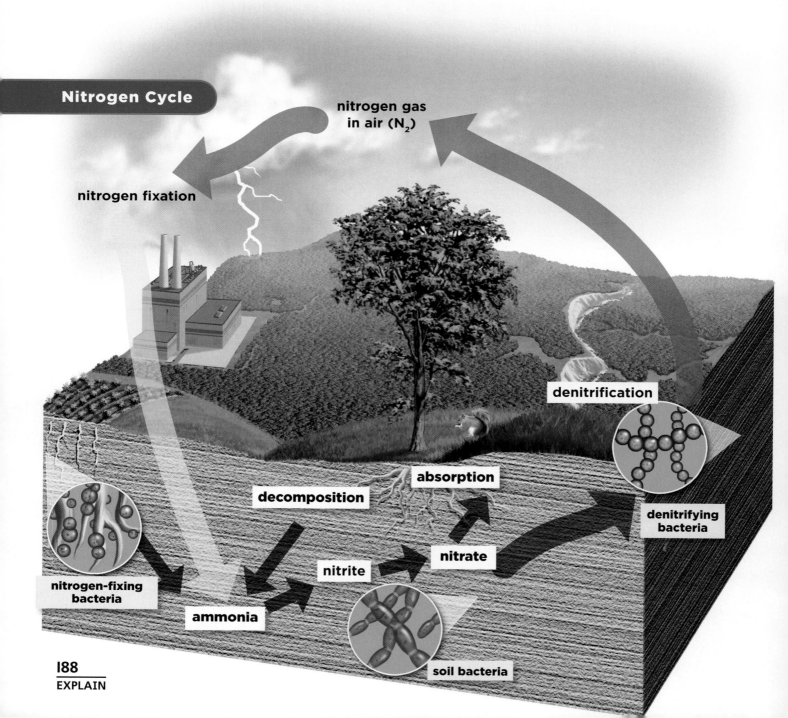

Nitrogen Cycle

nitrogen gas in air (N_2)

nitrogen fixation

denitrification

absorption

decomposition

denitrifying bacteria

nitrogen-fixing bacteria

nitrite

nitrate

ammonia

soil bacteria

Nitrogen can be fixed by volcanic activity, lightning, and combustion. It can also be fixed by certain bacteria. Different types of bacteria play important roles in the nitrogen cycle. *Nitrogen-fixing bacteria* live on the root nodules of *legumes* (LEG•yewmz), such as bean, pea, and peanut plants. These bacteria turn nitrogen gas into *ammonia* (uh•MOHN•yuh), a nitrogen-containing substance.

The ammonia is then changed into a form of nitrogen that can be used by plants. This is accomplished by two groups of bacteria that live in the soil. The first type of soil bacteria turn ammonia into a nitrogen-containing substance called *nitrite* (NYE•trite) A second type of soil bacteria turn the nitrite into *nitrate* (NYE•trayt), another substance that contains nitrogen.

As plants grow, they absorb nitrate from the soil. They use the nitrogen in nitrates to make proteins. Animals take in nitrogen when they eat plants or other plant-eating animals. Animals use the nitrogen to form compounds and then excrete substances containing nitrogen in their waste.

The nitrogen in animal waste and decayed plant and animal material returns to the soil. Decomposers then convert the nitrogen from these materials back into ammonia. How does nitrogen return to the atmosphere as a gas? *Denitrifying bacteria* (dee•NYE•truh•fye•ing) in the soil change some of the nitrates back into nitrogen gas and the cycle continues.

Quick Lab

Observe Legume Roots

1. Examine a legume plant. Clean off all dirt on the roots of the plant.

2. **Observe** Use a hand lens or microscope to examine the roots. What did you observe?

3. Use a hand lens to examine a carrot root. Compare these roots to the legume roots.

4. How are the legume roots similar to the other roots you observed? How are they different?

5. **Infer** Why are root nodules important in the nitrogen cycle?

✓ Quick Check

Summarize Write a summary of the nitrogen cycle.

Critical Thinking Why do humans need soil bacteria?

How is matter recycled?

Just as nature recycles water, carbon, and nitrogen, we need to conserve and recycle natural resources. Some natural resources are *renewable resources*. For example, trees, which are used for wood and paper, can be replanted. Other natural resources, such as oil and metals, are *nonrenewable resources*. These resources cannot be replaced in the environment once they are used. We can reduce our uses of natural resources by recycling them. We can make new objects or materials out of old materials.

Repeated planting sometimes uses up nitrogen in the soil. To replenish worn-out soil, farmers and gardeners have three choices. They can add nitrogen by planting legumes, using nitrogen-rich fertilizers, or creating compost. **Compost** (KOM•pohst) is a mixture of dead organic material that can be used as fertilizer. Composting is a way to recycle nitrogen. It also reduces the amount of trash we make.

How does compost enrich soil? Decomposers break down decaying plant and animal materials in the compost. One product of the decomposition is the nitrogen-containing substance ammonia. Soil bacteria change ammonia into nitrites and nitrates. Compost replenishes the nitrogen used by plants as they grow.

 Quick Check

Summarize Write a summary of how compost replenishes soil.

Critical Thinking Compost is useful but it often has a bad smell. What might give compost its smell?

Decomposers, like these bark beetles, break down decaying materials in the compost.

Lesson Review

Visual Summary

The **water cycle** moves water from its liquid form to its gas form through evaporation, condensation, and precipitation.

The **carbon cycle** moves carbon in an ecosystem by respiration, photosynthesis, and decomposition.

The **nitrogen cycle** moves nitrogen from a gas to living things and back to a gas again. Composting is a way to recycle nitrogen.

Make a FOLDABLES™ Study Guide

Make a Layered-Look Book. Use the titles shown. Tell about the cycle named on each layer.

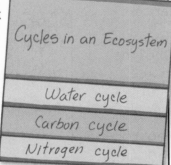

Cycles in an Ecosystem

Water cycle

Carbon cycle

Nitrogen cycle

Think, Talk, and Write

1. **Main Idea** What roles do plants play in the water, carbon, and nitrogen cycles?

2. **Vocabulary** A gas turns into a liquid during _____.

3. **Summarize** Write a summary of things that get recycled in an ecosystem.

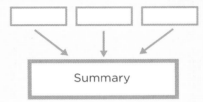

Summary

4. **Critical Thinking** A farmer's crops are less healthy than in previous years. What can the farmer do to get better crops?

5. **Test Prep** Which of the following processes release carbon dioxide?
   A photosynthesis, respiration
   B photosynthesis, burning oil
   C respiration, decomposition
   D photosynthesis, decomposition

6. **Test Prep** Animals add nitrogen into the ecosystem when they
   A eat plants.
   B excrete waste.
   C breathe.
   D burn sugars.

Writing Link

The Three Sisters
Research the "three sisters" planted by many Native American groups. These plants included corn, beans, and squash. Write a report about how this method of farming replenished soil naturally.

Art Link

Cycle Poster
Make a poster of one of the cycles described in this lesson. Use your imagination to illustrate the steps in the cycle.

Be a Scientist

Materials

spray bottle

4 plants in pots

water

4 plastic bags

string

equal pan balance

light source

How does water move in and out of plants?

Form a Hypothesis

Plants need water to survive. If a plant loses too much water it will wilt and eventually die. Plants lose water through transpiration, the evaporation of water from the leaves. As water evaporates, it pulls more water from the roots up through the xylem tissue. How does the amount of light a plant receives affect its transpiration rate? Write your answer as a hypothesis in the form "If the amount of light a plant receives is increased, then…"

Test Your Hypothesis

1. Use the spray bottle to water the four plants. Be sure to give all of the plants the same amount of water.

2. Place each of the plants' pots in a plastic bag and use the string to tie the bag securely around the stem of each plant.

3. **Measure** Weigh all four plants using the equal pan balance. Record their masses.

4. **Use Variables** Place two of the plants under the light source. Place the other two plants away from the light source.

5. **Record Data** After one hour weigh all four plants again. Record their masses and any changes you notice.

6. Return the plants to their original locations.

7. Repeat steps 5 and 6 after 24 hours and 48 hours. Record the masses and any observations.

Step 1

Step 2

Step 3

Draw Conclusions

1 What are the independent variables and dependent variables in the investigation? Which variables are controlled?

2 **Interpret Data** Did the mass of any of the plants change? Did your data show a connection between the transpiration rates and the amount of light?

3 Did your results support your hypothesis? Why or why not?

Guided Inquiry

How is the water loss in plants affected by changes in the environment?

Form a Hypothesis

You have seen how light affects the rate of transpiration. What other variables affect the rate of transpiration? How about wind? Write your answer as a hypothesis in the form "If wind increases, then the rate of transpiration . . ."

Test Your Hypothesis

Design a plan to test your hypothesis. Then write out the materials, resources, and steps you need. Record your results and observations as you follow your plan.

Draw Conclusions

Did your results support your hypothesis? Why or why not? Present your results to your classmates.

Open Inquiry

What other conditions in the environment can affect the rate of transpiration? Come up with a question to investigate. For example, how does humidity affect the rate of transpiration? Design an experiment to answer your question. Your experiment must be organized to test only one variable, or item being changed.

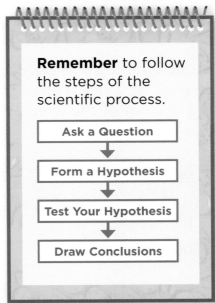

Remember to follow the steps of the scientific process.

Ask a Question

↓

Form a Hypothesis

↓

Test Your Hypothesis

↓

Draw Conclusions

Changes in Ecosystems

Ta Prohm Temple, Angkor, Cambodia

Look and Wonder

This stone building was once a magnificent temple built by kings. Today, trees and other plants grow out of the stone. What has changed in this ecosystem?

What happens when ecosystems change?

Make a Prediction

Each year a tree grows wider as a new layer of xylem forms an annual ring. Scientists often use tree rings to study changes in ecosystems. How did this tree's ecosystem change? Make a prediction.

Test Your Prediction

1. Count the number of tree rings on the tree-ring diagram. How old was this tree?

2. **Measure** Use the ruler to measure the width of each tree ring. Record your measurements.

3. **Interpret Data** Use the information provided in the chart to interpret your tree-ring data.

Draw Conclusions

4. Which years had the thickest rings? The narrowest rings?

5. **Predict** What most likely happened to the tree in its eighth year?

6. **Infer** What types of ecosystem changes did this tree experience? How can you tell?

Explore More

Have there been any fires, droughts, or floods in your community? Investigate using newspaper or Internet sources. Which parts of the environment have recovered better than others? Why?

Materials

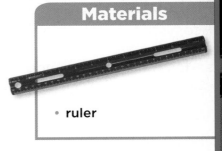

- **ruler**

Step 1

Step 3

Tree-Ring Data	
Type of Ring	**Event Affecting Tree**
thick ring	good growing conditions: warm, plenty of precipitation
narrow ring	poor growing conditions: cold, drought
dark scars	fire
long, light scars	insect infestation or disease

▶ Main Idea

Ecosystems can change naturally over time into a series of different living communities.

▶ Vocabulary

extinct species, p.198
endangered species, p.199
threatened species, p.199
succession, p.200
primary succession, p.200
pioneer species, p.200
pioneer community, p.200
climax community, p.201
secondary succession, p.202

LOG ON e-Glossary
at www.macmillanmh.com

▶ Reading Skill ✓

Cause and Effect

Cause	→	Effect
	→	
	→	
	→	
	→	

How can ecosystems change?

Most ecosystems are constantly changing. For example, an oak tree dies in a forest. Soon a new tree grows. Only this time the tree is a hickory, not an oak. But sometimes the balance of an ecosystem itself can change. Ecosystem changes can be caused by natural events or by human actions.

Natural events include natural disasters and changes caused by organisms. Earthquakes, floods, storms, volcanoes, droughts, and other natural disasters can drastically alter ecosystems. People can try to repair the damage from these disasters. But there is little or nothing anyone can do to prevent such events from occurring.

The second type of natural change is caused by organisms. Beavers, for example, build dams. They use mud, stones, and trees to create an artificial pond. Their dams can cause flooding, but the dams can also be beneficial by creating new habitats and food supplies. Large animals, like elephants, can cause changes by trampling trees and seedlings.

▼ A volcanic eruption causes lava to flow in Hawaii.

Beaver Dams

Read a Photo

How did this beaver change its ecosystem?

Clue: What is the beaver carrying?

Aquatic ecosystems are also changed by organisms. Recall that algae can sometimes rapidly reproduce, using all the nutrients and oxygen in the water. Aquatic organisms, such as coral, change their ecosystem by building reefs. Coral reefs create new habitats for many other organisms.

Humans cause ecosystem changes by shaping the environment to meet their needs. These changes often destroy or alter habitats. These changes affect the organisms that live in those habitats. Forests get cut down to build homes. Pesticides can harm organisms for which they were not intended. Pollution can damage water, soil, and air quality.

FACT All ecosystems are in a constant state of change.

Humans also change ecosystems by introducing new species or removing species. Introduced plant and animal species can threaten native species. Without natural predators, introduced species can threaten or even kill off local species. For example, zebra mussels were introduced into the Great Lakes. They began to reproduce rapidly and reduced the number of local mussel species. This drastically altered the food webs in the lakes.

Quick Check

Cause and Effect How do humans affect ecosystems?

Critical Thinking Can a natural ecosystem change cause more damage than a change caused by people? Give an example.

Extinction Game

Wild Atlantic sturgeon are endangered because of overfishing and pollution.

1 Count out 20 pennies to represent a school of sturgeon.

2 **Make a Model** Tape a piece of construction paper to your desk. Divide it into 6 sections. Sections 1 and 3 represent death. Sections 2, 4, and 6 represent life. Section 5 represents a new offspring.

3 Toss all 20 pennies onto the paper.

4 Remove any pennies that land in sections 1 and 3. Add a penny for any pennies that land in section 5. Record the new number of sturgeon in a data table.

5 Play the game for 20 rounds (years). After each round, record the number of sturgeon.

6 **Communicate** Did your school of sturgeon become extinct? If so, how many years did it take?

What happens when ecosystems change?

Some ecosystem changes are permanent. These changes affect the organisms within that ecosystem. Organisms must respond to changes in order to survive. Some respond by migrating to a place where they are more likely to survive. Recall that other organisms respond by adapting to the changes.

What happens when a species cannot respond to ecosystem changes? The individual members of that species begin to die. When the last member of a species dies, the species becomes an **extinct species** (ek•STINGT). Once a species becomes extinct it no longer exists on Earth. Some extinct organisms include all species of dinosaurs, the saber-toothed cat, and the dodo bird.

Thousands of species are expected to become extinct each year. Pollution, global warming, habitat destruction, and hunting all threaten organisms. The Tasmanian wolf, for example, became extinct about 65 years ago as a result of human actions. These wolves once lived in Australia. Farmers saw the Tasmanian wolf as a threat to their livestock and hunted the animal to extinction.

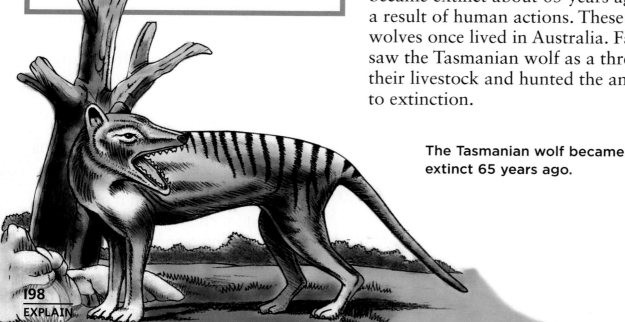

The Tasmanian wolf became extinct 65 years ago.

pitcher plant

habitat: pine forests, bogs, and stream banks in Alabama, Georgia, and North Carolina
status: endangered
main threats: overcollecting by humans, habitat loss

Karner blue butterfly

habitat: dry, sandy areas and open woods in Indiana, Wisconsin, Michigan, Minnesota, New York, and New Hampshire
status: endangered
main threats: habitat loss, over-collecting by humans

flying squirrel

habitat: coniferous and hardwood forests in Virginia and West Virginia
status: endangered
main threat: habitat loss

When a species is in danger of becoming extinct, it is called an **endangered species** (en•DAYN•juhrd). Endangered species today include the pitcher plant, the hawksbill sea turtle, the Karner blue butterfly, the flying squirrel, and many others. In some cases, such as the right whale, only a few hundred of these endangered organisms still exist.

Species with low numbers that could become endangered are called **threatened species**. The gray wolf, the manatee, and many others are threatened species. Each endangered and threatened organism is at risk for different reasons. Pollution, over-hunting or overcollecting, disease, and competition from newly introduced organisms can all cause a species to die out. The biggest threat to the survival of most organisms is habitat loss.

✔ Quick Check

Cause and Effect What can cause an organism to become threatened or endangered?

Critical Thinking Why are at least two members of an endangered mammal species needed for that species to survive?

hawksbill sea turtle

habitat: coral reefs and shallow coastal areas of Atlantic, Pacific, and Indian oceans; nests in Hawaii and Florida
status: endangered
main threats: hunting, loss of nesting habitats, water pollution

How do ecosystems come back?

Over time, ecosystems can gradually change. The process of one ecosystem changing into a new and different ecosystem is called **succession** (suhk•SESH•uhn). During succession, an area is changed by a certain species that is then replaced by other species over time. **Primary succession** takes place in a community where few, if any, living things exist, or where earlier communities were wiped out.

Primary succession occurs in barren, lifeless areas that have little or no soil. At first, the ecosystem is little more than solid, bare rock. Particles of dust and seeds blow in from neighboring environments. Lichens and plants, such as mosses, begin growing on the rock. These first organisms are **pioneer species**, the first species living in an otherwise lifeless area. They tend to be hardy organisms with short life cycles. Along with microorganisms, the pioneer species make up the **pioneer community**, the first living community in an otherwise lifeless area.

As they grow, lichens and mosses break down the rock and form soil. When lichens and mosses die, bacteria and other decomposers break them down as well. The decaying material adds nutrients to the soil. The soil can now support the growth of larger plants. In addition to plants, insects, spiders, and other small organisms begin to colonize. Decaying material from these organisms creates thicker, richer soil.

Grasses, ferns, and shrubs begin to sprout in the richer soil. Changes

Stages of Primary Succession

pioneer community		intermediate community	

| bare rock | lichens and mosses | small plants, lichens, grasses, and shrubs | shrubs and poplar, pine, and willow trees |

in the plant species cause changes in the animal species of the community. Flowering plants attract pollinators to the area, such as insects, birds, and small mammals. These animals attract larger predators to the community. After many years, this community may become a grassland or prairie.

If there is enough moisture, small trees, such as poplars and grey birch, begin to grow. The leaves of these fast-growing trees block the sun. This enables seedlings that need less sun to grow. Soon, pine trees that do not require as much sun start filling in the gaps. Eventually the pine trees give way to the hardwoods—maple and beech trees. As these trees fill in the forest, they form a climax community.

A **climax community** is the final stage of succession. Unless the community is disturbed by some natural disaster or human activity, the climax community will remain.

✔ Quick Check

Cause and Effect During succession, what causes larger plants to grow in place of mosses and lichens?

Critical Thinking A prairie is becoming a forest. Then a fire burns down the grassland. How does the fire affect the process of succession?

climax community

maple and beech trees

Read a Diagram

How do the earlier stages of succession compare to the climax community?

Clue: Look at the diagram and compare it to the photo.

What is secondary succession?

Secondary succession is the beginning of a new community where a community had already existed. Secondary succession can take place in a forest after a fire or logging has occurred. It can also take place in an abandoned farm field.

Secondary succession occurs faster than primary succession. The soil has already formed and some organisms might be present. For example, when a farm is abandoned, weeds and crabgrass begin to grow in the plowed field. After a couple of years, larger shrubs take over, as well as tree seedlings. For the next several years, the shrubs and seedlings compete for light, space, and resources. The tree seedlings finally win out and eventually the farm will become a pine forest.

From this point, the secondary succession process is similar to primary succession. After many years, the pine forest has a lower layer full of small hardwood seedlings. These seedlings take 40 or 50 years to take over and form the climax community.

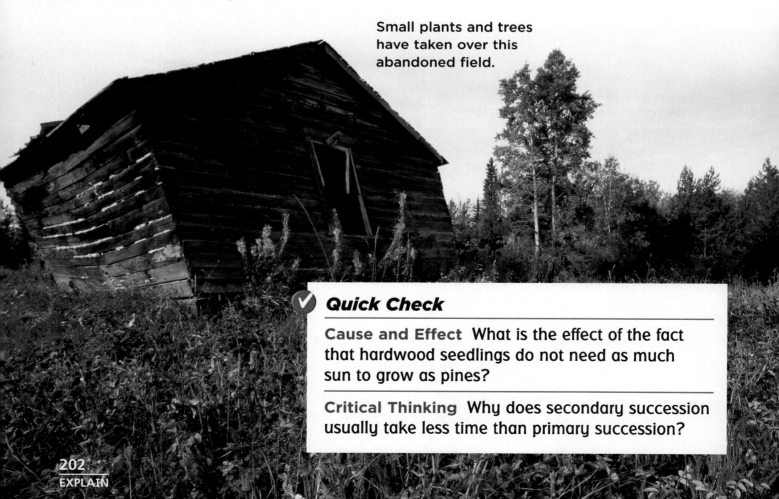

Small plants and trees have taken over this abandoned field.

> ### ✔ *Quick Check*
>
> **Cause and Effect** What is the effect of the fact that hardwood seedlings do not need as much sun to grow as pines?
>
> **Critical Thinking** Why does secondary succession usually take less time than primary succession?

Lesson Review

Visual Summary

Natural events and organisms can cause **ecosystem changes**. Human activity can also change ecosystems.

A variety of different things can cause organisms to become **extinct**. Most species become extinct from loss of habitat

Primary succession turns a barren, lifeless area into a living community. **Secondary succession** changes one living community into another.

Make a FOLDABLES™ Study Guide

Make a Trifold Book. Use the titles shown. Tell what you learned about each topic on the folds.

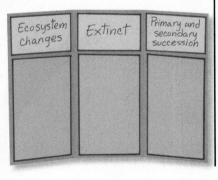

Ecosystem Changes | Extinct | Primary and secondary succession

Think, Talk, and Write

❶ **Main Idea** What can cause ecosystems to change?

❷ **Vocabulary** One of the first organisms to live in an area is a(n) _____.

❸ **Cause and Effect** Tell the causes and effects that result in a barren, lifeless ecosystem becoming a forest.

Cause → Effect	
	→
	→
	→
	→

❹ **Critical Thinking** How does primary succession affect food chains and food webs in an ecosystem?

❺ **Test Prep** Which list gives the correct order of primary succession?

A lichens, grasses, shrubs, pine, maple
B lichens, maple, grasses, shrubs, pine
C grasses, lichens, shrubs, pine, maple
D grasses, lichens, pine, shrubs, maple

❻ **Test Prep** Which of the following is a pioneer species?

A lichen
B pine tree
C owl
D soil

Writing Link

Endangered Species
Research an endangered species. Explain why this species is endangered. Tell how we might be able to prevent this animal from becoming extinct.

Math Link

Soil Collection
Soil in an ecosystem collects at a rate of $\frac{1}{6}$ of a centimeter every 10 years. At this rate, how long will it take for 2 centimeters of soil to collect?

Inquiry Skill: Interpret Data

Ecosystem changes can affect organisms. Scientists estimate that once there were more than 500,000 bald eagles in America. But by the 1960s, there were less than 450 nesting pairs. What happened? Scientists discovered particles of an insecticide called DDT in the eagles' eggshells. The United States outlawed the use of DDT in 1972. Did that help bring eagles back from the edge of extinction? Scientists learned the answer to that question by collecting and **interpreting data**.

▶ Learn It

When you **interpret data**, you use information that has been gathered to answer questions or solve problems. It is much easier to interpret data that has been organized and placed on a table or graph. Tables and graphs allow you to quickly see similarities and differences in the data.

The table below shows data gathered about bald eagle eggs. It lists the average number of eggs that hatched in the wild during a 16-year period. It also lists the levels of pesticide found in the eggs during that time.

Bald Eagle–Hatching Data		
Year	Average # Hatched	DDT in Eggs parts/million
1966	1.28	42
1967	.75	68
1968	.87	125
1969	.82	119
1970	.50	122
1971	.55	108
1972	.60	82
1973	.70	74
1974	.60	68
1975	.81	59
1976	.90	32
1977	.93	12
1978	.91	13
1979	.98	14
1980	1.02	13

Hatched Eagle Eggs

Average Number of Hatched Eggs

0

1966 1967 1968 1969 1970 1971 1972 1973 1974 1975 1976 1977 1978 1979 1980 1981

Year

▶ **Try It**

Study the table, then **interpret data** to answer these questions:

❶ In which year did the amount of pesticide in eggshells begin to decline? Why?

❷ Did the amount of pesticide continue in a steady decline?

❸ Does the data supply evidence that insecticide in eggs and the numbers of young hatched are related?

▶ **Apply It**

❶ Now use the data from the table to make two line graphs: one to show the average number of eggs that hatched and one to show the insecticide in the eggs. Do your graphs make it easier to **interpret data**? Why or why not?

❷ Lay one graph carefully on top of the other so the years across the bottom line up. Hold the pages up to the light. How would this help someone understand the relationship between the eagle eggs that hatched and the amount of insecticide in the eggs?

Biomes

Serengeti National Park, Tanzania, Africa

Look and Wonder

These gnu are grazing in an African grassland. Do the characteristics of an environment, such as soil, water, and temperature, affect the kinds of organisms that can live there?

How are soils different?

Make a Prediction

The nutrient content of soils can vary greatly. The amounts of nutrients in soils can influence the types of organisms that can live in certain places. Which type of soil has more nutrients? Make a prediction.

Test Your Prediction

① ⚠ **Be Careful.** Wear your goggles and apron. Place a spoonful of sand in the plastic cup.

② **Observe** Add hydrogen peroxide to the sand, drop by drop. Hydrogen peroxide is a chemical that bubbles when it reacts with nutrients.

③ **Communicate** Record the number of drops it takes until the sample starts bubbling.

④ Repeat steps 1 to 3 using soil instead of sand. Record your data.

Draw Conclusions

⑤ Which sample had more nutrients—sand or soil? Explain.

⑥ **Predict** Which would probably be better for growing plants—sand or soil? Explain.

⑦ **Infer** How might you classify the sand and soil—high in nutrients or low in nutrients?

Explore \ More

Collect other types of soil and test their nutrient levels using hydrogen peroxide. Which soil had the greatest amount of nutrients?

Materials

- goggles
- apron
- plastic spoons
- sand
- plastic cups
- hydrogen peroxide
- dropper
- soil

Step ①

Step ②

Read and Learn

▶ **Main Idea**

The six major land biomes on Earth are tundra, taiga, desert, rain forest, deciduous forest, and grassland.

▶ **Vocabulary**

biome, p.208

desert, p.209

tundra, p.210

taiga, p.211

tropical rain forest, p.212

temperate rain forest, p.212

deciduous forest, p.213

grassland, p.214

LOG ON e-Glossary
at **www.macmillanmh.com**

▶ **Reading Skill** ✔

Classify

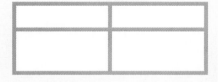

What are biomes?

Take a look out your window. If you live in Arizona you might see a dry desert. In Pennsylvania you might see a green forest. In Iowa the scene outside your window is likely to be a golden grassland. Each environment is a biome (BYE•ohm).

A **biome** is one of Earth's major land ecosystems with its own characteristic animals, plants, soil, and climate. A *climate* is an average weather pattern for a region. How is a biome different from other habitats? You can think of a biome as a set of habitats or ecosystems all grouped together into a kind of "super-ecosystem."

In all, there are six major land biomes: desert, tundra (TUN•druh), taiga (TYE•guh), rain forest, deciduous forest (di•SIJ•ew•uhs), and grassland. You can see how they are arranged on the map. For example, a desert stretches across the continent of Africa. A taiga covers the distance across Russia, a length of about 10,000 km (6000 miles). Each continent on Earth has a number of different biomes.

Read a Map

In which three places are most of the world's deciduous forests located?

Clue: Use the key to locate deciduous forests.

Global Biomes

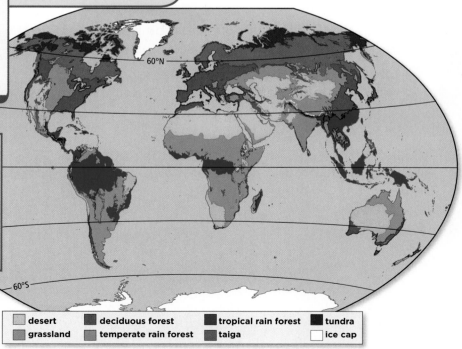

| desert | deciduous forest | tropical rain forest | tundra |
| grassland | temperate rain forest | taiga | ice cap |

Desert

A **desert** is a sandy or rocky biome, with little precipitation and little plant life. The main characteristic of a desert is the lack of water. Deserts are treeless and extremely dry biomes. Some deserts, such as the ones in Asia, are cold, but many deserts are brutally hot. Desert soil is often rich in minerals, but poor in animal and plant decay.

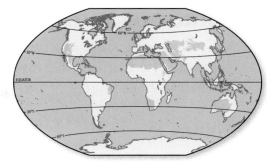

Plants and animals need special adaptations to survive in dry deserts. For example, cactus plants store water in their stems and branches. They also have thick, waxy cuticles that prevent water from evaporating. Other desert plants, like the mesquite, survive by having very long roots that reach moisture deep underground.

Only a few animals are active during the heat of the day in the desert. These include lizards and other reptiles that need the Sun's heat to warm their blood. Some insects and a few birds also are active during the day. Most desert animals are active at night. Animals such as kangaroo rats come out when the Sun goes down. When the Sun rises again, they move underground or find shelter to stay cool.

✓ Quick Check

Classify How would you classify a biome that is very dry and has cactus plants?

Critical Thinking Which biomes are found in the United States?

Scorpions are common in deserts.

FACT Some deserts have cold climates.

What are some harsh biomes?

Some biomes are characterized by having extremely cold weather. The tundra and taiga are cold biomes found in the northern hemisphere.

Tundra

The **tundra** is a large, treeless biome where the ground is frozen all year. The layer of *permafrost*, or permanently frozen soil, prevents trees from growing. The soil is also poor in nutrients. During the six- to nine-month winter, most tundra locations get very little sunlight. Temperatures can drop to −94°F (−70°C).

In the short summer, it stays light almost all day and temperatures rise to above freezing. Because of the permafrost, poor soil, and short summers, tundra plants have shallow roots and short growing seasons. Mosses, lichens, and some grasses and shrubs are common here.

Very few animals have adapted to living in the tundra. Caribou, polar bears, musk ox, and arctic hares and foxes make their homes in the tundra. In the spring and summer, shallow, boggy pools form from the melted ice. These are perfect breeding grounds for millions of mosquitoes and the birds that feed on them.

This caribou is grazing on lichens in an Alaskan tundra.

Wolverines are found in taiga forests.

Taiga

Just south of the tundra lies the world's largest biome— the taiga. The **taiga** is a cool forest biome of conifers found in northern regions. Unlike the frozen tundra, the taiga is full of coniferous evergreen trees. Temperatures in the taiga are cold, but not as cold as those in the tundra. Soil in the taiga is low in minerals.

Many plants have adapted to life in the taiga. The taiga has a longer growing season than the tundra. During the summer months the ice melts, which makes it possible for trees and other plants to grow. The taiga is able to support pines, firs, spruces, and other conifers.

Many taiga animals, like the snowshoe rabbit, look much like animals that live farther south. Their coats are thicker for extra warmth, and during the winter, lighter-colored to match the snow. Many taiga animals have thick layers of fat that protect against the cold. Many hibernate to avoid the coldest winter months.

Quick Check

Classify How would you classify a cold, treeless biome with very little sunlight?

Critical Thinking How would you expect a desert rabbit to be different from a rabbit that lives in the taiga?

What are some forest biomes?

Does a jungle in South America have anything in common with a tree-filled park near your home? Yes; both are examples of forest biomes.

Rain Forests

tropical rain forest

temperate rain forest

Read a Photo

How do tropical rain forests and temperate rain forests compare?

Clue: Look at the two photos. What similarities and differences do you see?

Rain Forest

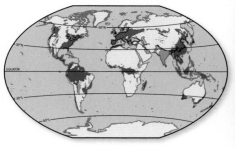

The **tropical rain forest** is a hot, humid biome near the equator, with heavy rainfall and a wide variety of life. Although its soil is nutrient-poor, the tropical rain forest supports more organisms than any other place on Earth.

Tropical rain forests have four layers. Few plants can grow on the *forest floor*, because there is very little light. The forest floor is home to many insects, frogs, and mice. The *understory* is made up of tree trunks, shrubs, and vines. Leopards, jaguars, and other large mammals are found in this layer.

The next layer, the *canopy*, is thick with plant life. The leaves of canopy plants prevent sunlight from reaching the lower levels. Monkeys, bats, toucans, tree frogs, snakes, and insects live among the tree branches. The highest layer is the *emergent layer*. It is made up of the upper parts of large trees. Many birds nest in this part of the rain forest.

A **temperate rain forest** is a biome with a lot of rain, fog, and a cool climate. Temperate rain forests have mild winters and cool summers. They are dominated by large evergreen trees and epiphytes (EP•uh•fites). *Epiphytes* can be lichens or plants, such as mosses and some ferns, that grow on trees. The animals in this biome include cougars, black bears, bobcats, owls, reptiles, and many amphibians.

Deciduous Forest

The **deciduous forest** is a forest biome with four distinct seasons and deciduous trees. Deciduous trees are hardwoods whose leaves change color and fall off every autumn. Examples are oak, maple, beech, and hickory trees. The soil is rich in nutrients. In winter, the forest is a snowy, empty place. Many birds have migrated to warmer places. Other animals are hiding or hibernating.

In the spring and summer, ferns, shrubs, and saplings shoot up. The leaves on the trees begin to grow. Common forest birds return, including robins, woodpeckers, owls, and hawks. Squirrels, mice, rabbits, raccoons, skunks, porcupines, and other small mammals also populate the forest. Larger animals include white-tailed deer, foxes, and bears. A variety of insects live in the forest, including ants, beetles, and butterflies.

Deciduous forests provide habitats for chipmunks.

213
EXPLAIN

What are grasslands?

The **grassland** is a biome where grasses, not trees, are the main plant life. American prairies are one kind of grassland. The African grassland is called a savanna. A temperate grassland has cold winters and hot summers. A grassland is wetter than a desert, but it is too dry for many trees to grow. Grassland soil is rich with nutrients. It is often used for growing crops such as wheat, oats, and corn.

Fires are common in this dry biome. The roots of the plants are deep, so the environment recovers quickly. New grass plants—and the grassland habitats—grow back.

The grasses are the producers of the ecosystem. They provide a rich food source for all the herbivores. In tall grass prairies, bluestem and switchgrass grows to heights of 3 meters (10 feet). Down below, the roots of these plants can grow just as deep or deeper.

Grasslands are also filled with animal life. Insects, like grasshoppers, crickets, butterflies, and moths, live among the wildflowers of American prairies. Low in the grass, toads, worms, insects, spiders, mice, prairie dogs, snakes, and other small organisms make their homes. These small organisms are prey for birds and other predators.

✓ Quick Check

Classify A grassland gets so dry that grasses cannot grow and soil blows away. How would you classify this biome?

Critical Thinking What do you think could happen to a grassland if fires were prevented?

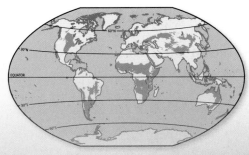

Bison roam the American prairie grasslands.

Visual Summary

Earth can be divided into six major **biomes**. Desert biomes are dry and hot.

The **tundra** and **taiga** are cold biomes.

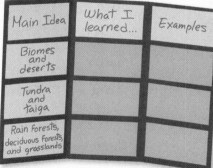

Forest biomes include **rain forests** and **deciduous forests**. Prairies and savannas are **grasslands**.

Make a FOLDABLES™ Study Guide

Make a Trifold Book. Use the titles shown. Tell about each topic on the folds.

Main Idea	What I learned...	Examples
Biomes and deserts		
Tundra and taiga		
Rain Forests, deciduous forests, and grasslands		

Think, Talk, and Write

❶ **Main Idea** What are Earth's six major land biomes?

❷ **Vocabulary** The biome just south of the tundra is the _____.

❸ **Classify** How would you classify the biome where you live? Give reasons to support your classification.

❹ **Critical Thinking** The Gobi Desert is dry, but its average temperature is around freezing. Should this region be classified as a different biome? Explain.

❺ **Test Prep** Which biome has permafrost?
   A deciduous forest
   B grassland
   C tundra
   D taiga

❻ **Test Prep** Which biome has many epiphytes?
   A grassland
   B rain forest
   C desert
   D tundra

 Writing Link

Global Warming
Research and write a report on global warming. How might global warming change some of Earth's biomes?

 Math Link

Temperature Ranges
Tundra temperatures may range from -70°C to 12°C. What is the total temperature range in degrees?

A Year in the Life of a Forest

Did you know that forests breathe?

Scientists can measure the gases in the forest air to gather data about the photosynthesis and respiration of the trees, animals, and other organisms that live there.

Take a look at the carbon dioxide data that scientists measured in the air from Howland Forest, a deciduous forest in Maine. Howland Forest has cold and snowy winters and hot and humid summers. How do these changes in seasons affect the amount of carbon dioxide in the air?

Spring

As the days become longer and warmer, activity in the forest increases. This increased activity results in higher levels of respiration, so the amount of carbon dioxide measured in the air starts to rise. The trees sprout new leaves and begin to photosynthesize.

Summer

Summer days are the longest and warmest of the year. Because the forest is so active, a lot of photosynthesis and respiration occurs. During the day, the amount of carbon dioxide is low. That's because the trees are taking in carbon dioxide and transforming it into food to store in their roots. During the night, the amount of carbon dioxide is high. All organisms in the forest, including the trees, are respiring and releasing carbon dioxide. These two processes together result in the different day and night carbon dioxide levels you see in the chart.

CO_2 Concentration (parts/million)		
month	minimum CO_2	maximum CO_2
Jan	378	388
Feb	377	385
March	377	384
April	376	388
May	371	393
June	362	413
July	356	427
Aug	355	424
Sept	362	418
Oct	358	386
Nov	366	379
Dec	368	377

These photos show Howland Forest during all four seasons.

Fall

Shorter days mean fewer hours of sunlight. Trees begin to lose their leaves and the forest becomes less active. The forest is photosynthesizing and respiring less. Day and night carbon dioxide levels are similar.

Winter

Winter days are the shortest and coldest of the year. The forest is much less active. Most of the trees have lost their leaves, and there is no photosynthesis. Day and night carbon dioxide levels are very similar as all the life-forms continue to respire.

Write About It

Main Idea and Details

1. Tell how the levels of carbon dioxide change in Howland Forest throughout the year.
2. Research other biomes and explain how they change during the year.

LOG ON e-Journal Research and write about it online at **www.macmillanmh.com**

Main Idea and Details

▶ Look for the central point of a selection to find the main idea.

▶ Details are important parts of the selection that support the main idea.

AMERICAN MUSEUM OF NATURAL HISTORY

Water Ecosystems

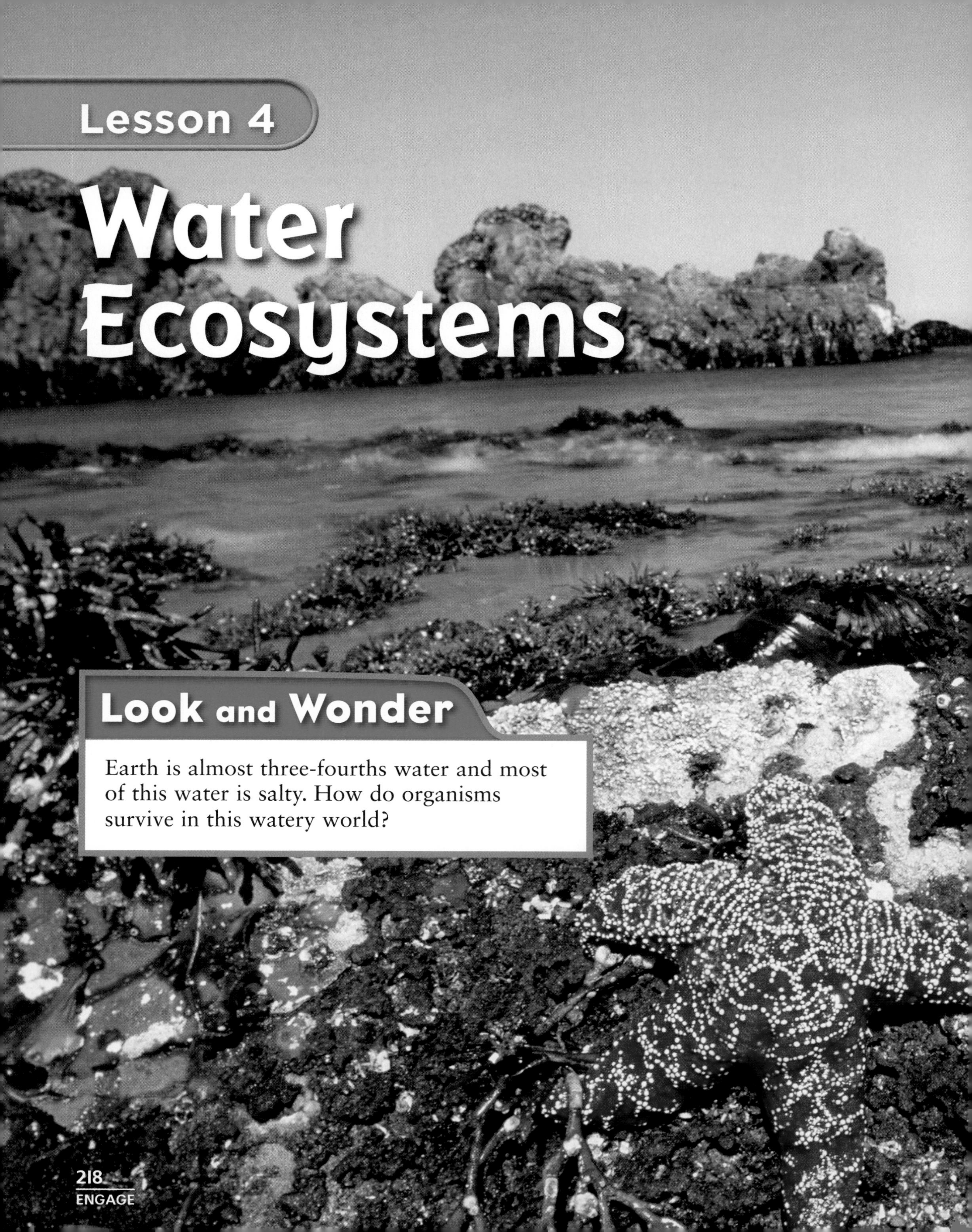

Look and Wonder

Earth is almost three-fourths water and most of this water is salty. How do organisms survive in this watery world?

How does the ocean get salty?

Purpose
To make a model that shows how ocean water becomes salty.

Procedure

1 Measure In the plastic cup, mix 2 tablespoons of salt and a few drops of food coloring. Use the spoon to stir until it's well-mixed.

2 Pour 2 cups of soil into one side of the shallow baking pan.

3 Mix the salt with the soil in the pan.

4 Tip the pan so the side with the mixture in it is slightly off the table. Try not to knock any of the mixture to the other side.

5 As you hold the pan slightly off the table, slowly pour some water onto the mixture.

6 Observe Note the color of the water when it reaches the other side of the pan.

Draw Conclusions

7 How does the color of the water compare to the color of the dyed salt?

8 Infer How does this model resemble what happens as fresh water flows to the ocean?

Explore More

Are some oceans saltier than others? Research Earth's oceans to find out if some have more salt than others. Write a report that explains how some oceans become saltier than others.

Materials

- plastic cup
- salt
- blue food coloring
- plastic spoon
- soil
- shallow baking pan
- container of water

Step 1

Step 5

What are water ecosystems?

Most of Earth's surface is covered by water, but all water is not alike. About 97 percent of the world's water is salty ocean water. The other three percent is fresh water. Fresh water is water that does not contain salt, or is very low in salt. Fresh water is the water you drink.

Freshwater and saltwater ecosystems are organized in similar ways. The organisms in water ecosystems are divided into three main categories. Plankton (PLANGK•tuhn) are creatures that drift freely in the water. They are not able to swim. Some plankton, such as diatoms, are producers, and others are consumers, such as some animal larvae. The second group includes the larger, active swimmers in a body of water called **nekton** (NEK•ton). Fish, turtles, and whales are all nekton.

The third group, organisms that live on the bottom of a body of water, are called **benthos** (BEN•thahs). Many benthos are scavengers or decomposers because they feed on material that floats down from shallower water. Some benthos, such as oysters, fix themselves to one spot. Others, like worms, burrow in the sand. Many, like lobsters, walk about on the bottom of a body of water looking for food.

Saltwater Organisms

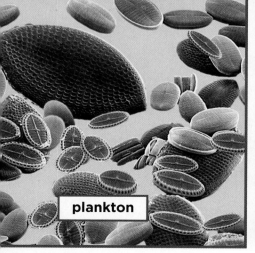
plankton

nekton

benthos

Like land ecosystems, biotic and abiotic factors determine the types of organisms that can survive in water ecosystems. Unlike land ecosystems, water is never a limiting factor. However, the amount of light, dissolved salt, and dissolved oxygen are important. They can all affect the types of organisms that can live in bodies of water.

Little sunlight is able to penetrate to the bottom of a body of water. The water gets cooler and darker away from the surface. Shallow water allows more light to penetrate. For this reason, producers that require light for photosynthesis are usually found on the surface or in very shallow water.

These same producers release oxygen during photosynthesis. Water can also pick up oxygen from the air. So surface and shallow water tend to have more dissolved oxygen than deeper water. The oxygen and nutrients from producers enable more organisms to live in the surface waters.

Quick Lab

Salt Water vs. Fresh Water

1. Fill a cup with fresh water. Fill a cup with salt water. Label each cup. Place flowers in each cup.

2. **Observe** Examine each flower after two hours.

3. **Communicate** Did you see any changes to either flower? Explain your observations.

Salt water   Fresh water

✔ Quick Check

Main Idea and Details How does the amount of fresh water compare to the amount of salt water?

Critical Thinking Suppose you had a bottle containing ocean water and another with fresh water. How could you tell the two apart?

Freshwater Organisms

plankton

nekton

benthos

This river is a running-water ecosystem.

What are freshwater ecosystems?

Not all freshwater ecosystems are alike. Rivers and streams are running-water ecosystems. Lakes and ponds are standing-water ecosystems. Swamps, bogs, and marshes are freshwater wetlands.

Running-Water Ecosystems

Moving bodies of fresh water can vary from small, fast-moving brooks to large, slow-moving rivers. Faster-moving bodies of water tend to have more oxygen, because air mixes in as the water flows. Other nutrients are washed into the water from the land. Organisms that live in fast-moving streams or rivers have adaptations to prevent them from being swept away. Some grow attached to rocks. Others, such as salmon, have strong muscles that enable them to swim against strong currents.

Slower-moving waters have less oxygen and are less dependent on the land for nutrients. More producers, such as algae, are able to survive in slow-moving water. Mussels, minnows, and other organisms can also live here.

▼ Wood ducks are common organisms in freshwater ecosystems.

Standing-Water Ecosystems

The typical freshwater lake or pond is divided into three zones. The shallow-water zone along the shore is where most of the organisms live. Cattails, sedges, arrowgrass, and other rooted plants grow here. These plants are home to a variety of microorganisms. They are eaten by animals such as insects, worms, tiny fish, insect larvae, and crustaceans. These animals are eaten by larger organisms that include frogs, birds, and small sunfish. Finally, top predators, such as herons, bass, and turtles, eat the smaller predators.

The open-water zone includes the water away from the shore. This zone may be too deep for rooted plants to survive. Algae and plankton float near the surface. Nekton, such as trout, whitefish, and pike are found here.

The third zone is below the open-water zone and includes the bottom. Very little light reaches the bottom, so producers cannot grow here. As a result, there is very little oxygen. Food must drift into this zone from other zones. Benthos, including worms and mollusks, are found in this zone.

Freshwater Wetlands

Wetlands, such as marshes, swamps, and bogs, are regions that are wet for most of the year. They are found in areas that lie between land and water. In order to survive here, plants must be adapted to water-soaked soil. Grasslike plants, moss, and some shrubs are found in wetlands. Beavers, muskrats, otters, birds, and fish live in wetlands.

 Quick Check

Main Idea and Details Describe the zones of standing water. Give details about the organisms in each zone.

Critical Thinking Why is fast-moving water dependent on the land?

Freshwater Zones

open water

shallow water

bottom

Read a Diagram

Which zone of fresh water contains the benthos?

Clue: Which zone contains the deepest water?

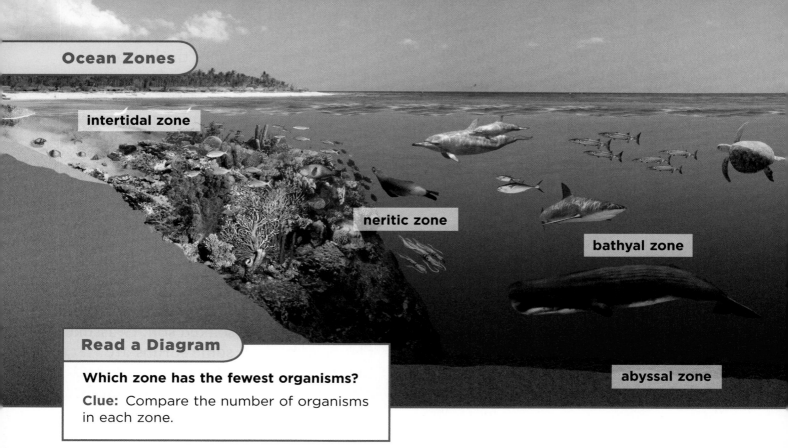

intertidal zone

neritic zone

bathyal zone

abyssal zone

Read a Diagram

Which zone has the fewest organisms?

Clue: Compare the number of organisms in each zone.

What are ocean ecosystems?

Like pond and lake ecosystems, oceans are divided into zones. The shallowest part of the ocean ecosystem is called the **intertidal zone**. Every day, the pull of the Moon's gravity causes ocean tides to rise and fall over the intertidal zone. It is covered by water at high tide. At low tide the organisms of the intertidal zone are exposed to the air.

Beyond the intertidal zone is the *neritic zone* (nuh•RI•tik). The key resource in this zone is sunlight. Algae, kelp, and other producers grow in huge numbers near the surface water where sunlight can penetrate. These producers attract herbivores including small fish, animal-like plankton, shrimplike

crustaceans, and some whales. These organisms in turn attract larger organisms: seals, sea turtles, jellyfish, and other large fish.

The third zone of the ocean is the *oceanic zone* (oh•shee•AN•ik). It is divided into the *bathyal zone* (BA•thee•uhl) and the *abyssal zone* (uh•BIS•uhl). The bathyal zone is home to many large consumers, such as sharks, but few producers.

Further down is the abyssal zone, where it gets darker and colder because the sunlight is completely blocked. Without light, producers cannot perform photosynthesis. Without producers, few consumers can survive. Organisms in this zone tend to be scavengers or decomposers. They live on nutrients that float down from other zones.

FACT Hydrothermal vents on the ocean floor support many organisms.

Intertidal Zone The intertidal zone is part of the sunlight zone. It can be rocky or sandy. Organisms such as crabs, clams, barnacles, sea stars, sea grasses, snails, and seaweeds live here. These organisms must be adapted to daily changes in temperature, moisture, and wave action.

Neritic Zone The neritic zone is also part of the sunlight zone. Coral reefs and kelp forests are important parts of the neritic zone. They provide habitats and food for many other organisms in this zone.

Bathyal Zone The bathyal zone can be divided into the twilight and dark zones. Little sunlight reaches these parts of the ocean. Many large organisms, like this octopus, make the dark zone their home. You will find sharks, squid, and many other large nekton in this zone.

Abyssal Zone The organisms in the abyssal zone, or the abyss, are well-adapted to its cold, dark conditions. Tube worms live in and around hydrothermal vents, cracks in the deep ocean floor. The heat and minerals from these vents support many organisms.

 Quick Check

Main Idea and Details Which zone is home to the most producers? What feature does this location have that helps organisms survive?

Critical Thinking At a depth of five meters, which would you expect to have more organisms living in it — clear water or murky water? Explain your answer.

Where do salt and fresh water meet?

The boundary where fresh water feeds into salt water is called an **estuary** (ES•chew•er•ee). Estuaries are unique ecosystems that are part salt water and part fresh water. Like intertidal zones, estuaries change with the tides. When the tide comes in, estuary water becomes more salty. The tide also brings in nutrients from the land. When the tide runs out, the estuary becomes mostly fresh water and wastes are flushed out.

Estuaries usually contain salt marshes. These boggy areas are covered with grasses and other marsh plants. The marsh grasses and surrounding water provide habitats for a wide variety of organisms. Large numbers of shrimp, oysters, clams, and fish live in estuaries. Many ocean fish return to estuaries to lay their eggs.

Countless insect larvae, young fish, and tiny crustaceans begin their lives in the calm, protected waters within an estuary. Larger organisms, including egrets, herons, frogs, turtles, muskrats, raccoons, otters, and bobcats feed on these smaller consumers.

Salt marshes within estuaries perform several special functions for a coastal region. They act like a sponge during storms to soak up excess water. The roots and stems of marsh plants also act as filters for river water. The grasses in salt marshes slow the water down and trap nutrients and pollution. The water that then flows into the ocean has been cleaned.

 ## Quick Check

Main Idea and Details What are estuaries? Use details to explain how they help the environment.

Critical Thinking A river empties into a freshwater lake. Is the place where the river and the lake meet an estuary? Why or why not?

Bobcats are one of the top carnivores in estuary ecosystems.

Lesson Review

Visual Summary

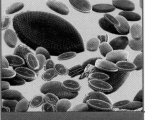

The organisms that live in fresh water and salt water are divided into **plankton**, **nekton**, and **benthos**.

Freshwater ecosystems include standing water, running water, or wetlands. **Saltwater ecosystems** are divided into the intertidal zone, neritic zone, and oceanic zone.

Estuaries are rich ecosystems that exist where fresh water and salt water meet.

Make a FOLDABLES™ Study Guide

Make a Trifold Book. Use the titles shown. Tell what you learned about each topic.

Main Ideas | What I learned... | Examples

Plankton, nekton, and benthos

Freshwater and saltwater ecosystems

Estuaries

Think, Talk, and Write

1. **Main Idea** Which water ecosystem contains most of the world's water?

2. **Vocabulary** The shallowest part of the ocean ecosystem is the _____.

3. **Main Idea and Details** Which freshwater ecosystems get a fresh supply of water every day? Which do not? Use details to support your answer.

Main Idea	Details

4. **Critical Thinking** Why are lakes and rivers more sensitive to pollution than the oceans?

5. **Test Prep** Which of the following organisms are nekton?
   A whales, turtles, sharks
   B sea stars, plankton, lobsters
   C crabs, tube worms, algae
   D rays, crabs, diatoms

6. **Test Prep** Which of the following helps to clean the environment?
   A neritic zone
   B benthos
   C plankton
   D estuary

 Math Link

Ocean Fractions
The top 200 meters of 3,000-meter-deep ocean water are penetrated by sunlight. What fraction of the water is dark?

 Art Link

Water Ecosystems
Make a poster that shows every water ecosystem, including streams, rivers, ponds, lakes, oceans, and estuaries.

Writing in Science

Keep Our Water Clean

Laura wanted people in her town to help clean up trash that had been dumped in the local stream. She wrote an e-mail to the mayor and asked him to put her message on the town Web site.

Laura_Email.com   Send   Cancel

Our supply of fresh water is limited to what we can get from streams, rivers, lakes, and under the ground. If we pollute these sources of water, we will run out of fresh water to use! I believe that adopting a local stream, creek, or watershed is one way you can protect your water.

Come this weekend and help clean up our stream. We will pick up the trash and plant trees and grasses that will keep the soil near the creek from washing away. We will also put up signs to remind people not to dump trash in our stream.

Don't take the water near you for granted. Take care of it!

Write About It

Persuasive Writing Write a letter to the mayor of your town. Explain a need that the students in your community have and why people should help. State your position clearly and support it with relevant facts and evidence organized in a logical way.

LOG ON e-Journal Research and write about it online at www.macmillanmh.com

Understanding Earth's Water

Percents and Fractions

To convert a percent

▶ **place the percent over a denominator of 100**

▶ **reduce the fraction to its simplest form**

$$15\% = \frac{15}{100} = \frac{3}{20}$$

To find a fraction of a fraction

▶ **multiply the two fractions. So if you ate $\frac{1}{3}$ of $\frac{1}{2}$ a pizza, you ate $\frac{1}{6}$ of the whole pizza**

$$\frac{1}{2} \times \frac{1}{3} = \frac{1}{6}$$

Most of the water on Earth is salt water. Just a small fraction of Earth's water is fresh. Most of that fresh water is frozen! It is trapped as polar ice. Polar ice is found around the North and South poles. If you understand percents and fractions, you can figure out what part of Earth's water is polar ice.

 Solve It

1. What fraction of Earth's water is salt water?
2. What fraction of Earth's water is fresh water?
3. Polar ice is about $\frac{3}{4}$ of the 3% of Earth's fresh water. What fraction of all of Earth's water is polar ice? Hint: Use your answer to question 2 to help you solve this. Then use a calculator to change your answer to a decimal. First divide the numerator by the denominator. Then multiply the answer by 100 to find out what percent of Earth's water is polar ice.

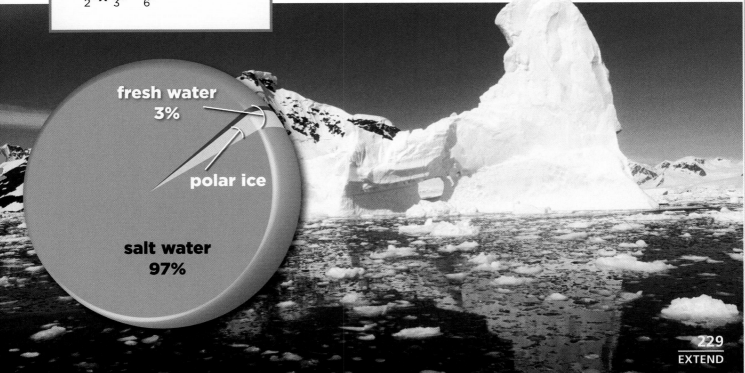

fresh water
3%

polar ice

salt water
97%

Visual Summary

Lesson 1 The important chemicals for life—water, carbon, nitrogen, and oxygen—are used and reused as they flow through ecosystems.

Lesson 2 Ecosystems can change naturally over time into a series of different living communities.

Lesson 3 The six major land biomes on Earth are tundra, taiga, desert, rain forest, deciduous forest, and grassland.

Lesson 4 Water ecosystems include bodies of fresh water, salt water, and mixed water.

Make a FOLDABLES™ Study Guide

Assemble your lesson study guide as shown. Use your study guide to review what you have learned in this chapter. Don't forget to include your Lesson 4 study guide in the back.

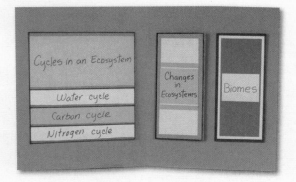

Fill in each blank with the best term from the list.

biome, p. 208

compost, p. 190

deciduous forest, p. 213

estuary, p. 226

extinct species, p. 198

intertidal zone, p. 224

secondary succession, p. 202

water cycle, p. 184

1. A desert ecosystem is an example of a(n) _____.

2. When its last member dies, a species becomes a(n) _____.

3. Trees such as oak and maple grow in a(n) _____.

4. The continuous movement of water between Earth's surface and the air is called the _____.

5. The beginning of a new community where another has already existed is called _____.

6. The Moon's gravity causes constant changes to the ocean's _____.

7. A fertilizer made from dead plant and animal material is called _____.

8. An ecosystem where fresh water meets salt water is called a(n) _____.

Answer each of the following in complete sentences.

9. **Cause and Effect** Why does burning fossil fuels create carbon dioxide?

10. **Sequence** In the process of primary succession, what three stages would come before the one shown below?

11. **Interpret Data** In which biomes would you expect to find animals that require little water to survive?

Biome Annual Precipitation	
Tundra	8 inches
Desert	0.1 inch
Rainforest	more than 100 inches
Deciduous Forest	32 inches

12. **Critical Thinking** Why would you not expect to find algae in the abyssal zone of the ocean?

13. **Persuasive Writing** Write a speech to persuade your community to recycle. Explain why recycling is important.

The Big Idea

14. **How are ecosystems different?**

Succession in Action

What to Do

Identify a place where primary or secondary succession is taking place.

1. Write a short paragraph describing primary and secondary succession.

2. Think of an area you have visited or read about where succession is taking place. Observe or research the types of plants and animals that inhabit this place. Draw a diagram of your observations or research.

3. Write a report based on your observations and/or research. In your report, list the evidence that succession is taking place in the area you chose.

Analyze Your Results

▶ Make a prediction about what will happen to this area if it is not disturbed for 20 years.

Test Prep

1. **Which process is shown below?**

A water cycle

B carbon cycle

C nitrogen cycle

D primary succession

Careers in Science

Gardener

Think about a garden or park. Besides walkways and some playground equipment, what comes to mind? Chances are you will think about plants, such as trees, shrubs, grass, and flowers. The beauty of gardens and parks depends on the work of gardeners. These are people who plant seeds and then take care of the plants that grow. Do you like to work outdoors? Do you have a "green thumb"? If both your answers are "yes," then you might like a career as a gardener. As a high school graduate, you can get a job where you learn gardening skills as you work. Among the rewards of being a gardener is the enjoyment of the beauty you help to create.

▲ **Gardeners work with many kinds of plants.**

▼ **This plant ecologist is observing pitcher plants.**

Plant Ecologist

Do you have a strong love of nature, especially plants? If so, then you might want to become a plant ecologist. As a plant ecologist, you would do much of your work outdoors as well as in a laboratory. Plant ecologists study the interrelationships of plants and their environments. Their concerns include natural resources, the protection of endangered species, and conservation issues. For example, plant ecologists are concerned with wetland ecosystems. Many plants within that ecosystem are endangered, such as the pitcher plant. A bachelor of science degree is needed for a beginner in this field, after which you might do graduate work.

LOG ON **e-Careers** at www.macmillanmh.com

Earth and Its Resources

Over 2.8 million liters (740,000 gallons)
of water flow over Niagara Falls per second!

This diamond mine is in the
Republic of Botswana, Africa.

THE MANY SIDES OF DIAMONDS

from *Time for Kids*

Diamonds are famed for their beauty. Diamonds are also the hardest and one of the most useful substances on Earth.

Most diamonds are shaped too strangely or are too small to be made into jewelry. These stones are still valuable. They are used in making thousands of products, from eyeglasses to computer chips. Workers cut, grind, or shape building materials using tools with diamond edges. Dentists use diamond-tipped drills.

Diamonds may be colorless, white, gray, blue, yellow, orange, red, green, pink, brown, or black.

 ## Write About It

Response to Literature This article describes the formation and use of diamonds. Research additional information about the history of industrial diamonds, how they are formed, and how they are used. Write a report about industrial diamonds. Include facts and details from this article and from your research.

 e-Journal Write about it online at **www.macmillanmh.com**

After diamonds are mined, they may be used in various industries or to make jewelry.

CHAPTER 5

Our Dynamic Earth

How does Earth's surface change?

Kilauea volcano, Hawaii

Key Vocabulary

mantle
a nearly melted layer of hot rock below Earth's crust (p. 246)

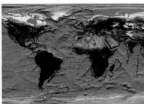

plate tectonics
a scientific theory that Earth's crust is made of moving plates (p. 254)

fault
a crack in Earth's crust whose sides show evidence of motion (p. 256)

hot spot
a stationary location in the mantle where magma melts through a tectonic plate (p. 266)

weathering
the process through which rocks are broken down (p. 284)

deposition
the process of dropping off bits of eroded rock (p. 288)

Earth's Landforms

Look and Wonder

Looking down from above Earth's surface, you can see oceans, mountains, and rivers below. What do these features look like?

Explore

What are Earth's features?

Purpose

To examine and classify Earth's features.

Procedure

1 Observe Examine the photos.

2 List the features of Earth's surface that you can identify in these photos.

3 Communicate Describe how the features are similar and different.

Draw Conclusions

4 Classify Identify groups into which you could sort Earth's features.

5 Infer What processes might have produced one or more of the features you have identified?

Explore More

Find photos of the Grand Canyon in Arizona. What do you think happens when water runs over rock for a long time? Form a hypothesis about how water was involved in forming the canyon. Design an experiment that would test your prediction.

Cannon Beach, OR

Alaska Range, AK

Marjorie Glacier, AK

Monument Valley, AZ

Ausable River, NY

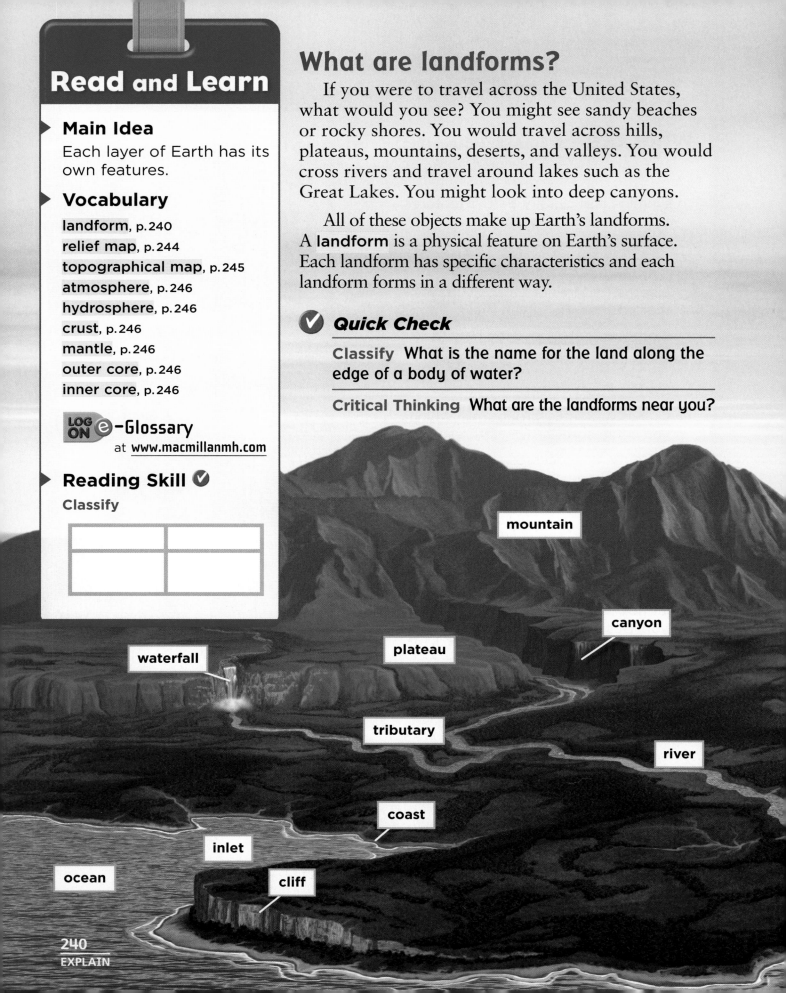

▶ **Main Idea**

Each layer of Earth has its own features.

▶ **Vocabulary**

landform, p. 240
relief map, p. 244
topographical map, p. 245
atmosphere, p. 246
hydrosphere, p. 246
crust, p. 246
mantle, p. 246
outer core, p. 246
inner core, p. 246

LOG ON **e-Glossary**
at www.macmillanmh.com

▶ **Reading Skill** ✓

Classify

What are landforms?

If you were to travel across the United States, what would you see? You might see sandy beaches or rocky shores. You would travel across hills, plateaus, mountains, deserts, and valleys. You would cross rivers and travel around lakes such as the Great Lakes. You might look into deep canyons.

All of these objects make up Earth's landforms. A **landform** is a physical feature on Earth's surface. Each landform has specific characteristics and each landform forms in a different way.

✓ **Quick Check**

Classify What is the name for the land along the edge of a body of water?

Critical Thinking What are the landforms near you?

mountain

canyon

plateau

waterfall

tributary

river

coast

inlet

ocean

cliff

Earth's Land Features

A mountain is a landform that rises high above the ground.

A hill is lower and rounder than a mountain.

A valley is low land between hills or mountains.

A canyon is a deep valley with high, steep sides.

A cliff is a high, steep section of rock or soil.

A plain is a wide, flat area.

A plateau is flat land that is higher than the land around it.

A desert is an area with very little precipitation.

A beach is the land along the edge of a body of water.

A dune is a mound or ridge of sand.

Earth's Water Features

An ocean is a large body of salt water.

A coast is where a body of water meets land.

A tributary is a small river or stream.

A river is a natural body of moving water.

A waterfall is a natural stream of water falling from a high place.

A lake is a body of water surrounded by land.

An estuary is where river water and ocean water meet.

A delta is the mass of land that forms at the mouth of a river.

An inlet is a narrow body of water off of a larger body of water.

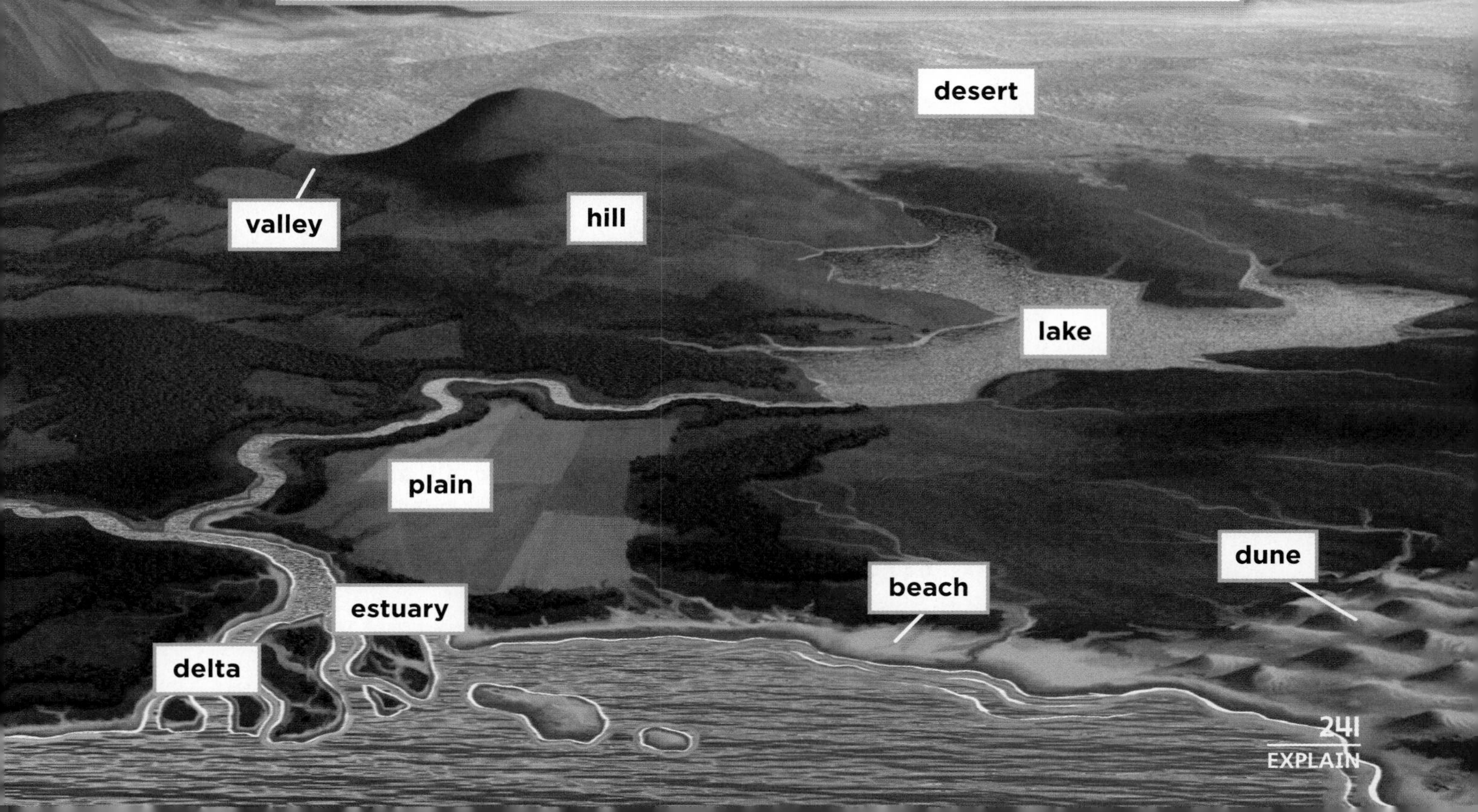

desert

valley

hill

lake

plain

dune

beach

estuary

delta

What are the features of the ocean floor?

Although waves move over the surface of the ocean, the ocean's surface is mostly flat. However, if you could travel beneath the ocean's surface, you would find features on the ocean floor that look like mountains and valleys.

An ocean basin is a large underwater area between continents. Along the coast of a continent, the ocean floor is called the *continental shelf*. Here the ocean floor is covered by shallow water and gradually slopes down. A continental shelf ends at a point where a sharp slope begins. This sharp slope is called the *continental slope*.

A *submarine canyon* is a steep-sided valley in a continental slope. Submarine canyons often are found near the mouths of large rivers. At the end of a continental slope is another gradual downward slope called a *continental rise*.

The *abyssal plain* (uh•BIS•uhl playn) is a wide, flat area of ocean floor. Abyssal plains cover about 40% of the ocean floor.

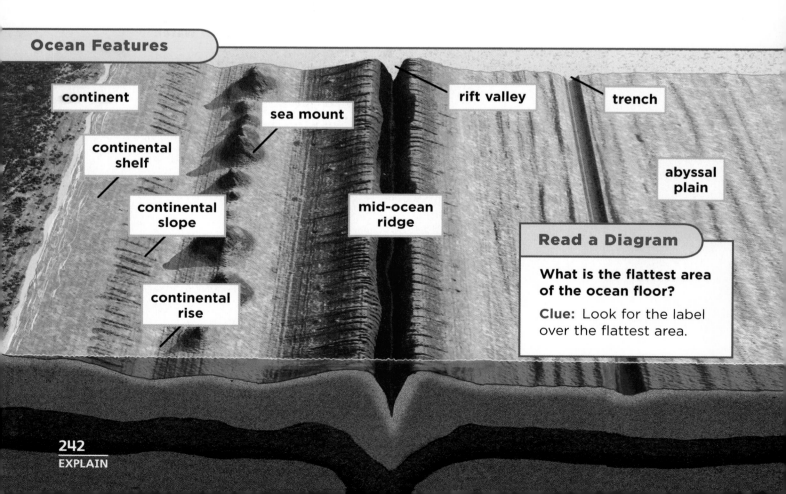

Ocean Features

continent

sea mount

rift valley

trench

continental shelf

continental slope

mid-ocean ridge

abyssal plain

continental rise

Read a Diagram

What is the flattest area of the ocean floor?

Clue: Look for the label over the flattest area.

▲ This underwater vehicle explored the ocean around Hawaii.

Trenches are the deepest parts of the ocean floor. They are usually long and narrow. A *seamount* is an underwater mountain that rises from the ocean floor but stops before it reaches the surface of the ocean.

Mid-ocean ridges are underwater mountain ranges. An indentation called a *rift valley* occurs along the top of these mountains.

Scientists can tell the depth of the ocean floor by sending sounds into the ocean and waiting for the echo to come back. How do scientists know what the ocean floor looks like? Scientists use underwater vehicles to observe the ocean floor. These vehicles may carry cameras, instruments to measure the underwater environment, or mechanical arms to gather samples.

✔ Quick Check

Classify Which ocean features are underwater mountains that do not reach the surface?

Critical Thinking How could rivers form submarine canyons?

≡ Quick Lab

Mapping the Ocean Floor

1. Place soft clay in the bottom of a container to form features of the ocean floor.

2. Cover your container.
3. Exchange your container with another classmate.
4. **Measure** Gently drop a probe into each hole and measure how much of the probe sticks out of the hole. Record the depth of each square of the grid.

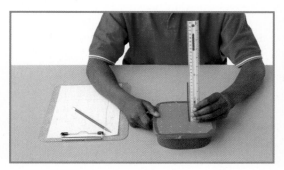

5. **Interpret Data** Use your probe measurements to figure out the height of the features, then draw and label them.
6. Remove the top of the container and compare your drawing to the ocean features.

How are Earth's features mapped?

Earth's surface is uneven. Hills rise in one place, while valleys dip in another place. While some places are higher or lower than other places, maps of the surface are flat. How can a flat map show an uneven surface?

Relief Maps

A *surveyor* is a person who takes measurements of land. As the first step in making a map, a surveyor measures the elevation in a specific location. *Elevation* is the height of land above sea level. The surveyor may leave an object as a *benchmark,* or permanent reference point, for that elevation.

Mapmakers then use the surveyor's measurements to show changes in elevation on a map. One way to do this is to draw a shaded picture of the land. The shading makes a map look as if it has three dimensions: length, width, and height. The map, of course, really has only two dimensions: length and width. A map that uses shading to show elevations is called a **relief map**.

Relief Map of Nunivak Island, AK

Cape Etolin

Etolin Strait

Cape Mohican

Nash Harbor

Cape Manning

Bering Sea

Elevation in feet
- over 1,600
- 1,400–1,600
- 1,200–1,400
- 1,000–1,200
- 800–1,000
- 600–800
- 400–600
- 200–400
- under 200

N
W E
S

0   10   20 miles

0   10   20 kilometers

Cape Corwin

Duchikthluk Bay

Cape Mendenhall

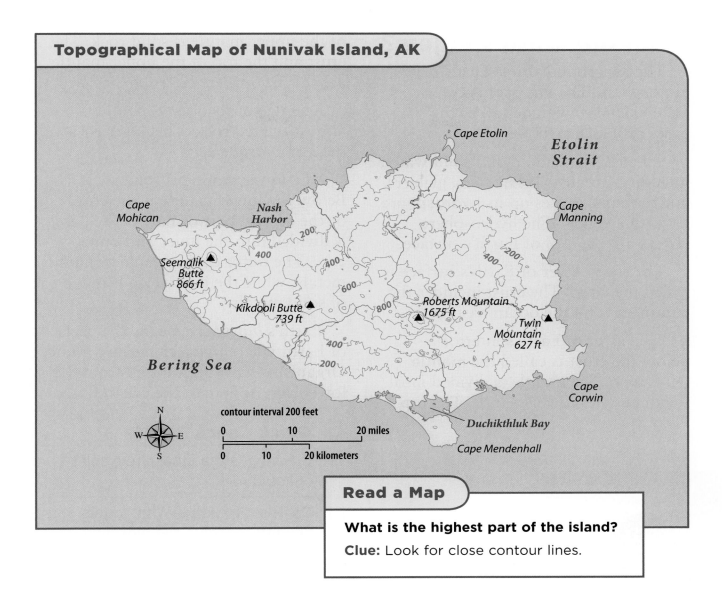

Topographical Map of Nunivak Island, AK

Cape Etolin

Etolin Strait

Cape Mohican

Nash Harbor

200

Cape Manning

400

Seemalik Butte 866 ft

400

400

200

400

600

Kikdooli Butte 739 ft

800

Roberts Mountain 1675 ft

Twin Mountain 627 ft

400

Bering Sea

200

Cape Corwin

contour interval 200 feet

N
W E
S

0   10   20 miles

0   10   20 kilometers

Duchikthluk Bay

Cape Mendenhall

Read a Map

What is the highest part of the island?

Clue: Look for close contour lines.

Topographical Maps

A **topographical map** (top•uh•GRAF•i•kuhl) uses lines to show elevation. Each *contour line* represents a different elevation. A number on the line gives the elevation. The number is usually in units of meters or feet.

In addition to elevation, contour lines can tell you how steep or gradual a slope is. Contour lines that are close together mean that elevation is changing rapidly and the slope of the land is very steep. Contour lines that are far apart mean that elevation is changing gradually.

 Quick Check

Classify What kind of map uses shading to show elevations?

Critical Thinking While hiking, why would you avoid a route where a map shows contour lines that are very close together?

What are Earth's layers?

The air around you is Earth's atmosphere. The **atmosphere** (AT•muhs•feer) includes all of the gases around Earth.

All of Earth's liquid and solid water, including oceans, lakes, rivers, glaciers, and ice caps, makes up its **hydrosphere** (HYE•druh•sfeer). The hydrosphere covers about 70% of Earth's surface.

The rocky layer of Earth's surface is called the **crust**. The crust includes the continents and the ocean basins.

The layer of Earth's interior below the crust is called the **mantle** (MAN·tuhl). The mantle is divided into the upper and lower mantle. The top of the upper mantle is solid rock. The crust and the top of the upper mantle are the *lithosphere* (LITH•uh•sfeer).

The rest of the upper mantle is almost-melted rock. This layer is called the *asthenosphere*.

The lower mantle is solid rock. Below the lower mantle is the core, or the central part of Earth. The core is divided into the inner and outer core. The **outer core** is made of liquid metals, while the **inner core** is made of solid metals.

Earth's *biosphere* means the parts of Earth where living things are found. Organisms have been found from the lower atmosphere to the ocean floor.

✔ Quick Check

Classify Is the lithosphere solid or liquid rock?

Critical Thinking What layers of Earth make up the biosphere?

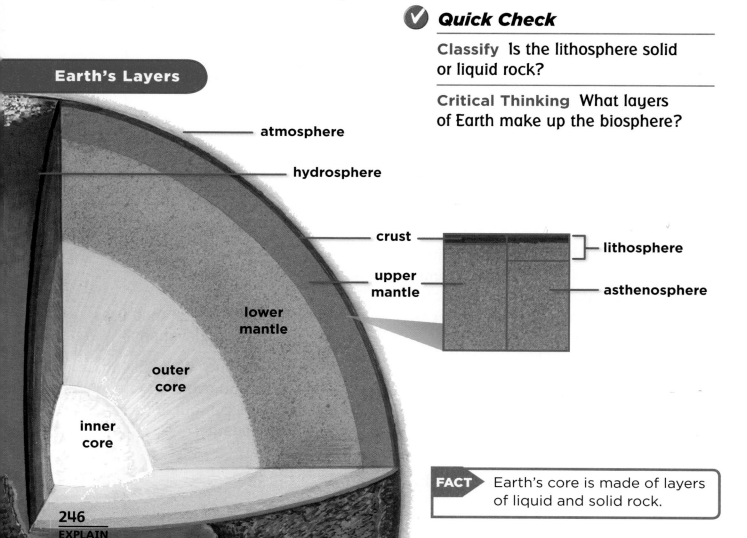

Earth's Layers

- atmosphere
- hydrosphere
- crust
- upper mantle
- lower mantle
- outer core
- inner core
- lithosphere
- asthenosphere

FACT Earth's core is made of layers of liquid and solid rock.

Lesson Review

Visual Summary

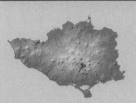

Earth's surface and ocean floor are covered by landforms.

Relief maps and topographical maps show elevations on Earth's surface.

Earth's layers include the atmosphere, hydrosphere, crust, mantle, and core.

Make a FOLDABLES™ Study Guide

Make a Three-Tab Book. Use the titles shown. Then write about what you have learned about those topics.

Earth's surface and ocean floor are...

Relief maps and topographical maps show...

Earth's layers include...

Think, Talk, and Write

1. **Main Idea** What are Earth's layers?

2. **Vocabulary** Mountains, valleys, deserts, and rivers are examples of _____.

3. **Classify** Which features of the ocean floor angle downward and which angle upward?

4. **Critical Thinking** How could a map showing elevation help you plan a hike?

5. **Test Prep** What is an abyssal plain?
   A an underwater mountain range
   B a steep-sided valley
   C a slope covered by shallow water
   D a wide, flat area of the ocean floor

6. **Test Prep** What is the outermost solid layer of Earth called?
   A hydrosphere
   B lithosphere
   C inner core
   D mantle

Writing Link

Explanatory Writing
Write about how a relief map is different from a topographical map. Give examples of circumstances under which you would want to use each type of map.

Social Studies Link

The Deepest Trench
The Marianas Trench in the Pacific Ocean is the deepest known trench in the ocean floor. Research the different depth measurements that have been taken and discuss how the depth measurements were made.

Focus on Skills

Inquiry Skill: Make a Model

Models show the basic features of a structure or process. When scientists **make a model**, they simplify a process or structure that would otherwise be difficult to see and understand. Many scientists use laboratory materials or computers to make a model so they can explain an idea, an object, or an event.

Mt. Rainier

▶ Learn It

What is the relationship between elevation and contour lines? Mapmakers use elevation measurements that are made by surveyors to make contour lines on maps. The surveyors use various instruments that can accurately measure distances. Telescopes, aerial photographs, and satellite images help them to get elevation measurements.

How can you understand the relationship between elevation and contour lines without those instruments? **Make a model** of a mountain and use it to make a topographical map to help you understand the relationship between contour lines and elevation.

▶ Try It

 Materials soft clay, pencil, ruler, dental floss, graph paper

1. **Make a Model** Form the clay into a mountain shape.

2. Using the pencil, poke a hole through the center of your mountain.

3. Measure the height of your mountain in centimeters.

4. Using the dental floss, cut a 1-centimeter slice off the top of the mountain.

Mt. Hood

5 Place the slice of mountain onto a piece of graph paper. Trace the edges of the slice and mark the location of the hole in the middle. Then put the slice to the side.

6 Cut the next 1-centimeter slice of mountain. Place it on the graph paper so the hole in the middle of the slice lines up with the spot you marked on the graph paper from the previous slice. Then trace the edges of this slice.

7 Cut, line up, and trace the rest of the slices of mountain. When you are finished with the bottom slice, put the clay mountain back together.

8 Is there a point on your topographical map where the lines are closer together? If there is, what does the mountain look like in this area?

▶ Apply It

Now that you understand the relationship between elevation and contour lines, you can apply your understanding when reading topographical maps. The topographical map below shows a student's neighborhood with the contour lines measured in feet. Use the map to answer the following questions.

1 What is the lowest elevation on the map?

2 What is the highest elevation on the map?

3 The purple circle represents a school. What is its elevation?

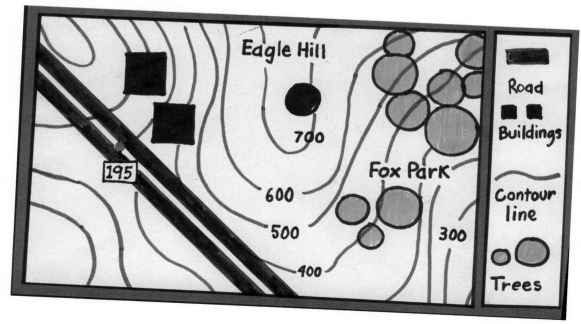

Plate Tectonics

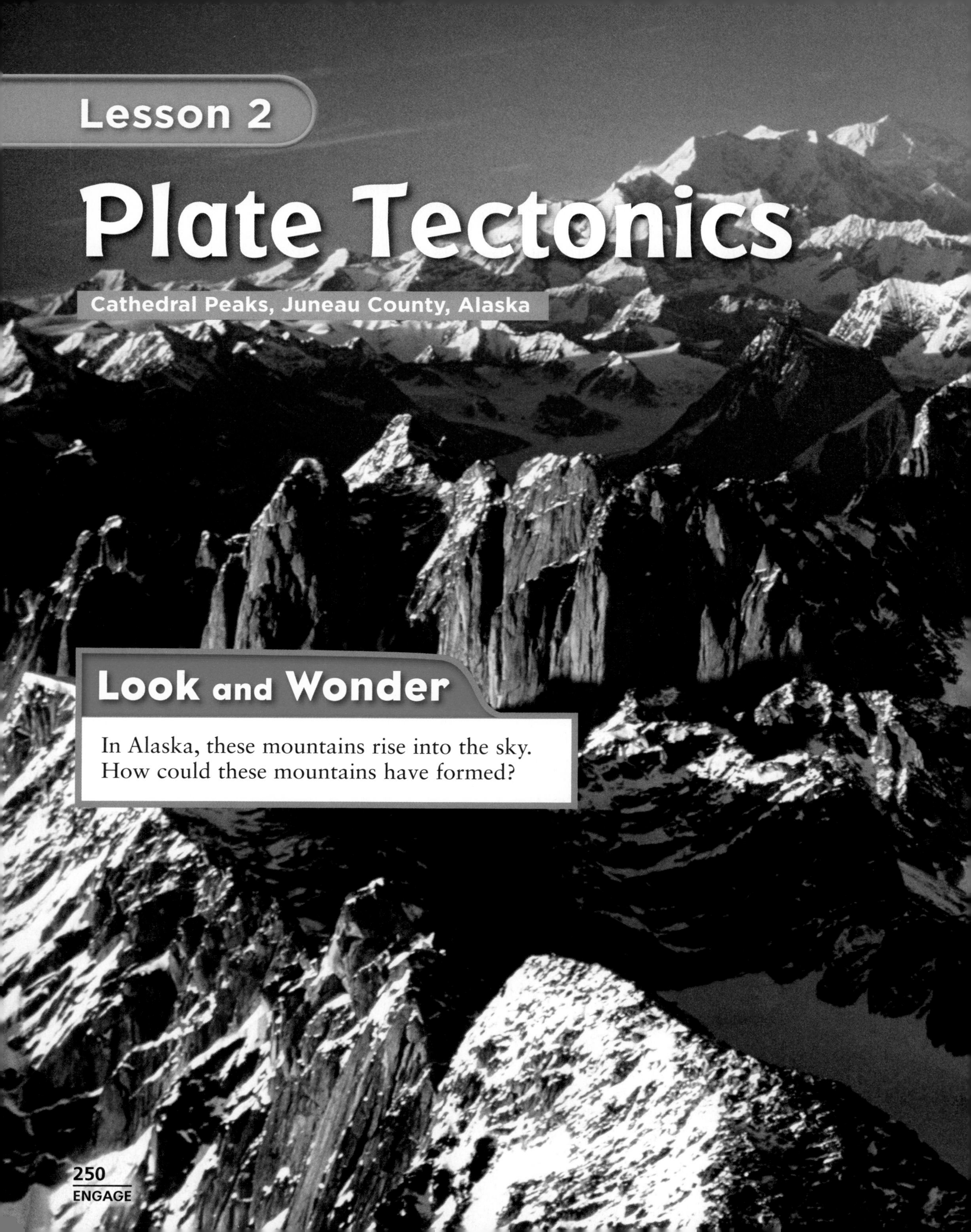

Cathedral Peaks, Juneau County, Alaska

Look and Wonder

In Alaska, these mountains rise into the sky. How could these mountains have formed?

How can mountains form?

Purpose
To model one way a mountain range can form.

Procedure

1 Lay the sheet of newspaper on your desk. In the middle of the sheet of newspaper, place one sheet of construction paper about 5 centimeters over the other sheet.

2 Spread sand over the overlapped pieces of construction paper. Then smooth the sand with your hand.

3 Make a Model Hold down the farther-away piece of construction paper with one hand. With your other hand, slowly push the closer sheet of construction paper underneath the other sheet.

4 Observe What happens to the sand?

Draw Conclusions

5 Infer In this model, the mountain range that formed was straight. How could you change the model to make the mountain range curve?

6 What would you expect to happen if you continued to push on the closer sheet?

Explore More

In this model of mountain range formation, you pushed one sheet of paper underneath another. What landforms would you model if you pulled the sheets of paper in other directions?

Materials

- sheet of newspaper
- 2 sheets of construction paper
- sand

Step 2

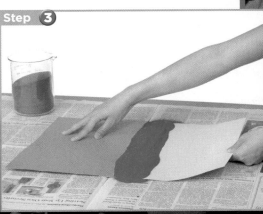
Step 3

Read and Learn

▶ **Main Idea**

Earth's surface is made of plates that constantly move.

▶ **Vocabulary**

geologist, p. 252

plate tectonics, p. 254

magma, p. 254

fault, p. 256

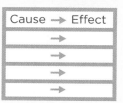

LOG ON ℮-Glossary

at www.macmillanmh.com

▶ **Reading Skill** ✔

Cause and Effect

Cause	→	Effect
	→	
	→	
	→	
	→	

Is Earth's crust moving?

About 100 years ago, Alfred Wegener, a German geologist (jee•OL•uh•jist), noticed something curious about Earth's continents. A **geologist** is a scientist who examines rocks to find out about Earth's history and structure. Wegener noticed that the continents looked like pieces of a jigsaw puzzle.

Wegener thought that millions of years ago Earth's continents were joined together like a completed jigsaw puzzle. Then, as time passed, some force pulled the puzzle pieces apart. The continents slowly moved to the positions they are in today. Wegener's idea became known as the *theory of continental drift.*

At that time, scientists did not know of any forces that could move continents. Over time, scientists found evidence that supported continental drift. One piece of evidence was that the mountains on the east coast of South America had the same types of rocks as the mountains on the west coast of Africa. Also, the rocks were the same age.

Theory of Continental Drift

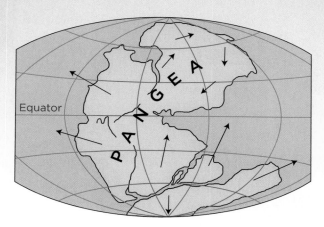

225 million years ago

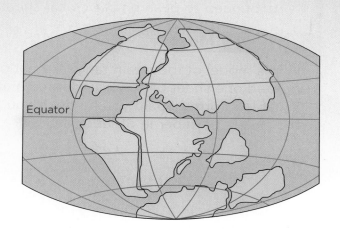

200 million years ago

FACT The continents are constantly moving.

Additional evidence that the continents had moved came from plant and animal fossils. *Fossils* are the remains or signs of a once-living thing. A fossil of a freshwater reptile called *Mesosaurus* has been found in very old rocks in South America and in Africa.

Mesosaurus could not have swum through the salt water in the Atlantic Ocean. However, if the continents had once been joined, *Mesosaurus* could have traveled across the larger continent through freshwater rivers.

What conclusion would you reach from this evidence? South America and Africa were once part of a supercontinent called Pangea. Over millions of years, South America and Africa have drifted apart. The water that filled the space between them formed the Atlantic Ocean.

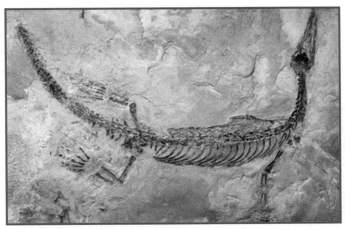
a *Mesosaurus* fossil

✔ Quick Check

Cause and Effect What caused scientists to conclude that Earth's continents were once joined?

Critical Thinking How would present-day animals and rocks on different continents compare since the continents have been separated for a long time?

Read a Diagram

How do the rock layers in Africa and South America provide evidence that the continents were once next to each other?

Clue: Compare the rock layers on both continents.

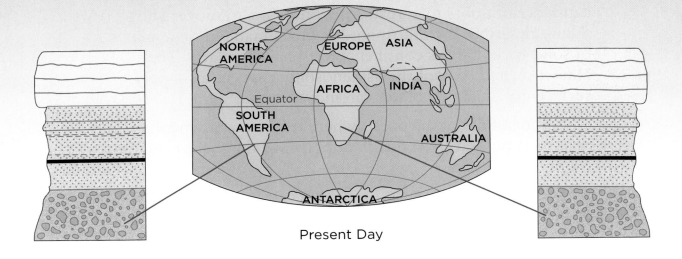

Present Day

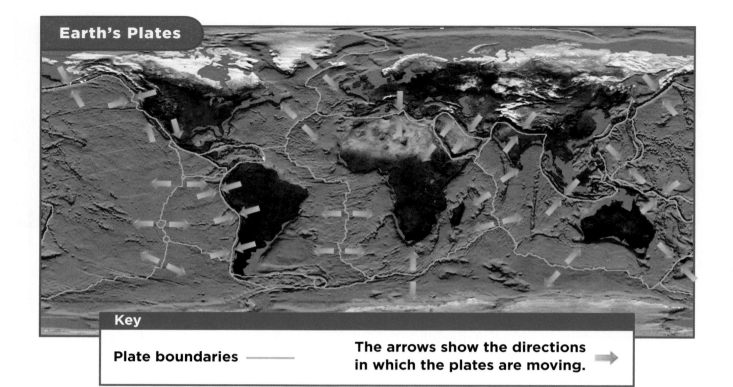

Earth's Plates

Key

Plate boundaries ———

The arrows show the directions in which the plates are moving. →

What causes the ocean floor to spread?

The ocean floor between South America and Africa is spreading at a rate of about 4 centimeters (1.5 inches) every year. If this rate of movement has happened for the past 130 million years, South America and Africa should be 520,000,000 centimeters (5,200 kilometers or 3,230 miles) apart. This is about the width of the Southern Atlantic Ocean. The ocean has filled up the space between the two continents.

Scientists have developed a theory called **plate tectonics** (tek•TAH•niks) to explain how forces deep within Earth can cause ocean floors to spread and continents to move. This theory describes the lithosphere as being made of huge plates of solid rock. Earth's continents rest on these plates. The almost-melted rock of the asthenosphere acts as a slippery surface on which the plates can move around.

In the middle of the ocean where the plates are moving apart, magma is pushed from the mantle toward the surface. **Magma** is hot, melted rock.

The upward movement of magma causes *tension*, or a stretch or push, on the plates. This push moves the ocean floor apart and separates the plates on either side of the mid-ocean ridge. Since the continents rest on these plates, they also move apart.

As the hot rock reaches the surface, it cools and builds up equally on both sides of the opening. The cooling rock forms the mid-ocean ridge and the rift valley along its top. The mid-ocean ridge runs roughly parallel to the separating continents. The ridge remains stationary while the ocean floor on both sides of the ridge grows wider. As the ocean floor grows wider, the continents move farther apart.

 Quick Check

Cause and Effect What happens as an effect of magma rising through the ocean floor?

Critical Thinking What causes mid-ocean ridges to have rift valleys along their tops rather than coming to a sharp peak as mountains do?

≡ *Quick Lab*

Spread of the Ocean Floor

❶ Place two stacks of books next to each other with a narrow space between them.

❷ Fold a sheet of paper and slip it into the space between the books.

❸ **Measure** Using a ruler, measure 3 cm of the paper. Fold that much paper down onto the books. Color this part of the sheet purple.

❹ **Make a Model** Move the sheets of paper up another 3 cm. Color the newly exposed areas of both sheets with another color. Repeat this step until you run out of paper.

❺ If the two sheets of paper represent the spreading ocean floor, which color represents the youngest rock?

Spread of the Ocean Floor

Age of ocean floor in millions of years

| 150-200 | 100-150 | 50-100 | 0-50 | 50-100 | 100-150 | 150-200 |

Read a Diagram

Where is the newest ocean floor?

Clue: Use the key to find the color that represents the newest ocean floor.

Folded Mountains

Fault-Block Mountains

How do mountains form?

You have learned that tension moves Earth's plates. What other ways do the plates move? Plates can be pushed into or pushed past each other.

When plates are pushed into each other, the force that occurs is called compression. *Compression* is a squeezing or pushing together of the crust. When a continental plate is compressed, the ground is forced upward, producing *folded mountains*.

The Himalayan Mountains are folded mountains. They formed millions of years ago as India and Asia collided. As the plates continue to push into each other, the Himalayan Mountains grow about 5 millimeters (0.2 inches) taller every year.

When one plate rubs past another plate, this movement causes *shear*, or a force that twists, tears, or pushes one part of the crust past another. In some places, there are deep cracks in Earth's crust called **faults** (fawlts). Shear can cause Earth's crust to break apart along a fault. When this happens, one side of the fault moves up and the other side moves down. This produces *fault-block mountains*. The Sierra Nevada in the western United States are fault-block mountains.

 Quick Check

Cause and Effect What process produces folded mountains?

Critical Thinking Looking at the location of Earth's tectonic plates, where else would you expect to find folded mountains?

Lesson Review

Visual Summary

Geologists have found evidence that chunks of Earth's crust have been and are moving.

Plate tectonics explains how forces deep within Earth move chunks of Earth's crust.

Mountains are produced where Earth's crust is compressed or being sheared.

Make a **FOLDABLES**™ Study Guide

Make a Folded Chart. Use the titles shown. Then summarize what you have learned.

Think, Talk, and Write

1 **Main Idea** Is Earth's crust stationary or moving?

2 **Vocabulary** A deep crack in Earth's crust is a _____.

3 **Cause and Effect** What causes the ocean floor to spread?

Cause	→	Effect
	→	
	→	
	→	
	→	

4 **Critical Thinking** Describe one piece of evidence that supports the theory of continental drift.

5 **Test Prep** What is magma?
- **A** water
- **B** solid rock
- **C** liquid rock
- **D** mountain

6 **Test Prep** When plates are compressed, they produce
- **A** folded mountains.
- **B** fault-block mountains.
- **C** mid-ocean ridges.
- **D** flat land.

Math Link

Measuring the Age of an Ocean
You measure an ocean and find that it is 100 kilometers wide. The ocean floor has been spreading 2 centimeters a year. How old is the ocean?

Art Link

Continent Location
You know that the continents are moving. Draw where you think the continents will be in 50 million years.

Writing in Science

Pangea and Other Supercontinents

This view of Earth shows one possible shape of the supercontinent Rodinia.

Expository Writing

A good exposition

▶ develops the main idea with facts and supporting details

▶ summarizes information from a variety of sources

▶ uses transition words to connect ideas

▶ draws a conclusion based on the facts and information presented

Based on fossils, rocks, and other geological evidence, scientists have concluded that all of the continents that exist today were once combined as one supercontinent. This supercontinent is called Pangea, which means "all continents." Around Pangea there was an ocean called Panthalassa, or "all seas."

Scientists have concluded that other supercontinents existed before Pangea. The oldest known supercontinent is called Rodinia, which means "homeland." Rodinia formed about 1.1 billion years ago and started to break up about 750 million years ago. The pieces formed another supercontinent called Pannotia. Then, about 550 million years ago, Pannotia broke apart. The fragments of

Write About It

Expository Writing Research the movement of Rodinia and Pannotia. Select a main idea. Write an expository essay with details that support your main idea.

LOG ON **e-Journal** Research and write about it online at **www.macmillanmh.com**

Math in Science

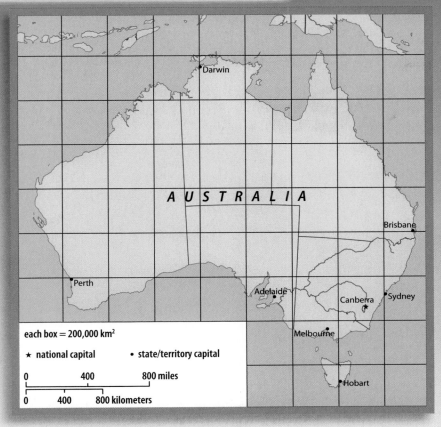

each box = 200,000 km²

★ national capital   • state/territory capital

0   400   800 miles

0   400   800 kilometers

Estimate the Area of a Continent

The area of a continent is the area of the continental plate that rises above the ocean. While continents have irregular shapes, area is measured in square units. One way to estimate the area of an irregular object is to place a grid over the object and then count the number of squares in the grid that cover the object.

 Solve It

1. How many boxes are full of land? How many are three-quarters full? Half-full? One-quarter full?

2. How many full boxes can you make from the quarter-full, half-full, and three-quarters-full boxes?

3. What is the total number of full boxes?

4. Now estimate the area of Australia in square kilometers.

5. Using reference materials, find an official measurement of the area of Australia. How does your estimate compare?

Estimate Area

To estimate area

▶ count the number of boxes that are one-quarter full, half-full, three-quarters full, and full

▶ multiply the number of each type of partly full boxes by the amount of land covered in that type of box

▶ add the total number of full boxes together

▶ multiply that number by the square kilometers that each full box equals to get the total estimated area

Volcanoes

Kilauea Volcano, Hawaii

Look and Wonder

When a volcano is about to erupt, hot rock inside the volcano pushes up and increases the steepness of the slope of the volcano. How can scientists predict whether a volcano will erupt?

How can you predict when a volcano will erupt?

Purpose

To build a tiltmeter, or an instrument that can measure changes in the slope of a volcano.

Procedure

1. ⚠ **Be Careful.** Using the tip of a pencil, punch a small hole into the side of a foam cup about 2.5 cm from the bottom of the cup. Push the coffee stirrer into the hole so it fits tightly.

2. Punch a small hole in the second cup. Push the other end of the coffee stirrer into that hole.

3. Mix 3 drops of food coloring into a container of water.

4. Put the cups and stirrer into a baking pan. Then pour each cup about half-full of the dyed water.

5. **Observe** Carefully tilt one end of the pan. What happens to the level of the water in the cups?

- pencil
- 2 foam cups
- coffee stirrer
- food coloring
- container of water
- baking pan

Draw Conclusions

6. **Experiment** Raise the tiltmeter to different heights and measure the change in the height of the water. How well does the tiltmeter record changes in height?

7. **Infer** What might it mean if a tiltmeter measured an increase in the steepness of a volcano's slope?

Step 1

Explore More

How would the tiltmeter work with different-sized cups or different lengths of straw? Pick a variable that you want to change. Design an experiment to test how well the model with the changed variable measures tilt.

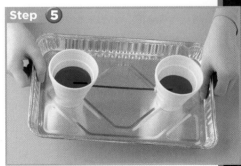

Step 5

Read and Learn

Main Idea

Volcanoes occur where magma reaches Earth's surface.

Vocabulary

volcano, p. 262
lava, p. 263
shield volcano, p. 264
cinder-cone volcano, p. 264
composite volcano, p. 265
island chain, p. 266
hot spot, p. 266
island arc, p. 266

 e-Glossary
at **www.macmillanmh.com**

Reading Skill ✓

Infer

Clues	What I Know	What I Infer

Where are volcanoes found?

Volcanoes (vol•KAY•nohz) form on land and on the ocean floor. A **volcano** is an opening in Earth's crust. Volcanoes are only located in certain places on Earth's surface. As you know, Earth's crust is broken up into a number of moving plates. Most volcanoes are found where plates meet.

For example, a circle of volcanoes called the *Ring of Fire* surrounds the Pacific Ocean. The Ring of Fire follows the boundaries of the plates that meet around the Pacific Ocean. Volcanoes are more likely to erupt at plate boundaries than anywhere else on Earth. An *eruption* is an outpouring of melted rock, ash, gases, or a combination of these.

Cross-Section of Volcano

vent

crater

vent

magma chamber

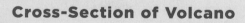

Read a Diagram

How many vents does this volcano have?

Clue: Look for the vent labels.

 Science in Motion Watch a volcano erupt at **www.macmillanmh.com**

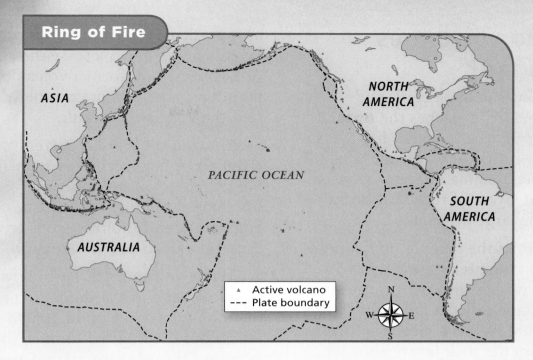

Ring of Fire

ASIA

NORTH AMERICA

PACIFIC OCEAN

AUSTRALIA

SOUTH AMERICA

▲ Active volcano
--- Plate boundary

However, volcanoes do not erupt at all plate boundaries. What makes some plate boundaries likely places for volcanoes to erupt? After collecting data about the directions in which plates moved, scientists concluded that volcanoes tend to erupt where one plate is pushed under another plate.

When rocks in the plate that is being pushed down reach the heat and pressure in the mantle, they melt. Magma forms and pools in a chamber underneath the crust.

The magma may rest quietly for hundreds or thousands of years. Sometimes, a crack forms above the chamber or the pressure in the chamber grows too great to be held in by the rock above it. Then the magma rushes upward toward Earth's surface.

All volcanoes have at least one *vent*, or opening. Once magma reaches Earth's surface, it is called lava (LAH•vuh). Lava, ashes, and gases erupt through vents. Over time, a cup-shaped depression may form around a vent. This depression is called a *crater*. Most craters are found at the top of a volcanic mountain.

Sometimes the magma chamber beneath a volcano is emptied. The volcano may then collapse inside itself. The hole that forms is called a *caldera* (kal•DER•uh).

✔ Quick Check

Infer What can you infer from the Ring of Fire map about volcano and tectonic plate locations?

Critical Thinking What makes a volcano more likely to erupt where one plate slides under another plate than where two plates collide?

FACT More than 80 percent of Earth's volcanic activity occurs on the ocean floor.

How do volcanoes build land?

Sometimes magma cools and hardens before it reaches the surface. A *dike* forms when magma hardens in vertical or nearly vertical cracks. A *sill* forms when magma hardens between horizontal layers of rock. Dikes and sills can be large or small.

Sometimes the magma pushed into a sill does not spread horizontally. Instead it pushes upward and forms a dome shape called a *laccolith* (LAK•uh•lith). When a laccolith forms, it may raise the rock layers above it.

The largest and deepest of all underground magma formations are batholiths (BA•thuh•liths). A *batholith* is huge and irregularly shaped. It reaches deep into the crust.

When lava comes out of a vent, it is liquid. Lava forms a solid layer of rock as it hardens. Over thousands of years, layers of lava may increase the height of a volcano and form a volcanic mountain. The mountain is new land that the volcano has built.

An *active volcano* is currently erupting or has recently erupted. Sometimes a volcano does not erupt for some time and is called a *dormant volcano*. A volcano that stops erupting is called a dead or *extinct volcano*.

Active volcanoes differ in how they form and in the shapes of the mountains they build. **Shield volcanoes** are built by thinner, fluid lava that spreads over a large area. These mountains have a broad base and gently sloping sides.

Cinder-cone volcanoes are built by thick lava that is thrown high into the air and falls as chunks or cinders. These mountains form as a cone shape with a narrow base and steep sides.

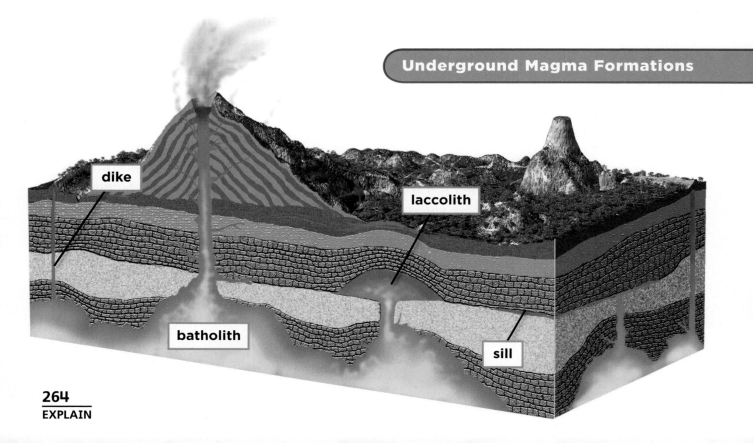

Underground Magma Formations

dike

laccolith

batholith

sill

Types of Active Volcanoes

Mt. Kilauea in Hawaii

shield

Mt. Capulin in New Mexico

cinder-cone

Stromboli in Italy

composite

Quick Lab

Types of Volcanoes

1. **Make a Model** Use materials of your choice to make models of the three kinds of volcanic mountains.

2. **Communicate** Write descriptions of each type of volcano and how it was formed. Place the description next to each model.

3. **Compare** How are the layers of the volcano models you made similar to or different from the layers of volcanic mountains?

Composite volcanoes (kuhm•POZ•it) are built by layers of ash and cinders sandwiched between layers of hardened lava. The shape on one side of a cone formed by a composite volcano usually looks the same as the shape on the opposite side.

 Quick Check

Infer How can a cone-shaped mountain form?

Critical Thinking Compare and contrast extinct, dormant, and active volcanoes.

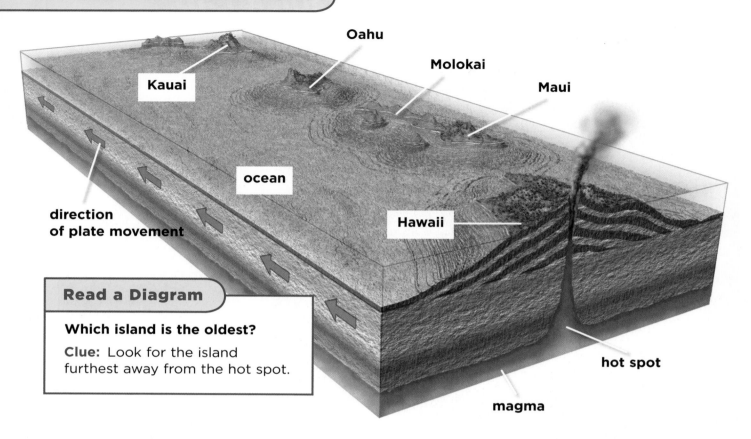

Formation of Hawaiian Islands

Kauai

Oahu

Molokai

Maui

ocean

Hawaii

direction
of plate movement

hot spot

magma

Read a Diagram

Which island is the oldest?

Clue: Look for the island
furthest away from the hot spot.

How do volcanoes build islands?

The Hawaiian Islands are an island chain, or a line of volcanic mountains. How did these islands form? Scientists know that the Hawaiian Islands rest on a slowly moving tectonic plate. As it moves, the plate passes over a stationary pool of magma called a hot spot.

Over millions of years, lava erupting from the hot spot formed a mountain. Eventually, the mountain grew taller than the ocean's surface and became a volcanic island. As the plate moved, that island moved away from the hot spot and a new island began to form.

In areas where an ocean-floor plate is pushed under another ocean-floor plate, an island arc forms. As the plate is pushed down, it melts. Magma forms, rises upward, and erupts through the ocean floor. The eruptions form a series of volcanic islands along the plate boundary. The Aleutian Islands in Alaska form an island arc.

✔ Quick Check

Infer What information would you need to figure out if islands formed from a hot spot?

Critical Thinking How can you distinguish between an island arc and an island chain?

Lesson Review

Visual Summary

Most volcanoes are found at plate boundaries where one plate is sliding under another plate.

Lava, rocks, and ashes that erupt on land build volcanic mountains.

Lava that erupts from the sea floor builds volcanic islands.

Make a FOLDABLES™ Study Guide

Make a Folded Table. Use the titles shown. Then summarize what you have learned.

Main Idea	What I learned...	Sketches
Most volcanoes are found...		
Magma, rocks, and ashes...		
Magma that errupts...		

Think, Talk, and Write

1 Main Idea What is a volcano?

2 Vocabulary A volcano that is not currently erupting but may erupt again is _____.

3 Infer Where will the next Hawaiian island be located?

Clues	What I Know	What I Infer

4 Critical Thinking Where do volcanoes occur?

5 Test Prep What kind of volcanic mountain is made from thin, fast-flowing lava?

  A cinder-cone volcano
  B shield volcano
  C composite volcano
  D crater

6 Test Prep What is a hot spot?

  A an erupting volcano
  B a stationary pool of lava
  C a stationary pool of magma
  D a location on the Ring of Fire

 Writing Link

Interviewing a Witness to an Eruption
Research a recent volcanic eruption. Write a newspaper article about the eruption. Include quotes from a witness who saw the eruption.

 Social Studies Link

Eruption of Mount Vesuvius
Research the eruption of Mount Vesuvius in 79 A.D. Identify the location of Mount Vesuvius and describe the eruption's effect on the people and the buildings in nearby communities.

Be a Scientist

measuring cup

plaster of paris

large container

spoon

funnel

squeeze bottle

1 piece of cardboard

tray

Structured Inquiry

How do volcanoes form islands?

Form a Hypothesis

If tectonic plates are moving over hot spots at different speeds, what do the islands that form look like? Write your answer as a hypothesis in the form "If one tectonic plate is moving over a hot spot faster than another tectonic plate, then . . ."

Test Your Hypothesis

1. **Measure** ⚠ **Be Careful.** Wear goggles. Place 750 mL of plaster of Paris into a large container. Add 250 mL of water and stir the mixture until a thin paste forms.

2. **Make a Model** Pour this mixture into the squeeze bottle. This mixture represents magma. The nozzle of the bottle represents the hot spot.

3. **Make a Model** Place the tip of the bottle at one end of the hole in the cardboard. The cardboard represents the tectonic plate.

4. Gently squeeze the bottle until lava starts to flow up through the hot spot. Continue to squeeze the bottle as you pull the piece of cardboard toward you. What happens?

5. Refill the bottle with the plaster of Paris and water mixture. Place the tip of the bottle into the end of a second hole in the cardboard. Slowly pull the piece of cardboard towards you as you squeeze the bottle. What happens?

Step 1

Step 2

Step 3

Draw Conclusions

6 Compare what happened as a result of steps 4 and 5. Do the results appear different? If so, why?

7 **Infer** How do volcanic islands appear if the tectonic plate is moving slowly over a hot spot?

8 **Infer** How do volcanic islands appear if the tectonic plate is moving rapidly over a hot spot?

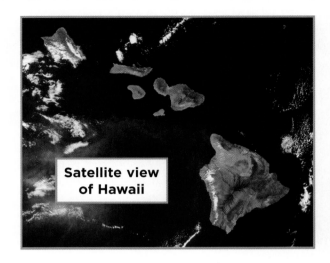

Satellite view of Hawaii

Guided Inquiry

How do eruptions of different types of lava affect the height of a volcano?

Form a Hypothesis

You know that the shapes of volcanoes are different when they form from lava of different thicknesses. How will thicker or thinner lava affect the height of a volcano? Write your answer as a hypothesis in the form "If a volcano forms from thicker lava, then . . ."

Test Your Hypothesis

Design an experiment to investigate effects of eruptions with different types of lava on the height of volcanoes. Write out the materials you need and the steps you will follow. Record your results and observations.

Draw Conclusions

Did your results support your hypothesis? Why or why not? Present your results to your classmates.

Open Inquiry

Does lava with bubbles of gas in it move differently from lava without bubbles? Think of a question and design an experiment to answer it. Your experiment must be organized to test only one variable. Keep careful notes as you do your experiment so another group could repeat the experiment by following your instructions.

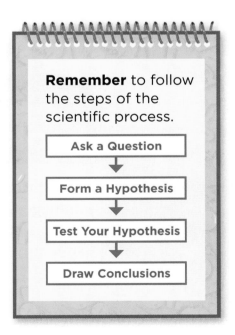

Remember to follow the steps of the scientific process.

Ask a Question

↓

Form a Hypothesis

↓

Test Your Hypothesis

↓

Draw Conclusions

Earthquakes

Look and Wonder

This bridge near Valencia, California, was damaged during an earthquake. What can cause something as solid as a bridge to suddenly break?

How does ground move during an earthquake?

Materials

- **cut pieces of foam**
- **pan**
- **soil**
- **wooden block**

Purpose

To model the movement of the ground during an earthquake.

Procedure

1. Place 2 pieces of foam in a pan so the cut surfaces touch each other.

2. Cover the foam with soil and smooth the soil over both pieces of foam.

3. Pull about 5 centimeters of the pan off of the edge of the table.

4. **Observe** ⚠ **Be Careful.** Gently tap the bottom of the pan with a block. What happened to the blocks and the soil?

5. What happens as you continue to tap the pan?

Draw Conclusions

6. **Infer** What would happen if you tapped the pan harder?

7. What do the foam blocks and the cut between the blocks represent?

Explore More

In this model, the cut between the foam blocks has a certain angle. How do you think the model would work if the blocks were cut at a different angle? Form a hypothesis about which angle will cause more dirt to fall. Make a model and test your hypothesis.

Step 2

Step 4

▶ Main Idea

Earthquakes occur when huge slabs of rock in Earth's crust suddenly move.

▶ Vocabulary

earthquake, p. 272
focus, p. 273
epicenter, p. 273
magnitude, p. 276
tsunami, p. 277

 **e-Glossary**
at **www.macmillanmh.com**

▶ Reading Skill ✓
Draw Conclusions

Text Clues	Conclusions

What is an earthquake?

The ground trembles. Dishes fall off shelves. Walls and buildings come crashing down. Water and gas pipes break. Highways buckle and bridges are torn apart. These events are caused by an earthquake (URTH•kwayk). An **earthquake** is a sudden movement of Earth's crust.

The rock on both sides of a fault is pushed and pulled by forces in the crust. Usually, rocks on both sides of a fault are stuck together. When layers of rock that are stuck together suddenly slip, an earthquake occurs. As a model, press your hands strongly together. Now try to move your hands past each other. It is not easy to do, but if you build up a strong enough sliding force, your hands will slip.

Slipping along a fault begins beneath Earth's surface. Most earthquakes occur at depths of less than 80 kilometers (50 miles). However, earthquakes can occur as deep as 644 kilometers (400 miles).

Focus, Epicenter, and Fault

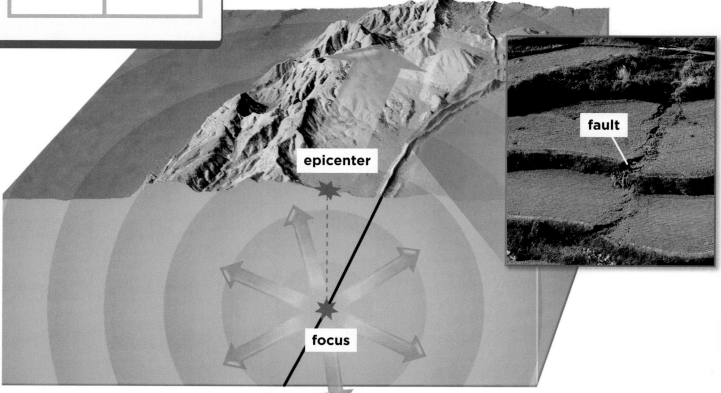

epicenter

fault

focus

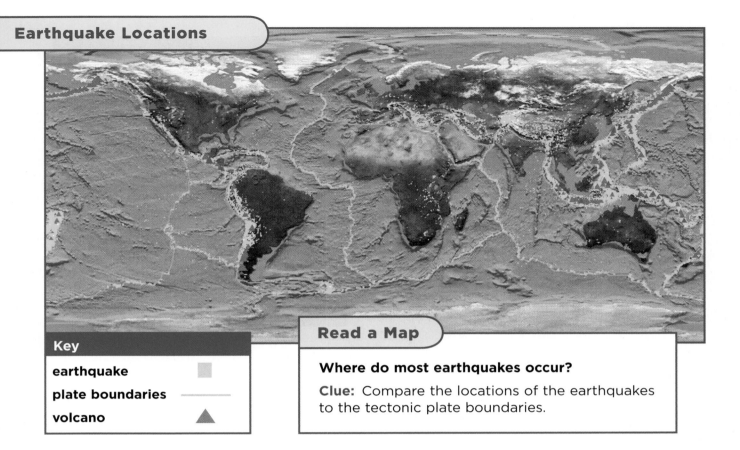

Key

earthquake

plate boundaries

volcano

Read a Map

Where do most earthquakes occur?

Clue: Compare the locations of the earthquakes to the tectonic plate boundaries.

The place where the slipping begins is called the earthquake's **focus** (FOHK•uhs). Waves of energy ripple outward from the focus. When they reach Earth's surface, the waves spread out from a point directly above the focus. That point is the earthquake's **epicenter** (EP•i•sen•tuhr). People first feel the ground shaking at the epicenter of the quake.

The waves move like ripples on a pond. They make the surface and anything on it move. The amount of damage an earthquake causes depends partly on the amount of energy released at the earthquake's focus. When more energy is released, more damage is likely to occur.

Earthquakes happen along the boundaries of tectonic plates because the pressure from the movement of the plates pushes on nearby faults. The map on this page shows the locations of earthquakes and the boundaries of Earth's tectonic plates. Alaska, Oregon, Washington, and California are near tectonic plate boundaries. Most of the earthquakes in the United States happen in these states.

 Quick Check

Draw Conclusions Why are there fewer earthquakes in the center of the United States than on the West Coast?

Critical Thinking How does the location of volcanoes compare to the location of earthquakes?

P   S   Lg

A seismometer records waves produced by an earthquake.

What waves do earthquakes make?

At 8:50 A.M. on October 8, 2005, a powerful earthquake struck the country of Pakistan in Asia. Within minutes, waves from that quake reached seismometers (size•MOM•uh•tuhrz) in the United States.

A *seismometer* is an instrument that detects and measures waves produced by an earthquake. Earthquakes produce three different kinds of waves: primary waves, or P waves, secondary waves, or S waves, and surface waves, or Lg waves.

Primary waves move the fastest. They pass through both the solid and liquid layers of Earth. Primary waves move back and forth as rocks squeeze together and spread apart.

Secondary waves travel about half as fast as primary waves. They move only through Earth's solid layers. Secondary waves move up and down.

Surface waves are the slowest-moving waves. They travel along Earth's surface like waves across an ocean or ripples across a pond. These waves cause the most damage.

Earthquake Waves

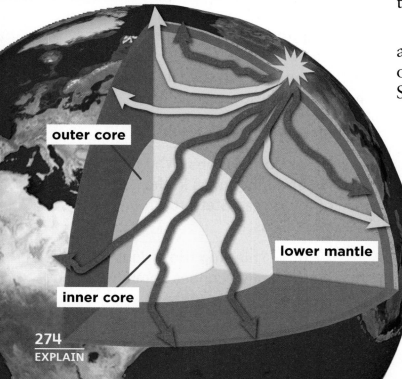

outer core

inner core

lower mantle

Key

S waves →

P waves →

Locating Earthquakes

How did scientists know that the earthquake of October 8, 2005, started in Pakistan? Seismometers can be used to pinpoint the location of an earthquake.

The difference in time between the arrival of P and S waves at a seismometer station gives the distance to the epicenter of the quake. However, say the distance from that seismometer station to the quake's epicenter is 1,000 kilometers. The epicenter could be anywhere on a circle with a radius of 1,000 kilometers from the seismometer station.

To pinpoint the exact location, scientists need data from at least three stations. This data is used to draw three circles on a map. The point where the three circles intersect is the epicenter of the earthquake.

≡ *Quick Lab*

Modeling P and S Waves

1. While your partner holds one end of a coiled toy, stretch the toy until it is fully extended.

2. **Make a Model** Using your thumb, pluck one of the coils. What happens? As you move the toy, draw or describe the motion that one coil makes.

3. Which type of wave does this model represent?

4. **Make a Model** Stretch the coiled toy again. Move your end up and down. What happens? As you move the toy, draw or describe the motion that one coil makes.

5. Which type of wave does this model represent?

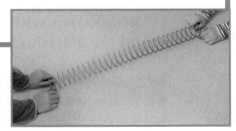

The epicenter of the earthquake is located where the circles around the seismograph station intersect.

Station 3

Station 2

Station 1

Epicenter

✓ *Quick Check*

Draw Conclusions How many seismometer stations are needed to find the epicenter of an earthquake?

Critical Thinking If S waves only pass through some of Earth's layers, what would that tell you about the composition of those layers?

How is an earthquake's energy measured?

Some earthquakes release more energy than others. Scientists measured the magnitude (MAG•NUH•tewd) of the 2005 earthquake in Pakistan to be 7.6. **Magnitude** is a measure of the amount of energy released by an earthquake.

The Richter Scale measures magnitude at the epicenter. It rates earthquakes from weakest to strongest, starting from 1. Each larger whole number indicates an earthquake that releases about 30 times more energy than the previous whole number. For example, a 7.6 earthquake releases 900 (30 x 30) times more energy than a 5.6 earthquake.

The Mercalli Scale measures what people felt and what happened to objects at a specific location. On the Mercalli Scale, earthquakes are rated in Roman numerals from I to XII, where XII is the most severe. The earthquake in Pakistan was a XII on this scale.

Aftershocks are sometimes felt after the main earthquake. *Aftershocks* are earthquakes with lesser magnitudes. Earthquakes with larger magnitudes have more aftershocks and the aftershocks have larger magnitudes.

Comparison of the Richter and Mercalli Scales

General Description	Richter Scale	Mercalli Scale	Estimated Number per Year	Effects
Micro earthquake	Less than 2.0	--	600,000	Not usually felt
Barely felt	2.0-2.9	I-II	300,000	May be felt by a few persons at rest.
Felt generally	3.0-3.9	II-III	49,000	Felt noticeably by people indoors.
Minor	4.0-4.9	IV-V	6,000	Felt indoors by many and outdoors by few. Furniture moves and books fall off shelves.
Moderate	5.0-5.9	VI-VII	1,000	Felt by nearly everyone. Windows, dishes, and glasses break.
Large	6.0-6.9	VII-VIII	120	Buildings partially collapse.
Major	7.0-7.9	IX-X	18	Buildings shift off of their foundations.
Great	8.0-8.9	XI-XII	1	Most buildings and their foundations are destroyed.

Effects of a Tsunami

Read a Photo

Which photo was taken after a tsunami?

Clue: Look for flooding.

Tsunamis

When an earthquake occurs underneath the ocean, a wave of water is pushed upward and outward in all directions. The water rushes across the ocean at speeds from 500 to 1,000 kilometers per hour (311 to 622 miles per hour). This wave carries tremendous amounts of energy.

Out at sea, the energy is spread through deep water, so the wave is not very tall. As the wave reaches shallow water near shore, the same amount of energy is spread through much less water and the height of the wave increases. This produces huge waves called **tsunamis** (TSEW•NAH•meez) in places such as beaches and harbors. Tsunami is a Japanese word that means "wave in port."

Tsunamis usually occur after underwater earthquakes that are greater than magnitude 6.5 on the Richter scale. Tsunamis can be predicted by detecting such earthquakes. In the Pacific Ocean, a system of instruments measures changes deep in the water. When a tsunami occurs, a warning is sent out to the areas at risk. A global tsunami warning system is being planned to detect future tsunamis.

 Quick Check

Draw Conclusions How much stronger is a 5.0 earthquake than a 3.0 aftershock?

Critical Thinking What causes a tsunami to grow taller as it reaches the shore?

Cross Section of a Tsunami

tsunami

epicenter

How can people prepare?

People cannot stop an earthquake from occurring. However, they can take steps to reduce the damage caused by shaking during an earthquake. Sometimes people can be warned about a future earthquake. If people are caught in an earthquake, they can seek a safe place and be aware of the dangers.

Damage to buildings is reduced by placing layers of rubber and steel between a building and its foundation. These layers cushion up-and-down motion but still allow the building to move from side to side as the ground moves. In an earthquake, a building on top of these layers will sway, but is much less likely to be damaged. Most newer buildings in U.S. cities such as Los Angeles and San Francisco have been built using these methods.

Before an earthquake, people should secure objects that might fall. People can identify safe spots to use as shelter during an earthquake, such as under a sturdy table or kitchen counter.

Buildings in San Francisco have been designed to resist damage from earthquakes. ▶

Predicting an earthquake is not easy to do. Instruments have been placed throughout California to detect possible signs of an earthquake, such as bulges or changes in the angle of the ground.

How soon will an earthquake happen after these changes are observed? It could be hours, days, weeks, or months before enough energy builds up for the ground to actually shift. This makes predicting earthquakes difficult. Some earthquakes have been predicted. When they can be predicted, people have time to seek safe shelter before the earthquake strikes.

Predicting a tsunami is easier because it follows an earthquake that can be felt using seismometers. People warned of a coming tsunami can seek shelter on high ground away from shorelines.

 Quick Check

Draw Conclusions Where is the safest place to be if you are warned that a tsunami is on the way?

Critical Thinking What makes it difficult for people to rely on earthquake predictions?

FACT ▶ If you are inside during an earthquake, the safest place to be is under a sturdy table.

Lesson Review

Visual Summary

 **Most earthquakes occur** near plate boundaries when huge slabs of rock move suddenly at a fault.

 **Earthquakes are detected** and measured using seismometers.

 **The effects of earthquakes and tsunamis** can be reduced by warning systems, building safer structures, and following personal safety practices.

Make a FOLDABLES™ Study Guide

Make a Three-Tab Book. Use the titles shown. Then summarize what you learned.

Think, Talk, and Write

1 **Main Idea** What is an earthquake?

2 **Vocabulary** The point on Earth's surface directly above the focus of an earthquake is the _____.

3 **Draw Conclusions** What can you conclude has happened if a tsunami strikes a shore?

Text Clues	Conclusions

4 **Critical Thinking** Contrast the Richter and Mercalli scales.

5 **Test Prep** What is a fault?
  A a type of earthquake wave
  B a layer of rock
  C a crack in Earth's crust
  D the shaking of land

6 **Test Prep** Which should you do to stay safe during an earthquake?
  A stay near a window
  B duck under a sturdy table
  C allow heavy furniture to fall
  D hide inside a chimney

Math Link

Measuring Earthquake Strength
How much stronger is an earthquake that measures 8.0 on the Richter Scale than one that measures 5.0?

Social Studies Link

 Earthquake Magnitude
Research the largest earthquake in your state. Include information about when and where it happened and what the effects were.

How Earthquakes Help Predict Volcanic Eruptions

What happens before a volcano erupts? First, magma moves into the magma chamber beneath the volcano. Then, the magma starts to rise to the surface. As the magma moves, it causes small earthquakes. The chance of an eruption increases as these earthquakes occur closer to the surface. Scientists use seismometers to detect this activity.

Scientists also look at the type of earthquake. Short-period earthquakes happen as magma breaks through rock on its way to the surface. This tells scientists that the amount of magma near the surface is increasing. Long-period earthquakes tell scientists that there is an ongoing movement of magma beneath the surface. This may mean that magma is flowing and moving toward the surface.

Mount St. Augustine, Augustine Island, Alaska

Explanatory Writing

A good explanation

▶ develops the main idea with facts and supporting details

▶ lists what happens in an organized and logical way

▶ uses time-order words to make the description clear

 Write About It

Explanatory Writing What are the differences between short-period and long-period earthquakes? Research these earthquakes. Write an explanatory essay with details that support your main idea.

LOG ON **e–Journal** Research and write about it online at **www.macmillanmh.com**

Using the Richter Scale

You have learned that an earthquake that measures 1.0 higher than another earthquake on the Richter scale releases 30 times more energy. How do you compare the strength of earthquakes that are less than 1.0 apart? An earthquake that is 0.1 higher releases about 1.4 times as much energy than the weaker earthquake does.

⊕ Solve It

1. How many times stronger was the Fort Tejon earthquake than the Kodiak Islands earthquake?

2. How many times stronger was the Kodiak Islands earthquake than the Owens Valley earthquake?

3. How many times stronger was the Pleasant Valley earthquake than the Denali earthquake?

Historic U.S. Earthquakes

location	date	magnitude
Fort Tejon, CA	Jan. 9, 1857	7.9
San Francisco, CA	Apr. 18, 1906	7.8
Kodiak Islands, AK	Oct. 9, 1900	7.7
Owens Valley, CA	Mar. 26, 1872	7.4
Pleasant Valley, NV	Oct. 3, 1915	7.1
Denali, AK	Oct. 23, 2002	6.7

source: U.S. Geological Survey

Santa Cruz, California

Multiplying Decimals

To multiply decimals

▶ multiply as with whole numbers

▶ add the decimal places in the factors

▶ put that many decimal places in the product

$$4.3 \leftarrow \text{one decimal place}$$

$$\times\ 2.5 \leftarrow \text{one decimal place}$$

$$10.75 \leftarrow \text{two decimal places}$$

Shaping Earth's Surface

Exit Glacier, Kenai Fjords, Alaska

Look and Wonder

Earth's surface is constantly changed by the movement of ice, water, and wind. How does the land change?

How does ice break up rocks?

Form a Hypothesis

Which takes up more space, liquid water or frozen water? Write your answer as a hypothesis in the form "If water is frozen in a confined space, then . . ."

Test Your Hypothesis

1 Using the marker and ruler, mark 15 centimeters on each container.

2 Mix 5 drops of food coloring into the water.

3 Fill both containers with water until the water reaches the 10-centimeter mark.

4 Put caps on both containers. Place one container in a freezer. Leave the other container at room temperature.

5 When the water in the container in the freezer has completely frozen, remove the container.

6 Observe Is there a change in the height of the water in either container? Is there a change in the shape of either container?

Draw Conclusions

7 Interpret Data What happens to the amount of space water takes up when it freezes?

8 Infer What do the results of your experiment indicate about what happens when water freezes in a crack in a rock?

Explore More

Other processes can change the surfaces of rocks. Observe the sidewalks in your neighborhood and pay special attention to cracks or changes in their surfaces. What might have caused these changes?

Materials

- marker
- metric ruler
- two identical plastic containers with caps
- food coloring
- water

Step **1**

Step **3**

Read and Learn

▶ **Main Idea**
Weathering and erosion change the shape of Earth's surface.

▶ **Vocabulary**
weathering, p. 284
erosion, p. 286
glacier, p. 287
deposition, p. 288
meander, p. 288
sediment, p. 288
floodplain, p. 290

-Glossary
at **www.macmillanmh.com**

▶ **Reading Skill** ✔
Problem and Solution

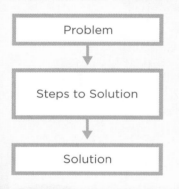

What is weathering?

As rain falls on the cliffs of the Palisades in New Jersey, water trickles into cracks in the rocks and collects there. If temperatures drop below the freezing point of water, the liquid water turns to ice. The ice takes up more space than the liquid water and pushes the crack further apart. Sometimes this causes bits of rock to break off.

Over many years, the cliff will slowly be worn away, or weathered. **Weathering** (WETH•uhr•ing) is the process through which rocks or other materials are broken down.

Physical Weathering

Physical weathering is caused by temperature changes, pushing, pulling, or rubbing. As water in cracks of a rock freezes, the frozen water pushes on the rock and can break pieces off.

Trees and other plants can grow out from cracks in a cliff. Their roots push against the walls of the cracks. This pressure can break off particles of rock.

Weathering is slowly wearing away the rocks of the Palisades, which rise above the New Jersey side of the Hudson River.

The Carlsbad Caverns in New Mexico formed as chemicals in groundwater broke up the rock.

Gravity pulls rocks down a slope. The rocks bump into other rocks on the slope as they roll downward. With each bump, parts of the rocks may break off.

Winds often blow on exposed rock. The winds pick up small particles of sand or dirt and rub it against the rock. As rocks are carried along in moving water, they may also bump or rub against each other. This rubbing wears away the surface of the rock.

Chemical Weathering

Chemical weathering occurs when chemicals break down rocks. When chemicals in groundwater break up underground rock, caves form.

In some parts of the eastern United States, you may see stone or metal statues whose features have been worn away or pitted. Sometimes the color on the statue is changed or damaged. If you see these changes, that statue may have been affected by acid rain.

Acid rain forms when gases from factories enter the air and combine with raindrops. Acid rain wears away stone and metals. The worn material may crumble and be swept away by winds and precipitation.

 Quick Check

Problem and Solution How does acid rain damage statues?

Critical Thinking How are physical and chemical weathering different?

Acid rain has discolored this bronze lion.

Landslides occur when loose soil and rock are pulled down by gravity.

What is erosion?

When it rains, some water soaks into the ground. Once the ground can no longer absorb water, the water mixes with the soil and forms mud. Eventually the mud contains so much water that it cannot stay on the slope. As the mud flows down the slope, it can knock down trees and destroy whatever is in its path. The movement of a large amount of wet soil and rocks down a slope is called a *mudslide*.

Erosion (i•ROH•zhuhn) is the process through which weathered rock is moved from one place to another place. Land can be eroded in one of five ways—by gravity, glaciers, running water, waves, or wind.

Gravity

When rocks and soil on a slope are loosened, gravity pulls them downward. Mudslides are pulled down by gravity. A *landslide* is the movement of a large amount of rock and soil down a slope. Landslides may occur after an earthquake or a volcanic eruption. These events cause the ground to move. This loosens rocks and soil enough that gravity can pull them downward.

Plant roots grow down into soil and around rocks. If many plants grow on a hill, their roots will hold on to the soil and rocks. Their roots hold on to soil that may otherwise move in a mudslide or a landslide.

Glaciers

A **glacier** (GLAY•shuhr) is a large mass of slowly flowing ice. Glaciers form in cold areas where snow piles up and freezes. Water freezes in cracks in the rock underneath the snow and weathers the rock. As the glacier moves, it carries away weathered pieces of rock.

The bits of rock wear away the ground at the beginning of the glacier, forming a steep bowl-shaped hollow called a *cirque* (SUHRK). The rocks and flowing ice also wear away dirt and rock along the sides of the glacier.

Eventually, the ice at the front of the glacier will melt. If the ice melts faster than the glacier is moving forward, the glacier will shrink. A valley that had a sharp V-shape before a glacier came through will now have a U-shape, with a flatter bottom and sides.

Formation of a Valley

Quick Lab

Rate of Erosion

1 **Form a Hypothesis** How does the speed of running water affect how fast soil erodes? Write your answer in the form of a hypothesis.

2 **Make a Model** Place dirt in two identical baking pans so the dirt is at the same level in each pan.

3 Place a wooden block underneath each pan.

4 Fill a watering can with a sprinkler head with 2 cups of water. Slowly pour the water into the pan. Record your observations.

5 Remove the sprinkler head and fill the watering can with 2 cups of water. Pour the water slowly into the pan. Record your observations.

6 **Draw Conclusions** Do your results support your hypothesis?

✓ Quick Check

Problem and Solution What could people do to reduce the chances of a mudslide happening on a hill?

Critical Thinking If a glacier completely melted, what landform would form in the cirque?

Read a Photo

What type of erosion might have formed this valley?

Clue: Look at the sides and bottom of the valley.

What is deposition?

When a glacier erodes dirt and rock, the eroded materials are pushed in front of it. When the glacier starts to shrink, the eroded materials are left behind. The process by which eroded materials are dropped off in another place is called **deposition** (de•puh•ZISH•uhn). Erosion and deposition work together to change the shape of Earth's surface.

Erosion and Deposition by Running Water

As water runs down hills, it can wash away soil and erode rock. The water, soil, and rocks will eventually flow into a larger body of water, such as a river. Rivers with fast-moving water tend to follow straight paths and have deeper channels and steeper banks. Rivers with slow-moving water tend to follow looping paths and have shallow channels and low banks. More deposition occurs in slow-moving than in fast-moving water.

Meanders (mee•AN•duhrz), or gentle loops, sometimes form in rivers with slow-moving water. Water moves slowly around the inside of a meander. Particles of soil and rock that are carried along in water are called **sediment**. Along the inside of a meander, sediment has time to settle out of the water. As sediment is deposited, it may eventually build new land.

Water moves more rapidly around the outside edge of a meander. The sediment in this part of the river is carried farther downstream. Sometimes additional sediment is eroded from land along the outer edge of the curve.

The loops in this river are called meanders.

Erosion of an Arch

Erosion and Deposition by Waves

Waves often either hit beaches at an angle or curve as they move into shallower water. This means that as they erode the shoreline, they move sand and rocks further down the beach or to the side.

When waves reach a *headland*, or an area of land that has water on three sides, they curve and wash away at the sides of the headland. As the waves continue to erode the sides, an arch forms.

When waves wash sand off beaches, the sand may be deposited in the water rather than back on the beach. Over time, enough sand has been deposited in the water that a strip of sandy land forms. This strip of land is called a *sandbar*. A sandbar may last for some time until movements of ocean water break it down.

Erosion and Deposition by Wind

Wind can wear away at rocks, smoothing them out. Wind also can move sand or sediment from one place to another. When the winds slow down, the sand and soil are deposited.

Sandbars form when waves pick up sand and deposit it in the water.

✔ Quick Check

Problem and Solution What makes the erosion of sand from a beach a problem for people?

Critical Thinking How does gravity deposit materials?

How are shorelines changed?

A shoreline, or the edge of a body of water, is changed by the erosion and deposition of sediment. Sediment is eroded and deposited along a shoreline by running water, waves, and winds.

Running Water

Water runs over the ground into streams and rivers. Sometimes, water enters a river faster than the river can carry it away. When water collects on land that is normally dry, it is called a *flood*.

Floods occur when water from a body of water overflows banks or beaches. A flood may also occur during a heavy rainfall. Natural wetlands can soak up water and reduce the chances of a flood. Draining wetlands or cutting down plants along a river bank may make floods more likely.

Flood waters carry and deposit sediments over the land. A **floodplain** (FLUHD•playn) is a place that is easily flooded when river water rises. Floods also can erode the shoreline of a body of water and change its shape or course.

Floodplain

Read a Photo

Which photo shows sediments being carried over a floodplain?

Clue: Flood waters carry sediments.

Wind blows sand away from the front of a beach.

Waves

When large sandbars are formed that stretch for hundreds of kilometers along a coastline, they are called *barrier islands*. Barrier islands protect beaches from erosion caused by large waves during storms. The waves hit the barrier islands first and erode the barrier islands rather than the beaches. After severe storms, barrier islands may be so completely eroded that they no longer appear above the water. Without the barrier islands, beach erosion will be worse during the next storm.

Wind

Some coastal areas have one or more sets of dunes running along the shoreline directly inland from the beach. A dune forms when wind erodes sand and deposits it along the back of the beach.

Dunes form in the direction that the wind usually blows. As the wind blows, it will pick up sand from the dunes closest to the water and blow it further inland. This can cause dunes to shift position.

Dunes protect areas further inland from the large waves that can occur during storms. Dunes also shelter inland areas from the wind. If severe winds or waves occur, dunes may be completely eroded.

 Quick Check

Problem and Solution What is more likely to happen if wetlands are drained?

Critical Thinking How do dunes protect the land?

How can shorelines be protected?

A shoreline may be damaged when more sand is being eroded than deposited. People may take steps to prevent further erosion or to reduce the rate of erosion.

People can protect river shorelines by changing the speed or the direction of running water. Dams can control the speed of the flow of water in a river.

Other structures change or block the direction in which water can travel. *Levees* (LEH•veez) are walls built to hold back water or prevent a flood. Canals or channels can be dug to carry away water that would otherwise cause floods.

Barricades can be built in the water along beaches to slow erosion. People can move sand from the water back onto the beach using pumps and hoses. Sometimes sand is even brought from other places to replace sand that was lost through erosion.

How do people prevent wind from eroding beaches and dunes? Fences are often put up near sand dunes to decrease the speed of the wind so less sand is blown away. Sometimes people plant grasses on dunes so the roots will grow into the sand and hold on to it.

▲ Some barricades may be built to prevent large waves from causing floods.

Fences and grasses help keep sand dunes from being eroded by wind.

 Quick Check

Problem and Solution How does planting grasses prevent wind from eroding dunes?

Critical Thinking When trying to prevent or reduce erosion, why should you consider the entire shoreline?

Lesson Review

Visual Summary

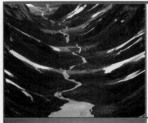

Weathering wears away land.

Erosion moves land from one place to another.

Moving water can cause flooding and build and break down land.

Make a FOLDABLES™ Study Guide

Make a Trifold Book. Use the titles shown. Then summarize what you have learned.

Weathering wears away...	Erosion moves...	Moving water can cause...

Think, Talk, and Write

1 **Main Idea** What processes break down and build up land?

2 **Vocabulary** As the speed of river water slows, _____ is deposited.

3 **Problem and Solution** If you needed to prevent waves from eroding a beach, what would you do?

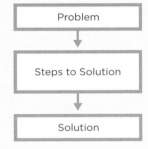

4 **Critical Thinking** How might you tell which type of weathering had worn away a cliff?

5 **Test Prep** What is a sandbar?
   A a strip of sandy land in water
   B land surrounded by water
   C sediment in the mouth of a river
   D rock moved by slowly moving ice

6 **Test Prep** What moves sand dunes from one place to another?
   A water
   B gravity
   C acid rain
   D wind

Writing Link

Fictional Writing
Write about a trip down the Mississippi River. Describe the different areas of the river that you would see.

Social Studies Link

Effects of Floods
Research how the ancient Egyptians used the annual flooding of the Nile River in Africa to water their crops.

Garrison Dam in North Dakota

Wrestling with the Big Muddy

The Missouri River is more than 4,000 kilometers (2,486 miles) long and flows through ten states. It is nicknamed the "Big Muddy" because it is full of sediment. The Missouri River is an important transportation route for boats and ships and a major source of water for millions of people and thousands of towns, industries, and farms.

The Missouri River used to be unpredictable. During droughts, it would dry up. Farmers couldn't water their crops, and transportation slowed. When it rained more than normal, the river would flood. Farmland and towns along the banks of the Missouri were swamped by water.

In 1944, the government decided to control the flow of the Missouri River. Engineers constructed a series of dams and levees that regulate the flow of water. This changed the waterway dramatically.

For example, the river once turned into a series of small channels that covered a wide flood plain. Now 1,127 kilometers (700 miles) of levees direct this water into one long, deep channel. However, when record rains fall, floods can still happen. In 1993, floods left 15 million acres of farmland along the river under water.

Sandpipers search the edge of the Missouri River for small insects.

Washington, Missouri, flooded in 1993.

Sturgeons are measured and tagged so changes in their population can be tracked.

Controlling the natural flow of a river solves some problems and causes others. Dams interfere with fish migration. In fact, many species of fish and other wildlife in and along the Missouri River are now threatened or endangered. Dams also interfere with soil fertilization. Nutrients are trapped behind dams and are no longer carried to farmlands along the river.

It takes different types of knowledge to balance the short- and long-term effects of controlling a river as complicated as the Big Muddy. Engineers, environmentalists, and government officials work together to develop solutions that meet the needs of people and the environment.

Problem and Solution

▶ Identify the problem by looking for a conflict or an issue that needs to be resolved.

▶ Think about how the conflict or issue could be resolved.

 Write About It

Problem and Solution

1. What problems did the Missouri River cause before 1944?
2. What did the government do to control the flow of the Missouri River? How did that change the river?

LOG ON e-Journal Research and write about it online at www.macmillanmh.com

AMERICAN MUSEUM of NATURAL HISTORY

Visual Summary

Lesson 1 Each layer of Earth has its own features.

Lesson 2 Earth's surface is made of plates that constantly move.

Lesson 3 Volcanoes occur where magma reaches Earth's surface.

Lesson 4 Earthquakes occur when huge slabs of rock in Earth's crust suddenly move.

Lesson 5 Weathering and erosion change the shape of Earth's surface.

Make a FOLDABLES Study Guide

Put your lesson study guides together as shown. Attach the Lesson 5 study guide to the back. Use your study guides to review what you have learned in this chapter.

Fill in each blank with the best term from the list.

fault, p. 256

hot spot, p. 266

landform, p. 240

magnitude, p. 276

outer core, p. 246

plate tectonics, p. 254

volcano, p. 262

weathering, p. 284

1. Magma flows through an opening in Earth's crust called a(n) _____.

2. The breakdown of rocks and other materials is called _____.

3. A measure of the amount of energy released by an earthquake is _____.

4. The layer of Earth's core that is made up of liquid metal is the _____.

5. A pool of magma called a(n) _____ formed the Hawaiian Islands.

6. In some places, there is a deep crack in Earth's crust called a(n) _____.

7. The theory that explains how continents move is called _____.

8. A physical feature on Earth's surface is a(n) _____.

Answer each of the following in complete sentences.

9. Problem and Solution How can people prevent damage from earthquakes?

10. Compare Describe the types of plate movement and what occurs as a result of their movement.

11. Make a Model Trace the shapes of the continents onto construction paper. Cut the continents out. Treat the continents as pieces of a jigsaw puzzle and see how they fit together. How does the shape of your model compare to the shape of Pangea?

12. Critical Thinking How do you think landforms affect the people who live near them? Give examples.

13. Explanatory Writing How are the locations of the focus and the epicenter of an earthquake related?

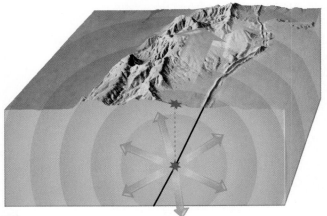

14. How does Earth's surface change?

Erosion or Weathering?

Your goal is to observe rock formations, buildings, statues, and other structures in your neighborhood.

What to Do

1. Look for evidence of erosion or weathering. Record details about what you saw.

2. Make a list of three examples of erosion and three examples of weathering.

Analyze Your Results

▶ Explain which type of weathering or erosion you think has taken place and the reasons for your decisions.

1. An earthquake occurred 200 kilometers from a seismograph station. What conclusion can be drawn from the map below?

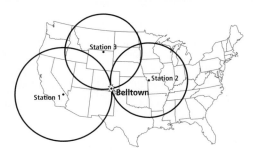

A The earthquake occurred 300 kilometers from station 3.

B The epicenter of the earthquake is in Hat City.

C The epicenter of the earthquake is in Belltown.

D The earthquake was only felt at stations 1 and 2.

CHAPTER 6

Protecting Earth's Resources

The **Big Idea**

What are Earth's resources?

Planting saplings in a tree nursery

Key Vocabulary

mineral
a solid material of Earth's crust with a distinct composition (p. 302)

rock cycle
a never-ending process in which rocks change from one kind into another (p. 306)

soil horizon
any of the layers of soil from the surface to the bedrock (p. 317)

fossil fuel
a fuel formed from the decay of ancient forms of life (p. 327)

renewable resource
a resource that can be replanted or replaced in a short period of time (p. 331)

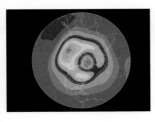

ozone
a gas that forms a layer in the atmosphere that screens out much of the Sun's ultraviolet rays (p. 348)

More Vocabulary

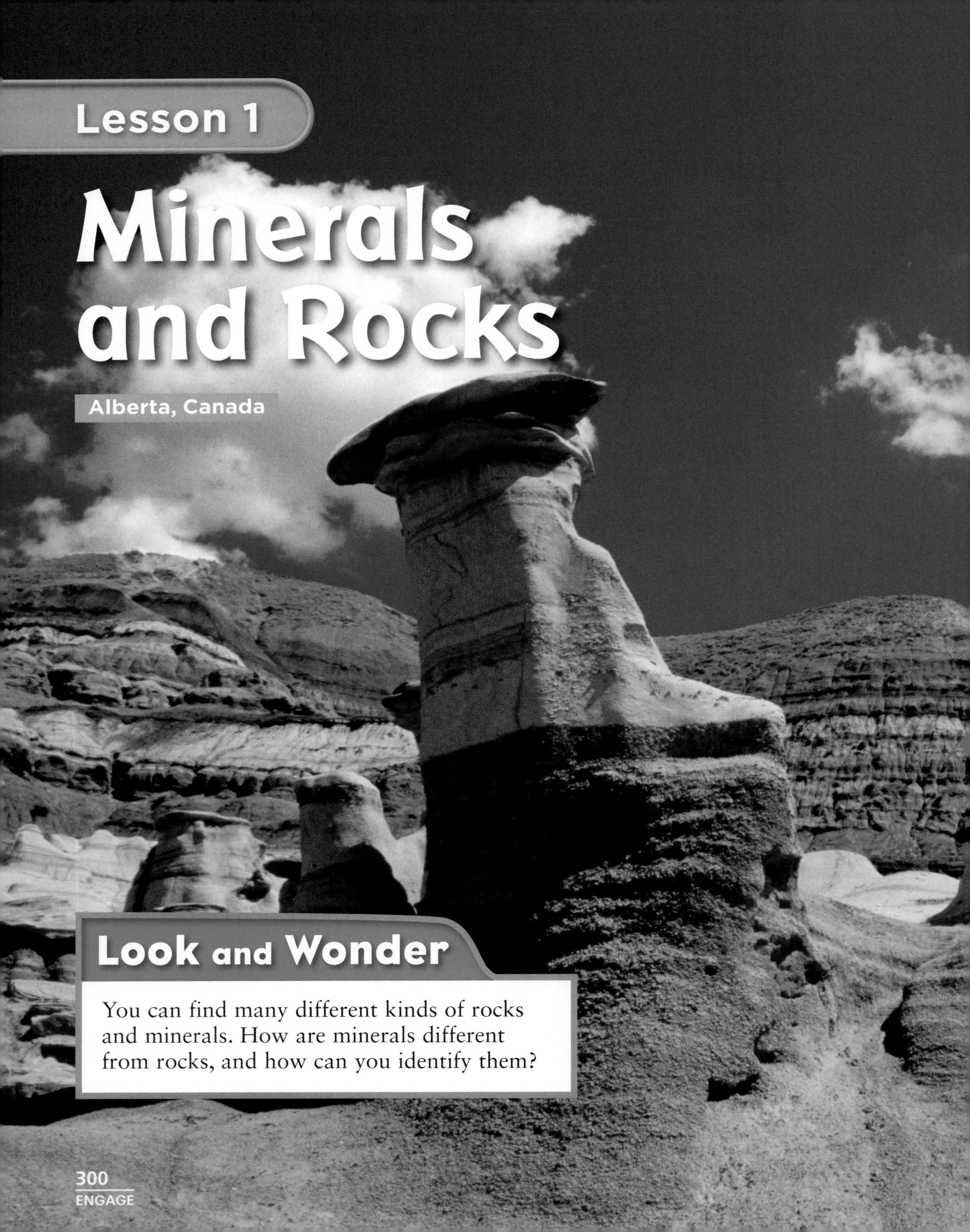

Minerals and Rocks

Alberta, Canada

Look and Wonder

You can find many different kinds of rocks and minerals. How are minerals different from rocks, and how can you identify them?

What are properties of minerals?

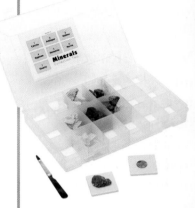

Purpose

To observe the properties of minerals.

Procedure

1. Use the clear tape and the marker to label each mineral with a different sample number.

2. Make a chart like the one shown below.

Sample Number	Mineral	Color	Shine (yes/no)	Streak	Scratch	Other
1						
2						

- clear tape
- marker
- mineral samples
- porcelain tile
- copper penny
- steel file

3. Fill in the columns of the chart for *color* and *shine* (like a metal).

4. **Observe** Rub the mineral across the porcelain tile. Record the color that you see on the tile.

5. **Observe** ⚠ **Be Careful.** Scratch the mineral on a copper penny and a steel file. Record whether the mineral scratches the penny or the file.

Draw Conclusions

6. **Infer** Examine your data. What can you say about the properties of different minerals?

7. How could the properties of minerals help you classify minerals?

Explore More

Using reference sources, identify these minerals. Then label and display them.

Step 4

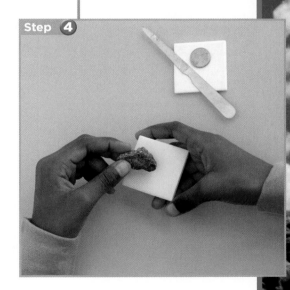

Main Idea

Rocks and minerals are produced in different ways and have different properties.

Vocabulary

mineral, p. 302
luster, p. 302
hardness, p. 303
sedimentary rock, p. 306
igneous rock, p. 306
metamorphic rock, p. 306
rock cycle, p. 306

LOG ON e-Glossary
at **www.macmillanmh.com**

Reading Skill ✓

Sequence

First

↓

Next

↓

Last

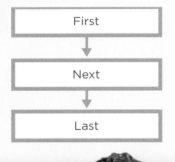

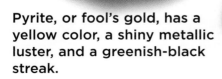

Pyrite, or fool's gold, has a yellow color, a shiny metallic luster, and a greenish-black streak.

What are minerals?

If you collect rocks, you might find a rock with red chunks or specks of yellow in it. The red chunks and specks of yellow are minerals (MIN•uhr•uhlz). A **mineral** is a solid, natural material made from nonliving substances in the ground.

Minerals are made up of elements. An *element* is a pure substance that cannot be broken down into a simpler substance. Gold is an element, as are aluminum, oxygen, hydrogen, and iron. Some minerals are made of a single element. Other minerals are made of two or more elements.

Minerals are found in nature. For example, iron is found in ore. *Ore* (awr) is a rock that contains a useful mineral. Steel is a combination of iron and carbon. Because steel is made by people, it is not a mineral.

Properties of Minerals

The color of a mineral is the color of the surface of the mineral. Streak is the color of the powder left when the mineral is rubbed on a rough surface. The color of a mineral and its streak are often different.

Luster (LUS•tuhr) is the way a mineral reflects light from its surface. There are two general kinds of luster. Minerals with a metallic luster look shiny, while minerals with a nonmetallic luster look duller. Minerals with a nonmetallic luster may look waxy, pearly, earthy, oily, or silky.

Mohs' Hardness Scale		
Hardness	Mineral	Can be scratched by
1	talc	
2	gypsum	fingernail
3	calcite	copper (penny)
4	fluorite	
5	apatite	steel (knife blade)
6	feldspar	porcelain (streak plate)
7	quartz	
8	topaz	
9	corundum	
10	diamond	

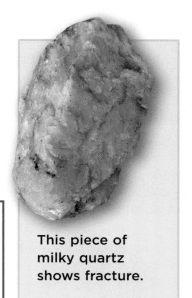

This piece of milky quartz shows fracture.

Read a Table

Which mineral is scratched by copper but not by a fingernail?

Clue: Find copper and fingernail on the table.

Hardness is a measure of how well a mineral resists scratching. A hard mineral resists scratching better than a softer mineral. On Mohs' hardness scale, minerals are ranked from 1, which is the softest, to 10, which is the hardest. A mineral with a higher number will scratch a mineral with a lower number. By scratching an unknown mineral with materials that have a known hardness, you can find the hardness of the unknown mineral.

When a mineral is broken, the appearance of the surfaces of the mineral can help identify it. When the surfaces are smooth, the property that it has is cleavage (KLEE•vij). *Cleavage* is described by the number of planes along which the mineral breaks. *Fracture* (FRAK•chuhr) is the property a mineral shows when it has uneven or rough surfaces.

This pink calcite shows cleavage along three planes.

 Quick Check

Sequence Rank feldspar, gypsum, and quartz from softest to hardest.

Critical Thinking If you found an unknown mineral, what properties could you use to identify it without damaging it?

FACT	Glass has a hardness of about 5.5 and can be scratched by any harder material.

What are the shapes of a mineral?

As minerals form, the elements that they are made from form patterns. These patterns cause minerals to form geometric shapes called crystals (KRIS•tuhlz). A *crystal* is a solid whose shape forms a fixed pattern.

Different minerals have different crystal shapes. Sometimes the larger structure of the mineral shows the same shape as the crystal structure. For example, if you look at crystals of table salt with a hand lens, you will observe that the salt crystals look like tiny cubes. In other minerals, the crystal structure is much smaller and can be seen only with a microscope.

Topaz is an example of a mineral with an orthorhombic structure.

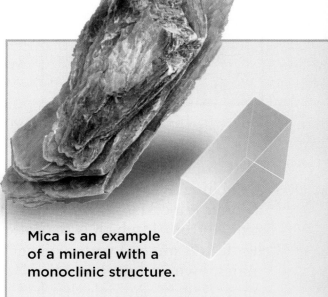

Mica is an example of a mineral with a monoclinic structure.

Amethyst is an example of a mineral with a hexagonal structure.

Emeralds are minerals that occur naturally in some rocks. Emeralds are cut and polished for use in jewelry.

Some minerals, such as emeralds, are also called gems (jemz). A *gem* is a mineral that is rare, beautiful, usually transparent, and glitters or shines in light. In their natural form, gems are often rough and uneven. Gems can be cut into shapes to show their color and to remove flaws. After they are cut and polished, gems may be used in jewelry.

Minerals may be more useful in a purer form than the one in which they are naturally found. For example, copper ore contains other materials. The copper is separated from the other materials by processes that include grinding the ore to a powder, rinsing it with chemicals, and then heating it.

At the end of these processes, the copper is 98 percent to 99 percent pure. For uses where a coating of copper is applied, such as for copper coins or copper utensils, this is pure enough. For other uses, such as wires in electronics, the copper may need to be purified further.

Quick Lab

Crystal Shapes

⚠ **Be Careful.** Wear goggles. Use a kitchen mitt if you need to hold or move the cup. Don't touch the very warm water.

1 Using a plastic spoon, slowly add small amounts of sugar to a cup of very warm water. Continue to add sugar and stir until you can see sugar in the water.

2 Tie one end of a 15 cm length of string around a sugar cube. Tie the other end to a pencil. Place the pencil across the cup so that the crystal hangs in the very warm sugar water without touching the sides or bottom.

3 **Observe** Watch the experimental setup for several days.

4 **Communicate** Describe what you observed in the cup.

✔ Quick Check

Sequence What processes are used to separate copper from copper ore?

Critical Thinking What shape is repeated to form a cube?

What is the rock cycle?

Nothing on Earth stays the same. The atmosphere, water, land, and living things on Earth all change over time. The same is true of rocks. A *rock* is a solid object made naturally in Earth's crust that contains one or more minerals. A rock may be as small as a grain of sand or larger than a house. Layers of rock beneath Earth's surface may be many kilometers thick.

Most rocks are made of mixtures of materials. Rocks have different mixtures because they are made in different ways. Scientists classify rocks into three types based on the way they are made: sedimentary rock (sed•uh•MEN•tuh•ree), igneous rock (IG•nee•uhs), and metamorphic rock (met•uh•MAWR•fik).

As you have learned, weathering and erosion on Earth's surface move particles of dust, sand, and soil. As time passes, layers of these sediments are deposited. The upper layers press down on the lower layers. The pressure compacts the sediments, or squeezes them together. Over time, the pressure cements the sediments, or makes the minerals stick together. A rock that forms from sediments is called **sedimentary rock**.

As sedimentary rocks are pushed underneath Earth's crust, heat and pressure melt them into magma. The magma may erupt through a volcano. If it does, an **igneous rock** forms as the lava cools and hardens.

As time passes, sedimentary and igneous rocks may become buried deep beneath Earth's surface. There they are under pressure from the weight of the rock above them. The temperature is also much hotter. Metamorphic rocks usually form no deeper than 20 kilometers (12 miles) below the surface and at temperatures between 200°C and 800°C (392°F to 1,472°F). A **metamorphic rock** is a rock that forms when sedimentary and igneous rocks change under heat and pressure without melting.

Sedimentary rocks can change into igneous or metamorphic rocks. Igneous rocks can change into sedimentary or metamorphic rocks. Metamorphic rocks can change into sedimentary or igneous rocks.

A change from one type of rock to another is caused by changes in conditions on and underneath Earth's surface. The changing of rocks over time from one type to another is called the **rock cycle**.

Quick Check

Sequence What are the steps by which an igneous rock turns into a sedimentary rock?

Critical Thinking Why is the rock cycle called a cycle?

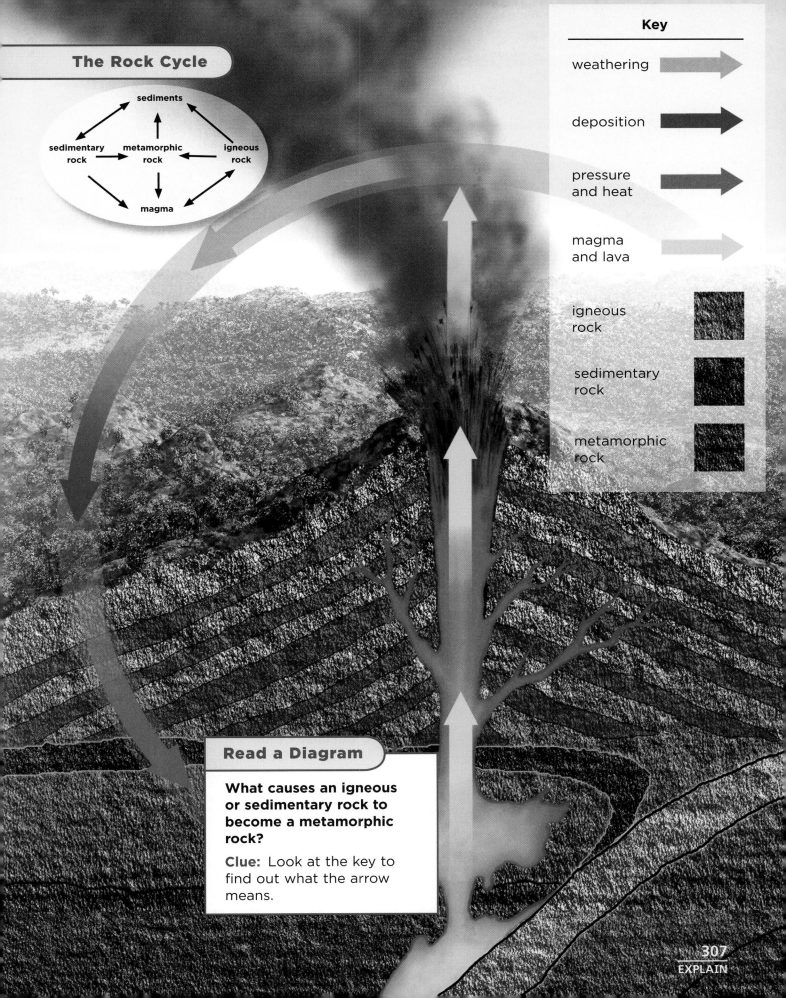

The Rock Cycle

sediments

sedimentary rock → metamorphic rock ← igneous rock

magma

Key

weathering

deposition

pressure and heat

magma and lava

igneous rock

sedimentary rock

metamorphic rock

Read a Diagram

What causes an igneous or sedimentary rock to become a metamorphic rock?

Clue: Look at the key to find out what the arrow means.

What are igneous and sedimentary rocks?

Since at least 50,000 years ago, people have used rocks to make weapons and to start fires. How did these rocks form? What other uses have people found for rocks?

Igneous Rocks

When an igneous rock forms from magma inside Earth, it is called an *intrusive rock* (in•TREW•siv). Below Earth's surface, intrusive rocks cool slowly. They may take 100 years or more to cool just a few degrees. The slow cooling often produces large crystals. If you find an igneous rock with large crystals in it, you can conclude that the rock is intrusive.

Granite is the most common intrusive rock. It is often used as a building material. The minerals that make up gems, such as rubies and sapphires, may form in intrusive rocks. These minerals can be used to make jewelry.

An igneous rock that forms from lava on Earth's surface is called an *extrusive rock* (ek•STREW•siv). On Earth's surface, lava is exposed to air or water, causing it to cool and harden very rapidly. Lava may cool in minutes when it spills into the sea or in a number of days as it flows over land. Large crystals do not have time to form. Basalt, the most common extrusive rock, is made of many small crystals.

Some extrusive rocks develop so quickly that they do not contain any crystals. Obsidian, which is also called volcanic glass, is an example of an extrusive rock that has no crystals. Its surface is smooth and glassy. People have used obsidian to make sharp tools and weapons.

Pumice is another type of extrusive rock. As pumice forms, gases bubble through the rock. The holes that are left behind make pumice light and rough. Because it is rough, pumice is often used for grinding or polishing.

Pink granite is an intrusive igneous rock, while obsidian is an extrusive igneous rock.

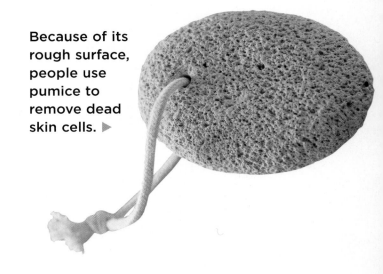

Because of its rough surface, people use pumice to remove dead skin cells. ▶

Sedimentary Rocks

Sedimentary rocks are made of different materials that have been compacted and cemented together. Some sedimentary rocks contain minerals that were once dissolved in water. The minerals formed crystals among the sediments that came together to form the rock.

Some sedimentary rocks are made from smaller rounded stones that have been cemented together. This type of rock is called a *conglomerate rock* (kuhn•GLOM•uhr•it).

Sedimentary rocks are often used in buildings. Limestone is used on the outside of a building or for decorations on buildings. Sandstone also is used on the outside of buildings and for decoration. Ground limestone is an ingredient in concrete.

An example of a conglomerate rock.

 Quick Check

Sequence Place these rocks in order from fastest cooling to slowest cooling: granite, obsidian, pumice.

Critical Thinking You are driving along a highway cut through walls of rock made up of layers. What type of rock is this?

City Hall in Buffalo, NY, is covered with sandstone and limestone (below).

What are metamorphic rocks?

If sedimentary and igneous rocks are put under heat and pressure, the shape or the size of the crystals within them can change. The crystals may also change position to form layers. Heat and pressure may even change one of the minerals in the rock into another mineral. The high pressure also squeezes the particles in the original rock more tightly together.

If you look closely at limestone, you can often see fossil fragments in the rock. As limestone changes into marble under heat and pressure, the fossils are usually crushed. Marble is a more compact rock than limestone, with crystals that are locked together like pieces of a jigsaw puzzle. The color in marble comes from minerals in the original piece of limestone.

Slate is a type of metamorphic rock in which the minerals are tightly packed together, making it waterproof. When slate is broken, it shows cleavage as it breaks into thin sheets. This makes slate useful as a roofing material as well as for stepping stones and outside floors.

Marble is a shiny metamorphic rock that contains minerals that give it brilliant colors. Marble is easy to carve or shape, and thus is used for making statues, floors, kitchen counters, and monuments.

▲ The color in this marble was caused by mineral impurities in the limestone from which it formed.

The Taj Mahal in India was built using white marble (below).

 Quick Check

Sequence What happens to limestone as it turns into marble?

Critical Thinking Describe the ways rocks are used in your neighborhood.

Lesson Review

Visual Summary

Minerals have properties by which they can be identified.

Rocks can be classified into three groups: igneous, sedimentary, and metamorphic.

During the rock cycle, rocks form and change into other types of rocks.

Make a FOLDABLES™ Study Guide

Make a Three-Tab Book. Use the titles shown. On the inside of each tab, summarize what you have learned about that topic.

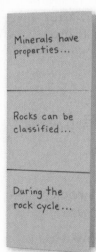

Minerals have properties...

Rocks can be classified...

During the rock cycle...

Think, Talk, and Write

1. **Main Idea** How are minerals and rocks similar and different?

2. **Vocabulary** When a mineral is rubbed on a rough surface, the color of the powder that is left is the _____.

3. **Sequence** Rank topaz, talc, and calcite from hardest to softest on Mohs' scale.

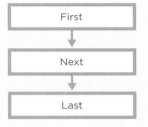

First

↓

Next

↓

Last

4. **Critical Thinking** Which steps of the rock cycle involve heat?

5. **Test Prep** From what material does an extrusive rock form?
   - **A** magma
   - **B** lava
   - **C** mineral
   - **D** sediment

6. **Test Prep** Which property of a mineral is that it breaks into smooth surfaces?
   - **A** hardness
   - **B** color
   - **C** cleavage
   - **D** fracture

Math Link

Layers of Rock
If a hill has 50 layers of sedimentary rock and each layer is 30 meters tall, how high is the hill?

Art Link

Rocks in Art
Select a sculpture and describe the rock from which it was made. Give reasons why the artist might have chosen that type of rock to make the sculpture.

Focus on Skills

Inquiry Skill: Classify

As you just read, rocks are naturally formed solids made up of one or more minerals. Each mineral adds its own properties to a rock. There are billions of different rocks on Earth. Scientists group, or **classify**, rocks into three groups based on the way they form. In order to determine how they form, scientists observe the properties of the rocks. These properties include color, weight, texture, and whether the rocks float or sink.

▶ Learn It

When you **classify**, you group objects that share properties. You need to compare and contrast the objects in order to find out what properties they share. Remember, to compare you look at how things are alike, while to contrast you look at how things are different.

Classifying is a useful tool for organizing and analyzing. It can help you understand why things belong in the same group and how some things can belong to several different groups. It is important to keep notes as you **classify**. Your notes can help you figure out how to classify other things.

▶ Try It

| Materials | 8 different rocks, water, small bowl

1. Make a table as shown. List the properties that you want to look for in the first column.

2. Examine the first rock carefully.

3. **Classify** Mark an X in the appropriate box if this rock can be classified by the property listed in the rows.

4. Fill the bowl with water. Place the rock in the bowl to test whether the rock floats.

5. Repeat using the remaining rocks.

Classifying Rocks by Properties

	#1	#2	#3	#4	#5	#6	#7	#8
color: dark								
color: light								
several colors								
heavy								
light								
rough								
smooth								
sharp								
has holes								
has layers								
floats								
sinks								

▶ Apply It

Now that you have classified rocks by their properties, look for rocks on the ground and in buildings. Make a table that lists all of the rocks that you have seen. List the properties by which these rocks can be classified. Finally, mark the chart to show which rocks have the same properties and can be classified together.

6 **Classify** How many rocks would you classify as smooth?

7 **Classify** How many rocks would you classify as having layers?

8 Which property is shared by the most rocks?

9 Decide whether each rock is an igneous, sedimentary, or metamorphic rock.

Soil

Look and Wonder

These young plants are growing in a field in Manitoba, Canada. Plants grow well in some types of soil but don't grow well in others. What is in soil that helps plants grow?

What is in soil?

Materials

- toothpicks
- hand lens
- soil sample

Purpose

To examine the contents of a soil sample.

Procedure

1. **Observe** Use the toothpicks and hand lens to separate the contents of the soil sample.

2. **Record Data** Identify and list the different materials in the soil sample.

Draw Conclusions

3. **Classify** Does your soil sample contain nonliving things? What about once-living things?

4. Based on your observations, what are the contents of soil?

Explore More

Collect and examine samples of soil from different places in your neighborhood. How do the contents of these samples compare with the one you studied in this activity? Do the additional samples change the conclusion you drew about the contents of soil?

Step 1

▶ **Main Idea**

Soil is a natural resource made of a mixture of nonliving and once-living materials.

▶ **Vocabulary**

soil, p. 316

soil horizon, p. 317

humus, p. 317

topsoil, p. 317

pollution, p. 319

conservation, p. 320

e-Glossary

at www.macmillanmh.com

▶ **Reading Skill** ✔

Summarize

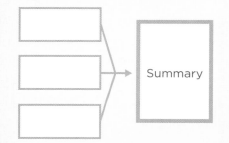

What is soil?

If you watched the same rock over many years, you would see that as time passed, the rock weathered. Microscopic organisms would grow among the bits of rock. Some of these organisms would break down the rocks into chemicals that could nourish plants.

As the rock weathers, grasses would grow, followed by bushes and trees. Animals would come to eat the plants, and other animals would feed on the animals that fed on the plants. When the animals and plants die, their bodies add organic nutrients back to the soil. *Organic* means having to do with or coming from living things.

Soil is a mixture of bits of rock and bits of once-living parts of plants and animals. Soil covers most of Earth's landmasses. Without it, plants and animals would not be able to live on land.

Soil covers the ground in rain forests, grasslands, and deserts. The soils in these places look different, but they all started from rocks. As rocks weather, the soil forms in layers. If you dig a hole in the ground, you will see the layers as you dig deeper.

Soils in different locations look different, but they form in similar ways.

Soil Horizons

Each layer of soil is called a **soil horizon** (huh•RYE•zuhn). No matter where it is found, soil is divided into three horizons called A, B, and C.

The **A** horizon, which holds the most nutrients, contains humus (HYEW•muhs). **Humus** is the part of the soil that is made up of decayed organic materials. These materials are the remains of dead plants and animals that are decayed by microscopic organisms. Humus contains nutrients that feed plants. Humus also soaks up and holds water more easily than the bits of rock do.

The soil in this horizon is called **topsoil**. Most plant roots grow in this soil. The roots absorb nutrients and water from humus.

The **B** horizon is called *subsoil*. You will find less humus in subsoil and lots of fine particles of rock, such as the particles that make up clay.

Next is the **C** horizon, which is made mostly of larger pieces of weathered rock. These soil horizons rest on solid, unweathered bedrock.

Different areas will have different depths of soil horizons. Some areas may not have one of these soil horizons.

✔ Quick Check

Summarize What are the main steps in the formation of soil?

Critical Thinking How could erosion change soil horizons and how plants grow in that soil?

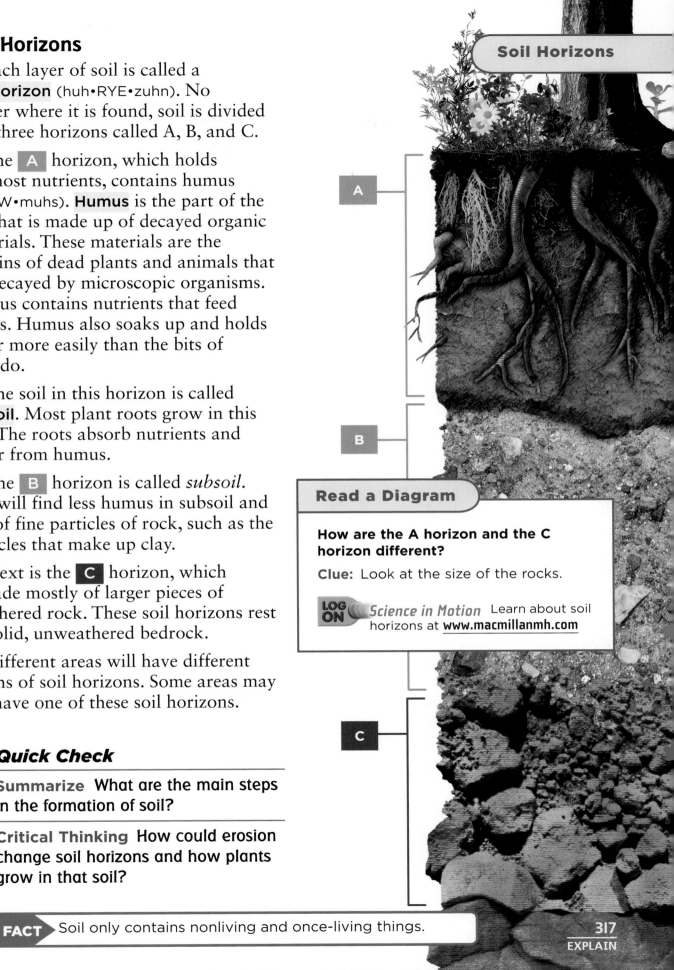

Soil Horizons

A

B

Read a Diagram

How are the A horizon and the C horizon different?

Clue: Look at the size of the rocks.

 LOG ON *Science in Motion* Learn about soil horizons at **www.macmillanmh.com**

C

FACT Soil only contains nonliving and once-living things.

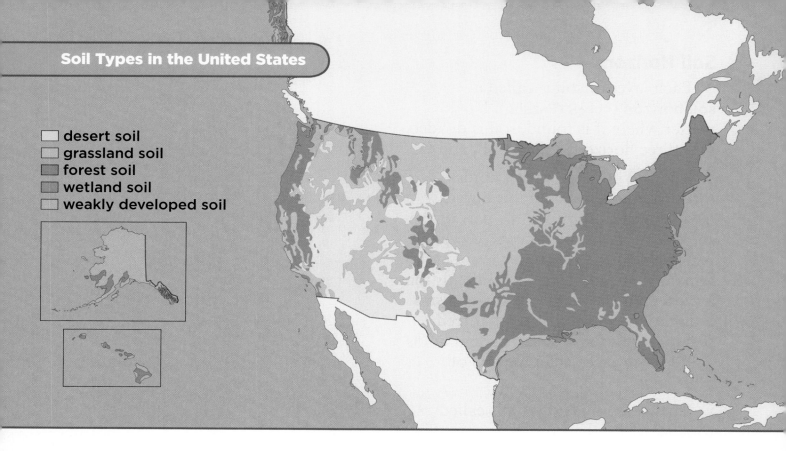

Soil Types in the United States

- ☐ desert soil
- ☐ grassland soil
- ■ forest soil
- ■ wetland soil
- ☐ weakly developed soil

How is soil used?

Soils in different places have different properties. Each type of soil supports different plant and animal life. Most of the United States is covered by three types of soil: forest soil, desert soil, and grassland and prairie soil.

The soil in a forest has a thin layer of topsoil with little humus. Frequent heavy rainfall carries minerals deep into the ground. Plants with shallow roots cannot reach these minerals. Crops with shallow roots do not grow well in such soil. Most of the forest soil in the United States is in the eastern third of the country.

Desert soil is sandy and does not hold much humus. Because desert areas receive little rain, plants have special adaptations to grow there. However, desert soil is rich in minerals since the minerals are

not washed away by rain. Animals can sometimes be raised in areas with desert soil. Crops can only be grown if water for the plants is artificially piped to the area.

The grasslands and prairies of the United States are found between the Rocky Mountains and the country's eastern forests. Crops such as corn, wheat, and rye grow on land from Texas to North Dakota. The soil is rich in humus, which provides nutrients for crops. The humus holds water so minerals are not washed deep into the ground. Animals eat the grasses that grow naturally in this soil.

Soil is a resource. Like other resources, it can be used up, wasted, or spoiled. Soil can be eroded by flowing water and wind. Plant roots hold soil in place. If plants are removed, more soil may be eroded. This may change the type of plants that can grow in an area or make it difficult for any plants to grow.

The nutrients in soil are naturally removed by plants. The plants use the nutrients to grow and to build their body parts. The nutrients are normally replaced when plants die, fall to the ground, and decay. What happens when a farmer completely removes a crop from the land? Then there are no plants left behind to die and decay. The land becomes less able to support the growth of new crops.

Pollution (puh•LEW•shuhn) is the addition of harmful materials to soil, air, or water. Soil can be polluted by chemicals released onto the ground. It can also be polluted by chemicals used to kill insects and weeds. When people dump garbage on the ground, the garbage can also pollute soil.

✓ Quick Check

Summarize What properties of soil are best for farming?

Critical Thinking How might insect pests be controlled without using chemicals that may pollute the soil?

Farmers use chemicals to kill insects that eat crops, but these chemicals can pollute the soil that the crops need to grow.

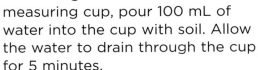

≡ Quick Lab

Soil Soaks Up Water

1. In a bowl, measure out topsoil and sand to make a soil mixture that you predict will hold water well.

2. ⚠ **Be Careful.** Using the point of a pen, punch an equal number of small holes in the bottom of three cups.

3. Fill one cup with topsoil, one cup with sand, and one cup with your mixture.

4. **Experiment** While holding the cup with the topsoil over a large measuring cup, pour 100 mL of water into the cup with soil. Allow the water to drain through the cup for 5 minutes.

5. Measure the water that passed through the soil.

6. Repeat steps 4 and 5 with the sand and with your mixture.

7. Calculate the amount of water that the soil soaked up.

8. **Interpret Data** Which type of soil holds the most water?

How is soil conserved?

The preservation or protection of natural resources, including soil, is called **conservation** (kon•suhr•VAY•shuhn). Here are some methods of conserving soil:

Fertilizing Fertilizers containing one or more nutrients can be added to soil to replace nutrients used up by previous crops.

Crop Rotation Farmers can plant different crops on the same land in different years. They can choose crops that add the nutrients that have been removed by other crops.

Conserving Soil

Read a Photo

How does the method shown in the photograph conserve soil?

Clue: Look for patterns in the fields.

Strip Farming Plant roots help prevent soil from being washed or blown away. For this reason, farmers may plant grasses between rows of other crops.

Contour Plowing Rain water flows swiftly down hills and can carry away rich topsoil. Farmers can slow the speed with which water flows down the hill by contour plowing, or plowing furrows across the slope of a hill instead of plowing up and down the slope of the hill.

Terracing Terraces are flat shelves that are cut into a hillside. Crops are planted along each terrace. This also slows the speed of water flowing down a slope.

Wind Breaks Farmers plant tall trees along the edges of farmland to slow the speed of wind across the ground. Where there are trees, the wind is less likely to blow away topsoil.

Laws Governments may pass laws to stop the pollution of soil.

Individual Efforts You can avoid polluting soil with trash and help clean up land that has already been polluted.

Education You can help inform people of the value of soil and how to conserve it.

 Quick Check

Summarize What methods are used to conserve soil?

Critical Thinking What might cause mountaintops to have little or no topsoil?

Lesson Review

Visual Summary

Soil is a mixture of bits of rock and bits of once-living parts of plants and animals.

Soil supports plant and animal life and can be polluted.

Soil can be conserved in many different ways.

Make a **FOLDABLES**™ Study Guide

Make a Three-Tab Book. Use the titles shown. On the inside of each tab, summarize what you learned about that topic.

Soil is a mixture...

Soil supports...

Soil can be conserved...

Think, Talk, and Write

① **Main Idea** What is soil?

② **Vocabulary** The part of soil that is made up of decayed organic materials is called _____ .

③ **Summarize** Describe the methods used to protect soil from erosion by water.

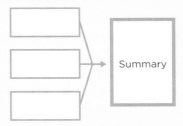

Summary

④ **Critical Thinking** Compare and contrast forest soil with desert soil.

⑤ **Test Prep** What is the C horizon of soil made of?
- **A** clay
- **B** humus
- **C** bedrock
- **D** large rocks

⑥ **Test Prep** What is strip farming?
- **A** adding fertilizer to soil
- **B** cutting shelves in hills
- **C** planting grasses between crop rows
- **D** planting trees around crops

 Writing Link

Conserve Soil!
Make a leaflet or flyer designed to persuade your neighbors to conserve soil. Explain why the soil in your area should be conserved and suggest ways people can conserve it.

 Social Studies Link

The Dust Bowl
Research the Dust Bowl of the 1930s. Write an essay describing the causes of the Dust Bowl and how it affected people in the United States.

Be a Scientist

Materials

2 pans

potting soil

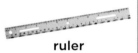

ruler

sand

grass seeds

2 measuring cups with water

Which soil is better for plant growth?

Form a Hypothesis

Different types of soil are made of different materials. Sand is a type of soil made from small pieces of rocks. Potting soil is made from bits of sticks and leaves. How fast will grass seeds grow in potting soil compared to sand? Write your answer as a hypothesis in the form "If grass seeds are planted in potting soil and in sand, then . . ."

Step 1

Test Your Hypothesis

1 Fill one pan with potting soil until the soil is 1 inch deep. Fill the other pan with sand until the sand is 1 inch deep.

2 Evenly scatter grass seeds over each pan.

3 Place the pans in the sunlight.

4 Every other day, pour the same amount of water on the seeds in both pans.

5 **Observe** What do the pans look like after three days? After one week?

Step 2

Draw Conclusions

6 Why is it important to make sure the pans get the same amount of light and water?

7 **Infer** What differences between the potting soil and the sand may have affected the plant growth?

Step 4

What effect does pollution have on plants?

Form a Hypothesis

You now know the type of soil in which plants will grow faster. How fast will plants grow in polluted soil? Write your answer as a hypothesis in the form "If grass seeds are planted in soil and polluted soil, then . . ."

Test Your Hypothesis

Design an experiment to investigate how fast plants will grow in soil compared to polluted soil. Write out the materials you need and the steps you will follow. Record your observations and results.

Draw Conclusions

Did your results support your hypothesis? Why or why not? Present your results to your classmates.

How efficient are conservation methods that slow down the flow of water over soil? Think of a question and design an experiment to answer it. Your experiment must be organized to test only one variable. Keep careful notes as you do your experiment so another group could repeat the experiment by following your instructions.

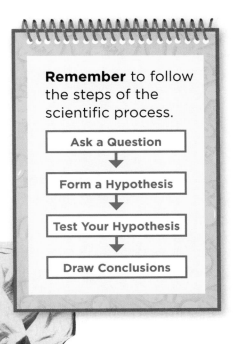

Remember to follow the steps of the scientific process.

Ask a Question
↓
Form a Hypothesis
↓
Test Your Hypothesis
↓
Draw Conclusions

Fossils and Energy

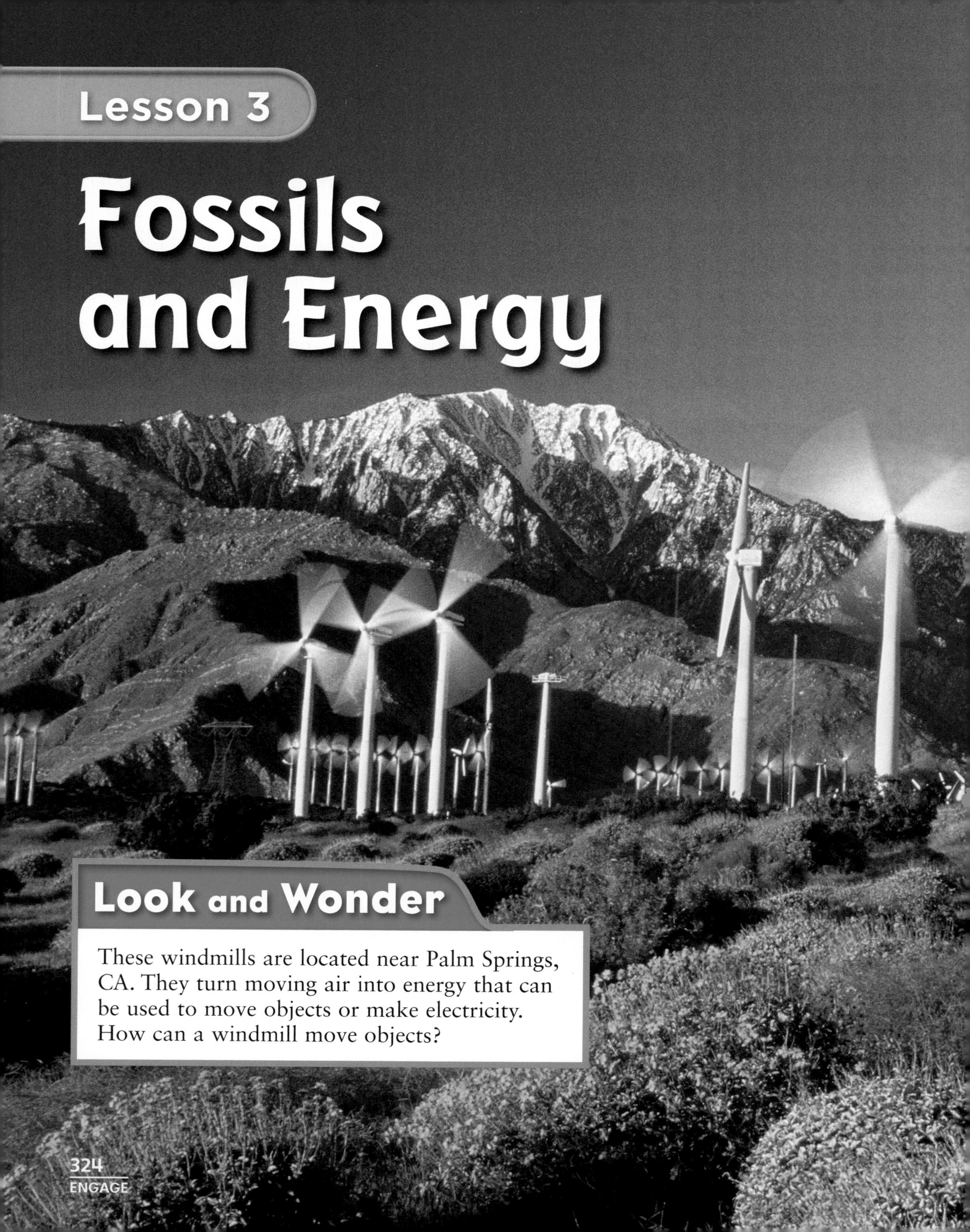

Look and Wonder

These windmills are located near Palm Springs, CA. They turn moving air into energy that can be used to move objects or make electricity. How can a windmill move objects?

How can wind move objects?

Form a Hypothesis

How many paper clips do you think you can move with your breath using a windmill? Write your answer as a hypothesis in the form "If the speed of the wind against a windmill blade increases, then . . ."

Test Your Hypothesis

1 Wrap the 8 cm by 15 cm strip of paper around the pencil. Have a partner tape the edges of the paper together to form a tube.

2 Tape the 5 cm side of the 8 cm by 5 cm strips to the tube of paper near one end of the tube to make blades for the windmill. Space the strips so they are equally far apart.

3 Tie one paper clip to the string. Tape the other end of the string to the paper tube.

4 Hold the ends of the pencil and blow on the paper strips. What happens to the paper clip?

5 **Experiment** Now attach more paper clips to that one. How many paper clips did you add before your breath could no longer lift them?

Draw Conclusions

6 How is the energy from your breath used to raise the paper clip?

7 **Infer** If you used larger rectangles for windmill blades, what do you think would happen to the number of paper clips you could lift?

Explore More

What result do you think you would get with different-shape blades? Think of a shape to test and come up with a design. Then experiment to find out whether your shape works better than a rectangle.

Materials

- **8 cm by 15 cm strip of paper**
- **new pencil**
- **tape**
- **four 8 cm by 5 cm strips of paper**
- **paper clips**
- **string**

Step **1**

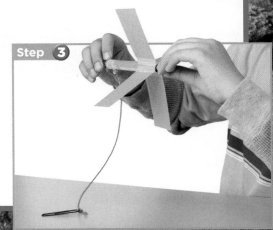

Step **3**

▶ Main Idea

Ancient organisms became fossils and fossil fuels. Fossil fuels are nonrenewable resources, so renewable sources of energy are also used.

▶ Vocabulary

fossil, p. 326
fossil fuel, p. 327
relative age, p. 328
absolute age, p. 328
era, p. 329
nonrenewable resource, p. 331
renewable resource, p. 331
alternative energy source, p. 332

LOG ON ⓔ-Glossary
at www.macmillanmh.com

▶ Reading Skill ✓

Fact and Opinion

Fact	Opinion

What are fossils?

Fossils (FOS•uhlz) are the remains or traces of ancient organisms preserved in soil or rock. Sometimes when an organism died its remains were covered with soil, sand, or some other sediment. Over many centuries, these sediments hardened over and around the organism's remains. Almost all fossils are found in sedimentary rock.

Scientists also can use fossils to learn more about ancient environments. Scientists know that modern animals that live in a warm climate have certain characteristics. What could scientists conclude if they examined fossils from Antarctica and saw that these fossils have characteristics of warm-climate animals? The scientists could conclude that Antarctica, which is now cold, was once much warmer.

Sometimes different characteristics are shown by comparisons of the bones of modern animals to fossilized bones. In the 1880s, the fossilized bones of an organism named *Smilodon* were discovered. *Smilodon* is a type of large cat popularly known as a saber-toothed cat because of its large fangs. The fossilized fangs are larger than the fangs of modern lions and tigers. Scientists generally agree that *Smilodon* used its fangs for hunting. However, scientists are not sure whether *Smilodon* used its fangs to grab or to bite prey.

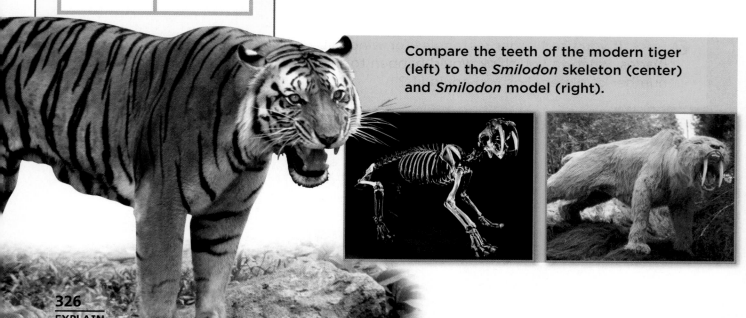

Compare the teeth of the modern tiger (left) to the *Smilodon* skeleton (center) and *Smilodon* model (right).

Fossil Fuels

Millions of years ago, plants used the energy in sunlight to build their bodies. In the process, they stored the Sun's energy as sugars and starches. The plants eventually died and fell to the ground. Layers of sediment built up on top of this layer of dead plants.

Over millions of years, pressure from the weight of the layers of sediment pressed the dead plants together and formed *peat*. As the peat hardened, it turned into a sedimentary rock called *bituminous coal* (bye•TEW•muh•nuhs kohl), or soft coal. As it was buried even deeper, this coal was changed into a metamorphic rock called *anthracite* (AN•thruh•site), or hard coal.

Bituminous coal and anthracite are types of fossil fuels. A **fossil fuel** is a material that formed from the decay of ancient organisms and is used today as a source of energy.

Sometimes, partly decayed parts of ocean organisms were buried deep under the ocean. There, a combination of the weight of rock, heat, and the action of bacteria turned the decayed materials into oil and natural gas. Oil and natural gas also are fossil fuels.

Over time, natural gas and oil fill up connected spaces between rocks. Natural gas is often found above oil. Today, in many parts of the world, oil and natural gas are pumped from deep beneath the ocean floor. Oil and natural gas also are now found beneath land that used to be covered by oceans.

coal

Dead organisms in a swamp form peat.

oil and gas

Dead organisms fall to the ocean floor.

coal

Peat is covered with layers of sediment.

oil and gas

The dead organisms are buried in sediment.

coal

Pressure turns peat into coal.

oil and gas

Pressure forms oil and gas.

 Quick Check

Fact and Opinion Most people like to hunt fossils. Is this an opinion?

Critical Thinking What happens to igneous rocks that makes finding fossils in them unlikely?

How old are fossils and fossil fuels?

Fossils can provide information about whether an organism is older or younger than other organisms. In general, each layer of rock is older than the layer above it and younger than the layer below it. This is called the *law of superposition*.

If you found one fossil in a top layer of rock and another fossil in a lower layer, the relative age of the fossil in the lower layer is older than the fossil in the higher layer. The **relative age** of a rock is how old it is compared to another rock.

Although you know that the bottom fossil is older, how can you tell whether it is 50,000 years old, 500,000 years old, or 5,000,000 years old? The **absolute age** is the age of a fossil in years. To find the absolute age of a fossil, you have to find out the absolute age of the rock in which the fossil was found. Since the fossil formed when the rock formed, the fossil is the same age as the rock.

How can you tell how old a rock is? All rocks contain different elements. Some of these elements change into other elements in constant ways.

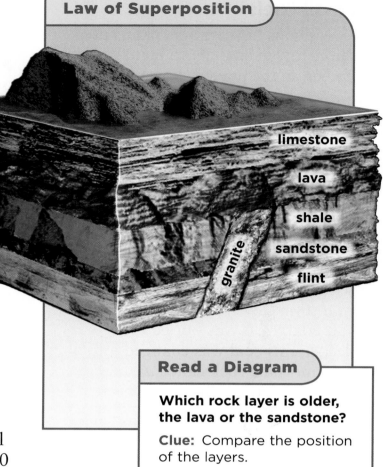

Law of Superposition

limestone

lava

shale

sandstone

flint

granite

Read a Diagram

Which rock layer is older, the lava or the sandstone?

Clue: Compare the position of the layers.

For example, element A changes into element B over time. If half of element A changes into element B every 1 million years, after 1 million years have passed, the rock holds equal amounts of elements A and B.

The time it takes for the amount of an element to be cut in half is called the element's *half-life*. The half-life of element A is 1 million years. Different elements have different half-lives.

Geological Eras

PRECAMBRIAN	PALEOZOIC – AGE OF ANCIENT LIFE					
	CAMBRIAN	ORDOVICIAN	SILURIAN	DEVONIAN	CARBONIFEROUS	PERMIAN

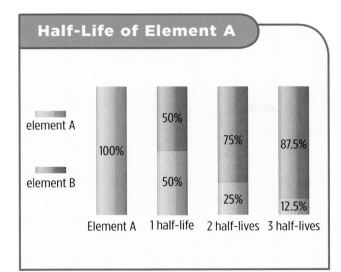

Half-Life of Element A

element A

element B

100%	50% / 50%	75% / 25%	87.5% / 12.5%
Element A	1 half-life	2 half-lives	3 half-lives

Scientists have used the relative and absolute ages of fossils to develop a history of Earth since its formation about 4.6 billion years ago. When describing the age of Earth, scientists use units called eras (EER•uhz). An **era** is a unit of time measured in millions of years. A geologic period divides an era into a smaller unit of time.

Fossil fuels formed during the Carboniferous period, which was between 350 million and 280 million years ago. During this time, the land was covered with swamps filled with large leafy plants. Over millions of years, these plants were buried and turned into fossil fuels. As the tectonic plates moved, the fossil fuels moved. Coal is now found in northern Europe, Asia, and midwestern and eastern North America.

≡Quick Lab

Half-Life of a Penny

1. **Record Data** Count the total number of pennies that you were given.

2. Place all of the pennies in a box so they are heads up.

3. **Experiment** Close the box and shake it to mix the pennies.

4. Open the box and remove all the pennies that have turned tails up. Set them aside.

5. Record the number of pennies remaining in the box.

6. Repeat steps 2–4 until one or no pennies remain in the box.

7. How many pennies were removed after each shake?

8. What is the "half-life" of a penny?

✓ Quick Check

Fact and Opinion A fossil in a 2-million-year-old rock is 2 million years old. Is this a fact or an opinion?

Critical Thinking Could you tell the relative age of a fossil if layers of rock have been shifted by earthquakes?

MESOZOIC – AGE OF DOMINANT REPTILES	CENOZOIC – AGE OF DOMINANT MAMMALS

TRIASSIC	JURASSIC	CRETACEOUS	TERTIARY	QUATERNARY

How are fossil fuels used?

When fossil fuels are burned, they release energy from the sunlight that was stored in the dead plants and in the animals that ate them. People can either turn the stored energy into a different kind of energy or use it to do work.

For example, in a car, gasoline releases energy that runs the car's engine. The energy released from burning oil also is turned into heat energy and used to warm buildings. Natural gas can be burned in a stove to cook food or in a furnace to heat a home.

Electricity is a form of energy that people use every day. Electricity is used to light homes, schools, office buildings, and streets. It is used to run equipment from clocks and elevators to DVD players and computers.

Most of the electricity that people use is made in power plants. In a power plant, energy is used to make an electric generator move. As the generator moves, electricity is produced. The electricity then travels through wires to places where it is used, such as your home.

Coal power plants burn coal to produce electricity.

Drilling rigs extract oil and natural gas from underneath the ocean floor.

Power plants get the energy to run electric generators from sources such as coal, oil, and natural gas. However, these are nonrenewable resources (non•ri•NEW•uh•buhl). A **nonrenewable resource** is one that can be used up faster than it is made. It took millions of years to make the oil, natural gas, and coal that we use today, but we may use up these resources in only hundreds of years.

Renewable Resources

Earth also has renewable resources. A **renewable resource** is a living or nonliving resource that can be replaced naturally. Wind, water power, and sunlight are considered nonliving renewable energy sources.

Living renewable resources include such things as fish and forests. Living renewable resources must be treated with care. It is possible to use populations of living things faster than they can reproduce. Once completely gone, a living thing cannot be replaced.

Extracted oil and natural gas are purified in refineries such as this one near Port Huron, MI.

 Quick Check

Fact and Opinion Fossil fuels are made from decayed plants and animals. Is this statement a fact or an opinion?

Critical Thinking Why are wind, water, and sunlight renewable energy sources?

How can the Sun, wind, and water make energy?

Our planet provides other sources of energy that could be used to make electricity, keep us moving, and keep us warm. Any source of energy other than fossil fuels is called an **alternative energy source** (awl•TUR•nuh•tiv EN•uhr•jee sawrs). Alternative energy sources include wind, moving water, and *solar energy*, or energy from the Sun.

The energy from these sources can be used to do work. Sunlight can heat water, air, or other materials. It can also be turned into electricity. The energy in wind and moving water can be used to move machines and make electricity.

Energy from Wind

Wind is simply air that is moving. The wind moves the blades of a windmill. The blades are attached to gears and shafts. The gears and shafts are attached to an electric generator.

When the blades of the windmill turn, the parts of the generator move and electricity is produced. Windmills are used to produce electricity in parts of California and Hawaii and in countries such as Denmark, Germany, Spain, and India.

Wind energy does not pollute the air we breathe. However, it can only be used where winds blow almost all the time. Some people are concerned that windmills might interfere with the habitats of migrating birds.

Energy from Moving Water

Water running through streams and rivers has energy. Waterwheels use energy from moving water to do work. Running or falling water turns a wheel,

Alternative Energy Sources

This windmill turns wind into wind power.

The dam helps make hydroelectric power from water.

and the turning wheel moves an axle. The axle can be connected to one of a number of different machines.

In a mill, the axle moves two large, round stones. When grain is put between the stones, the motion of the stones grinds the grain into a powder.

In a hydroelectric plant, running, falling, or flowing water spins a generator. The prefix *hydro* means "water." A hydroelectric plant is one that uses water to produce electricity.

Hydroelectric power plants do not pollute air or water. However, they can only be used where there is moving water. They may also disrupt the lives of animals that live in the water.

Solar Energy

Energy from the Sun is solar energy. Solar energy is a resource that will last as long as the Sun shines. How are light and heat energies from the Sun changed into energy people can use?

Fields of solar cells turn the sunlight that strikes them into electricity for homes. Some calculators are powered by solar cells that also change sunlight into electricity. Solar energy also can heat water for a house.

Solar energy will not run out. It does not cause pollution of any kind and it is available wherever the Sun shines. In order to be most effective, solar cells need to be located in areas where cloud-free days occur most of the year.

 Quick Check

Fact and Opinion Solar energy will last as long as the Sun shines. Is this statement a fact or an opinion?

Critical Thinking If fossil fuels ran out, what would be the effect on the world's people?

Read a Photo

Which of these alternative energy sources uses water?

Clue: Look for water in the photos.

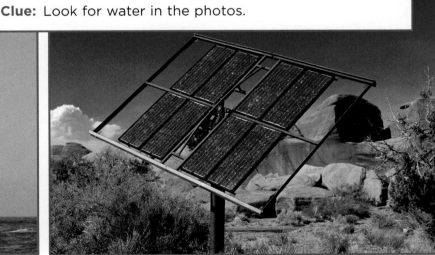

The turbine makes hydroelectric power from tides.

A solar panel captures solar power.

What are other sources of alternative energy?

Alternative energy sources also include energy in the nucleus of atoms, heat from inside Earth, and plant and animal materials that are used as fuel.

Nuclear Energy

Certain elements are made of atoms that can change into atoms of other elements. The change occurs in the center, or nucleus, of the atoms.

▼ A view inside a nuclear power plant.

When these changes occur, a great amount of heat energy is given off. This energy is called *nuclear energy* (NEW•klee•uhr). The heat makes steam and the steam moves the parts of an electric generator.

Electricity made by using nuclear energy is generally less expensive than electricity made by burning fossil fuels. Nuclear power plants do not pollute the atmosphere unless an accident occurs. However, accidents have occurred that have caused radioactive material to pollute the air.

Geothermal Energy

Heat produced inside Earth, or *geothermal energy* (jee•oh•THUR•muhl), causes volcanoes and hot springs. Hot springs contain water at Earth's surface that has been heated underground. Steam from hot springs can be piped to machines that spin generators. The hot water also can be piped into homes and buildings to heat them. Nearly all of the homes in the country of Iceland are heated by geothermal energy.

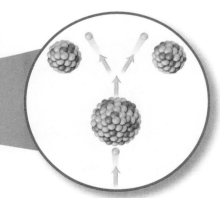

nuclear chain reaction

Geothermal power plants heat almost all the homes in Iceland.

Geothermal energy sources have to be tapped from underground. Only a few places on Earth are located where this kind of energy is easy to reach.

Biomass

Anything that burns can be used as an energy source. However, when nonrenewable resources such as coal, oil, and natural gas are burned, they are used up. *Biomass* (BYE•oh•mas) is made up of materials from living things. Wood, animal waste, and plant materials, such as cornstalks that are leftover parts of crops, are biomass.

Because trees and crops are living renewable resources, biomass does not run out. Trees can be replanted and crops can be grown again. However, biomass does not release as much energy as the burning of coal, oil, or natural gas. Growing plants for fuel also decreases the amount of land available for growing crops.

This biodiesel is made of 20 percent soybean oil and 80 percent diesel fuel. ▼

 Quick Check

Fact and Opinion State one fact and one opinion about nuclear energy.

Critical Thinking How is location a drawback that is shared by wind, solar power, flowing water, and geothermal heat in terms of their uses as sources of energy?

How can we conserve energy?

You and your family use energy every day. You may not even realize that a certain activity uses energy. For example, when you turn on a light, you are using power. You also are using coal, oil, or natural gas because many power plants burn these fuels to produce electricity.

When you ride in a car, the gasoline that is burned in the car's engine comes from oil. When you take a shower, you may be using up natural gas or oil because many hot-water boilers burn natural gas or oil. If you feel cold at home, you may turn up a thermostat. When you do, a signal goes to a furnace, turning it on and burning more oil or natural gas.

For every way you use up energy, there is a way you can conserve it. How do you think you and your family can conserve energy?

Houses in this community in England are designed to be energy efficient.

Quick Check

Fact and Opinion Give your opinion about ways you could conserve energy.

Critical Thinking Explain how you could be using fossil fuels while watching a television show.

Guidelines for Energy Conservation

 Turn off lights when you leave the room.

 Turn off electronic equipment when you aren't using it.

 Use water-conserving showerheads and take shorter showers.

 Turn hot water off when you aren't using it.

 Carpool or use public transportation whenever you can.

 Turn the heat down or air conditioning up when you aren't home. Insulate windows and doors to prevent heat loss.

Lesson Review

Visual Summary

Fossils are the remains or evidence of past life and are used to shed light on Earth's history.

Fossil fuels have been produced from decayed living things, and are nonrenewable resources.

People can use sources of renewable energy or conserve nonrenewable energy resources.

Make a FOLDABLES™ Study Guide

Make a Trifold Book. Use the titles shown. In each column, write down a fact that you know.

Think, Talk, and Write

1. **Main Idea** What do people get from ancient organisms?

2. **Vocabulary** The main energy source for heating homes in Iceland is heat from inside Earth, or _____.

3. **Fact and Opinion** Describe one solution to the problem of decreasing oil supplies due to the use of cars.

Fact	Opinion

4. **Critical Thinking** How are relative age and absolute age different?

5. **Test Prep** Which of the following is an alternative source of energy?
   A oil
   B biomass
   C natural gas
   D coal

6. **Test Prep** Which of the following is NOT a renewable resource?
   A plants
   B solar energy
   C coal
   D animals

Math Link

Half-Life of a Jam Sandwich
The half-life of a jam sandwich is 5 minutes. How much of the sandwich is left after 15 minutes?

Art Link

Carboniferous Period
Research the plants and animals that lived during the Carboniferous period. Then draw a picture of their environment.

Writing in Science

Scientists dig for fossils in Dinosaur National Monument in Colorado and Utah.

So You Want to Be a Fossil Hunter

You should look for fossils in locations where fossils will be on the surface, such as craggy mountains and hills or steep canyons. Fossils can also be found along tall riverbanks or ocean shores.

You should look for areas where the rocks are in layers. You may have to dig to find fossils in the layers. Look for patterns or different colors in the rocks that could be a plant leaf or an animal shell.

If you think you have found a fossil, record your location and the type of rock in the area. Next, take a photo of the fossil. After that, you may carefully use a stiff brush or small trowel to clean the dirt or rock away from the fossil without damaging it. To identify your fossil, compare the patterns and details in your fossil to reference materials.

Descriptive Writing

A good description

▶ uses sensory words to describe how something looks, sounds, smells, tastes, or feels

▶ includes vivid details to help the reader experience what is being described

Complete fossils may be put in museums.

 Write About It

Descriptive Writing Select a fossil discovery and write a description of it. Use sensory words and specific details.

LOG ON e-Journal Research and write about it online at **www.macmillanmh.com**

Waterbury 8 KM
Burlington 50 KM

Converting Units

When people shop for a car, they often consider how much gas it uses to drive a certain distance. A fuel-efficient car travels a longer distance on a smaller amount of gas. This saves the driver money and also helps conserve oil. In the United States, we measure the amount of gas that a car uses in miles per gallon (mpg). Scientists and people in most other countries get the same information in kilometers per liter (kmpl).

Converting Units

To convert kmpl to mpg

▶ Multiply the number of kmpl by 2.352

12 kmpl
x 2.352
─────────
28.224 mpg

To convert mpg to kmpl

▶ Multiply the number of mpg by 0.425

40 mpg
x 0.425
─────────
17.000 kmpl

 Solve It

1. Sam's car gets 25 mpg. Jasmine's car gets 29 mpg. Which car is more fuel-efficient?

2. Lori's car gets 36 mpg. Henry's car gets 9 kmpl. How much gas does Henry's car use in miles per gallon? Whose car is more fuel-efficient?

3. Maria drove 64 kilometers on 4 liters of gas. Jerry drove 80 miles on 4 gallons of gas. How many kilometers per liter of gas can each car drive? Whose car is more fuel-efficient?

Air and Water

Seljalandsfoss Waterfall, Iceland

Look and Wonder

Every day, fresh water flows from this waterfall. How much fresh water do you use in the same day?

How much fresh water do you use?

Make a Prediction

How much water do you use in a day for a particular activity, such as brushing your teeth or washing your hands?

Test Your Prediction

1 Put the container in the sink.

2 Turn the water on and pretend to brush your teeth or wash your hands. Run the water as long as you would if you were really doing that activity. Once you are done, turn the water off.

3 **Measure** Using the measuring cup, scoop water out of the container into the sink. Keep track of each cup that you pour so you can estimate the total amount of fresh water you use for that activity.

Draw Conclusions

4 **Use Numbers** On a chart, figure out how many gallons of fresh water you use for the activity in a week, a month, and a year.

5 **Communicate** Discuss how much water you used with your classmates. Exchange data for the amount of water you used for your chosen activity. Whose use of water was closest to their prediction?

6 Design and complete tables or graphs to display the results of all of the data collected by the other students.

Materials

- container
- sink
- measuring cup

Step **2**

Step **4**

Activity: _____

Cups	6 6×day
1 week	252 c
1 month	1008 c
1 year	12,096 c 756 gal.

Explore More

Think of a way you can reduce the amount of water that you used. Predict how much water you can save. Redo the activity you chose using your new idea. Were you able to save water? Discuss your idea and its result with your classmates.

▶ **Main Idea**

Air and water are resources that support life on Earth.

▶ **Vocabulary**

reservoir, p. 343
aquifer, p. 343
smog, p. 348
ozone, p. 348

▶ **Reading Skill** ✔

Main Idea and Details

Main Idea	Details

What are sources of fresh water?

Many organisms on Earth need fresh water to survive. About 70 percent of Earth's surface is covered with water. However, about 97 percent of the water on Earth is salt water in the oceans. Roughly 2.3 percent of the fresh water on Earth is frozen at or near the North Pole and South Pole. Another 0.6 percent is liquid fresh water. Finally, 0.1 percent of Earth's water is present in the air as water vapor. If all the water on Earth were the size of this page, the amount of fresh water on Earth would be the size of this square.

What causes so much of the water on Earth to be salty? Water that falls as rain or snow is fresh water. As rain runs downhill, it picks up salts that are in soil and rocks. The flowing water runs into rivers. River water does not taste salty because it contains a very small amount of salts.

Usable Sources of Fresh Water

snow

streams

reservoir

well

aquifer

Rivers carry these salts into the ocean. Waves pick up salts from rocks and sand. Erupting volcanoes also add salts to the ocean. Each of these sources adds only a small amount of salt to the ocean. Since salts have been added for many millions of years, over time, the amount of salt in the ocean has slowly increased to its current concentration of 3.5 percent.

Because so much of the water on Earth is salty, fresh water is a limited resource. Most of the fresh water that people use is obtained from running water, standing water, and groundwater.

Running Water

Many cities and towns are built next to sources of running water, such as streams or rivers. Running water provides a source of fresh water for homes, farms, and businesses.

Standing Water

Bodies of standing fresh water, such as lakes and reservoirs (REZ•uhr•vwahrz), fill holes in the

ground. A **reservoir** is an artificial lake that is built to store water. Reservoirs are usually made by building a dam across a river. Water is stored behind the dam and is released when it is needed.

Groundwater

Groundwater seeps into the ground through aquifers (AK•wuh•fuhrz). An **aquifer** is an underground layer of rock or soil that has pores and is capable of absorbing water. As water seeps through an aquifer, it eventually reaches a layer of rock that does not absorb water. Over the years, fresh water builds up on top of this rock.

Groundwater is most useful to people when it is close enough to the surface that it can be reached by drilling or digging into the ground. Groundwater is then pumped up through a well. As water is removed, the level of the water underground drops. In order for the water to reach that height again, more water must seep down to replace it.

 Quick Check

Main Idea and Details What makes fresh water a limited resource?

Critical Thinking What are some reasons why one area of an ocean might be more salty?

Read a Diagram

What are ways people use artificial construction to get water?

Clue: Look for artificial construction.

dam

river

How do we use water?

In the United States, about 408 billion gallons of water are used every day. How do we use this water?

The largest category of water use in the United States is thermoelectric power. In this case, water is mostly used to cool the equipment that makes the electricity. As this water is not directly used by people, these power plants often use salt water.

The largest use of fresh water in the United States is irrigation (ir•ri•GAY•shuhn). *Irrigation* means to supply with water by artificial means. Unless crops and plants are supplied by natural sources of fresh water, they need irrigation in order to grow.

Most of the time, households and businesses use water that is supplied by a water company. Some of this

▲ Farmers need water for their crops.

water also is used for fighting fires, in swimming pools, and in public buildings, such as schools.

Some businesses have their own water sources. They may use water as part of making, washing, cooling, or transporting their product.

Some people have their own supply of fresh water. The source is usually a well. They can use this water in their house or for animals.

As water condenses in the sky or runs across the ground, it may pick up substances such as chemicals or harmful organisms that can pollute it. Polluted water cannot be used by people.

What are some signs that water might be contaminated? It might smell. It might be cloudy, have a strange color, or have dead fish in it. However, you can't always tell that water is polluted by looking at it.

Water Use in the United States

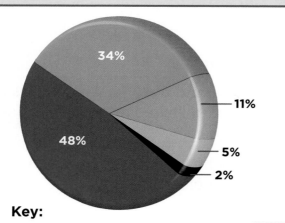

34%

11%

48%

5%

2%

Key:
- ■ – thermoelectric power
- ■ – irrigation
- ■ – water company
- ■ – business
- ■ – wells

Read a Graph

What is the second-largest use of water in the United States?

Clue: Look for the second-largest piece of the pie chart.

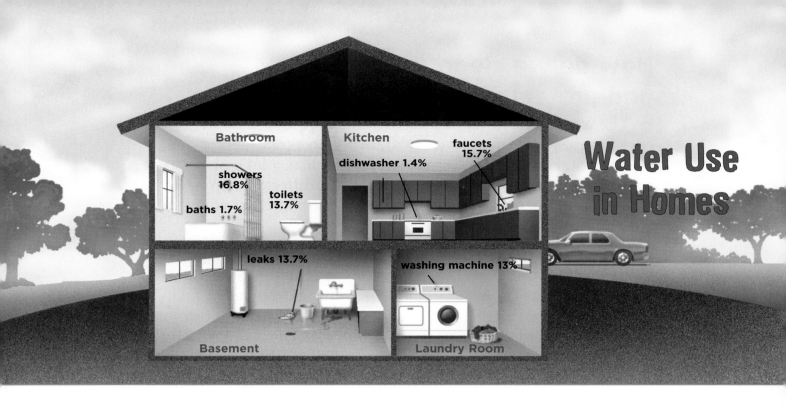

Water Use in Homes

Bathroom
showers 16.8%
baths 1.7%
toilets 13.7%

Kitchen
dishwasher 1.4%
faucets 15.7%

leaks 13.7%

washing machine 13%

Basement

Laundry Room

Farmers and homeowners need to water plants so the plants will grow. Sometimes they also use chemicals to help their crops and lawns grow, or to kill specific types of plants. People also use chemicals to kill organisms that are harmful to plants. Some factories and mines produce wastes. If these chemicals or wastes reach a source of water, they can pollute it.

Polluted water from this sewage pipe is pouring into a reservoir.

Water washes over streets and driveways. The flowing water can pick up contaminants, such as salt used to melt ice and snow, spilled motor oil, and trash.

Watering lawns and gardens may use 50 percent to 70 percent of a household's water. Inside their homes, people use the largest amounts of water to take showers, flush toilets, wash clothes, and wash dishes. When they are done with these tasks, the water flows back into the pipes. However, the water now has wastes and household chemicals in it and will need to be cleaned before it can be used again by people.

✅ Quick Check

Main Idea and Details How is water used and polluted?

Critical Thinking Why do farmers and homeowners use products that can contaminate fresh water?

How do we clean, conserve, and protect water?

The water that flows to houses and businesses in most communities is treated, or cleaned, in a water treatment plant. There, water from a freshwater source, such as a lake or reservoir, runs through several tanks. In each tank, a different step takes place. The steps may vary depending on the source of your community's water.

First, sticky particles are added to the water to attract any dirt in it. This step is called *coagulation* (koh•ag•yuh•LAY•shuhn). In the next tank, as *sedimentation* (sed•uh•muhn•TAY•shuhn) takes place, the clumps of dirt and sticky particles fall to the bottom of the tank. Then the water passes through a series of filters, which are layers of sand, gravel, and charcoal. These filters remove remaining bits of soil or other particles from the water. After water leaves this tank, chlorine and other chemicals are added to the water to kill harmful bacteria. This step is called *disinfection* (dis•in•FEK•shuhn). The clean water is kept in a storage tank until it is released to the community.

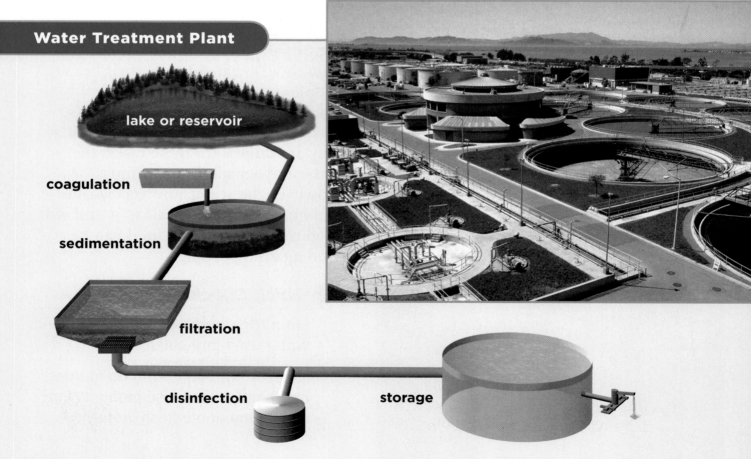

Water Treatment Plant

lake or reservoir

coagulation

sedimentation

filtration

disinfection

storage

Conserving Water

People can conserve water by reducing their use of water. Water conservation can be done by entire communities and by individuals. Sometimes water conservation efforts are aimed at larger areas, such as a river's watershed.

Protecting Water

In communities that have a limited supply of water, the local government may make regulations that stop people from watering lawns, filling swimming pools, and washing cars. People who break such regulations can be fined. These restrictions usually go into effect when natural sources of water, such as rain or snow, fall below the amount of water needed for people to use.

Local and state governments, as well as the United States government, have passed laws to protect our water supply. In 1974, Congress passed the Safe Drinking Water Act, which sets rules that communities must follow to keep drinking water clean and safe.

Three years later, Congress passed the Clean Water Act, which made it illegal to throw pollutants into surface waters. People or businesses that break this law can be fined.

 Quick Check

Main Idea and Details What can you do to conserve water?

Critical Thinking How would you change the steps in a treatment plant if the water were heavily polluted?

Rules of Water Conservation

 Use water-conserving showerheads and take shorter showers.

 Don't leave water running when you aren't using it.

 Wash dishes by hand. If you use a dishwasher, use a water-saving model and don't run it unless it is full.

 Fix leaking pipes or faucets.

 Use a water-saving washing machine and wash full loads of clothes.

 Grow plants that don't require frequent watering and water your plants after dark so the water does not evaporate.

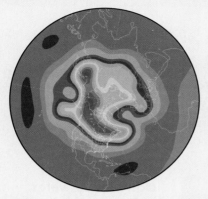

thicker ozone

thinner ozone

Northern Hemisphere (1986)

Northern Hemisphere (2006)

How do we use and pollute air?

You cannot see air, and if air is clean, you cannot smell it or taste it. If air does not move, you cannot feel it or hear it. However, air is all around you.

Earth's atmosphere holds the gases that living things need to stay alive. These gases are oxygen, carbon dioxide, and nitrogen. Plants and animals use oxygen to produce the energy they need. Plants take in carbon dioxide to make the foods they need to survive. Bacteria in soil turn nitrogen into chemicals that plants use to grow.

Sometimes particles of a pollutant build up in the air. These particles are produced when fossil fuels or trash are burned. Dust from plowed fields, construction sites, and mines also can put polluting particles into the air. Chemicals that are produced by factories can cause air pollution.

On certain days, you might see a yellow haze in the air over cities. This is a sign of a type of air pollution called smog. **Smog** is a type of air pollution that is caused by particles that are produced when fossil fuels are burned. Smog irritates eyes and can make breathing difficult. It is especially dangerous for people who have breathing problems such as asthma.

Air pollution does not only cause problems near the ground. About 30 kilometers above Earth's surface, there is a layer of a gas called **ozone** (OH•zohn). Ozone is a form of oxygen that protects living things on Earth's surface from dangerous energy from the Sun.

In recent years, chemicals in aerosol cans and air conditioners have escaped into the atmosphere. When these chemicals rise high up into the atmosphere, they set in motion chemical reactions that destroy ozone. When ozone is destroyed, harmful radiation from the Sun reaches Earth's surface more easily. Evidence suggests that this radiation can increase the chances of getting skin cancer.

Smoke from cars and trucks is a source of air pollution.

An *ozone hole* is a thinner area in the layer of gas. Compare the images of the ozone over the North Pole in 1986 and in 2006. What has happened to the thickness of the ozone layer?

In industrial areas, smoke and gases that pour into the air from factories can combine with rain to form acid rain. Acid rain has many harmful effects. It can kill trees and wear away stone buildings and statues.

Air can also be polluted by events in nature. Volcanic eruptions of gases, dust, and ash may stay for days, weeks, or months.

≡ Quick Lab

Dirty Air

1. Using a plastic knife, smear a thin layer of petroleum jelly on an index card.

2. Holding the edges of the index card, carefully place the card in a corner of the room.

3. **Observe** What does the index card look like after one day? After one week?

4. **Infer** How does the petroleum jelly help you track air pollution?

5. **Form a Hypothesis** Would you expect more air pollution near a road or away from a road? Why?

✔ Quick Check

Main Idea and Details How is air pollution produced?

Critical Thinking Discuss how forest destruction would affect air pollution.

Ash from erupting volcanoes is a source of air pollution.

▲ Planting trees can lower carbon dioxide levels and reduce air pollution.

How do we protect air?

What can be done to prevent air pollution? The best way to prevent air pollution is to keep the pollutants from getting into the air.

Congress passed the Clean Air Act in 1963 and has added several updates to it. As a result of these laws, many pollutants are now banned and other pollutants are filtered out before they get into the air.

The chemicals that destroy ozone can no longer be used in aerosol cans. Factories must equip smokestacks with devices that trap pollutants before they can be set loose in the air. The exhaust systems of cars, buses, and trucks also are fitted with devices that limit the gases and soot that come out of exhaust pipes.

A park ranger measures air pollutants in Yellowstone National Park, WY.

 Quick Check

Main Idea and Details List three causes of air pollution.

Critical Thinking List a benefit and cost of controlling air pollution.

Lesson Review

Visual Summary

Most organisms on Earth need clean fresh water to live.

Air contains gases, such as oxygen, carbon dioxide, and nitrogen, that are needed by living things.

Water and air can be polluted or conserved.

Make a FOLDABLES™ Study Guide

Make a Folded Chart. Use the titles shown. In each box, write the main idea of what you learned.

Think, Talk, and Write

1. **Main Idea** Why are air and water important resources?

2. **Vocabulary** Dangerous radiation from the Sun is blocked by _____ in Earth's atmosphere.

3. **Main Idea and Details** List three ways you can conserve fresh water.

Main Idea	Details

4. **Critical Thinking** Some whales feed on krill, a small sea animal. Krill feed on green organisms called algae, which produce oxygen. Explain how killing the whales might affect Earth's atmosphere.

5. **Test Prep** An aquifer is a(n)
   - **A** body of surface water.
   - **B** body of underground water.
   - **C** form of precipitation.
   - **D** ocean.

6. **Test Prep** Which gas do plants release into the air?
   - **A** nitrogen
   - **B** carbon dioxide
   - **C** oxygen
   - **D** nitrates

Writing Link

The End of Trees
Write a science fiction story in which all the trees on Earth are destroyed by an event of your choosing. Describe how the end of trees affects the environment and all living things on Earth.

Health Link

Waterborne Diseases
Do research to identify a disease that is caused by polluted water. Write a report describing the type of pollution, the effects of the disease, and ways to prevent the pollution.

Getting the Salt Out

Salts are removed from ocean water inside the Santa Catalina Island desalination plant.

Why does California have water shortages when it is next to the Pacific Ocean? People cannot drink ocean water because of the salts in it.

The island of Santa Catalina lies off the coast of Southern California. It is completely surrounded by the Pacific Ocean. However, people on the island use water from the ocean all the time—to water crops, to take showers, and even to drink. How can they drink and use the salty ocean water? The water is transformed from salty to fresh at the Santa Catalina desalination plant. *Desalination* means "to remove salts."

At the desalination plant, ocean water is taken from an ocean water well. Once it is moved into the plant, salt and other impurities are removed from the water. The fresh water that is produced can now be used by people.

The Santa Catalina plant is one of the few desalination plants in the United States that produces water for public use. Desalination is an expensive process that uses a lot of energy. Despite its cost, there are desalination plant projects all over the world, including places like Saudi Arabia and Japan. Desalination is generally used when a community has so little access to fresh water that it is willing to pay a high price to get it. Scientists continue to research cheaper and more efficient ways to produce fresh water from ocean water.

Problem and Solution

▶ Identify the problem by looking for a conflict or an issue that needs to be resolved.

▶ Think about how the conflict or issue could be resolved.

Write About It
Problem and Solution

1. What is in ocean water that prevents people from using it directly from the ocean?
2. How do the people of Santa Catalina get fresh water?

LOG ON **e-Journal** Research and write about it online at **www.macmillanmh.com**

AMERICAN MUSEUM OF NATURAL HISTORY

Visual Summary

Lesson 1 Rocks and minerals are produced in different ways and have different properties.

Lesson 2 Soil is a natural resource made of a mixture of nonliving material and once-living things.

Lesson 3 Ancient organisms became fossils and fossil fuels. Fossil fuels are nonrenewable resources, so renewable sources of energy are also used.

Lesson 4 Air and water are resources that support life on Earth.

Make a FOLDABLES™ Study

Assemble your lesson study guide as shown. Use your study guide to review what you have learned in this chapter.

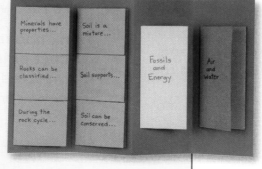

Fill in each blank with the best term from the list.

aquifer, p. 343

conservation, p. 320

era, p. 329

igneous rock, p. 306

mineral, p. 302

renewable resource, p. 331

smog, p. 348

topsoil, p. 317

1. Water power can be replaced naturally, so it is a(n) _____.

2. A unit of time measured in millions of years is called a(n) _____.

3. Most plant roots grow in _____.

4. The protection of natural resources is called _____.

5. An underground layer of rock or soil that can absorb water is a(n) _____.

6. A solid natural material made from nonliving substances in the ground is a(n) _____.

7. When fossil fuels are burned, the resulting particles can cause _____.

8. As lava or magma cools, a(n) _____ is formed.

LOG ON **e-Review** Summaries and quizzes online at **www.macmillanmh.com**

Answer each of the following in complete sentences.

9. **Fact and Opinion** Some minerals contain crystals. Is this statement a fact or an opinion? Explain your answer.

10. **Summarize** Write a description of the soil horizons.

11. **Classify** Identify whether each of the following is a renewable or nonrenewable fuel resource: wind, oil, sunlight, coal, natural gas, tides, and waves.

12. **Critical Thinking** Why do you think that most households and businesses get water from water companies?

13. **Explanatory Writing** Explain how you can tell that this is an extrusive rock and not an intrusive rock.

Big Idea

14. What are Earth's resources?

Alternatives for the Future

Create a brochure about an alternative energy source.

What to Do

1. Choose one alternative energy source. Review its advantages and disadvantages.

2. Do research to learn how this alternative energy source is used today.

3. Brainstorm ideas about how this alternative energy source may be used in the future.

Analyze Your Results

▶ Create a brochure to teach people about your alternative energy source. Use the information you have collected.

Test Prep

1. **What method of soil conservation is shown below?**

   A terracing

   B contour plowing

   C crop rotation

   D strip farming

Careers in Science

Cartographer

Do you like making maps and charts? Do you have good math and computer skills? If so, then you might become a cartographer, or mapmaker. Whenever families go on road trips or truckers do their jobs on the highways, they depend on maps or a global positioning system (GPS). Other people who depend on maps or GPS include airplane pilots, ship captains, hikers—most anyone who is going anywhere. For starters, after high school, you'll need a college degree in geography and cartography. After that, it's on to further studies for a fascinating career.

▲ **These cartographers are analyzing maps.**

▼ **Oil rig drillers at work on a wellhead.**

Oil Rig Driller

Do you like working outdoors? Do you think you also would like to operate heavy machinery? Then you might have a career as an oil rig driller. Oil rigs are used to drill for petroleum and gas found deep in Earth—under the sea as well as the land. Oil rig drillers use power-driven derricks and other equipment to drill for oil and gas. To get started in this career after high school, you would work as a roustabout, or helper, with an oil or gas field service. You would train and advance on the job. In time, you could become a professional oil rig digger—a position with a high degree of responsibility and job satisfaction.

Weather and Space

Jet streams are powerful enough to push airplanes backward!

Red Sea, Egypt

STRONG STORMS

Corte Madera Creek,
San Anselmo, California

from *Time for Kids*

LOS ANGELES, JANUARY 21, 2005
Severe weather pounded the western region of the United States last week. Part of a coastal town in California was buried in sliding soil, rocks, and mud. Thirteen homes were crushed. Rainstorms caused flooding in many areas. According to the National Weather Service, these were the wettest days in a row on record for California.

A flood washed this California home into a river.

 Write About It

Response to Literature This article describes the damage caused by severe rainstorms in California. Research the damage that severe rainstorms can cause. Write a report about the effects of severe rainstorms. Include facts and details from this article and your own research.

 **e-Journal** Write about it online at **www.macmillanmh.com**

CHAPTER 7

Weather Patterns

The
Big
Idea

How can we tell what the weather will be?

Stovepipe tornado over New Mexico

Key Vocabulary

weather
what the lower atmosphere is like at any given place and time (p. 366)

air pressure
the force put on a given area by the weight of the air above it (p. 367)

humidity
the amount of water vapor in the air (p. 369)

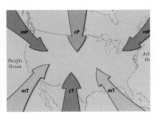

air mass
a large region of the atmosphere in which the air has similar properties (p. 384)

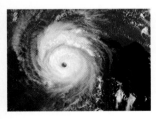

hurricane
a large rotating storm with low pressure at its center (p. 400)

climate
the average weather pattern of a region (p. 408)

More Vocabulary

insolation, p. 364

troposphere, p. 366

global wind, p. 370

barometer, p. 374

front, p. 384

weather map, p. 388

thunderstorm, p. 394

blizzard, p. 396

tornado, p. 398

storm surge, p. 401

cyclone, p. 401

current, p. 410

rain shadow, p. 411

El Niño, p. 412

The Atmosphere and Weather

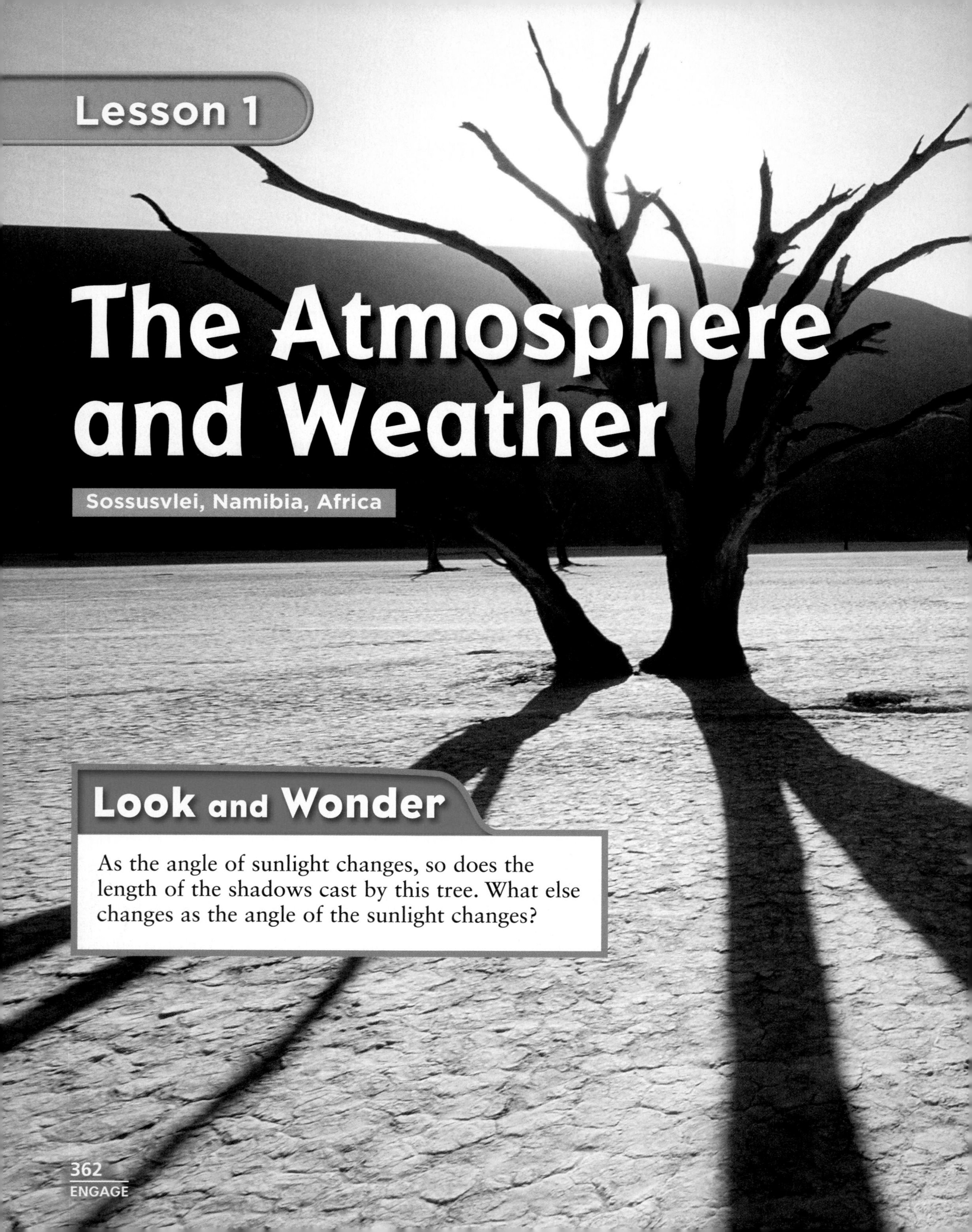

Sossusvlei, Namibia, Africa

Look and Wonder

As the angle of sunlight changes, so does the length of the shadows cast by this tree. What else changes as the angle of the sunlight changes?

How does the angle of sunlight affect temperature?

Materials

Form a Hypothesis

What happens to the temperature of Earth when sunlight reaches it at different angles? Write your answer as a hypothesis in the form "If the angle of the sunlight increases, then …"

- scissors
- 3 thermometers
- 3 sheets of black construction paper
- masking tape
- 3 pieces of cardboard
- protractor

Test Your Hypothesis

1. ⚠ **Be Careful.** Cut a slot for the thermometers in the middle of each piece of construction paper.

2. Tape each sheet of construction paper to one of the pieces of cardboard.

3. Place a thermometer into each slot so the bulb is between the construction paper and the cardboard.

4. Tape the thermometers in place. Place the thermometers in the shade until they read the same temperature. Record this temperature.

5. ⚠ **Be Careful.** Do not look directly at the Sun. Put the thermometers in the sunlight as shown.

6. **Record Data** Every two minutes, record the temperature shown on each thermometer.

Draw Conclusions

7. What are the independent and dependent variables in this experiment?

8. **Interpret Data** Graph the change in temperature over time for each thermometer setup. Which thermometer's temperature rose fastest?

Step 3

Explore More

You know that sunlight warms Earth's surface. Which is warmed faster by sunlight—soil or water? Form a hypothesis, design an experiment to test it, record your data, and communicate your results.

Step 5

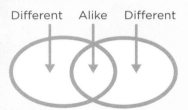
Read a Diagram

How does the angle of the sunlight affect the area of the ground that the sunlight covers?

Clue: Look at the arrows that indicate where the sunlight is hitting Earth. Then compare that to the boxes that show how much area the sunlight covers.

How does the Sun warm Earth?

On the first day of spring, the thermometer outside your window in New York City, NY, reads 41°F (5°C). Your friend lives in Miami, FL, about 2,080 kilometers (1,290 miles) to the south. She says that the temperature there is 72°F (22°C). What causes the temperature to be warmer in Miami than in New York City?

When sunlight shines on Earth, energy from the Sun warms Earth's surface. The solar energy that reaches a planet is called **insolation** (in•suh•LAY•shuhn). However, insolation does not warm all places on Earth equally.

One reason for the difference in temperature has to do with Earth's shape. Earth is shaped like a sphere, or a ball. An imaginary line called the *equator* (ee•KWAY•tuhr) runs around Earth's middle.

Angles of Sunlight on the First Day of Spring

180°

150°

120°

90°

equator

120°

150°

Sunlight strikes with the most vertical angle at or near the equator.

If you think of sunlight as a beam of light, the beam shines on Earth in a circle at the equator. Since Earth's surface is curved, the same beam will strike Earth at a wider angle above or below the equator.

The beam of sunlight always has the same amount of heat energy. However, a beam that warms Earth in an oval covers a greater area of Earth's surface than a beam that warms Earth in a circle. The heat energy of the sunlight is spread over a greater area.

Because the area is larger but the heat energy in the sunlight is the same, each part of that area receives less energy. Areas that are farther north or south of the equator receive less heat energy from sunlight than areas that are closer to the equator. Since New York City is farther away from the equator than Miami, it receives less of the Sun's heat energy.

 Quick Check

Compare and Contrast What is one reason why Miami is warmer than New York?

Critical Thinking On what part of Earth is sunlight the least concentrated?

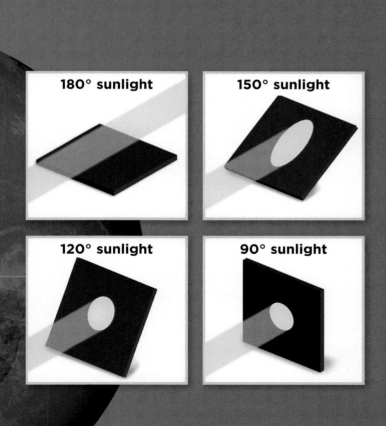

180° sunlight

150° sunlight

120° sunlight

90° sunlight

▲ New York (top) is generally cooler than Miami (bottom). Part of the reason for this is that New York receives less heat energy from sunlight than Miami.

What are the layers of the atmosphere?

When energy from the Sun hits Earth, about 50 percent of the insolation is absorbed by Earth's surface and about 5 percent of energy from insolation is reflected by Earth's surface. What happens to the rest of the energy?

The atmosphere forms five layers of gases around Earth. The layer of gases closest to Earth's surface is called the **troposphere** (TROP•uh•sfeer). The troposphere is between 8 and 18 kilometers (5 to 11 miles) thick. The depth of the troposphere is greatest at the equator and smallest at the poles.

Weather is the condition of the troposphere at a particular time and place. Almost all weather occurs in the troposphere. Weather can be hot or cold, wet or dry, calm or stormy, and sunny or cloudy. Clouds may absorb or reflect about 45 percent of the Sun's energy.

Above the troposphere are the stratosphere, mesosphere, thermosphere, and exosphere. As the height above Earth increases, the number of particles of gas in the atmosphere decreases. The exosphere begins at about 640 kilometers (400 miles) and ends at about 10,000 kilometers (6,200 miles) above Earth's surface. The particles of gas in the exosphere are very far apart.

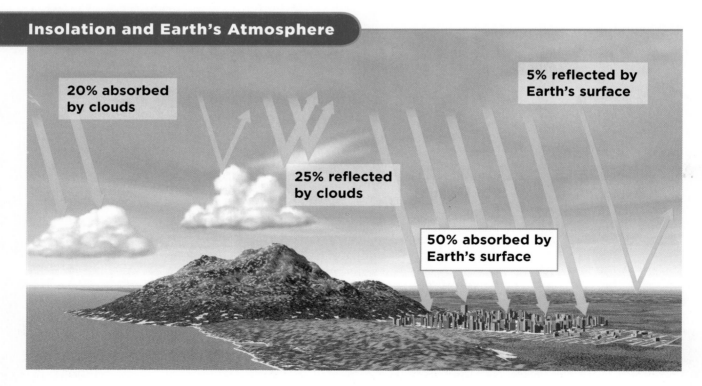

Insolation and Earth's Atmosphere

20% absorbed by clouds

5% reflected by Earth's surface

25% reflected by clouds

50% absorbed by Earth's surface

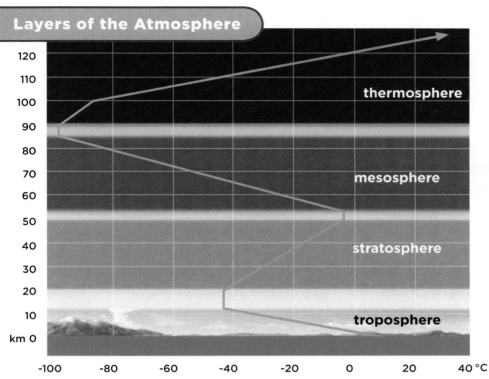

thermosphere

mesosphere

stratosphere

troposphere

Key

change in
temperature

The particles of gas press on Earth's surface and on everything they surround. The force put on a given area by the weight of the air above it is called **air pressure** or atmospheric pressure. At sea level, the average air pressure is 1.04 kilograms per square centimeter (1.04 kg/cm^2), or 14.7 pounds per square inch (14.7 lb/in.^{2}).

You can think of this as the weight of a column of air pressing on a patch of Earth's surface about the size of your thumbnail. You do not feel this weight because atmospheric pressure pushes in all directions and these pushes balance each other.

Even though air looks empty, it contains a mixture of gases, such as nitrogen and oxygen. You can tell that air takes up space because it fills up blimps, balloons, and car or bike tires.

Air takes up space in a basketball.

✔️ *Quick Check*

Compare and Contrast How much of the Sun's energy is absorbed or reflected by Earth's surface?

Critical Thinking Are there particles of gas in space?

FACT Air exerts air pressure in all directions.

What changes air pressure?

Many variables affect air pressure. These include volume, temperature, height above Earth's surface, and amount of water vapor.

Volume

Volume is a measure of how much space an object takes up. Think about a container with a bag taped over it. If you pull up on the bag, you are increasing the total volume. Because the bag is taped over the container, the amount of air cannot change. As you pull, more space is available for the same amount of air. The air pressure in the container decreases. Now, the air outside pushes in harder than the air inside pushes out. As you pull the bag up, you pull against that difference in air pressure.

Temperature

Air pressure also depends on temperature. When air is heated, the gases speed up and spread out into a larger space. There are now fewer particles of gas in the original space. The air pressure decreases and the air weighs less.

Height Above Earth's Surface

The column of air above a mountain is shorter than the column of air above sea level. The column above the mountain weighs less and pushes with a lower pressure.

Atmospheric pressure decreases with higher altitude (AL•ti•tewd). *Altitude* is the height above Earth's surface. Altitude is measured from sea level.

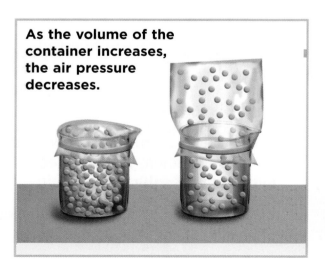

As the volume of the container increases, the air pressure decreases.

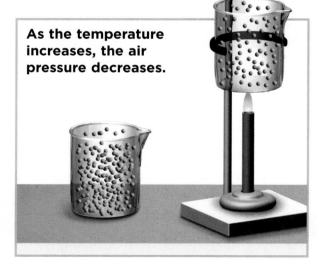

As the temperature increases, the air pressure decreases.

lower air pressure
at top of mountain

higher air pressure
at sea level

Amount of Water Vapor

Air is a mixture of gases. Water vapor weighs less than most of the other gases in air. When water vapor is added to air, the mixture of gases becomes lighter, and so exerts less pressure than dry air. The **humidity** (hyew•MID•i•tee) is the amount of water vapor in air.

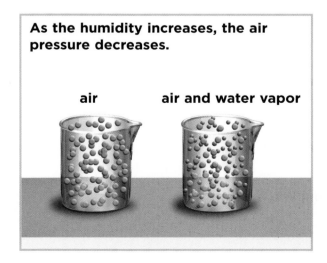

As the humidity increases, the air pressure decreases.

air   air and water vapor

 Quick Check

Compare and Contrast Would you expect higher atmospheric pressure on a dry day or a rainy day?

Critical Thinking What happens to air pressure when air is cooled?

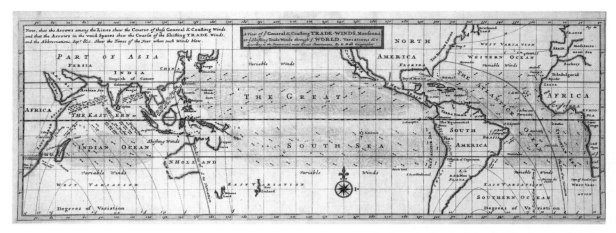

A map of trade winds from 1775.

What are global winds?

Hundreds of years ago, sailing ships carried trade goods around the world. When sailing from Europe to the Americas, the captains wanted to travel from northeast to southwest. They found that winds blow in this direction between the equator and 30°N latitude (LAT•i•tewd). *Latitude* is a measure of how far north or south a place is from the equator. Winds that blow between about 30°N latitude and about 30°S latitude became known as *trade winds*.

Trade winds are part of a system of winds called global winds (GLOH•buhl windz). A **global wind** blows steadily over long distances in a predictable direction. Global winds blow because sunlight heats areas near Earth's equator more than it heats areas near Earth's poles.

Global Winds

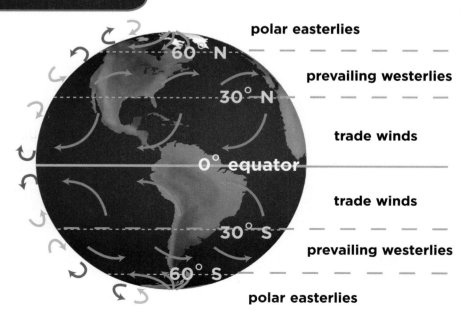

polar easterlies

60° N

prevailing westerlies

30° N

trade winds

0° equator

trade winds

30° S

prevailing westerlies

60° S

polar easterlies

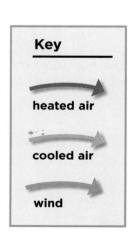

Key

heated air

cooled air

wind

Coriolis Effect

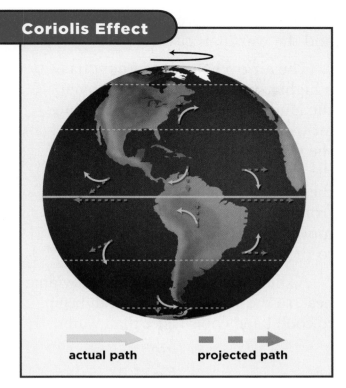

actual path   projected path

As Earth rotates, places near the poles travel a shorter distance than places near the equator. This causes winds away from the equator to curve. In the Northern Hemisphere, winds curve to the right. In the Southern Hemisphere, winds curve to the left. The *Coriolis effect* (kawr•ee•OH•lis i•FEKT) describes the change in direction of something on or above Earth's surface because of Earth's rotation.

If the Coriolis effect was the only factor in the movement of winds from the North Pole to the equator, you might predict that all of the global winds would blow in the same direction. However, between latitude 30°N and 60°N, the global winds do not blow in the predicted direction.

What causes these global winds to change direction? If you measure the number of particles in a specific volume of cold air and compare it to the number of particles in a specific volume of warm air, the cold air will have more particles. The cold air is more *dense*, or packed with particles, than the warm air. A cold, more dense volume of air will sink. A warm volume of air is less dense and will rise.

As air is heated at the equator, it rises up within the troposphere. As the air goes higher, the temperature decreases and the air begins to cool. Because heated air continues to rise, the cooled air above it is pushed away from the equator to the north toward the pole.

You might think that the cooling air would flow all the way to the North Pole. Actually, air warms and cools in three different bands as it moves north from the equator. In the middle band, the direction of the cooling air is reversed and the winds blow in the opposite direction.

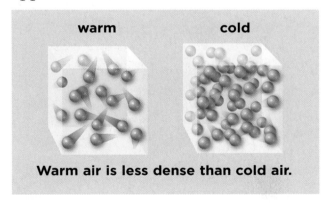

warm   cold

Warm air is less dense than cold air.

Quick Check

Compare and Contrast In which direction do winds in the Northern and Southern Hemispheres curve?

Critical Thinking What would happen to the global winds if no more heat energy reached Earth?

What are local winds?

About 50 percent of the Sun's energy reaches Earth's surface, and about three quarters of Earth's surface is covered in water and about one quarter in land. What happens when the Sun's energy heats land and water?

During the day, the Sun warms the water and the land. The temperature of the land increases faster than the temperature of the water. As the land warms up, it heats the air above it. Air over the land becomes warmer than air over the sea. The warm air over the land becomes less dense, the atmospheric pressure decreases, and the warm air rises.

The air over the ocean now has a higher pressure than the air over the land. The cooler, higher-pressure air over the ocean moves toward the warmer, lower-pressure air over the land. If you visit a beach during the day, you feel the wind blowing from the ocean onto the land. This movement of air from the water to the land is called a *sea breeze*.

As night falls, the Sun stops providing heat, so the land and water both begin to cool. Land cools faster than water,

---

Air Movement in Sea and Land Breezes

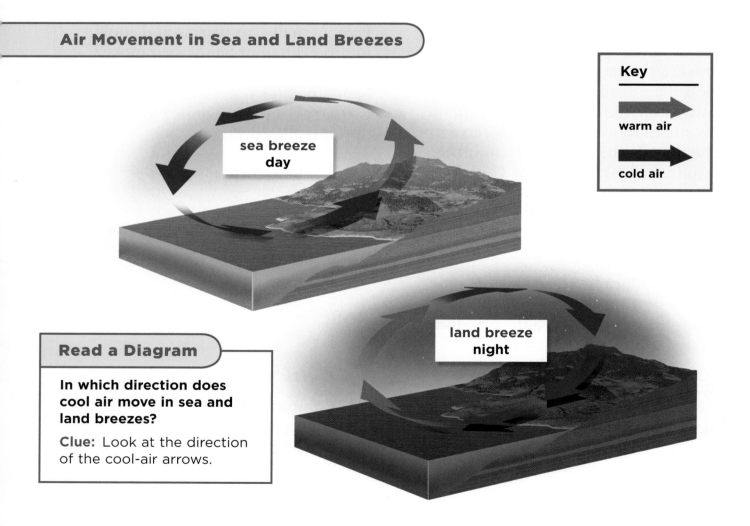

sea breeze
day

land breeze
night

Key

warm air

cold air

Read a Diagram

In which direction does cool air move in sea and land breezes?

Clue: Look at the direction of the cool-air arrows.

so the air over the water stays warmer and has a lower pressure than the air over the land. The air over the water rises, and cooler, higher-pressure air over the land moves toward the water. If you are at a beach in the evening, you feel the wind blowing from the land toward the ocean. This movement of air from the land to the water is called a *land breeze*.

In the morning, sunlight reaches mountain slopes before it reaches valleys. As the mountain slopes are warmed, the warm air begins to rise. Cool air moves up out of the valleys to replace the rising warm air, forming a *valley breeze*.

In the late afternoon or evening, the valley receives more heat from the Sun, while the mountain slopes cool from the morning's heat. As warm air rises over the valley, cool air from the mountains flows down into the valley, forming a *mountain breeze*.

 Quick Check

Compare and Contrast Compare the air pressure during a sea breeze and a land breeze.

Critical Thinking How does the Sun's energy cause local winds?

Air Movement in Mountain and Valley Breezes

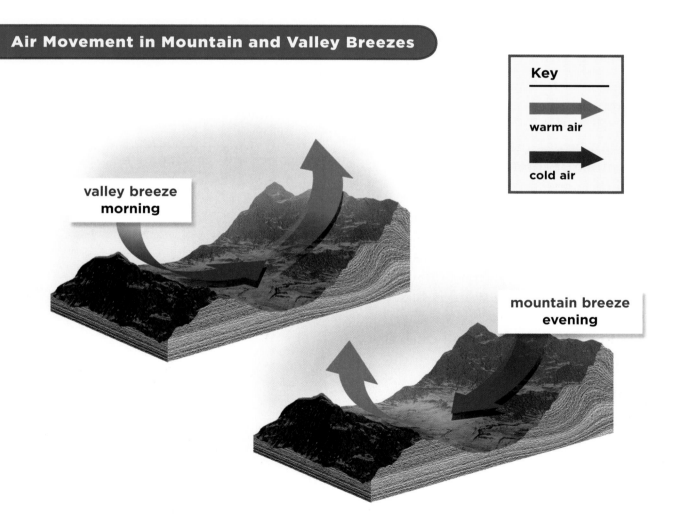

Key

warm air

cold air

valley breeze morning

mountain breeze evening

How do we measure air pressure and wind?

A **barometer** (buh•ROM•i•tuhr) measures atmospheric pressure. A *mercury barometer* measures the pressure of air on an airtight tube of mercury. The atmospheric pressure is the height to which the mercury rises.

An *aneroid barometer* measures changes in the size of an airtight container. As air pressure increases, the container is squeezed and becomes smaller. As air pressure decreases, the container expands.

A windsock is a large open-ended tube of cloth attached to a pole. When the wind blows, the sock fills and blows away from the wind.

An *anemometer* (an•uh•MOM•i•tuhr) measures wind speed using cups that rotate when the wind blows. The wind speed can be calculated by measuring the number of rotations during a period of time.

A *weather vane* is an instrument that shows which way the wind is blowing. One side is larger to catch the wind. The other side points in the direction in which the wind is blowing. For an accurate reading, a weather vane must be away from tall objects that might interfere with the wind.

✓ Quick Check

Compare and Contrast Which parts of mercury and aneroid barometers are airtight?

Critical Thinking Why does part of a barometer need to be airtight?

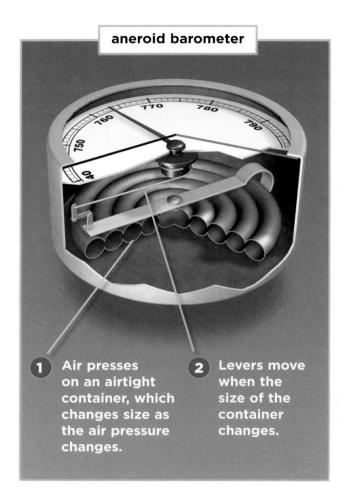

aneroid barometer

1 Air presses on an airtight container, which changes size as the air pressure changes.

2 Levers move when the size of the container changes.

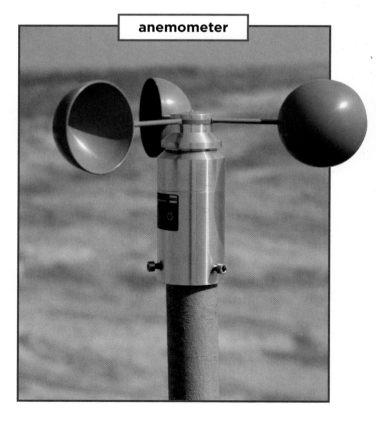

anemometer

Lesson Review

Visual Summary

The Sun heats Earth's atmosphere unequally, which changes air pressure.

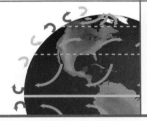
Air moves from areas of high pressure to areas of low pressure.

Changes in air pressure cause winds.

Make a FOLDABLES™ Study Guide

Make a Three-Tab Book. Use the titles shown. On the inside of each tab, compare and contrast information about the topic.

The Sun heats Earth's atmosphere

Air moves From...

Changes in air pressure Cause...

Think, Talk, and Write

❶ **Main Idea** What causes winds?

❷ **Vocabulary** The force put on a given area by the weight of the air above it is called _____.

❸ **Compare and Contrast** How does the heat energy over an area at the equator compare to an area at the poles?

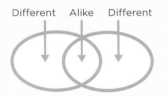
Different   Alike   Different

❹ **Critical Thinking** How does solar energy cause global and local winds?

❺ **Test Prep** Global winds
  A move air in a loop.
  B have atmospheric pressure.
  C heat Earth unequally.
  D blow over long distances.

❻ **Test Prep** What does a barometer measure?
  A air pressure
  B wind speed
  C wind direction
  D humidity

 Writing Link

Fictional Narrative

If you were a sailing captain traveling from Europe to the Americas, what information about the winds would you record? How could you tell wind speed or direction while on your ship?

 Art Link

Weather Vanes

Design a weather vane. Include a way for the weather vane to rotate. Decorate your weather vane and then test how well it shows wind direction.

Focus on Skills

Inquiry Skill: Communicate

When scientists complete an experiment, they **communicate** their results. When you **communicate**, you share information with others. You may do this by speaking, writing, drawing, singing, or dancing.

▶ Learn It

In the following activity, you will test whether air can lift a notebook off the table. Keep notes that include your hypothesis, materials, observations, and conclusion.

Scientists often try new experiments based on work that other scientists have done. If you **communicate** the details of your experiment, other students can do experiments based on yours. Writing down exactly what you did also lets you plan more experiments with different materials and different variables. If you get an unexpected result or disprove your hypothesis, you should **communicate** that as well.

▶ Try It

Materials notebook, balloons, tape, ruler

1. You know that air has weight and takes up space. Do you think air in a balloon will be able to lift a notebook off a table? If it can, how high will the notebook rise?

2. Tape two balloons to a notebook so the ends of the balloons stick out. Flip the notebook over so it is on top of the balloons.

How strong is air?

Hmmm?

WOW!

3 Blow into one of the balloons. What happens to the notebook? Fill both balloons with as much air as you can.

4 Using a ruler, measure the height between the table and notebook.

5 **Communicate** Exchange data about the height to which air was able to raise your notebook.

6 Using the data from your classmates, figure out the average height that your class was able to lift the notebooks. Make a chart to compare your results.

7 **Communicate** Who was able to raise their notebooks the highest? Was anyone unable to lift it? Discuss any problems that occurred or improvements that could be made to lift the notebook higher.

▶ **Apply It**

Think about how you could use air to lift the book even higher. What would happen if you used bigger balloons? If you placed smaller balloons under each corner of the notebook? How heavy of a book could you lift using these materials?

Plan a new experiment. Test your hypothesis and draw conclusions about using the power of air to lift objects. Then communicate to the class the results of your experiment by writing a report, drawing a cartoon strip, or composing and singing a song!

Clouds and Precipitation

Look and Wonder

When storm clouds roll over an area such as these fields in Montana, rain often follows. How much rain falls during a storm?

How much rain falls in your community?

Purpose

To measure the amount of rainfall in your community.

Procedure

1. ⚠ **Be Careful.** Use the scissors to cut the top off the carton.

2. Using tape, attach the carton to the baking pan. Then put the pan on the ground outside in an open area.

3. **Measure** Check the carton at the same time every day. If there is water in it, measure the height of the water in centimeters.

4. **Record Data** Write down the daily results on a table. Then empty the carton and put it back in the same spot outside.

Draw Conclusions

5. **Interpret Data** Design and complete a graph to display your results.

6. **Use Numbers** Convert your total measurement of cubic centimeters of precipitation into liters.

- scissors
- carton
- masking tape
- baking pan
- ruler

Step 2

Explore More

How close were your results to an official rain measurement for your area? Were there any problems that you ran into with the experiment? How could you improve your data collection?

Step 4

Main Idea

Water vapor in the air can form clouds, fog, rain, hail, sleet, or snow. Air masses and fronts change weather as they move.

Vocabulary

air mass, p. 384

front, p. 384

weather map, p. 388

LOG ON e-Glossary
at www.macmillanmh.com

Reading Skill ✓

Draw Conclusions

Text Clues	Conclusions

Technology

Explore air currents and winds with Team Earth.

How do clouds form?

Sometimes you look up at the sky and see nothing but blue. First one fluffy, white cloud appears, then another, and another. The clouds grow larger and darker. Now the sky is covered by a gray blanket of clouds. How did these clouds form in what looked like an empty sky?

Water vapor is one of the gases in the atmosphere. As water vapor particles are carried higher, they begin to lose heat energy and become colder. As they become colder, they lose speed and come closer together. When water vapor particles have slowed enough, they condense around tiny particles of dust.

You may have seen condensed drops of water on shower doors, on the sides of cold glasses of liquid, and on grass early in the morning. These drops of water condense from water vapor in the air in the same way that cloud droplets do.

The appearance of a cloud depends on how high it forms in the atmosphere and what the temperature there is. *Cirrus clouds* (SIR•uhs) form at the highest altitudes. They usually form from ice crystals when liquid water is cooled below its freezing point of 0°C (32°F).

Lower clouds form from water droplets. These clouds may appear gray or black. This occurs when the water droplets are so dense that sunlight is unable to pass through.

Two of the kinds of clouds that form from water droplets are cumulus clouds and stratus clouds. *Cumulus clouds* (KYEW•myuh•luhs) are puffy clouds that form at middle altitudes. *Stratus clouds* (STRAY•tuhs) are layered and form at low altitudes.

When the temperature near the ground is cold, water vapor forms fog. *Fog* (fawg) is a cloud that forms near the ground.

cirrus clouds

Cirrus clouds are wispy with fuzzy edges.

cumulus clouds

Cumulus clouds are individual puffy clouds.

stratus clouds

Stratus clouds form in blanketlike layers.

≡ **Quick Lab**

Types of Clouds

1 **Observe** Look for clouds in the sky. How many different types of clouds do you see?

2 **Classify** Do the clouds that you see look like cirrus, cumulus, or stratus clouds?

3 Continue your data collection for one week.

Cloud Observation

	Cirrus	Cumulus	Stratus
Day 1			
Day 2			
Day 3			

4 Which type of cloud did you see most frequently?

5 Write a report about the types of clouds that you saw. Do you think you would get different results at a different time of year? Explain.

✔ *Quick Check*

Draw Conclusions If you observed wispy clouds high in the sky, what would you conclude these clouds are?

Critical Thinking What can you learn about wind direction from watching clouds?

FACT Visible clouds are made of either tiny droplets of water or ice.

How does precipitation form?

While clouds are made of liquid water, the water is in the form of very small drops. As time passes, some of these drops will collide with each other and form larger drops.

As the drops collide, the cloud becomes thicker and grayer. Soon the drops become too heavy for winds to keep them in the air. They fall to the ground as precipitation.

When the air temperature is warm, water vapor forms liquid precipitation, or raindrops. When the air temperature is below the freezing point of water, water vapor forms solid precipitation. The types of solid precipitation are sleet, hail, and snow.

When raindrops fall through a layer of very cold air near the ground, they freeze, changing into tiny bits of ice called *sleet*. Sleet only reaches the ground when air temperatures are cold near the ground.

Hail forms in severe thunderstorm clouds when raindrops in the cloud collide with bits of ice. The drops freeze, forming a hailstone. Winds

How Precipitation Forms

rain

sleet

hail

push the hailstones back up into the cloud and the hailstones grow larger. The largest hailstones are about the size of a softball.

Snow forms when air temperature is close to or below freezing. Then the temperature is so cold that water vapor turns directly into a solid crystal. If you look carefully at snowflakes landing on a window, you may see the shape of the snowflake before it melts.

Precipitation is measured in units of depth. The measurements tell how deep the rain would be in a container or how deep the snow is on the ground.

To measure the depth of rain, the rain is caught in a rain gauge (rayn gayj). A *rain gauge* is a container that has a scale marked on it in millimeters or inches. To measure the depth of rainfall, someone must read the level of the water in the rain gauge.

The depth of snowfall can be measured using a meterstick or yardstick by pushing the stick through the snow to the ground and reading the mark at the top of the ruler. Drifting snow may interfere with accurate measurements.

✔ Quick Check

Draw Conclusions
If the air temperature was above freezing, which type of precipitation would fall?

Critical Thinking
Which type of precipitation occurs when rain falls and the air temperature near the ground is cold?

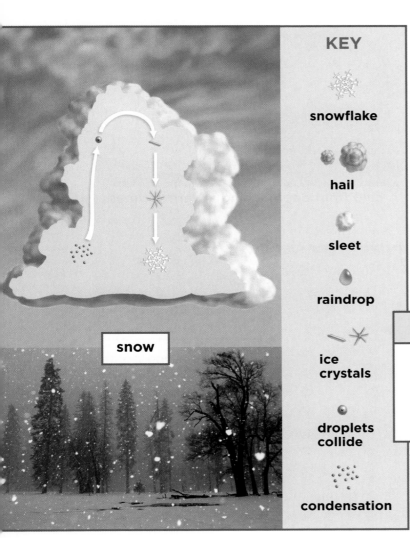

KEY

snowflake

hail

sleet

raindrop

ice crystals

droplets collide

condensation

snow

Read a Diagram

What is the first step in each type of precipitation?

Clue: Look at what happens first in the clouds.

What are air masses and fronts?

The weather in an area is affected by the air mass that is passing over the area. An **air mass** is a large region of air that has a similar temperature and humidity. Air masses can cover thousands of square kilometers of land and water.

Depending on where they form, air masses can be cold or warm and dry or humid. An air mass that forms above a warm area of water will be warm and humid. An air mass that forms over a cold area of land will be cold and dry. The area from which an air mass gets its characteristics is called its source region.

When one air mass meets a different air mass, this meeting place between air masses is called a **front**. A front marks the front edge of the oncoming air mass.

A warm front approaching a cold air mass.

A cold front approaching a warm air mass.

Air Mass Source Regions

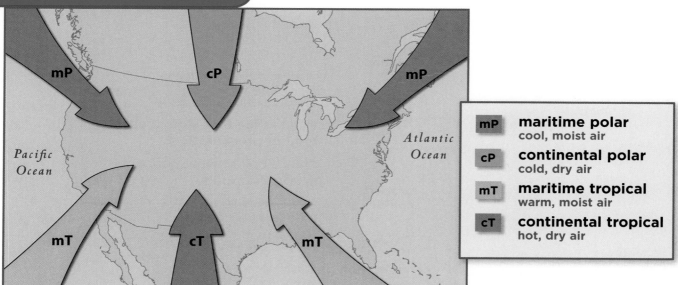

Pacific Ocean

Atlantic Ocean

mP	**maritime polar** cool, moist air
cP	**continental polar** cold, dry air
mT	**maritime tropical** warm, moist air
cT	**continental tropical** hot, dry air

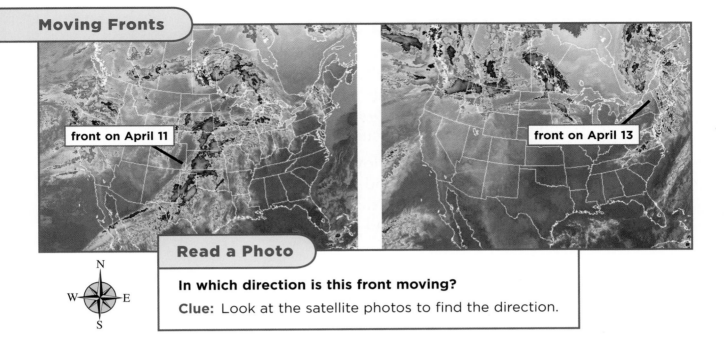

front on April 11

front on April 13

N
W E
S

Read a Photo

In which direction is this front moving?

Clue: Look at the satellite photos to find the direction.

When a cold and dry air mass runs into a warm and moist air mass, the dense, heavy cold air pushes the lighter warm air up. As the warm air rises, the moisture in it condenses. Towering clouds form and storms may follow.

Sometimes the temperature and humidity of the air masses behind a front are similar. When this happens, the air masses do not move, and a stationary front forms. A stationary front is a boundary between two different air masses where neither one is strong enough to replace the other.

What moves air masses? You learned about global winds such as trade winds. Another type of global wind is the jet stream. The jet stream is a high-altitude wind caused by large temperature differences between air masses. The jet stream winds can exceed speeds of 240 kilometers per hour (150 miles per hour).

Over North America, the jet stream blows from west to east. Since the jet stream winds push air masses, almost all weather fronts in North America move from west to east as well.

Warm and cold fronts help you predict what the weather is going to be in the future. Today's weather is likely to be similar to whatever yesterday's weather was further west. As the jet stream winds blow one front to the east, the fronts behind it are pushed east as well.

 Quick Check

Draw Conclusions What happens when a cold air mass moves into an area where the air mass is warm?

Critical Thinking What would you need to know to estimate how long it would take for a front to move across the United States?

What are highs and lows?

Knowing the location of low-pressure systems and high-pressure systems can tell you about local weather conditions. A *low-pressure system* is a large mass of air with low air pressure in the center. A *high-pressure system* is a large mass of air with the highest air pressure in the center.

Since warm, humid air has low pressure, low-pressure systems usually bring warm and stormy weather. Moisture that was held in a low-pressure air mass condenses as it rises and cools, bringing clouds, rain, and other types of precipitation.

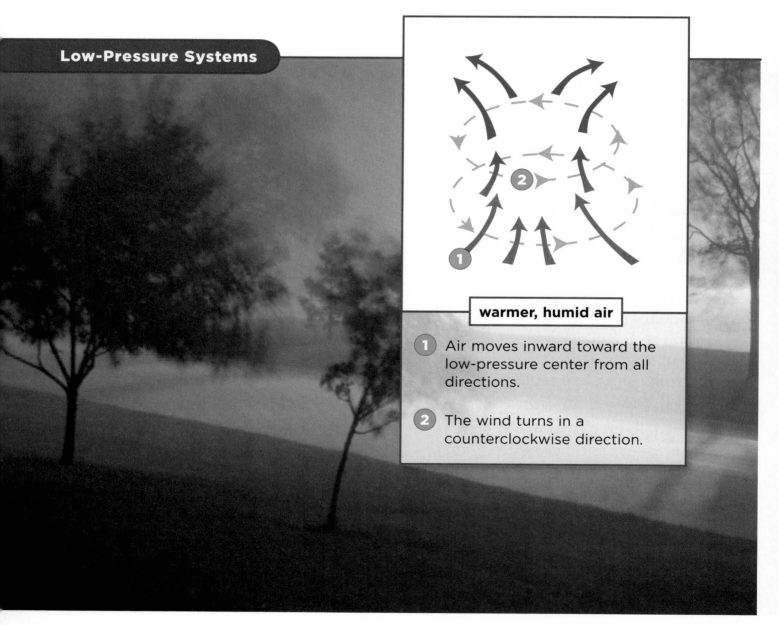

Low-Pressure Systems

warmer, humid air

1 Air moves inward toward the low-pressure center from all directions.

2 The wind turns in a counterclockwise direction.

Since cool, dry air has high pressure, high-pressure systems usually bring dry, clear weather. Any moisture carried in a high-pressure system tends to evaporate, clearing the sky of clouds and precipitation.

To figure out where high- and low-pressure systems are at a particular time, scientists plot the air pressure of different areas on a map. They then connect all the places that have the same air pressure with a line called an *isobar* (EYE•suh•bahr).

 Quick Check

Draw Conclusions What causes winds in low-pressure and high-pressure systems to rotate in different directions?

Critical Thinking How does knowing the location of low- and high-pressure systems help you forecast the weather?

High-Pressure Systems

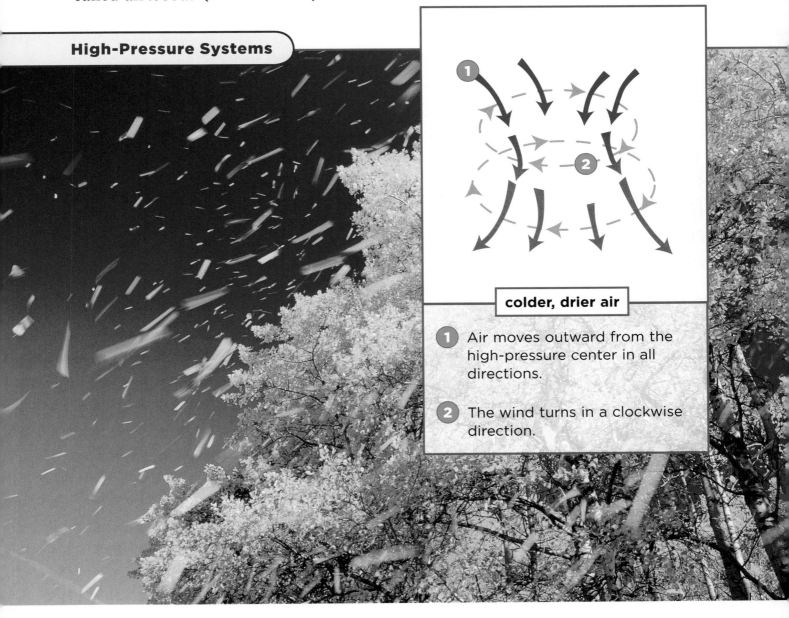

colder, drier air

1 Air moves outward from the high-pressure center in all directions.

2 The wind turns in a clockwise direction.

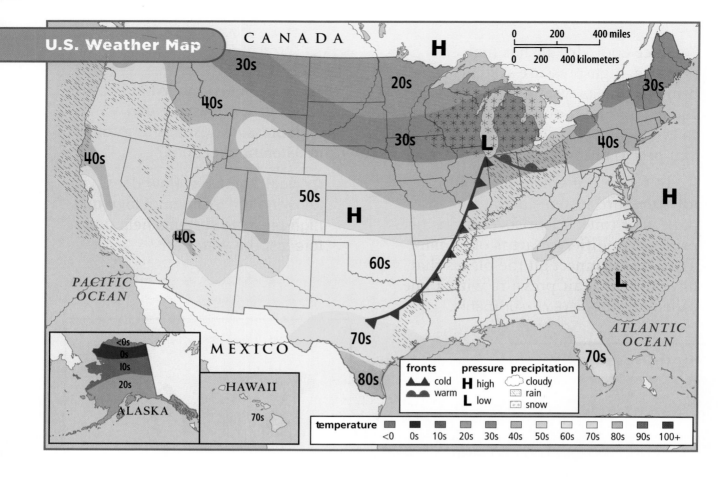

fronts
▲▲ cold
◓◓ warm

pressure
H high
L low

precipitation
◌ cloudy
▨ rain
▨ snow

temperature
<0 0s 10s 20s 30s 40s 50s 60s 70s 80s 90s 100+

What do weather maps tell you?

How could you find out where rain is falling? A **weather map** shows the weather in a specific area at a specific time. Weather maps can show one variable, such as air pressure, or many different variables.

On a weather map, a symbol is used to represent each variable. For example, cold fronts are shown by a blue line with blue triangles. The triangles point in the direction that the cold front is moving.

Meteorologists are scientists who study Earth's atmosphere and weather. Meteorologists track variables that might affect the weather so they can make forecasts, or predictions,

about what weather will occur. Meteorologists may forecast what the weather will be for the next day, for the next five days, or even for the next few months.

Meteorologists constantly measure variables because sudden changes in one variable can change the weather. This is why it may rain when a clear sky was predicted.

Quick Check

Draw Conclusions If a large high-pressure system was moving toward your area, what would you predict for the next day's weather?

Critical Thinking Which weather variables do you want to know about before you go outside? Why?

Lesson Review

Visual Summary

Clouds and precipitation form from water vapor in the air.

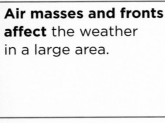

Air masses and fronts affect the weather in a large area.

Knowing where high- or low-pressure systems are can help you predict the weather.

Make a FOLDABLES™ Study Guide

Make a Trifold Book. Use the titles shown. Then draw a conclusion about each topic.

Think, Talk, and Write

1 **Main Idea** What can water vapor form?

2 **Vocabulary** Clouds that form close to the ground are called _____.

3 **Draw Conclusions** What type of cloud is a blanketlike layer of low-altitude clouds?

Text Clues	Conclusions

4 **Critical Thinking** If you built your own weather station, what instruments would you include?

5 **Test Prep** Which of the following is a variable that meteorologists track to make weather forecasts?
   - **A** traffic
   - **B** air pressure
   - **C** barometer
   - **D** population

6 **Test Prep** Which of the following is NOT a type of solid precipitation?
   - **A** rain
   - **B** snow
   - **C** hail
   - **D** sleet

Math Link

Predict Rainfall

During a 5 A.M. weather report, rain is reported to be falling at a rate of 2 inches per hour. If the next report is at 7 A.M., how much rain will have fallen?

Art Link

Local Weather Map

Draw a map of your local area with your weather prediction for tomorrow. Come up with symbols to include in a key.

Be a Scientist

Materials

2 bottles

scissors

tape

cobalt chloride paper

2 plastic cups

sheet of paper

How can you tell that water vapor is in the air?

Form a Hypothesis

Cobalt chloride paper is blue in dry air and turns pink in air that has water vapor in it. Write a hypothesis in the form "If water evaporates, then cobalt chloride paper near or above the water will . . ."

Test Your Hypothesis

1 ⚠ **Be Careful.** Cut the tops off the 2 bottles.

2 Tape 1 strip of cobalt chloride paper in the bottom of each bottle.

3 Place 1 bottle upside down over an empty plastic cup. Fill the other plastic cup half-full of water. Place a bottle upside down over that cup.

4 Tape a third strip of cobalt chloride paper to a sheet of paper. Leave it in open air.

5 **Observe** Examine the color of the strips of cobalt chloride paper.

6 **Record Data** Write down any changes in the color of the cobalt chloride paper.

Step 1

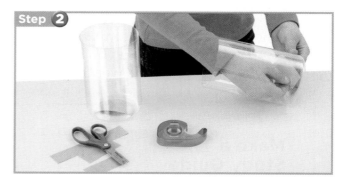
Step 2

Step 3

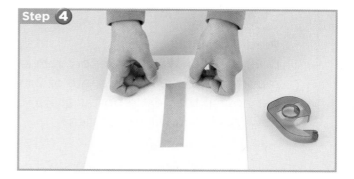
Step 4

Draw Conclusions

1 **Use Variables** Identify the variables in this experiment. What purpose does the cobalt chloride paper that is taped to the piece of paper serve?

2 **Draw Conclusions** Does the evidence from your observations support your hypothesis?

Does surface area affect rate of evaporation?

Form a Hypothesis

You have already learned water vapor can be detected in the air. Does water evaporate faster from a body of water with a bigger surface area? Write your answer as a hypothesis in the form "If you increase the surface area of water, then the evaporation rate will . . ."

Test Your Hypothesis

Design a plan to test your hypothesis. Write out the materials, resources, and steps that you need to take. Record your results and observations as you follow your plan.

Draw Conclusions

Did your test support your hypothesis? Why or why not? Present your results to your classmates.

What effect does wind have on the evaporation rate of water? Come up with a question to investigate. Design an experiment to answer your question. Your experiment must be organized to test only one variable or one item being changed. Your experiment must be written so that another can complete it by following your instructions.

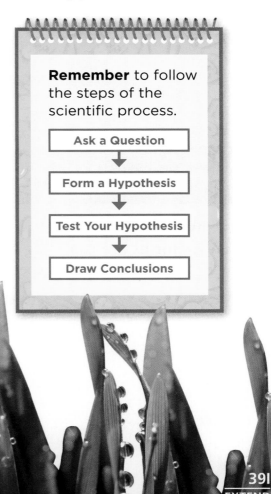

Remember to follow the steps of the scientific process.

Ask a Question

↓

Form a Hypothesis

↓

Test Your Hypothesis

↓

Draw Conclusions

Severe Storms

Tucson, AZ

Look and Wonder

On any given day, more than 40,000 thunderstorms are rumbling somewhere on Earth. What causes these spectacular storms?

What happens when air masses of different temperatures meet?

Form a Hypothesis

What happens to an air mass when it meets another air mass of the same temperature or of a cooler temperature? Write your answer as a hypothesis in the form "If an air mass meets another air mass of the same or of a cooler temperature, then . . ." Like air, water flows and carries heat. Using water as a model for air can help you test your hypothesis.

Test Your Hypothesis

1. ⚠ **Be Careful.** Using scissors, cut the cardboard so it fits tightly in the clear box. Wrap the cardboard in aluminum foil.

2. Pour 4 cups of cold water into one container and 4 cups of warm water into the other one. Place a few drops of blue food coloring into the cold water and red into the warm water.

3. Hold the cardboard tightly against the bottom of the box. Pour the cold water on one side and the warm water on the other.

4. **Observe** Watch the box from the side as you remove the cardboard.

5. Now repeat the same test with warm water in both containers and food coloring in only one.

Draw Conclusions

6. What are the variables in this experiment?

7. **Infer** Which test looked like it formed storms? Why?

Explore More

Will a greater difference in temperature between the warm and cold water increase the observable effects? Form a hypothesis and test it.

Materials

- scissors
- cardboard
- clear plastic box
- aluminum foil
- cold water
- 2 containers
- warm water
- food coloring

Step 3

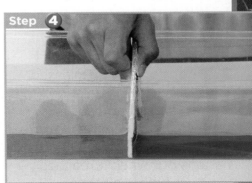

Step 4

▶ Main Idea

Storms are caused by the collision of air masses.

▶ Vocabulary

thunderstorm, p. 394
blizzard, p. 396
tornado, p. 398
hurricane, p. 400
storm surge, p. 401
cyclone, p. 401

LOG ON ℮-Glossary
at **www.macmillanmh.com**

▶ Reading Skill ✓

Cause and Effect

Cause	→	Effect
	→	
	→	
	→	
	→	

▶ Technology **SCIENCE QUEST**

Explore weather with Team Earth.

(1) **Fronts** A cold front moves in and pushes warm, humid air upward. As the air rises, it cools and water vapor condenses.

(2) **Thunderheads** Released energy from condensation warms the air and causes updrafts. A thunderhead forms. The top of the thunderhead flattens out when it reaches winds at higher altitudes.

(3) **Precipitation** Rain falls.

What are thunderstorms?

Lightning flashed through the sky and thunder rumbled over the city. Heavy rain fell during the storm, flooding streets and sewer systems. Thunderstorms similar to this one happen all over the world. A **thunderstorm** is a rainstorm that includes lightning and thunder.

In order for a thunderstorm to occur, warm air must rise, carrying moisture with it. Any upward movement of air is called an *updraft*. Updrafts in a thunderstorm cause a cloud to increase in height, forming a tall cloud called a *thunderhead*. As rain falls, a downdraft may occur. A *downdraft* is a sudden downward movement of cool or cold air.

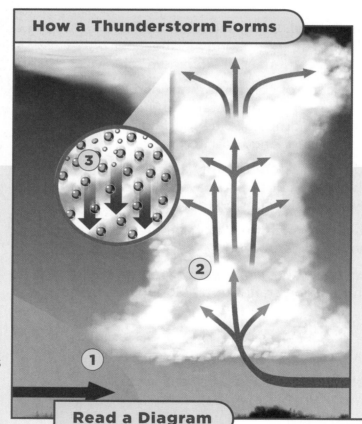

How a Thunderstorm Forms

Read a Diagram

What happens to the temperature of air in a thunderhead?

Clue: Red represents hot and blue represents cold.

LOG ON *Science in Motion* Watch how thunderstorms form at **www.macmillanmh.com**

Lightning and Thunder

Lightning is the spark caused when electricity that has built up in a thunderhead discharges. Lightning can jump between parts of the same cloud, between different clouds, or between a cloud and the ground.

One theory about the cause of lightning is that air in motion rubs particles of ice and rain moving downward against particles moving upward. As the particles rub, they become charged with static electricity.

This is similar to what happens when you shuffle your feet across a carpet and a charge of static electricity builds up in your body. When you then touch your finger to a metal object, you feel a tingle as a spark jumps between you and the object. That spark is a discharge of electricity.

Lightning raises the temperature of the air around it to more than five times the temperature of the surface of the Sun. This burst of heat makes the air expand violently. *Thunder* is the sound of the rapidly expanding air.

The dangers in thunderstorms are lightning, windblown objects, and flash floods. Stay indoors and away from windows. If you are outside, stay away from tall objects.

 Quick Check

Cause and Effect What happens as a thunderstorm forms?

Critical Thinking How is thunder similar to the "pop" produced by a pricked balloon?

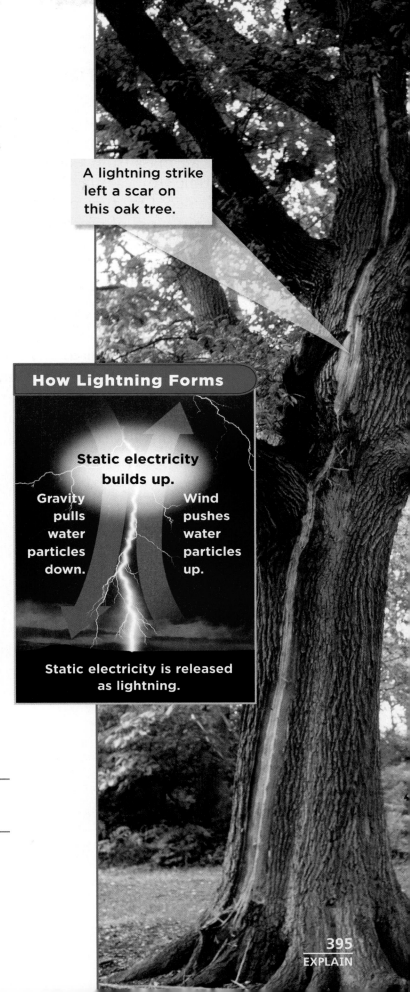

A lightning strike left a scar on this oak tree.

How Lightning Forms

Static electricity builds up.

Gravity pulls water particles down.

Wind pushes water particles up.

Static electricity is released as lightning.

What are winter storms?

Winter storms occur when two air masses of different temperatures and moisture levels meet. In the United States, winter storms often form when a continental polar air mass from Canada moves south and meets a maritime tropical air mass moving north from the Gulf of Mexico. One type of winter storm is a storm in which the main types of precipitation, such as snow or sleet, occur with cold air temperatures. Another type of winter storm is a rainstorm where ground temperatures are cold enough that ice forms on outside surfaces.

Blizzards

People often refer to any severe snowstorm as a blizzard (BLIZ•uhrd). In the United States, the official definition of a **blizzard** is a snowstorm with 35-mile-per-hour winds and enough snowfall that you can only see up to one quarter of a mile.

Winter Storms

North Dakota

New York

Read a Photo

Which photo shows an ice storm?

Clue: Look for ice in the photo.

Blizzards often drop several feet of snow on an area over a few days. Wind can move the snow into drifts that can be significantly deeper than the total amount of snow that has fallen.

When the snow that falls is light and flaky, strong winds may continue to blow it around after the snowfall has stopped. When weather conditions for a blizzard are met after snow has stopped falling, meteorologists refer to the storm as a ground blizzard.

Sometimes heavy snowfall during a blizzard combines with strong updrafts and downdrafts. Snow blows rapidly around and makes it difficult to distinguish the ground from the air. People can quickly lose their sense of direction. These conditions are called a whiteout.

Ice Storms

When a warm front approaches a cold air mass, the warm air mass usually pushes the cold air mass away. Sometimes, in areas with a lower altitude such as valleys, a thin layer of cold air remains behind. If rain falls as the warm air mass cools, the rain will freeze as it hits the cold air near the ground.

If the ground is also cold, the sleet or freezing rain will coat everything it lands on with a thin layer of ice. An *ice storm* is a storm in which freezing rain forms a layer of ice on outside surfaces.

The weight of ice and snow on power lines and tree branches can cause them to break. Ice and snow can also make streets slippery and dangerous for walking or driving.

Another danger from winter storms is exposure to cold. To stay safe, stay indoors. Keep warm clothes and blankets available in case you lose power.

 Quick Check

Cause and Effect Under what conditions does a ground blizzard occur?

Critical Thinking If you were stuck in a car during a blizzard, how would tying a colored object to the antenna help?

Traveling during winter storms can be difficult.

What are tornados?

Under the right conditions, a thunderstorm can turn into a tornado (tawr•NAY•doh). A **tornado** is a rotating, funnel-shaped cloud with wind speeds up to 500 kilometers (300 miles) per hour. People refer to these storms as "twisters."

Tornados begin to form when warm air moves upward in a thunderhead, creating an area of low pressure. The low air pressure draws air inward and upward. As air flows into the low-pressure area, it rotates faster and faster.

From the ground, the shape of the cloud looks like a funnel. Warm air rises up the center of the spinning funnel cloud. Rain falls outside of the cloud. When the tip of the funnel cloud touches the ground, it becomes a tornado.

Because only a relatively small section of the tornado actually touches the ground, tornados have been known to destroy houses on one side of a street while leaving houses on the other side untouched.

The dangers during a tornado are flying objects and powerful winds. If you hear a tornado warning, seek shelter in a tornado shelter or in a basement, bathroom, or closet on the lowest floor. If you are in a car or a mobile home, get out and seek shelter.

 Quick Check

Cause and Effect What causes tornado winds to rotate?

Critical Thinking Why might differences in air pressure cause a closed building to explode outward as a tornado passes over it?

How a Tornado Forms

1 Warm air moves upward in a thunderhead.

2 A funnel cloud forms when the air starts rotating.

Tornado in a Bottle

1. Fill a 2-liter plastic bottle one-third full of water.

2. Place an empty 2-liter plastic bottle upside down over the mouth of the first bottle. Use masking tape or a connector to join the two bottles together.

3. **Make a Model** Holding the bottles by the necks, flip them upside down so the bottle with the water in it is now on top. Place the bottles on a desk.

4. **Observe** What do you see?

5. How is this model similar to the movement of wind in a tornado?

3 **The funnel cloud becomes a tornado when it touches the ground.**

What are hurricanes?

Thunderstorms can turn into a tropical storm. A *tropical storm* has rotating winds with low pressure at its center. It forms near the equator where the ocean is warm. Water vapor evaporates from the warm water. As the warm, moist air rises, cooler air flows toward the space where the warm air had been.

Water continues to evaporate, which lowers the air pressure even more. Surrounding high-pressure air moves into the area of low air pressure and causes rotating winds.

A tropical storm turns into a **hurricane** (HUR•i•kayn) when the wind speed of the storm reaches more than 119 kilometers per hour (74 miles per hour). From space, hurricanes look like a spiral of clouds with a hole in the middle. The hole is the center of the low-pressure area, which is called the "eye" of the hurricane. Clouds form a border around the eye and spread out beyond it.

A hurricane's fastest winds and heaviest rains occur next to the eye. Winds near an eye can reach speeds of almost 300 kilometers per hour (190 miles per hour). The area inside the low-pressure area of the eye is calm, with no precipitation or wind.

Hurricanes can pull nearby thunderstorms into the growing storm. Large hurricanes may be up to 2,000 kilometers (1,200 miles) in diameter and may cover many states.

tracking wind speed

tracking location

Storm surges can damage shorelines as well as buildings near the water.

Storm Surges

Hurricane winds whip up large waves in the ocean. These waves cause a bulge of water in the ocean known as a **storm surge** (stawrm surj). As a storm moves over a coast, the storm surge can cause water levels to suddenly rise, or surge, several meters.

The main dangers in a hurricane are flying objects, strong winds, and flooding. If possible, leave areas threatened by a hurricane. Take prescription medicines with you.

If you stay in a building, board up windows and stay away from windows and doors. Store food, bottled water, flashlights, and battery operated radios. Stay away from beaches and areas that may become flooded.

Cyclones

Any storm with a low-pressure center that causes a circular pattern of winds to form is called a **cyclone** (SYE•klohn). Tropical storms, hurricanes, and tornados have low-pressure centers and spinning winds and are all cyclones.

 Quick Check

Cause and Effect At what point does a tropical storm turn into a hurricane?

Critical Thinking Is a thunderstorm a cyclone?

Doppler radar

weather balloons

airplane

From inside the eye of a hurricane, this airplane gathers weather data.

How are storms tracked?

Meteorologists use different kinds of instruments to collect data about the variables that can affect a storm. Weather stations around the world use instruments such as weather vanes, barometers, and rain gauges to gather data about local weather conditions.

Many local weather stations have Doppler radar. Doppler radar tracks the speed and direction of winds and rain by measuring changes in the motion of an object as the object moves toward or away from the radar detector.

Scientists use weather balloons to gather data about weather conditions in the upper atmosphere. The instruments in the balloons use radio transmitters to send air pressure, temperature, and humidity data back to the ground station. By monitoring the balloon's movement, meteorologists gather data about the winds at its altitude.

Weather satellites take pictures of the atmosphere from space. One kind of camera takes pictures of the heat of the ground and the ocean. Another type of camera takes pictures of cloud cover and can track the size and location of a hurricane.

Satellite photos cannot tell the wind speeds in a hurricane. Meteorologists fly into hurricanes and drop instruments to collect this data. By gathering accurate information about the wind speed and pressure, meteorologists are better able to predict the path of a hurricane.

 Quick Check

Cause and Effect What instruments could measure the wind speed of a storm?

Critical Thinking What would tracking the ocean temperature during the year tell you?

Lesson Review

Visual Summary

Thunderstorms and winter storms occur when two air masses of different temperatures and moisture levels meet.

Cyclones, such as hurricanes and tornados, are storms with a low-pressure center and circular winds.

Meteorologists use different kinds of instruments to gather data about weather variables.

Make a FOLDABLES™ Study Guide

Make a Trifold Book. Use the titles shown. Discuss the topic on the inside of each fold.

Main Ideas | What I learned... | Sketches and Examples

Thunderstorms and winter storms occur when...

Cyclones, such as hurricanes and tornados, are...

Meteorologists use...

Think, Talk, and Write

❶ **Main Idea** What causes storms?

❷ **Vocabulary** Tornados and hurricanes are examples of a(n) _____.

❸ **Cause and Effect** What causes a hurricane to form?

Cause	→	Effect
	→	
	→	
	→	
	→	

❹ **Critical Thinking** What is the reason why most thunderstorms do not become cyclones?

❺ **Test Prep** What is a storm surge?
  A a circular pattern of winds
  B a bulge of water in the ocean
  C a winter storm with freezing rain
  D a large region of cold air

❻ **Test Prep** Which of the following is a storm with a low-pressure center?
  A thunderstorm
  B ice storm
  C tornado
  D blizzard

 Writing Link

Fictional Narrative
Write about what it would be like to work as a meteorologist. Discuss the daily tasks that you would do.

 Social Studies Link

Tornado Frequency
Research and write a report about Tornado Alley. Include information about how people that live in Tornado Alley prepare for tornados.

Living Through a Mudslide

April 15

Last night it rained very hard. As I was getting ready to leave for school this morning, I heard loud booming and cracking sounds. My whole family ran out of the house in our pajamas. Dad said, "It must be a mudslide. We have to get to higher ground."

First, Dad got us all in the van and drove to the top of a nearby hill. Then we turned on the car radio. We heard that about a half mile below us the mudslide was carrying cars and trees down the road. A few people were taken to the hospital with minor injuries, but fortunately no one was badly hurt.

We could not go back home for hours. Finally, we were told that it was safe to return. We were lucky that our house was only damaged a little.

Water and dirt from a mudslide blocked the road.

Rocks and a tree washed downhill by a mudslide blocked the road.

Personal Narrative

A good personal story

▶ tells a story from personal experience

▶ expresses the writer's feelings by using the first person point of view

▶ uses time-order words to connect ideas and show the sequence of events

Write About It

Personal Narrative Write a personal narrative about a storm, mudslide, or other severe weather condition that you have experienced. Use a clear sequence of events to tell what happened and what you did.

 LOG ON e-Journal Research and write about it online at **www.macmillanmh.com**

Math in Science

How Far Away Is Lightning?

Multiply Fractions

To multiply a fraction by a whole number

▶ write the whole number as a fraction by placing it over the denominator 1

▶ then multiply numerators and denominators

▶ then reduce the fraction

$$\frac{1}{5} \times 20 = \frac{1}{5} \times \frac{20}{1} = \frac{20}{5} = 4$$

When you see a lightning bolt, a few seconds will pass before you hear thunder. The sound of thunder travels at about $\frac{1}{5}$ of a mile per second. Once you see the lightning bolt, count the seconds until you hear thunder. If you know how to multiply fractions, you can use this information to find out how far away the lightning bolt was.

 Solve It

1. A rumble of thunder takes 15 seconds to reach your ears. How far away is it?

2. You see a bolt of lightning and 25 seconds later you hear thunder. How far away was the lightning?

3. If a lightning flash is seen 8 seconds before thunder is heard, how far away was the lightning?

Climate

Igloolik Island, Nunavut, Canada

Look and Wonder

People around the world live in areas with very different temperatures. In some places, the temperature is cold throughout the year, while in other places, it is hot. What causes these differences?

How does distance from an ocean affect temperature?

Make a Prediction

San Francisco, CA, is closer to the Pacific Ocean than Stockton, CA. Make a prediction about how distance from an ocean affects the temperature of a city.

Test Your Prediction

❶ Use the temperature data in the charts to compare the monthly high temperatures of the two cities.

❷ Use the temperature data in the charts to compare the monthly low temperatures of the two cities.

Draw Conclusions

❸ **Interpret Data** Which city has the greater change in temperature during the year? Which city has the smaller change in temperature during the year?

❹ **Infer** How might the ocean affect the temperature changes in these cities?

❺ **Communicate** Write a report explaining how the data for these two cities either support or do not support your prediction. Would examining data for more cities improve the accuracy of your prediction?

Explore More

Write a prediction explaining how being near an ocean will affect another weather variable. Collect and compare weather data for both cities. Write a report explaining how the data support or do not support your prediction.

Average High Temperature (°F)		
Month	San Francisco	Stockton
Jan.	55.7	53.4
Feb.	59.1	60.5
Mar.	61.3	65.9
Apr.	63.9	72.9
May	66.8	81.0
Jun.	70.0	88.4
Jul.	71.4	94.1
Aug.	72.1	92.5
Sep.	73.5	88.2
Oct.	70.2	78.4
Nov.	62.9	64.2
Dec.	56.4	53.7

Average Low Temperature (°F)		
Month	San Francisco	Stockton
Jan.	42.4	37.7
Feb.	44.9	40.5
Mar.	46.1	42.6
Apr.	47.6	46.1
May	50.1	51.6
Jun.	52.6	57.0
Jul.	53.9	60.4
Aug.	54.9	59.8
Sep.	54.7	57.2
Oct.	51.8	50.2
Nov.	47.3	42.2
Dec.	43.1	37.5

Read and Learn

▶ **Main Idea**

Average weather patterns determine an area's climate.

▶ **Vocabulary**

climate, p. 408
current, p. 410
rain shadow, p. 411
El Niño, p. 412

–Glossary
at www.macmillanmh.com

▶ **Reading Skill** ✓

Classify

What is climate?

Weather changes from day to day. However, the weather in any area tends to follow a pattern. For example, in Seattle, WA, rain usually falls between November and March, and daytime temperatures throughout the year are mild.

Seattle has a wet and mild climate (KLYE•mit). **Climate** means the average weather of a place. Although other weather information is considered, average temperature and average rainfall are important variables in determining climate. Climate can be determined for different areas, different periods of time, and different locations.

Because of the effect of insolation on temperature, latitude has the strongest effect on climate. On a global scale, the United States is in the temperate zone. The United States can also be divided into more specific local climate zones.

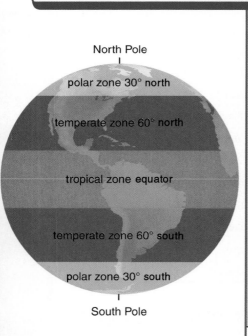

North Pole

polar zone 30° north

temperate zone 60° north

tropical zone equator

temperate zone 60° south

polar zone 30° south

South Pole

Climate Zones

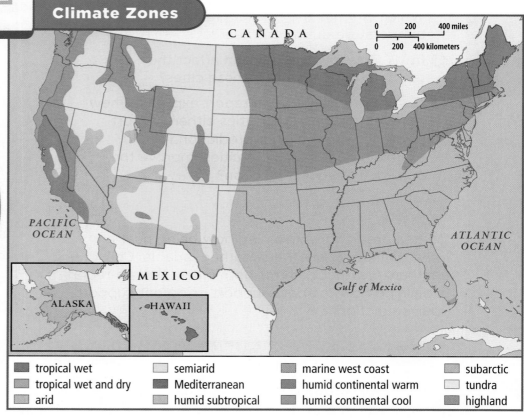

- tropical wet
- tropical wet and dry
- arid
- semiarid
- Mediterranean
- humid subtropical
- marine west coast
- humid continental warm
- humid continental cool
- subarctic
- tundra
- highland

408
EXPLAIN

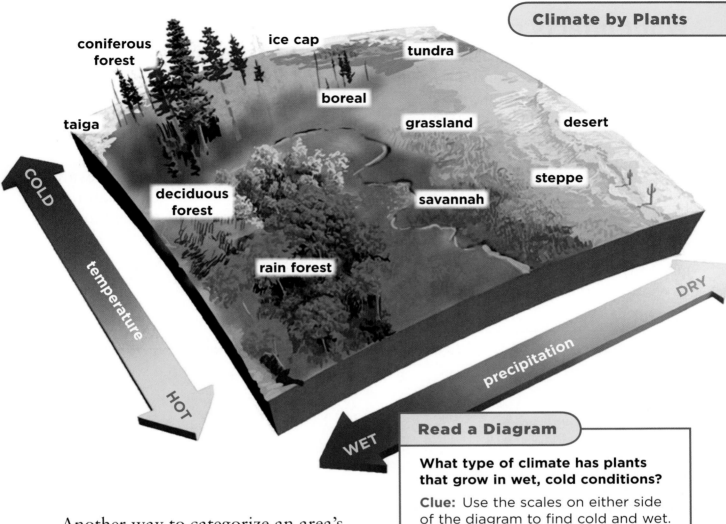

coniferous forest

ice cap

tundra

boreal

taiga

grassland

desert

COLD

deciduous forest

steppe

savannah

temperature

rain forest

DRY

HOT

precipitation

WET

Read a Diagram

What type of climate has plants that grow in wet, cold conditions?

Clue: Use the scales on either side of the diagram to find cold and wet.

Another way to categorize an area's climate is to describe the plants that grow there. Each kind of plant has its own conditions for growth. Since these conditions include precipitation, sunlight, and temperature, plants can be used to determine an area's climate.

Many scientists are concerned that the global climate is warming. Heat energy from sunlight absorbed by Earth's surface radiates back into the atmosphere as the ground cools at night. Much of this radiated heat is absorbed by a layer of greenhouse gases, which include water vapor, carbon dioxide, and ozone. Some of the heat absorbed by greenhouse gases then radiates back and warms Earth.

When fossil fuels are burned, they release greenhouse gases. Burning trees also increases the amount of carbon dioxide in the atmosphere. As the amount of greenhouse gases in the atmosphere increases, more heat radiates back toward Earth. This may be slowly causing the global temperature to increase.

 Quick Check

Classify What is the climate of the states near the Gulf of Mexico?

Critical Thinking What is the climate like where you live?

What affects climate?

In addition to latitude, factors such as distance from water, ocean currents, winds, altitude, and mountain ranges may affect the climate of an area.

Distance from Water

Most of Earth's surface is covered by water. However, some places can be located far from a large body of water. The temperature of an inland city is generally warmer in summer and cooler in winter than a city near the ocean.

Ocean Currents

A **current** is a constant movement of ocean water. The Gulf Stream, which runs along the East Coast of the United States and across the Atlantic Ocean, carries warm water from near the equator toward the poles. Other currents carry cold water from the poles toward the equator. The temperature of the current affects the climate of the land nearby. For example, the warm water in the Gulf Stream causes mild temperatures in the British Isles.

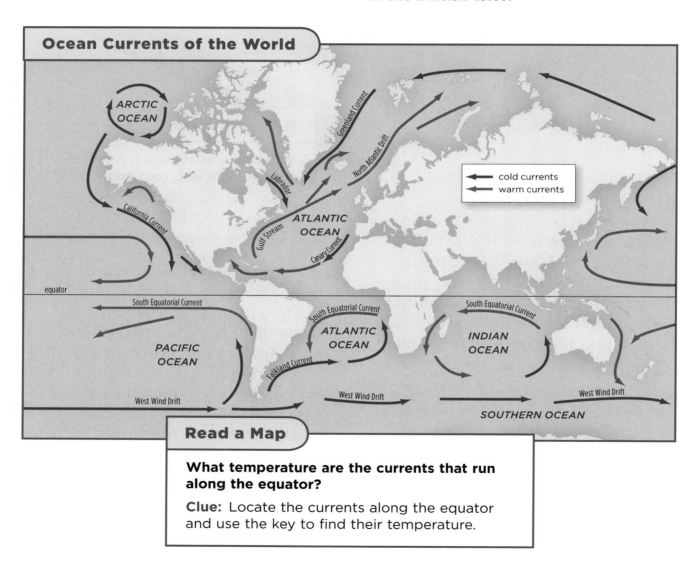

Ocean Currents of the World

ARCTIC OCEAN

Greenland Current

North Atlantic Drift

Labrador

California Current

ATLANTIC OCEAN

Gulf Stream

Canary Current

cold currents
warm currents

equator

South Equatorial Current

South Equatorial Current

ATLANTIC OCEAN

South Equatorial Current

INDIAN OCEAN

PACIFIC OCEAN

Falkland Current

West Wind Drift

West Wind Drift

West Wind Drift

SOUTHERN OCEAN

Read a Map

What temperature are the currents that run along the equator?

Clue: Locate the currents along the equator and use the key to find their temperature.

Rain Shadow

windward side

leeward side

warm, moist air | cold, moist air | hot, dry air

Mountain Ranges

Mountain ranges affect precipitation patterns. As warm, moist air moves up a mountain, it gets colder, water vapor condenses, and precipitation falls on the windward side of the mountain. The air that moves down the leeward side of the mountain is dry and hot. The dry area on the leeward side of the mountain is called a **rain shadow**.

Winds

As water vapor evaporates from warm ocean currents around the equator, winds carry the water vapor away from the equator to cooler regions. There the water vapor condenses and releases heat into the atmosphere. Global winds also move air masses and fronts.

Altitude

The higher a place is above sea level, the cooler its climate is. Along the base of a mountain near the equator, you may find tropical plants growing. At the peak of the mountain, you may find permanent ice and snow.

 Quick Check

Classify If one side of a mountain range is hot and dry, is that the leeward side or the windward side?

Critical Thinking How much rainfall would you expect on coastal areas near the Gulf Stream?

What is El Niño?

Most of the time, a cold current along the coast of Peru keeps the ocean water there cool. This causes high air pressure in the eastern Pacific Ocean. In the western Pacific Ocean, the water is warm, which causes low air pressure. The winds that blow across the Pacific Ocean usually blow from east to west.

Every two to seven years, the cold current sinks and stops pushing cold water up to the surface. This causes a change in weather conditions known as an **El Niño** (el NEEN•yoh).

During an El Niño, the warmer temperature at the surface of the water near Peru causes the air pressure there to decrease. The change in air pressure may cause the winds to reverse direction and blow from west to east. The winds move ocean water and cause higher tides on the coasts of North and South America. The winds also move moist air, causing heavy rains and storms.

Sometimes the cold current moves closer to the surface. When this happens, the air pressure in the western Pacific Ocean decreases and the air pressure in the eastern Pacific Ocean increases. The weather is dryer in South America and wetter in Australia. This change in weather conditions is called La Niña.

✔ Quick Check

Classify What weather patterns could be occuring if the air pressure along the coast of Peru is low?

Critical Thinking What conditions occur in Australia during an El Niño?

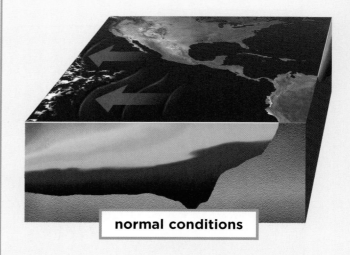

normal conditions

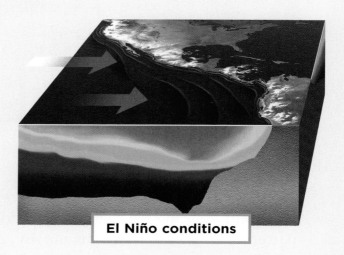

El Niño conditions

La Niña conditions

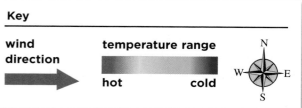

Key

wind direction

temperature range

hot   cold

N W E S

Lesson Review

Visual Summary

Plant growth is one way to define the type of climate.

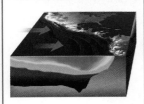
Areas in the rain shadow of a mountain range are usually dry.

El Niño causes climate changes in North America, South America, and Australia.

Make a FOLDABLES™ Study Guide

Make a Three-Tab Book. Use the titles shown. On the inside of each tab, classify the titles shown.

Plant growth is one way...

Areas in the rain shadow...

El Niño causes...

Think, Talk, and Write

1. **Main Idea** How is climate determined?

2. **Vocabulary** The average weather of a place is the _____.

3. **Classify** If data showed the cold current along the coast of Peru moved closer to the surface, which weather condition is happening?

4. **Critical Thinking** Draw the normal condition of the Pacific Ocean and then draw the changes that occur during an El Niño and a La Niña.

5. **Test Prep** What is a constant movement of ocean water called?
   - **A** precipitation
   - **B** El Niño
   - **C** high tide
   - **D** current

6. **Test Prep** Which of the following is NOT a variable that may affect the climate of an area?
   - **A** temperate zone
   - **B** altitude
   - **C** latitude
   - **D** ocean currents

Math Link

Average Temperature
Research the temperature in your area for the last year. Calculate the average monthly temperature and then make a graph or chart to compare the data.

Social Studies Link

Effects of El Niño and La Niña
Research and write about how people in North and South America are affected by El Niño and La Niña.

Museum Mail Call

Scientists at the American Museum of Natural History study the natural world and the people who live in it. They collect stories and objects from people around the world. Read these letters to find out how weather affects children in different countries at the same time of year.

June 13

Dear Museum Scientists,

Hola! (That's "hello" in Spanish.) It's the dry season here in Palmdale right now and it's *muy caliente*—very hot! We haven't had rain in weeks.

It's usually hot and dry here from May to November. We don't have a lot of water, so it has to be piped in from other areas. People have to watch how much water they use. Restaurants only serve water to people who ask for it.

Some people plant cactuses and shrubs around their homes. These plants need a lot less water than a thick, green lawn. I planted jalapeño peppers and pumpkins with *mi hermana*, my sister. We water the plants in the evening. That way the hot sunlight won't dry up all of the water.

Carlos

Carlos and his sister Alicia plant pumpkins.

June 23

Dear Museum Scientists,

   The *gio mua*, or monsoons, have brought wet weather to our land. Everything here is soaked! Our monsoon season lasts from May to October. Many inches of rain can fall during heavy storms. But the storms only last for about an hour each day. It's very hot, so we don't mind getting wet. It's actually a lot of fun, and we dry off right away.

   Our farm is near the Mekong River. Water floods our rice fields and helps the rice grow. It's hard work walking through the swampy ground. We carry the rice with *quang ganh*. These are baskets that we balance on the end of a pole.

   People here are used to a lot of water. We build our homes on stilts so the water won't get in. We ride boats down the river and sell our rice on a floating market. Some years, there is more water than we expect!

                                    Vang

Rice is carried in *quang ganh*.

Write About It
Compare and Contrast

1. How does the weather in Palmdale compare to the weather near the Mekong River?
2. What activity do both Carlos and Vang do?

LOG ON **e-Journal** Research and write about it online at **www.macmillanmh.com**

Compare and Contrast

▶ To compare, look for similarities, or things that are the same.

▶ To contrast, look for differences, or things that are not the same.

AMERICAN MUSEUM of NATURAL HISTORY

Visual Summary

Lesson 1 Heat energy from the Sun changes air pressure and causes winds.

Lesson 2 Water vapor in the air can form clouds, fog, rain, hail, sleet, or snow. Air masses and fronts change weather as they move.

Lesson 3 Storms are caused by the collision of air masses.

Lesson 4 Average weather patterns determine an area's climate.

Make a FOLDABLES™ Study Guide

Assemble your lesson study guides as shown. Use your study guides to review what you have learned in this chapter.

Fill in each blank with the best term from the list.

climate, p. 408   **humidity**, p. 369

current, p. 410   **storm surge**, p. 401

cyclone, p. 401   **weather**, p. 366

front, p. 384   **weather map**, p. 388

1. Weather in a specific area at a specific time is shown on a(n) _____.

2. Waves caused by a hurricane can lead to a(n) _____.

3. Air masses meet at a(n) _____.

4. The average weather of a place is its _____.

5. The amount of water vapor in the air is called _____.

6. A constant movement of ocean water is a(n) _____.

7. The condition of the atmosphere at a particular time and place is called the _____.

8. A circular pattern of winds with a low-pressure center is a _____.

Answer each of the following in complete sentences.

9. Main Idea and Details How does the Sun affect Earth's weather?

10. Classify Which type of severe weather is shown below? Explain your answer.

11. Communicate Suppose that you are giving a friend a windsock. Write a note that explains how the windsock works and what it measures.

12. Critical Thinking Why might people be concerned about an El Niño?

13. Personal Narrative Write a personal narrative about how you or your family prepared for a storm. How did you stay safe? Did you need any special supplies? If you have never prepared for a storm, write about how you could prepare for future storms.

14. How do scientists predict the weather?

Be a Weather Tracker

What to Do

1. Record the temperature, precipitation, and cloud cover at the same time each day for one week.

2. Write the highest and lowest temperatures you recorded each day in a chart. Include the precipitation and cloud cover in your chart as well.

3. Create a line graph showing the high and low temperatures you recorded in your chart.

Analyze Your Results

▶ Use your graph to draw conclusions about the weather patterns in your area during the week.

1. What type of cloud is shown in this photograph?

A stratus

B cumulus

C cirrus

D nimbostratus

The Universe

The Big Idea

What is in outer space?

Whirlpool Galaxy (M51) in Canes Venatici constellation

Key Vocabulary

inertia
the tendency of a moving object to keep moving in a straight line or of any object to resist a change in motion (p. 423)

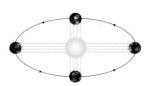

revolution
one complete trip around the Sun. Earth completes one revolution in 365 days (p. 424)

tide
the regular rise and fall of the water level along a shoreline (p. 436)

satellite
a natural or artificial object in space that circles around another object (p. 448)

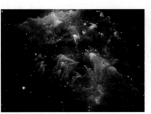

star
an object in space that produces its own energy, including heat and light (p. 458)

galaxy
a collection of billions of stars. Our Sun belongs to the Milky Way galaxy (p. 465)

More Vocabulary

gravity, p. 422

orbit, p. 423

rotation, p. 426

phase, p. 432

lunar eclipse, p. 435

telescope, p. 442

planet, p. 444

moon, p. 448

comet, p. 450

asteroid, p. 451

nebula, p. 458

white dwarf, p. 459

supernova, p. 460

black hole, p. 460

constellation, p. 462

light-year, p. 463

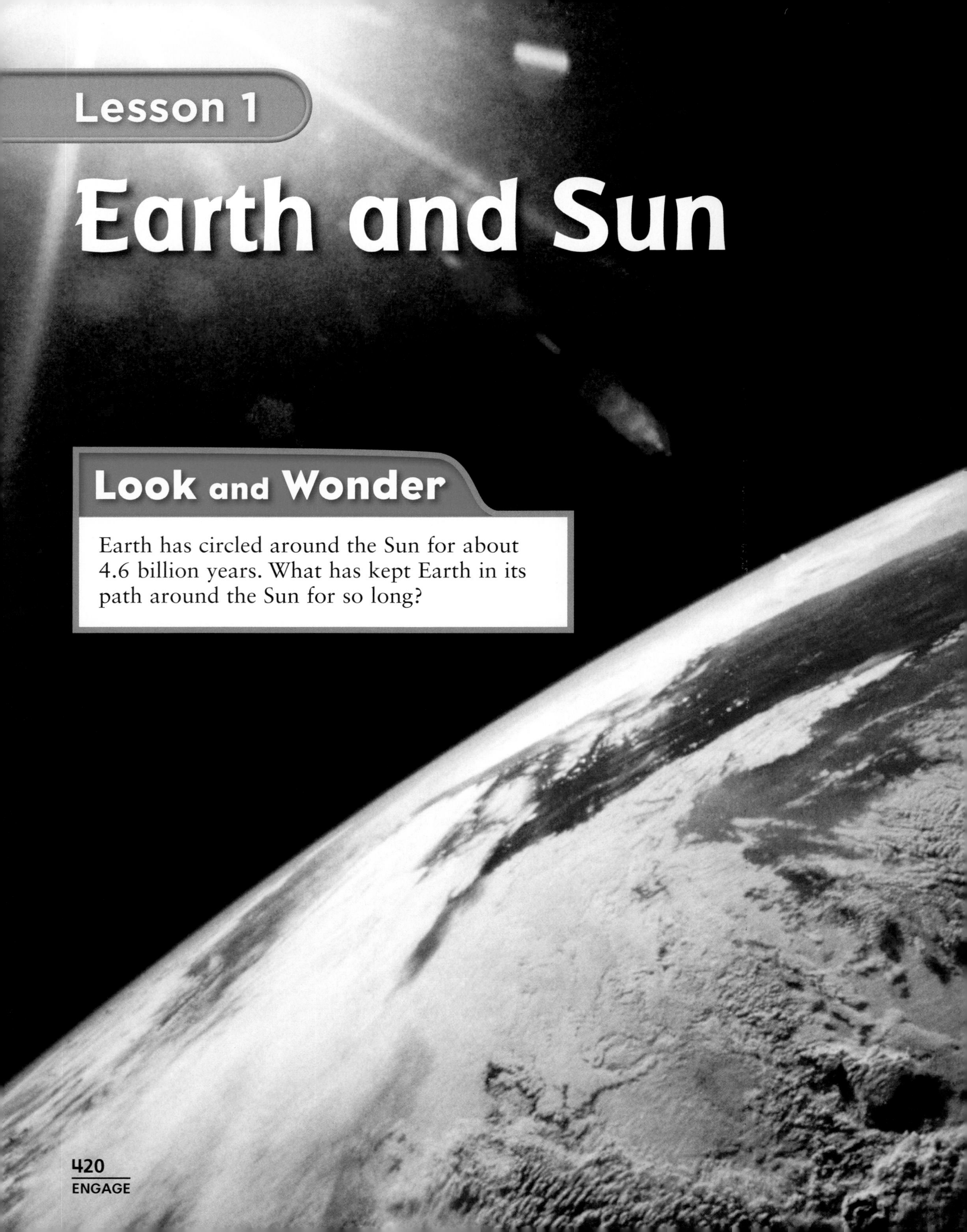

Earth and Sun

Look and Wonder

Earth has circled around the Sun for about 4.6 billion years. What has kept Earth in its path around the Sun for so long?

What keeps Earth moving around the Sun?

Form a Hypothesis

If you let go of a ball being swung in a circle, in what direction will the ball travel? Write a hypothesis in the form "If I let go of a ball being swung in a circle at a particular point, then . . . "

Test Your Hypothesis

1 Place the tennis ball on the fabric and bring the four corners of the fabric together so they cover the ball. Then tie string around the four corners to form a pouch.

2 △ **Be Careful.** While holding the other end of the string, lean forward and slowly spin the ball in a circle near your feet.

3 Observe Let go of the string. Watch the path that the ball takes.

4 Record Data Draw a diagram to show the path the ball took when you let it go.

5 Repeat the experiment, letting go of the ball at three different spots on the circle. Where does the ball go?

Draw Conclusions

6 Did the experiment support your hypothesis? Why or why not?

7 If this activity models the movement of Earth around the Sun, what do you, the ball, and the string represent?

Explore \ **More**

What results would you expect if you repeated the experiment using a lighter ball? Form a hypothesis, do the experiment, record your data, and write a report.

Materials

- **tennis ball**
- **fabric**
- **string**
- **graph paper**

Step **1**

Step **2**

What is gravity?

Each planet in the solar system is drawn toward the Sun by gravity. **Gravity** is a force of attraction, or pull, between any two objects. The strength of the pull of gravity is affected by the total mass of the two objects and by the distances between the objects. The strength of the pull of gravity decreases when the total mass of the two objects decreases and when the objects are farther apart.

Compare the pull of gravity you feel on Earth to the pull of gravity you would feel on the Moon. Your mass stays the same no matter where you are. Earth's mass is greater than the Moon's mass. This means that the total mass of you and Earth is greater than the total mass of you and the Moon. The pull of gravity between you and Earth is stronger than the pull between you and the Moon. In fact, the Moon's gravity is about one sixth of Earth's gravity.

Two objects do not have to touch each other to produce a force of gravity between them. The pull of gravity between Earth and the Sun acts across about 150 million kilometers (93 million miles) of space. Gravity also acts across roughly 6 billion kilometers (4 billion miles) of space between the Sun and Pluto. Since the distance is farther between the Sun and Pluto, the pull of gravity between the Sun and Pluto is weaker than the pull of gravity between the Sun and Earth.

In this photo, you can see the height of astronaut John Young's jump on the Moon. He can jump higher on the Moon than on Earth because the Moon's gravity is about one sixth of Earth's gravity.

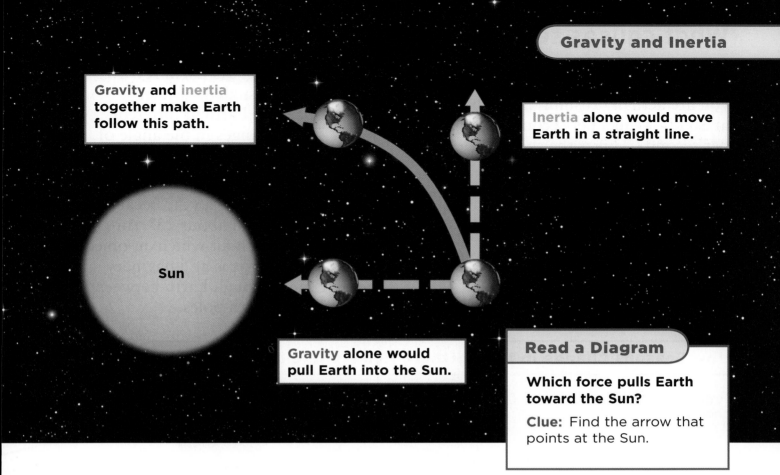

Gravity and inertia together make Earth follow this path.

Inertia alone would move Earth in a straight line.

Gravity alone would pull Earth into the Sun.

Sun

Read a Diagram

Which force pulls Earth toward the Sun?

Clue: Find the arrow that points at the Sun.

The planets are held in their orbits around the Sun by the force of gravity between each planet and the Sun. An **orbit** is a path one object takes around another object.

If gravity was the only force acting on a planet, the planet would be pulled into the Sun. What prevents this from happening? All objects have a property called *inertia*. **Inertia** is the tendency of a moving object to keep moving in a straight line.

As Earth orbits the Sun, it is pulled toward the Sun because of gravity. At the same time, Earth's inertia makes it move away from the Sun. As a result of the effects of gravity and inertia, Earth moves in a nearly circular orbit called an *ellipse*.

When Earth is closest to the Sun, it is about 147,000,000 kilometers (91,000,000 miles) away. When Earth is furthest from the Sun, it is about 152,000,000 kilometers (94,000,000 miles) away. This 5-million-kilometer (3-million-mile) difference shows that Earth's orbit is close to, but not quite, a perfect circle.

 Quick Check

Fact and Opinion Astronauts can jump higher on the Moon than on Earth. Is this a fact or an opinion?

Critical Thinking In what direction would the planets travel if the Sun suddenly disappeared? Explain.

What causes seasons?

You probably feel like you are sitting still as you read this page, but you are actually rushing through space at 30 kilometers per second (19 miles per second). This is the speed at which Earth is moving around the Sun.

Earth's orbit is about 924,000,000 kilometers (574,000,000 miles) long. How long does it take Earth to make one revolution? A **revolution** is one complete trip around the Sun. Earth makes this trip in one year, or in $365\frac{1}{4}$ days.

During a year, you observe seasons changing on Earth. What causes the seasons to change? As Earth revolves around the Sun, sunlight strikes different parts of Earth at different angles. These changes in the angle of the sunlight cause the seasons.

The angle at which sunlight hits Earth changes during a year because Earth's axis is tilted about 23°. An *axis* is a straight line about which an object rotates. Earth's axis is an imaginary line that runs through Earth between its North and South poles.

How Seasons Change in the Northern Hemisphere During a Year

This diagram is not to scale.

N

spring begins

S

summer begins

N

S

Sun

N

autumn begins

S

Read a Diagram

In the beginning of summer, which hemisphere receives more sunlight?

Clue: Find the diagram of Earth when summer begins.

As Earth revolves around the Sun, the tilted axis always points in the same direction. When the Northern Hemisphere is tilted away from the Sun, the ground does not receive much heat energy and temperatures are low. In the Northern Hemisphere, this is the beginning of winter.

At the same time, summer begins in the Southern Hemisphere. The Southern Hemisphere is angled toward the Sun, so the heat energy of the sunlight is more concentrated. The ground receives more heat energy and temperatures are warmer.

Because the tilt of Earth's axis always points in the same direction, the seasons in the Northern and Southern hemispheres are always opposite. In spring and in autumn, both hemispheres receive equal warmth from sunlight which makes temperatures mild.

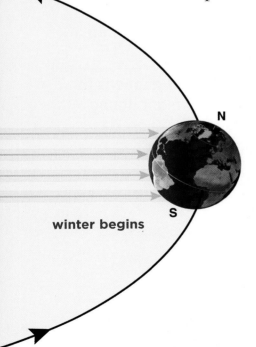

winter begins

FACT ▶ Earth is closer to the Sun in January than in July.

Quick Lab

Seasons and Earth's Tilt

1 Using modeling clay, make a sphere to represent Earth. Then make a base for the sphere.

2 ⚠ **Be Careful.** Push a toothpick through the sphere to represent Earth's axis. Use a pencil to draw a line around the center to represent the equator.

3 Hold the sphere so the toothpick is straight up and down, then tilt the sphere so the top of the toothpick is at an angle of about 23° and push the bottom of the toothpick into the base.

4 **Observe** Aim a flashlight at the sphere so the end of the toothpick points away from you. Describe how the light spreads over the sphere. What would the seasons be in the Northern and Southern hemispheres?

5 **Observe** Now shine the flashlight so the end of the toothpick points toward you. Describe how the beam of light spreads over the sphere. What would the seasons be in the Northern and Southern hemispheres?

✔ Quick Check

Fact and Opinion Write a fact about why Earth's seasons change.

Critical Thinking When the season in the Northern Hemisphere is autumn, what season occurs in the Southern Hemisphere?

What causes day and night?

You are moving through space at 30 kilometers per second (19 miles per second) as Earth revolves. You are also spinning in a circle at about 1,600 kilometers per hour (1,000 miles per hour) as Earth rotates. One **rotation** is a complete spin on the axis. Earth makes one rotation every day, or 24 hours.

At any point in time, half of Earth's surface faces the Sun and is in daylight. The other half of Earth's surface faces away from the Sun and is in darkness.

The tilt of Earth's axis affects the length of the day. If the axis was not tilted, day and night would each be 12 hours long. Instead, there are more hours of daylight and fewer hours of night during the summer.

Shouldn't you feel movement as Earth revolves and rotates? You don't feel these motions because you are carried along with Earth. It is as if you had your eyes closed as you sped down a perfectly smooth highway in a car. As you sat in the car, you would not feel the motion of the car and would not be able to tell that you were moving.

If you watch objects in the sky, such as the Sun, they appear to rise in the east and set in the west. This is the apparent motion of these objects, but is not the real motion. As Earth rotates from west to east, objects in the sky appear to move in the direction opposite of Earth's movement.

✔ Quick Check

Fact and Opinion The Sun rises in the east and sets in the west. Explain whether this is a fact or an opinion.

Critical Thinking What happens to the length of day and night during winter?

Earth's Rotation

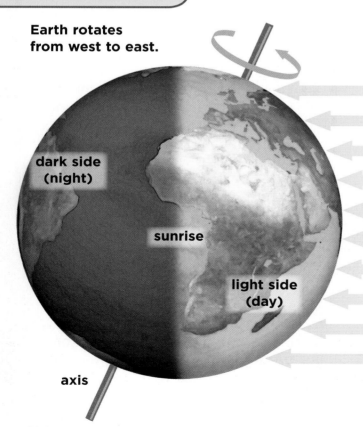

Earth rotates from west to east.

dark side (night)

sunrise

light side (day)

axis

sunlight

Lesson Review

Visual Summary

The pull of gravity depends on the masses of two objects and the distance between them.

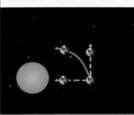

An object stays in orbit because of gravity and inertia.

Because of Earth's rotation, half of Earth's surface is in daylight and half is in darkness.

Make a FOLDABLES™ Study Guide

Make a Three-Tab Book. Use the titles shown. On the inside of each tab, write a fact about the title.

Think, Talk, and Write

1 **Main Idea** What factors affect the strength of the pull of gravity between two objects?

2 **Vocabulary** The tendency of a moving object to keep moving in a straight line is called _____.

3 **Fact and Opinion** Earth's day is 24 hours long. Is this a fact or an opinion?

Fact	Opinion

4 **Critical Thinking** Which movement of Earth causes day and night?

5 **Test Prep** What keeps Earth in its orbit around the Sun?
  A lift and pressure
  B thrust and inertia
  C gravity and inertia
  D pressure and gravity

6 **Test Prep** How much time does Earth take to complete one revolution?
  A one day
  B one week
  C one month
  D one year

Math Link

Calculating Earth's Mass
The Sun's mass is roughly 330,000 times Earth's mass. If you made a model of the Sun with a mass that was 1,000 kilograms, what would Earth's mass be?

Health Link

Weightlessness
Research and write about how human beings in space are affected by the lack of gravity and how astronauts deal with these effects.

Focus on Skills

Inquiry Skill: Use Numbers

When scientists **use numbers**, they add, subtract, multiply, divide, count, or put numbers in order to explain and analyze data.

The orbits of each planet in the solar system have different radii. This means each planet takes a different amount of time to revolve around the Sun. As the radius of the planet's orbit increases, the revolution time increases. What would your age be if you lived on a different planet?

▶ Learn It

The diagram of the planets shows the time each planet takes to revolve around the Sun in Earth days or years. Scientists **use numbers** to compare the revolution time of the other planets in our solar system to that of Earth. You can do that by dividing the revolution time of a planet by the revolution time of Earth.

For example, it takes Earth $365\frac{1}{4}$ days to travel around the Sun. Mars takes 687 days to complete its revolution. If you divide the length of time it takes Mars to make a revolution by the length of time it takes Earth to make a revolution, you get 1.88. Mars takes almost twice as long as Earth to complete one revolution.

If you were 62 years old in Earth years, how old would you be in Mars years? The ratio of Mars's revolution to Earth's is 1.88. Divide your age by the Earth-planet ratio to calculate your age on a specific planet.

Planet	Revolution (days)	Earth-Planet Ratio	Age on Planet
Mercury			
Venus			
Earth	365	1	62
Mars	687	1.88	33
Jupiter			
Saturn			
Uranus			
Neptune			

Number of Days Each Planet Takes to Revolve Around the Sun

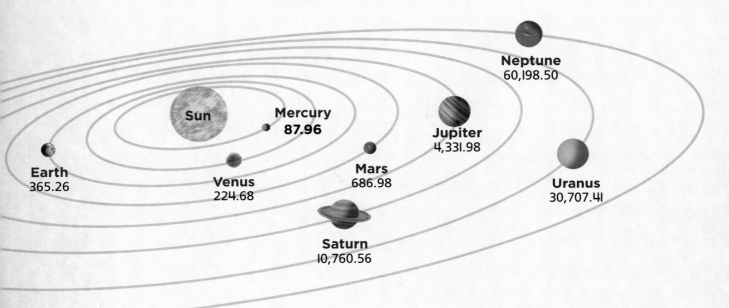

Neptune
60,198.50

Sun

Mercury
87.96

Jupiter
4,331.98

Earth
365.26

Venus
224.68

Mars
686.98

Uranus
30,707.41

Saturn
10,760.56

▶ **Try It**

❶ Make a chart with titles like the one shown. Record the revolution data from the diagram on your chart.

❷ **Use Numbers** Calculate the Earth-planet ratio for all of the planets.

❸ If you were 6 years old in Earth years, how old would you be in Mars years?

▶ **Apply It**

❶ **Use Numbers** Now calculate how old you would be if you lived on each of the planets.

❷ On which planet would you be the oldest in that planet's years? On which planet would you be the youngest?

❸ What can you infer about the revolution time of the planet and the age you would be on that planet?

Earth and Moon

Zion National Park, Utah

Look and Wonder

If you watched the Moon every night for a month, you would see that the shape of the Moon looks like it is changing. What makes the Moon appear to change shape?

What makes the Moon appear to change shape?

Purpose

To model changes in the appearance of the Moon as seen from Earth.

Materials

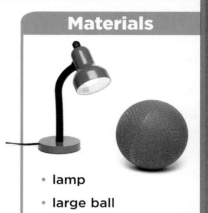

- lamp
- large ball

Procedure

1 **Make a Model** You represent an observer on Earth. A classmate uses a lamp to represent the Sun. A classmate with a ball represents the Moon.

2 Face the classmate with the lamp. Have your other classmate hold the ball directly between you and the lamp.

3 **Observe** Have your classmate turn the lamp on. How much of the surface of the ball is lit? Record what you see.

4 **Observe** Have your classmate with the ball move one eighth of the way around you. Turn to face the ball and record what you see.

5 Repeat step 4 until your classmate returns to the starting position.

Draw Conclusions

6 What causes the changes in the Moon's appearance as seen from Earth?

7 What happened to the shape of the ball that represented the Moon during this experiment?

Explore More

Mark one side of the ball. Repeat the procedure while keeping the marked side toward the light. How much sunlight do the marked and unmarked sides receive?

Step 3

Main Idea

The Moon is Earth's natural satellite.

Vocabulary

phase, p. 432
solar eclipse, p. 435
lunar eclipse, p. 435
tide, p. 436

LOG ON ℮-Glossary
at www.macmillanmh.com

Reading Skill ✓

Sequence

First

↓

Next

↓

Last

Craters, rills, maria, and mountains can be seen on the Moon's surface.

How does the Moon appear?

On July 20, 1969, astronaut Neil Armstrong was the first person to walk on the Moon. Armstrong sent this message back to Earth: "The surface is fine and powdery . . . like much of the high desert of the United States."

The Moon has no atmosphere. Because there is no atmosphere, there are no winds and there is no weather on the Moon. There is no air to block radiation from the Sun or for astronauts to breathe. Temperatures can reach as high as 253°F (123°C) and drop below −451°F (−233°C).

How did Earth's visitors to the Moon survive? They wore spacesuits to protect them from the changes in temperature and from radiation. They also carried containers of oxygen to breathe.

The craters on the Moon's surface were made over billions of years as rocks traveling through space hit the Moon. Vast plains cover other parts of the Moon. Early astronomers called these plains *maria*, a Latin word meaning "seas." Valleys, or *rills*, cut grooves in the Moon's surface. In other places, mountains rise thousands of meters.

Phases of the Moon

From Earth, you can only see the parts of the Moon's surface that are lit by sunlight. If you looked at the Moon from out in space, you would see that the side of the Moon that faces the Sun is always lit by sunlight. As the Moon revolves around Earth, different amounts of light reflect from the Moon's surface and the Moon appears to change shape.

During a full-Moon phase, an observer on Earth can see the entire half of the Moon that is lit by sunlight. During a new-Moon phase, the lit side of the moon is facing away from an observer on Earth. A phase of the Moon is the appearance and shape of the Moon as you see it at a particular time. The phase depends on the location of the Moon in relation to Earth and the Sun. The time from one phase of the Moon until the next time the Moon reaches the same phase is 29.5 days.

as seen from space

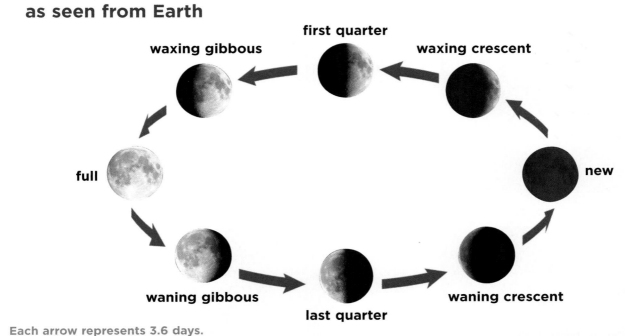

first quarter

waxing gibbous

waxing crescent

Earth

full

new

waning gibbous

waning crescent

last quarter

Each arrow represents 3.6 days.

as seen from Earth

first quarter

waxing gibbous

waxing crescent

full

new

waning gibbous

waning crescent

last quarter

Each arrow represents 3.6 days.

 Quick Check

Sequence Starting with a full Moon, what are the phases of the Moon during a month?

Critical Thinking How does the Moon change during the waxing and waning phases?

FACT The phases of the Moon are caused by the revolution of the Moon around Earth.

What causes eclipses?

What is happening when a dark shadow seems to move in front of the Sun or when the Moon dims or changes color? These events are called *eclipses*. An eclipse occurs when one object moves in front of another object in space.

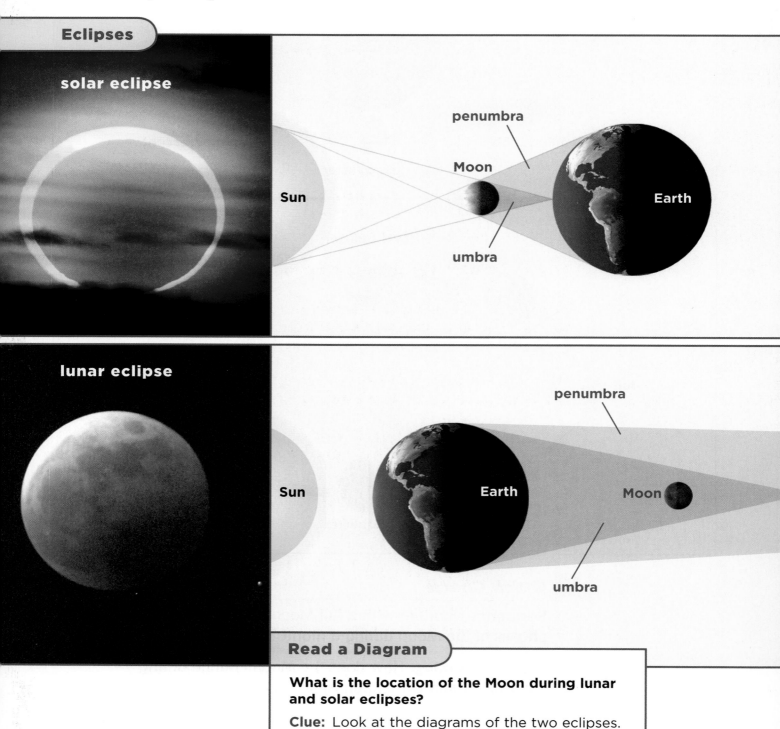

Eclipses

solar eclipse

penumbra

Moon

Sun

Earth

umbra

lunar eclipse

penumbra

Sun

Earth

Moon

umbra

Read a Diagram

What is the location of the Moon during lunar and solar eclipses?

Clue: Look at the diagrams of the two eclipses.

A **solar eclipse** occurs when the Moon passes directly between the Sun and Earth. When this happens, the Moon casts a shadow on Earth. People on Earth see darkness move across the Sun. A solar eclipse can only occur during a new-Moon phase.

A **lunar eclipse** occurs when the Moon moves into Earth's shadow and is no longer reached by direct sunlight. This happens when Earth is between the Sun and the Moon. Lunar eclipses happen only during full-Moon phases.

A full Moon and a new Moon occur once a month, but eclipses happen more rarely. What makes eclipses occur less frequently? The Moon's orbit moves above and below a straight line between Earth and the Sun. Because of these differences in its orbit, the Moon rarely travels exactly between Earth and the Sun.

Look at the way the Sun, the Moon, and Earth line up in lunar and solar eclipses. In both cases, there is an area where the light from the Sun is completely blocked. This area is called the *umbra*. Around the umbra, there is an area where light is not completely blocked. This is called the *penumbra*.

In solar eclipses, the shadow of the Moon causes the umbra and penumbra. If you are on Earth in the umbra during a solar eclipse, darkness covers the entire face of the Sun. This is called a *total solar eclipse*. If you are on Earth in the penumbra during a solar eclipse, darkness covers only part of the Sun. This is called a *partial solar eclipse*.

Quick Lab

Eclipses

1. **Make a Model** You will represent Earth using a large ball. One classmate will use a flashlight to model the Sun. Another classmate will use a tennis ball to represent the Moon.

2. Begin by facing the classmate with the lamp. Rotate the Moon through one month of revolution around Earth.

3. What positions of the Moon, Earth, and Sun produce eclipses?

In lunar eclipses, Earth's shadow causes the umbra and penumbra. Lunar eclipses may also be total or partial depending on whether or not the Moon is in the umbra or penumbra.

 Quick Check

Sequence In a solar eclipse, what are the positions of the Sun, the Moon, and Earth?

Critical Thinking What would you see if you were on the Moon's surface during a lunar eclipse?

What causes the tides?

The pull of gravity between the Moon and Earth and between the Sun and Earth causes a bulge in the surface of Earth. On the part of Earth's surface that is rocky, we do not notice this pull. However, the pull can be seen in the oceans and other large bodies of water. This pull causes the **tide**, or the rise and fall of the ocean's surface.

The pull of gravity causes ocean water to bulge on the side of Earth facing the Moon. A matching bulge occurs on the side of Earth that is opposite the Moon. This causes high tides at both locations. Low tides occur halfway between high tides. As Earth rotates, these bulges move across the oceans. Most oceans have two high tides and two low tides every day.

About twice a month, near the new- and full-Moon phases, the Sun and Moon line up and pull in the same direction. This causes higher high tides and lower low tides, called *spring tides*. The tides with the smallest range between high and low tides occur between these two spring tides. These more moderate tides are called *neap tides*. They take place when the Sun and Moon pull in different directions and their pulls partly cancel each other. Neap tides occur during the first–quarter and last–quarter Moon phases.

 Quick Check

Sequence When do the weakest tides occur?

Critical Thinking What makes spring tides and neap tides occur twice a month?

low tide

high tide

Tides During One Month

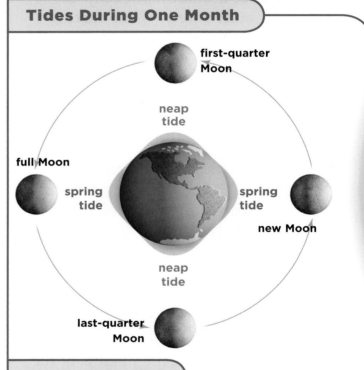

first-quarter Moon

neap tide

full Moon

spring tide

spring tide

new Moon

neap tide

last-quarter Moon

Read a Diagram

During which phases of the Moon do the strongest tides occur?

Clue: The strongest tides are caused when gravity from the Sun and from the Moon pull together on Earth.

 *Science in Motion* Watch how gravity causes tides at **www.macmillanmh.com**

Lesson Review

Visual Summary

Earth's Moon goes through eight phases during one month.

Eclipses are a darkening or hiding of the Sun, a planet, or a moon by another object in space.

Tides on Earth are caused by the pull of gravity from the Moon and the Sun.

Make a FOLDABLES™ Study Guide

Make a Trifold Book. On each page, summarize what you have learned.

Think, Talk, and Write

1. **Main Idea** What makes Earth's Moon a natural satellite?

2. **Vocabulary** During a(n) _____ the Moon passes through Earth's shadow.

3. **Sequence** Draw the phases of the Moon beginning with the new Moon.

   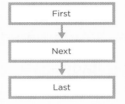
   ```
   First
     ↓
   Next
     ↓
   Last
   ```

4. **Critical Thinking** How did astronauts survive as they explored the Moon?

5. **Test Prep** When the Moon is completely in the umbra during a lunar eclipse, this is called a _____
   A total solar eclipse.
   B partial lunar eclipse.
   C total lunar eclipse.
   D partial solar eclipse.

6. **Test Prep** Which pair of Moon phases are opposites?
   A new and full
   B waxing crescent and first quarter
   C full and waning gibbous
   D waxing gibbous and last quarter

 Math Link

Diameter of Earth and the Moon
The diameter of Earth is 12,760 kilometers across, while the diameter of the Moon is 3,475 kilometers across. Convert these distances into miles and feet.

 Social Studies Link

What Is a Blue Moon?
"Once in a blue Moon" is an expression meaning an event that rarely happens. Research the meaning of the phrase "blue Moon" and explain how it could relate to this expression.

Writing in Science

What Would Happen If Gravity Went Away?

Explanatory Writing

A good explanation

▶ develops the main idea with facts and supporting details

▶ lists what happens in an organized and logical way

▶ uses time-order words to make the description clear

If gravity went away in my room, the first thing I would do would be to tie my furniture to the floor to keep it from floating around. Then, I would gather all my small toys and put them in a box so I'd know where to find them. Finally, I would practice somersaults and walk on the ceiling.

Write About It

Explanatory Writing You know that gravity keeps everything on Earth from floating off into space. Look at the picture, and explain what would change if gravity suddenly stopped working.

LOG ON e-Journal Research and write about it online at **www.macmillanmh.com**

Math in Science

0.38

0.91

1.00

0.38

Weight on Other Planets

2.53

1.07

0.90

1.14

Multiply Decimals

To multiply decimals

▶ multiply as with whole numbers

▶ count the number of decimal places in each factor

▶ add the total number of decimal places

▶ move the decimal point that many places to the left in the product

The force of gravity depends on the masses of the objects involved and the distance between them. On other planets, the weight of an object would not be the same as it is on Earth. Weight is measured in Newtons (N). One Newton is equal to 0.225 pounds. On some planets, such as Mercury, the pull is much weaker, while on others, such as Jupiter, it is much stronger. You can use the gravity of each planet to find the weight an object would have on that planet.

Solve It

1. *Spirit,* the Mars Exploration Rover, weighs 1700 Newtons on Earth. How much would it weigh on Mars?

2. An astronaut weighs 910 Newtons on Earth. How many Newtons would the astronaut weigh on Venus?

3. If a dog weighs 290 Newtons on Earth, how much would the dog weigh on each planet?

The Solar System

Stonehenge, England

Look and Wonder

How many planets do you see? Mars, Saturn, and Venus are in a triangle above the central stone. Mercury is below them to the left. Jupiter is much higher to the right. How far away are these planets from Earth?

How far apart are the planets?

Purpose

To make a model that shows the distances between the planets using Astronomical Units (AU), where one AU equals the average distance between Earth and the Sun. This distance is about 149,591,000 kilometers (92,960,000 miles).

Procedure

1. Let the length of each paper towel equal 1 Astronomical Unit. Using the information on the chart, roll out the number of paper towels you need to show the distance from the Sun to Neptune.

2. **Make a Model** Mark the location of the Sun at one end. Then measure the distance that each planet would be from the Sun and draw the planet on the paper towel.

- **paper towels**
- **markers**
- **ruler**
- **masking tape**

Step 1

Draw Conclusions

3. **Interpret Data** Compare the distances between Mercury and Mars, Mars and Jupiter, and Jupiter and Neptune. Which are farthest apart?

4. **Infer** What can you conclude about the distances between the planets in the solar system?

Distances of the Planets from the Sun

planet	distance in AU
Mercury	0.4
Venus	0.7
Earth	1
Mars	1.5
Jupiter	5.2
Saturn	9.5
Uranus	19.2
Neptune	30

Explore More

Your model has all of the planets in a line. How could you make a model to show the positions of the planets at a specific time? Write instructions that others can follow to make the model.

Step 2

Read and Learn

▶ **Main Idea**

Our solar system is made up of the Sun, the eight planets and their moons, and comets, asteroids, and meteoroids.

▶ **Vocabulary**

telescope, p. 442
planet, p. 444
moon, p. 448
satellite, p. 448
comet, p. 450
asteroid, p. 451
meteor, p. 451

LOG ON ⊕ -Glossary
at www.macmillanmh.com

▶ **Reading Skill** ✓

Infer

Clues	What I Know	What I Infer

▶ **Technology** SCIENCE QUEST 💿

Explore gravity and orbits with Team Earth.

How do we observe objects in space?

Until January 7, 1610, people observed the night sky using only their eyes. On that date, an Italian astronomer named Galileo Galilei looked at the sky through a telescope for the first time. A **telescope** is an instrument that makes distant objects seem larger and nearer.

Optical Telescopes

Galileo used an optical telescope, which uses lenses or mirrors to see objects by gathering visible light. Among the objects Galileo saw were four moons revolving around the planet Jupiter. At that time, most people believed that all of the objects in the solar system revolved around Earth.

Looking through an optical telescope makes a dim object such as a star seem brighter. It can also make objects appear larger so more details can be seen. When the diameter of the light-gathering lens or mirror is increased, more light is gathered and planets appear larger. Today's optical telescopes have lenses and mirrors many times larger than those of Galileo's telescope. Modern optical telescopes can magnify images of more distant planets and look farther into space.

However, observers on Earth have to look into space through Earth's atmosphere. As you learned, the air in Earth's atmosphere has different densities. As light from distant stars travels through the air, the changes in density make the faint light of stars appear fuzzy.

◀ This large globular cluster was seen through an optical telescope.

Telescopes in Space

In 1990, the Hubble Space Telescope was placed into orbit around Earth. Objects that are billions of trillions of kilometers from Earth can be seen through the Hubble Space Telescope. The Hubble Space Telescope was named after astronomer Edwin Powell Hubble, who studied galaxies.

Placing telescopes in space allows scientists to see into space while avoiding Earth's atmosphere. The Hubble Space Telescope and other space telescopes gather more than visible light from objects in space. For example, they can detect the heat that is given off by a star.

◄ This supernova was seen by the Hubble Space Telescope.

Radio Telescopes

Back on Earth, radio telescopes record data from radio waves given off by objects in space. Groups, or arrays, of dishes focus the radio waves so the radio data can be recorded. Computers then turn the data into an image. Radio waves can pass through Earth's atmosphere without interference.

 Quick Check

Infer What reasons could you give for placing an optical telescope in orbit instead of a radio telescope?

Critical Thinking How would Galileo's observation of Jupiter's moons affect the idea that everything revolved around Earth?

This image of an irregular galaxy was made by a radio telescope. ▶

What are planets?

A *solar system* is a star and the objects that orbit around it. In our solar system, there are eight planets orbiting the Sun. A **planet** is a large object that orbits a star.

From nearest to farthest from the Sun, the planets are Mercury, Venus, Earth, Mars, Jupiter, Saturn, Uranus, and Neptune. The planets travel in elliptical, nearly circular orbits around the Sun.

The *inner planets* are closer to the Sun than the asteroid belt and have surfaces made of rock. These planets are Mercury, Venus, Earth, and Mars. The *outer planets* are beyond the asteroid belt and have surfaces made of gases. These planets are Jupiter, Saturn, Uranus, and Neptune.

Pluto was once known as the ninth planet. Pluto's elongated orbit and small size were different from the other planets. Because of this, scientists debated whether Pluto should be classified as a planet.

In August 2006, the International Astronomical Union officially reclassified Pluto as a dwarf planet. Other dwarf planets include Ceres, which is found in the asteroid belt, and 2003 UB313, which is larger than Pluto and even farther from the Sun.

 ## Quick Check

Infer How do the surface materials of the inner and outer planets differ?

Critical Thinking What other objects are in a solar system?

The Solar System

Earth

asteroid belt

Saturn

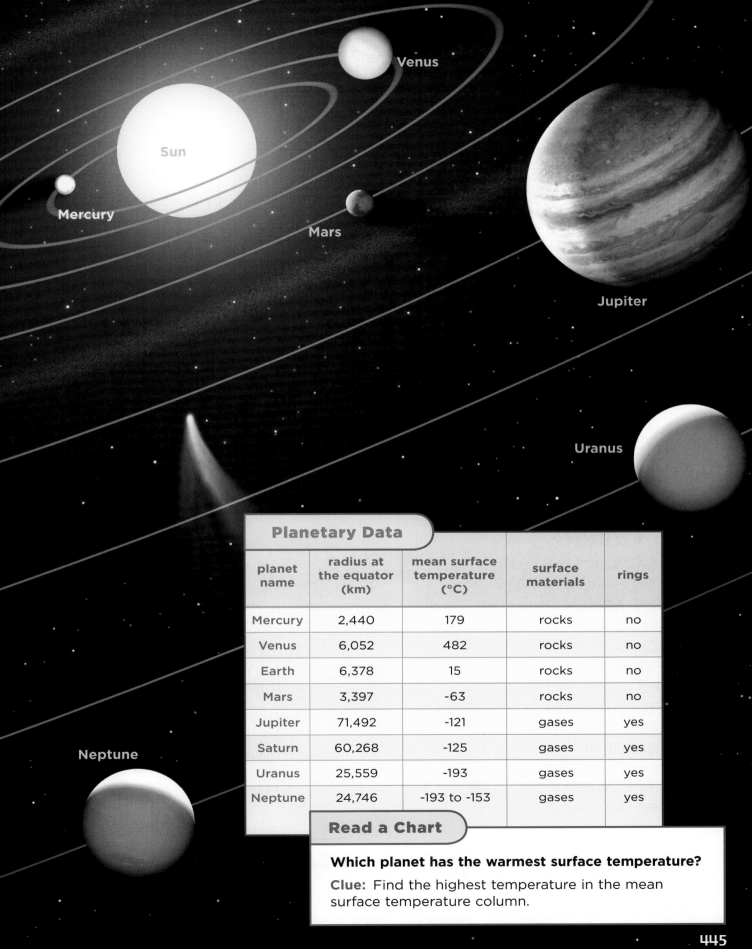

Planetary Data

planet name	radius at the equator (km)	mean surface temperature (°C)	surface materials	rings
Mercury	2,440	179	rocks	no
Venus	6,052	482	rocks	no
Earth	6,378	15	rocks	no
Mars	3,397	-63	rocks	no
Jupiter	71,492	-121	gases	yes
Saturn	60,268	-125	gases	yes
Uranus	25,559	-193	gases	yes
Neptune	24,746	-193 to -153	gases	yes

Read a Chart

Which planet has the warmest surface temperature?

Clue: Find the highest temperature in the mean surface temperature column.

How do the planets compare?

Each planet has unique features. By studying these features, you can learn more about the differences in the surfaces and atmospheres of the planets.

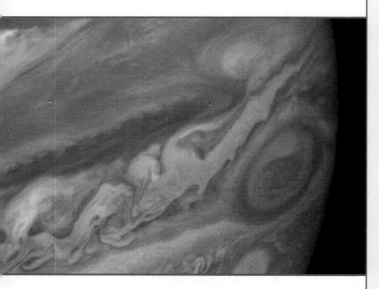

Jupiter's Great Red Spot

The Great Red Spot is a huge storm that has been blowing continuously for over 400 years. Its winds can reach speeds of about 435 kilometers per hour (270 miles per hour). This storm has a diameter of 24,800 kilometers (15,400 miles), which is almost twice the diameter of Earth. Scientists believe that a combination of sulfur and phosphorus in the atmosphere gives this storm its color.

Saturn's Rings

Saturn's rings were first observed by Galileo in 1610. The rings are made of pieces of ice and rock. Some of these pieces are as small as a grain of sand, while others are as large as a house. Scientists think the rings may be pieces of comets, asteroids, or moons that broke apart near Saturn and were pulled into orbit around it.

Until 1977, scientists thought Saturn was the only planet with rings. As scientists observed the outer planets, they also found faint rings around Jupiter, Uranus, and Neptune.

Venus's Surface

The surface of Venus shows evidence of violent volcanic activity in the past. Venus has shield and composite volcanoes similar to those found on Earth. Long rivers of lava have been seen on Venus.

Mars's Rocks

These dark boulders are volcanic rock fragments that have been found on Mars. These rocks look similar to rocks found near lava flows on Earth. On Earth, these types of rocks only form in the presence of water.

≡ *Quick Lab*

Planet Sizes

1. **Use Numbers** Using a scale of 2,000 kilometers = 1 centimeter, find the diameter of each of the planets in centimeters.

2. **Make a Model** Using a ruler and scissors, cut circles out of poster board to show the sizes of the planets. Then label each planet.

3. Arrange the planets in order from nearest to farthest from the Sun.

4. How do the sizes of the inner and outer planets compare?

✓ *Quick Check*

Infer What could the fact that there are volcanoes on Venus mean about the interior of the planet?

Critical Thinking How do winds form on Jupiter?

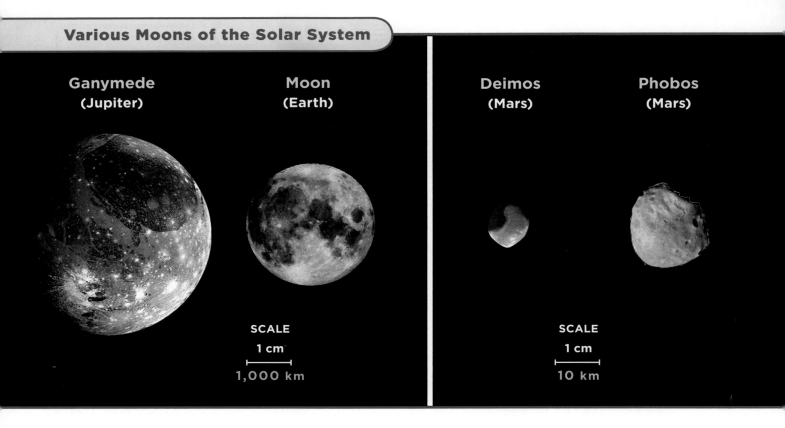

Ganymede
(Jupiter)

Moon
(Earth)

Deimos
(Mars)

Phobos
(Mars)

SCALE
1 cm
1,000 km

SCALE
1 cm
10 km

How do the moons compare?

A **moon** is a natural object that orbits a planet. Different planets have different numbers and sizes of moons.

The inner planets have fewer moons than the outer planets. Mercury and Venus do not have moons. Earth has one moon and Mars has two.

With at least 63 moons, Jupiter has the most moons of any planet in the solar system. Saturn has 47 moons. Astronomers have discovered 27 moons around Uranus and 13 moons orbiting Neptune. As astronomers observe the outer planets with better telescopes and with space probes, they continue to find more moons.

Moons are also called *satellites*. A **satellite** is an object in space that circles around another object. While moons are natural satellites, people also put objects into space that orbit Earth or other planets. These objects are called *artificial satellites*. They include weather and communications satellites and space probes that orbit planets to observe their surfaces.

The size of the moons in the solar system varies. Some of the moons are only a few kilometers wide. Jupiter's Ganymede is the largest moon in the solar system. Ganymede is larger in diameter than Pluto and Mercury. Earth's Moon is also larger than Pluto, and is visible without a telescope. Ganymede is the only other moon that may be seen without a telescope.

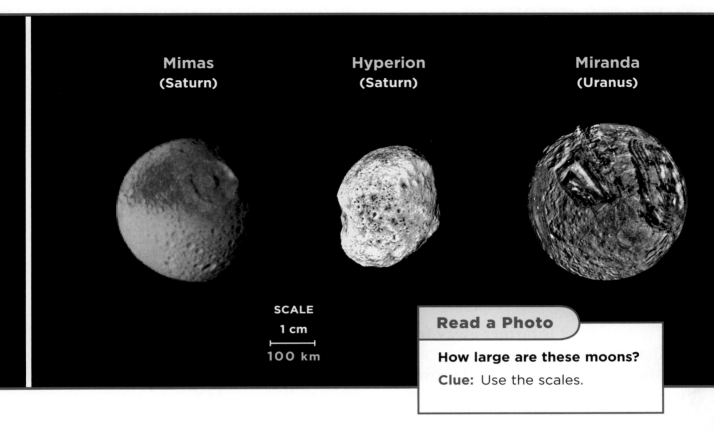

Mimas
(Saturn)

Hyperion
(Saturn)

Miranda
(Uranus)

SCALE
1 cm
⊢——⊣
100 km

Read a Photo

How large are these moons?

Clue: Use the scales.

Forming Craters

Sometimes small objects in space collide with large objects. When this happens, the impact often forms a crater, or a bowl-shaped hole, on the large object. Many moons have craters on their surfaces. Craters vary in size because the objects that hit a moon are different sizes and travel at different speeds.

On Earth's Moon, the impact of an object knocks the surface material away so the rock underneath is exposed. The surface material piles around the edges of the crater. This makes the Moon's craters distinct and easy to see from Earth.

Ganymede's surface is made of ice and rock. The dark rock is about 4 billion years old. The light-colored rock is somewhat younger. Craters are seen on both types of rock, which means that objects have been hitting Ganymede for at least 4 billion years. Unlike those on the Moon, craters on Ganymede are flat. This may be because flowing ice on Ganymede's surface smooths out their edges.

Deimos, Mars's smaller moon, is composed of carbon-rich rock and ice. Deimos's surface has craters that have been partially filled in by loose rock.

 Quick Check

Infer How are a moon and an artificial satellite different?

Critical Thinking What happens when objects in space collide with Earth?

What are asteroids, comets, and meteors?

Different types of small objects are present in space. These objects include comets, asteroids, and meteors.

Comets

A **comet** is a mixture of frozen gases, ice, dust, and rock that moves in an elliptical orbit around the Sun. Comets are thought to be bits of material left over from the formation of the solar system about 4.6 billion years ago.

When a comet is farther away from the Sun, the gases and ice in the comet are frozen. As the comet moves toward the Sun, the core of the comet, or the *nucleus*, warms up. Some of the ice and dust in the core form a cloud, or *coma*, around the nucleus. Together, the nucleus and coma make up the *head* of the comet.

As the comet gets closer to the Sun, radiation from the Sun pushes some of the coma away from the comet. This material forms a glowing tail that may stretch millions of kilometers behind the head. Sometimes two tails will form. One tail is made of ice and one is made of gases.

Heat energy moves out from the Sun in every direction. As a comet moves around the Sun, the head stays closest to the Sun and the tail trails out behind it. No matter where the comet is in its path around the Sun, the comet's tail always points away from the Sun.

Comets orbit around the Sun, but the amount of time that their orbits take is different. Halley's Comet was the first comet whose return was predicted. It gets close to Earth about every 76 years, most recently in 1986. The next time it will be near Earth is in 2061.

Comets have tails of ice and gases.

The Leonids are a meteor shower that occurs every year in mid-November.

Asteroids

An **asteroid** is a rock that revolves around the Sun. Most of the thousands of asteroids in the solar system are located between Mars and Jupiter in the asteroid belt. Many asteroids have irregular shapes and look like potatoes. Some asteroids are less than 2 kilometers (1 mile) wide, while others can be up to 800 kilometers (500 miles) wide!

Meteors

The solar system is full of other small objects. In space, these objects are called *meteoroids*. If an object crosses paths with Earth and enters Earth's atmosphere, it is called a **meteor**. Most meteors burn up before they reach the ground. When a meteor lands on the ground, it is called a *meteorite*.

Ida is a heavily cratered, irregularly shaped asteroid.

✓ Quick Check

Infer As scientists identify materials in comets, what might they infer about the materials that existed as the solar system formed?

Critical Thinking Draw the location of the tail and head of a comet as it moves around the Sun.

How do we explore the solar system?

Exploration of other worlds started in 1959 when a Soviet rocket carrying scientific instruments landed on the Moon. Since then, we have sent space probes to orbit and land on all of the planets in the solar system. A *space probe* is a vehicle carrying instruments that is sent from Earth to explore an object in space.

The first space probe to visit a planet arrived at Venus in 1965. In 1969, the United States sent the first astronauts to the Moon. An astronaut is a person who travels in a space vehicle. The Moon is the only place in space that astronauts have explored.

In 2004, two small robot cars, or rovers, landed on Mars. The rovers, named *Spirit* and *Opportunity*, drove over the Martian surface. Cameras took pictures of soil, pebbles, and rocks. Instruments aboard the rovers examined the Martian surface and found evidence that liquid water may once have existed on Mars. As far as

we know, liquid water is required to support life.

NASA is planning a mission to use rovers to collect Martian soil and bring it back to Earth. NASA may also use airplanes and balloons to study the atmosphere on Mars.

Other space probes have observed comets and asteroids. The *New Horizons* space probe launched in January 2006 and should reach the dwarf planet Pluto in 2015. This space probe will analyze Pluto's surface, geology, and atmosphere. Another space probe called *Dawn* will explore Ceres.

 Quick Check

Infer What might be inferred from the discovery that liquid water may once have existed on Mars?

Critical Thinking Design your own spaceship that is capable of carrying astronauts to the Moon.

This art shows what one of the Mars rovers might look like as it travels across the Martian surface.

Visual Summary

The solar system is made up of the Sun, the planets and their moons, asteroids, meteoroids, and comets.

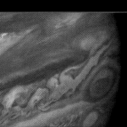

The planets and moons in the solar system vary in size and surface material.

Space probes are sent to objects in the solar system to gather information about them.

Make a FOLDABLES™ Study Guide

Make a Three-Tab Book. Use the titles shown. On each tab, summarize what you have learned.

The solar system is made up of...

The planets and moons...

Space probes are sent to...

Think, Talk, and Write

1. **Main Idea** What are planets?

2. **Vocabulary** A mixture of frozen gases, ice, dust, and rock that orbits the Sun is a(n) _____.

3. **Infer** If you knew that probes that traveled to Venus were crushed after a few hours, what could you infer about atmospheric pressure on Venus?

Clues	What I Know	What I Infer

4. **Critical Thinking** If you land on the closest planet to the Sun, where are you?

5. **Test Prep** What is the largest planet in the solar system?
   - **A** Earth
   - **B** Mars
   - **C** Saturn
   - **D** Jupiter

6. **Test Prep** Which was the first planet to be explored by a space probe?
   - **A** Mercury
   - **B** Venus
   - **C** Jupiter
   - **D** Mars

 Writing Link

Science Fiction
Read *The War of the Worlds* by H. G. Wells. Write a report about the novel. Discuss how much of the story is based on fact and how much is fiction.

 Art Link

Planet Surfaces
Research the surface features of one of the inner planets. Then draw an illustration of what the planet's surface might look like.

Voyager Discoveries

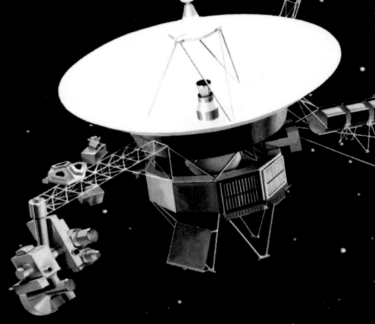

In 1977 NASA launched the Voyager Interstellar Mission to explore Jupiter, Saturn, Uranus, Neptune, and their moons. Each of the mission's trips had to be very precisely planned. Speeds and distances had to be accurately calculated. The two Voyager spacecraft had to be close enough to each planet to collect data and to get a boost from that planet's gravity in order to be propelled toward their next destination. At the same time, the spacecraft had to be far enough away from the planets that they would not go into orbit around them. All of NASA's careful planning worked. The Voyager mission has provided scientists with new and closer looks at our farthest neighbors.

Jupiter—1979

Images show Jupiter's rings. Volcanic activity is observed on Io, one of Jupiter's moons. Europa, another moon, may have an ocean under it's icy crust.

Saturn—1980

Scientists get a close look at Saturn's rings. The rings contain structures that look like spokes or braids.

Uranus—1986

Scientists discover additional dark rings around Uranus. They also see ten new moons. Voyager sends back detailed images and data on the planet, its moons, and dark rings.

Neptune—1989

Large storms are seen on the planet. One of these storms is Neptune's Great Dark Spot. Scientists thought Neptune was too cold to support this kind of weather.

After observing these planets, the Voyager spacecraft kept traveling. They are the first human-made objects to go beyond the heliosphere. The heliosphere is the region of space reached by the energy of our Sun. It extends far beyond the most distant planets in the solar system.

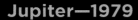

Write About It

Cause and Effect

1. What caused the Voyager spacecraft to be propelled from one planet toward the next?
2. How did scientists benefit from the Voyager missions?

LOG ON e-Journal Research and write about it online at **www.macmillanmh.com**

Cause and Effect

▶ Look for the reason why something happens to find a cause.

▶ An effect is what happens as a result of a cause.

AMERICAN MUSEUM of NATURAL

Stars and the Universe

Look and Wonder

If you look out into space from Earth, you would see stars such as these in the Carina Nebula. What makes some of these stars appear brighter than others?

How does distance affect how bright a star appears?

Form a Hypothesis

If one star gives off more light than another star, but they appear to be the same brightness to an observer, what does this mean about the distance of the stars from the observer? Write your answer as a hypothesis in the form "If one star gives off more light than another star but both appear to be the same brightness to an observer, then . . ."

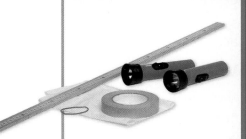

Materials

- **2 flashlights**
- **tissue paper**
- **rubber band**
- **masking tape**
- **meterstick**

Procedure

1 Cover the front of one flashlight with a few layers of tissue paper. Place a rubber band around the paper.

2 **Make a Model** Let each flashlight represent a star. Place a strip of masking tape on the floor. Have two classmates stand behind the tape and turn the flashlights on.

3 Select the flashlight that appears brighter and have the classmate holding it slowly move away from you. When do the two stars appear to have the same brightness? Measure the distance.

Draw Conclusions

4 **Infer** What factors affect how bright a star looks to an Earth observer?

5 **Communicate** Did your classmates see the stars as having the same brightness at different distances? What might that mean about individual observations of stars?

Explore More

Two stars give off the same amount of light, but one star looks dimmer. Form a hypothesis and use models to test your prediction. Collect data and communicate your results.

Step 2

Read and Learn

▶ **Main Idea**

Stars are organized into systems such as galaxies and solar systems.

▶ **Vocabulary**

star, p. 458
nebula, p. 458
white dwarf, p. 459
supernova, p. 460
black hole, p. 460
constellation, p. 462
light-year, p. 463
galaxy, p. 465

LOG ON ℮-Glossary
at www.macmillanmh.com

▶ **Reading Skill** ✓

Problem and Solution

Problem

↓

Steps to Solution

↓

Solution

How do stars form?

Stars form when matter comes together and starts to give off energy. A **star** is an object that produces its own energy, including heat and light. Stars can go through stages, or cycles, between their beginning and ending. Different kinds of stars have different cycles. The cycle of a star depends on how much hydrogen the star contains. A star's cycle ends when it stops giving off energy.

All stars form out of a nebula. A **nebula** is a huge cloud of gases and dust. Gravity pulls the mass of the nebula, most of which is hydrogen atoms, closer together. As the hydrogen atoms move closer, they collide with each other.

The collisions produce heat and the temperature in the cloud begins to rise. When the temperature reaches at least 10,000,000°C (18,000,000°F), hydrogen atoms begin combining together to form atoms of helium. This process gives off tremendous amounts of heat and light. The spinning cloud has become a *protostar*, or beginning star.

Stages of a Medium-Sized Star

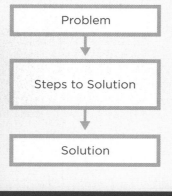

1 nebula

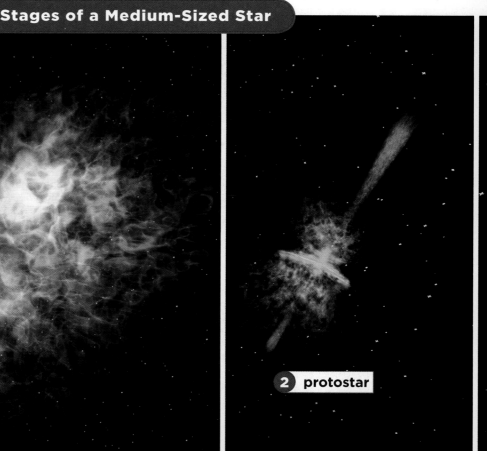

2 protostar

The Sun, and other stars like it, started with a medium amount of hydrogen. That hydrogen is the fuel that produces energy in the Sun. For a few billion years, hydrogen continues to combine to form helium and the star increases in temperature.

Eventually the heat forces the hydrogen on the edge of the star to expand into space. As the expanding hydrogen moves further away from the center of the star, it cools and turns red.

At this stage in its cycle, the star has become a *red giant*. A red giant is many times larger than the original star. In the star's core, the temperature has risen to about 100,000,000°C (180,000,000°F). Helium atoms now combine to form atoms of carbon.

Eventually all the helium is gone and the star can no longer combine helium to form carbon. Now the star begins to cool and shrink, becoming a white dwarf. A **white dwarf** is a small and very dense star that shines with a cooler white light. The white dwarf stage is the end of a medium-sized star's cycle.

About 10 billion years pass during this cycle. Since the Sun is about 5 billion years old, it is about halfway through its cycle.

✅ Quick Check

Problem and Solution What data could you use to find out the stage of a star's cycle?

Critical Thinking Contrast a protostar and a white dwarf.

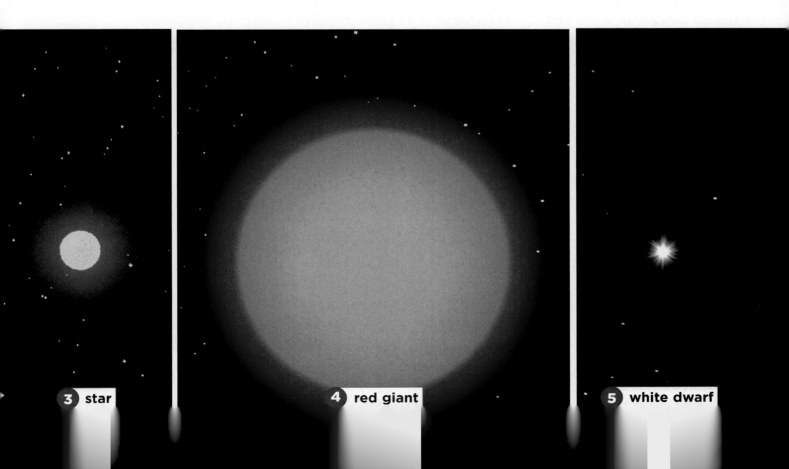

3 star   4 red giant   5 white dwarf

What happens to larger stars?

Stars that start off with greater amounts of hydrogen end their cycle differently. After they become red giants, the temperature of the core of these stars increases to about 600,000,000°C (1,080,000,000°F). At this temperature, their atoms combine to form atoms of iron.

Eventually the iron core produces more energy than gravity can hold together and the star explodes. The exploding star is called a **supernova**. Supernovas shine brightly for days or weeks and then fade away. A supernova will form a new nebula.

If a star is very massive, it may end its cycle as a black hole. A **black hole** is an object that is so dense and has such powerful gravity that nothing can escape from it, not even light. Stars are characterized by their size, color, and temperature. The Sun is a medium-sized yellow star whose surface temperature is about 6,000°C (11,000°F).

Giant stars have diameters that are 10 to 100 times that of the Sun. Super giants may have diameters that are 1000 times that of the Sun. Neutron stars are the smallest stars.

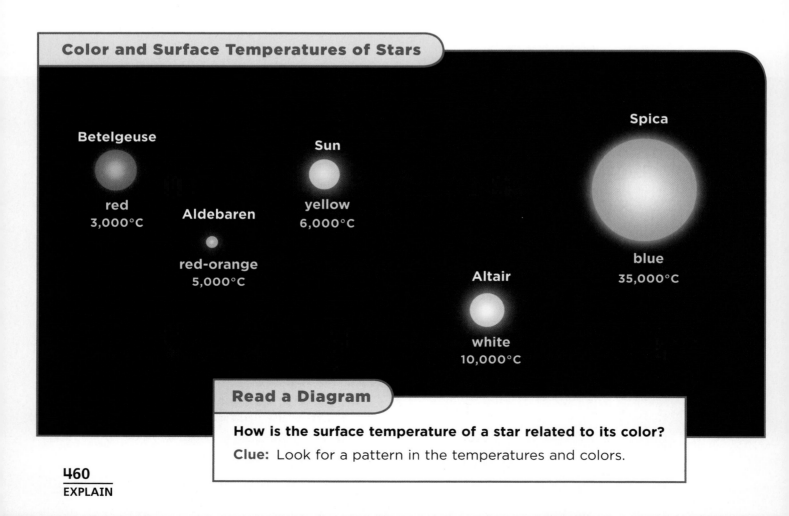

Color and Surface Temperatures of Stars

Betelgeuse
red
3,000°C

Aldebaren
red-orange
5,000°C

Sun
yellow
6,000°C

Altair
white
10,000°C

Spica
blue
35,000°C

Read a Diagram

How is the surface temperature of a star related to its color?

Clue: Look for a pattern in the temperatures and colors.

▲ Stars with planets might look like this.

Planets around distant stars are too dim, small, and far away to be seen even through a telescope. How are these planets discovered? Remember that gravity causes all objects to pull on all other objects. When scientists observe a star whose motion is not smooth, they infer that another source of gravity is present.

By measuring the motion of the star, astronomers can calculate the mass and distance from the star of the possible planet. Using such methods, astronomers have discovered what may be more than 160 planets beyond our solar system.

Most of these planets are probably gas giants. However, scientists have reported finding what may be a rocky planet orbiting a red dwarf. Scientists calculated that the planet was five times more massive than Earth and three times farther from its star than Earth is from the Sun. Temperatures on its surface were thought to be about −364°F (−220°C).

Astronomers found this planet by analyzing data of the star's brightness for changes that indicated that a planet passed in front of the star. This method is called *gravitational microlensing*.

 Quick Check

Problem and Solution How can astronomers discover planets around distant stars that they cannot see?

Critical Thinking What may happen to massive stars at the end of their cycles?

What are constellations?

When people in ancient cultures looked at the night sky, they saw patterns called **constellations** in the stars. Constellations were often named after animals, characters from stories, or familiar objects. Some constellations have been very useful to both ancient and modern travelers.

For example, if you can see either the Big Dipper or the Little Dipper in the night sky, you can follow the line that their stars make to find Polaris, the North Star. If you travel in the direction of Polaris, you will be moving north. If you ever become lost in the woods or at sea, look for Polaris in the night sky. It will help guide you to safety.

The ancient Greeks divided the sky into 12 sections. They named some constellations after characters from Greek myths, such as Orion, a hunter, and Hercules, a hero.

Finding Polaris

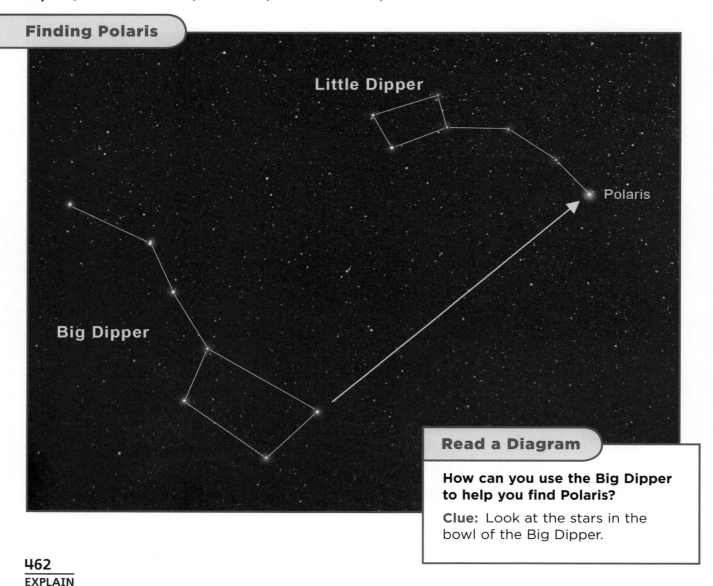

Little Dipper

Polaris

Big Dipper

Read a Diagram

How can you use the Big Dipper to help you find Polaris?

Clue: Look at the stars in the bowl of the Big Dipper.

The Greeks named other constellations after animals, such as Taurus, the bull, or Ursa Major, the big bear.

The ancient Chinese divided the sky into four major regions. The name of each region included a color, an animal, and a direction. For example, the western region was called the White Tiger of the West. Within each of these four major regions are seven smaller areas.

Star Distance

How far away are the stars in the constellations? After the Sun, the next closest star to Earth is called Proxima Centauri. It is about 40,000,000,000,000 kilometers (24,800,000,000,000 miles) away. Because stars are so far from Earth, writing their distance in kilometers becomes awkward.

To simplify the writing of such distances, astronomers use a unit called a *light-year*. A **light-year** is the distance that light travels in a year, or about 9.5 billion kilometers (5.9 billion miles). Proxima Centauri is 4.2 light-years from Earth.

Greek constellations

Chinese constellations

Quick Check

Problem and Solution How could you travel using the stars as a guide?

Critical Thinking What animals, characters, and objects would you choose to name constellations after?

Messier 81
spiral galaxy

NGC 1300
barred spiral galaxy

M32
ovoid elliptical galaxy

ESO 510-G13
disc-shaped elliptical galaxy

NGC 1427A
irregular galaxy

NGC 7673
irregular galaxy

What are star systems?

If you look at the night sky through a small telescope, you can see individual stars and some of the planets in our solar system. If you look carefully, you might see hazy patches of faint light. These hazy patches are galaxies. A **galaxy** is a huge, very distant collection of stars.

Each galaxy holds billions of stars. The universe is full of galaxies, and each galaxy varies in size and shape. The three basic shapes of galaxies are spiral, elliptical, and irregular.

Spiral galaxies are shaped like a pinwheel with many arms. They are fairly flat with a bulge in the middle.

Our solar system is in an arm of a spiral galaxy called the *Milky Way*. The individual stars that you see in the sky are part of the Milky Way galaxy.

Some spiral galaxies only have two arms. These are called *barred spiral galaxies*. The arms of this kind of galaxy spread out from a bar of stars that cut across the center of the galaxy.

Elliptical galaxies are rounded. They can be shaped like an egg or a thick pancake. They do not have arms.

Irregular galaxies do not have distinct shapes. They may look like clouds or blobs.

Star Clusters

Some stars in a galaxy form clusters. These clusters range in size from a few hundred stars to more than 100,000 stars. *Globular clusters* are shaped like a sphere. They hold 100,000 or more stars. If you looked at a star cluster without a telescope, the star cluster would look as if it were a single star.

Binary Stars

Sometimes when you aim a telescope at what looks like a single star, you discover two points of light instead of one. This happens when two stars form near each other and rotate around each other. These two stars are called *binary stars*. The prefix *bi-* means "two."

How would a blinking star indicate a binary star? An apparently single star blinks because it has a dim partner star that regularly comes between it and an observer on Earth. When this happens, the dimmer star blocks the brighter star's light from reaching Earth. It's as if you repeatedly passed your hand between your eyes and a lighted bulb. The bulb would appear to blink each time your hand passed in front of it.

 ## Quick Check

Problem and Solution You want to demonstrate a way of identifying a binary star without using a telescope. What do you do?

Critical Thinking What would a spiral galaxy look like if viewed from the side?

How did the universe form?

Astronomers have found evidence that the universe is expanding like ripples made by a stone dropped in a pond. The universe includes all matter and energy, including everything from the tiniest parts of atoms to the tremendous explosions of dying stars.

If the universe is expanding in all directions, it had to have started at a single point. That point was like the spot where a rock drops into a pond, sending ripples outward. The *big bang theory* states that the universe started with a big bang at a single point and has been expanding ever since. Evidence suggests that the big bang happened about 13.7 billion years ago.

The big bang theory states that all of the matter and energy in the universe exploded outward from a single point.

Quick Lab

Expanding Universe

1. **Make a Model** Blow the balloon up a little. The balloon represents the universe shortly after the big bang. Place stickers on it to represent galaxies.

2. **Observe** Blow the balloon half-full of air. What happens to the size of the stickers? To the distance between the stickers?

3. **Observe** Blow the balloon full of air. What happens to the size of the stickers? To the distance between the stickers?

✔ Quick Check

Problem and Solution How can you tell from the direction of the universe's expansion that the universe began at a single point?

Critical Thinking Which is older, the universe or the Sun?

Lesson Review

Visual Summary

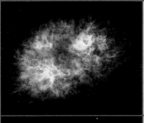

Stars produce their own light and heat.

The Sun is one of many different kinds of stars that make up the Milky Way galaxy.

Ancient cultures saw different constellations in the night sky.

Make a **FOLDABLES**™ Study Guide

Make a Trifold Book. Use the titles shown. Then summarize what you have learned.

Think, Talk, and Write

1. **Main Idea** How are stars organized?

2. **Vocabulary** An object so dense that nothing can escape from it, not even light, is a(n) _____.

3. **Problem and Solution** What could you test to determine the stage of a star's cycle?

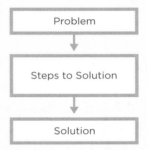

4. **Critical Thinking** What can you infer about the mass of a star whose cycle ends as a white dwarf?

5. **Test Prep** What is a supernova?
   A a small star   C an exploding star
   B a giant star   D a nebula

6. **Test Prep** What is a huge, distant family of stars?
   A a galaxy   C a universe
   B a planet   D a constellation

Writing Link

Discovering Galaxies
Edwin Powell Hubble was an American astronomer who discovered galaxies outside of the Milky Way. Research and write about Hubble's Law and how it relates to the big bang theory.

Social Studies Link

Supernovas
Research the supernovas that have been seen in the last thousand years. Pick one supernova and describe the last time it was seen.

Be a Scientist

Materials

newspaper

pan

cocoa powder

ruler

plastic spoon

white flour

3 rubber bands

3 marbles of different sizes

How do craters form?

Form a Hypothesis

You know that craters form when an object in space hits another object. Does the size of an object affect the size of the crater it forms? Write your answer as a hypothesis in the form "If a larger object hits, then . . ."

Test Your Hypothesis

1. Cover the floor with newspaper and place a pan on the paper.

2. **Make a Model** Fill the pan with cocoa powder to about 1 cm. Gently tap the pan until the cocoa powder is smooth. Using the spoon, shake white flour on top to represent topsoil.

3. By wrapping a cut rubber band around each marble, measure the diameter of three marbles of different sizes.

4. Drop the largest marble from 20 cm above the pan. Measure the diameter of the crater and record your data.

5. Repeat step 4 for the other 2 marbles. Make sure each marble falls in a different area of the pan.

Step 2

Step 3

Step 4

Draw Conclusions

6 **Analyze Data** How does the diameter of the crater compare to the diameter of the marble?

7 What did you see at the crater sites? Why did this happen?

8 How is this model similar to what happens when an object hits the surface of the Moon?

9 What are the controlled, independent, and dependent variables?

This crater shows what happens when an object hits a surface.

How does height affect crater size?

Form a Hypothesis

You now know the effect that objects of different sizes have on crater formation. What happens when similar-sized objects hit from different heights? Write your answer as a hypothesis in the form "If an object hits from a greater height, then . . ."

Test Your Hypothesis

Design an experiment to test your hypothesis. Write out the materials you need and the steps you will take. Record your results and observations.

Draw Conclusions

What were your independent and dependent variables? Did your experiment support your hypothesis?

What effect does the surface material have on crater formation? Think of a question and design an experiment to answer it. Your experiment must be organized to test only one variable. Keep careful notes as you do your experiment so another group could repeat the experiment by following your instructions.

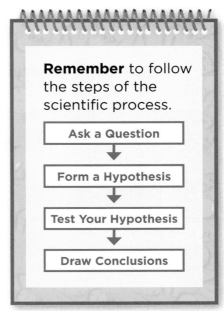

Remember to follow the steps of the scientific process.

| Ask a Question |
| Form a Hypothesis |
| Test Your Hypothesis |
| Draw Conclusions |

Visual Summary

Lesson 1
Gravity and inertia keep Earth in orbit around the Sun.

Lesson 2
The Moon is Earth's natural satellite.

Lesson 3
Our solar system is made up of the Sun, the eight planets and their moons, and comets, asteroids, and meteoroids.

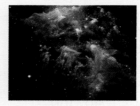

Lesson 4
Stars are organized into systems such as galaxies and solar systems.

Make a FOLDABLES™ Study Guide

Assemble your lesson study guide as shown. Use your study guide to review what you have learned in this chapter.

Fill in each blank with the best term from the list.

asteroid, p. 451

galaxy, p. 465

lunar eclipse, p. 435

nebula, p. 458

orbit, p. 423

planet, p. 444

rotation, p. 426

tide, p. 436

1. Gravity and inertia keep Earth moving around the Sun in its _____.

2. When the Moon moves into Earth's shadow, a(n) _____ occurs.

3. A large object that orbits a star is called a(n) _____.

4. All stars form from a huge cloud of gases and dust called a(n) _____.

5. Every 24 hours, Earth makes a complete _____, or spin on its axis.

6. A rock that revolves around the Sun is called a(n) _____.

7. A huge, distant collection of stars is called a(n) _____.

8. The rise and fall of the ocean's surface is called the _____.

Answer each of the following in complete sentences.

9. Infer What is the relationship between the universe, galaxies, solar systems, stars, and planets? How do Earth, Moon, and Sun fit into this picture?

10. Fact and Opinion Write a paragraph using facts to explain how the revolution of Earth causes the seasons.

11. Use Numbers The Islamic calendar is based on the revolution of the Moon around Earth, which takes 29.5 days. An Islamic year is made up of 12 complete Moon orbits. Calculate how much a year on an Islamic calendar differs from that on a solar calendar.

12. Critical Thinking You observe a small, white, very dense star with a telescope. Identify this star and explain whether this star is older or younger than our Sun.

13. Explanatory Writing What two planets are shown below? Explain how you know which planets they are.

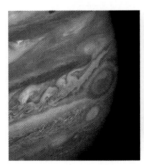 

14. What is in outer space?

Moon Music

Your goal is to analyze facts and opinions in songs about the Moon.

1. Research and write down the lyrics of a song about the Moon.

2. Write down any statements made about the Moon in your song. Identify each as either fact or opinion.

Analyze Your Results

▶ Why do you think so many songs have been written about the Moon? Why is the Moon's true nature often misunderstood?

1. Which type of galaxy is shown below?

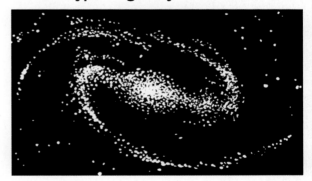

A spiral galaxy

B ovoid elliptical galaxy

C irregular galaxy

D barred spiral galaxy

Careers in Science

Weather Observer

There is an old saying that if you do not like the weather now, wait and it will change. How do people know what the weather will change to? After you finish high school, you might enjoy working as a weather observer. As a weather observer, you would collect data about weather conditions. You would be trained to use instruments that measure temperature, humidity, and air pressure. You would read radar scans and satellite photographs. The weather forecasts meteorologists make depend on data collected by weather observers.

Weather observers collect data.

Astronomer

Are you interested in looking for planets around distant stars or watching solar systems form? Astronomers study the stars and learn about the other planets and suns in the universe. As an astronomer, you would use telescopes and satellites to gather data about other solar systems. Then you would interpret that data to find out what the stars and planets are made of and how old they are. To be an astronomer, you need to be good at math and physics, have strong computer skills, and obtain a doctoral degree in astronomy. After that, you might say the sky's the limit!

▲ Astronomers observe stars and planets.

Matter

The barnacles and the rusty chain
are both forms of matter.

Scientists can use plants to help remove dangerous metals from soil.

Alpine pennycress can clean soils contaminated with zinc and cadmium.

Green and Clean

Plants as Pollution Control

Heavy metals—for example, lead (Pb), mercury (Hg), cadmium (Cd), and arsenic (As)—are toxic. They can make people and animals sick. They occur in small quantities in Earth's crust. They only become a problem when they build up in larger amounts. This often happens when people use them to make products or dump them as waste. For example, arsenic was once used in insecticides. In the past, lead was added to gasoline and paint. Over time, these toxic metals have accumulated in our soil.

Cleaning up heavy metals isn't easy. Scientists have discovered that certain plants can remove toxic elements from the soil or water. They take up heavy metals through their roots and store them in leaves. Then people can collect the contaminated plant parts, leaving behind soil that is cleaner and safer. They might even be able to remove the metals from the plants and reuse them!

Brake fern soaks up toxic arsenic, safely cleaning polluted soils.

Sunflowers have an unusual appetite for toxic metals, such as uranium.

Write About It

Response to Literature This article describes how plants are used to help clean polluted soil. Research additional information about cleaning up waste. Write a report about the cleaning process. Include facts and details from this article and from your research.

LOG ON e-Journal Write about it online at **www.macmillanmh.com**

CHAPTER 9

Comparing Kinds of Matter

The
Big Idea

How can you classify matter?

Key Vocabulary

matter
anything that has mass and takes up space (p. 481)

density
the amount of matter in a certain volume of a substance; found by dividing the mass of an object by its volume (p. 482)

buoyancy
the upward push of a liquid or gas on an object placed in it (p. 483)

element
a pure substance that cannot be broken down into any simpler substances through chemical reactions (p. 490)

atom
the smallest unit of an element that retains the properties of that element (p. 491)

malleability
the ability to be bent, flattened, hammered, or pressed into new shapes without breaking (p. 504)

More Vocabulary

Properties of Matter

Look and **Wonder**

This hot air balloon is pretty big, yet it easily floats in the air above the Red Sea in Egypt. How is this possible?

478
ENGAGE

Which has more matter?

Form a Hypothesis

Which do you think has more matter—the balloon or the tennis ball? Does having more matter make an object larger? Does it make it heavier? Write your answer as a hypothesis in the form "If matter increases, then an object will . . ."

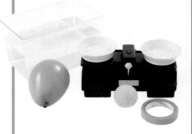

Materials

- inflated balloon
- tennis ball
- container of water
- tape
- equal pan balance

Test Your Hypothesis

1. **Measure** Place one object under water. Record how high the water rises with some tape, then remove the object. Do not spill any water! Next place the second object in the water and record the water level; this is your dependent variable.

2. Place the objects on either side of the equal pan balance. Which is heavier? This is your independent variable.

3. Repeat all your measurements to verify your answers.

Draw Conclusions

4. **Analyze** Did the heavier object also raise the water higher? Why or why not? Which object has more matter? Did the test support your hypothesis?

Step 1

Explore More

What if you were given a large bag of popped popcorn and a small bag with an equal number of popcorn kernels. Which do you think would have more matter? Form a hypothesis and test it. Then analyze and write a report of your results.

Step 2

▶ **Main Idea**

Matter can be described by many different properties.

▶ **Vocabulary**

mass, p. 480
weight, p. 481
volume, p. 481
matter, p. 481
density, p. 482
buoyancy, p. 483

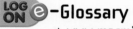

e-Glossary
at www.macmillanmh.com

▶ **Reading Skill** ✔

Classify

How can you describe matter?

Pretend you have lost your backpack. You want to describe your backpack to your friends so they can help you look for it. How could you describe it so that they know which one is yours? You could say how large the backpack is, or how many things there are inside. You could talk about its color, texture, or hardness. In each case you are describing a *property* (PROP•uhr•tee) of the backpack. A property is something that can be observed about an object or a group of objects. Properties of matter can describe the amount of matter or what that matter is like.

Mass

One way to describe the backpack would be its mass. **Mass** (mas) is the amount of matter in an object. This property compares the amount of matter in a sample to standard amounts. Mass can be measured on an equal pan balance. Gravity pulls on the standard pieces and on the sample. When the balance levels out, the amount of matter in the sample and the standard pieces must be the same. Mass is measured in kilograms or grams. One kilogram is equal to 1000 grams. Adding up the grams of mass in the standard pieces gives the mass of the sample in grams. The mass of an object is always equal to the sum of the masses of the pieces of the object.

The sum of all the masses in the backpack, plus the backpack itself, is equal to the mass of the full backpack.

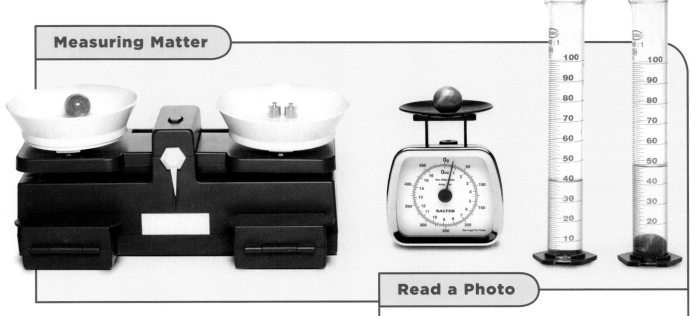

What is the volume of the marble?

Clue: What is the water level before and after the marble is in the cylinder?

Weight

Try holding a marble in one hand and a basketball in the other—they feel different. It is harder to hold up the basketball. You are feeling the weight of each ball. **Weight** (wayt) is how strongly gravity pulls on an object. If an object has more mass, it will also have more weight. Weight is measured in Newtons (N). One Newton is equal to 0.225 pounds (lbs) in the English system. Gravity is greater on planets with more mass, so an object's weight depends on the planet it is on. An object's mass, in contrast, is the same on different planets.

Volume

Mass and weight describe the amount of matter in an object, but what about its size? **Volume** (VOL•yewm) measures how much space matter takes up. A marble's volume makes the water level rise when you place it in a graduated cylinder. Matter in the marble takes up space and pushes water out of the way. The

change in the water level when an object is placed under water tells us the object's volume. Volumes of liquid are often measured in milliliters (mL) by using a graduated cylinder, a beaker, or a measuring cup. The volume of solids is usually measured in cubic centimeters (cm^3). A volume of 1 cm^3 equals 1 mL.

Now that you understand mass and volume, we can use the best definition for matter. **Matter** (MAT•uhr) is anything that has mass and volume. That is a broad definition! Almost everything in the world around you is matter.

Quick Check

Classify Which units are used to measure weight, mass, and volume?

Critical Thinking How could you change an object's volume without changing its mass?

What is density?

Think of a large empty box with a lid. It has a large volume, but little mass. If you place a few marbles in it, it has more mass but the same volume. Add more marbles and you get more mass, but the volume stays the same. As you add more marbles, you are increasing the box's density. **Density** (DEN•si•tee) is the amount of mass for each cm³ (or mL) of a substance. To calculate a sample's density, you divide its mass by its volume.

$$\text{Density} = \frac{\text{mass}}{\text{volume}}$$

A marble and a rubber ball can be about the same size. Which do you think is denser? The marble has more particles, or more mass, inside of it than the ball. The ball has air inside of it. Air is less dense than both glass and rubber. These observations tell you that the marble is denser than the ball.

With boxes and other containers you can change density by adding more mass inside. You could also change the density by changing the volume. You could stretch, crush, or bend the container to let air out or in. Most matter, however, acts like the marble. You cannot add more mass to the inside, nor can you stretch it. Its density will always be the same.

Density and Water

Read a Diagram

How would you represent the density of water?

Clue Look at the density of the marble that sinks and the ball that floats.

Floating and Sinking

When an object is placed in a fluid, gravity pulls the object down. In order for the object to go down, the fluid must move out of the way. What would happen if the fluid was denser and did not get out of the way? The object would stop sinking . . . it would float!

Objects can float as a result of buoyancy (BOY•uhn•see). **Buoyancy** is the resistance to sinking. It occurs because the fluid that is being pushed out of the way pushes back on the object. If an object is denser than the fluid, the object can push harder and it sinks. If the fluid is denser than the object, the fluid can push harder and the object floats.

Buoyancy depends on density. So, if you change the mass or volume of an object, you can change whether or not it will float. If you have a toy boat and keep adding mass to it, it will sink. Buoyancy also depends on shape. A block of aluminum will sink, but an aluminum canoe will float. Why? The canoe's shape holds in air so that it acts like it has a density less than that of water. But a canoe full of water instead of air would sink.

Many fluids also have a property called *surface tension* (SUR•fis TEN•shuhn). In water, every particle pulls itself toward the other particles. This attraction creates a "skin" on the surface. This skin is what surface tension means. If an object is spread over the surface it can rest on this skin even if it would not normally float. If an object is not spread out enough, it will break the skin and sink.

Quick Check

Classify Which properties depend on shape, volume, or mass?

Critical Thinking How would you design a boat to carry large masses across the ocean?

FACT The density of an object depends on the type of material.

What forms can matter have?

A *state of matter* is one of the three common forms that matter can take: solid, liquid, or gas. You interact with these states of matter every day. Desks and books are solids. The water you drink is a liquid. The sky and the air you breathe are made of gases. Each state has its own properties.

The particles in a solid have very little freedom to move. The particles just vibrate in place. A solid stays in a definite shape with a definite volume no matter what container it is in. A rock is a good example. The shape and volume will not change unless something changes the solid, such as if it is broken or heated. Many solids, like table salt, have regular shapes like cubes. In general, a solid is the densest state of matter.

In a liquid, particles move more freely than in a solid. The particles are close together but they can flow past each other. This flowing allows a liquid to take the shape of its container. So liquids have a definite volume but not a definite shape. In general a liquid is the second densest state of matter. Water, however, is denser as a liquid than as a solid.

Gas particles are not close together and can move past each other very easily. A gas has no definite volume or shape. At room temperature, gases move around to fill their container. If the volume of the container increases, the gas expands to fill it. This property allows gases to inflate things like balls and tires. In general, a gas is the least dense state of matter.

Quick Check

Classify Which states of matter have a definite shape? Which states of matter have a definite volume?

Critical Thinking What would happen to the particles in each state if you tried to squeeze them?

gas

solid

liquid

Lesson Review

Visual Summary

Samples of **matter** can be described by properties such as **mass, volume,** and **weight**.

Density measures the mass per volume and determines **buoyancy.**

Matter can exist in one of three common states: **solid**, **liquid** or **gas.**

Make a FOLDABLES™ Study Guide

Make a Trifold Book. Use the titles shown. Summarize what you have learned in each box. Draw sketches to illustrate your understanding.

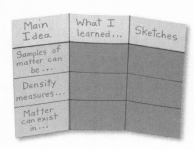

Think, Talk, and Write

1. **Main Idea** How could you predict if a block of plastic will float in water?

2. **Vocabulary** How strongly gravity pulls on an object's mass is the object's _____.

3. **Classify** Which properties do solids and liquids share? Which are different?

4. **Critical Thinking** Would a balance on the Moon, which has one-sixth the gravity of Earth, still read masses correctly? Why or why not?

5. **Test Prep** Which property measures the space taken up by an object?
   A hardness
   B mass
   C volume
   D weight

6. **Test Prep** Which sample most likely has the lowest density?
   A liquid mercury
   B oxygen gas
   C water
   D gold metal

Writing Link

Descriptive Writing
Pick a piece of matter inside your classroom. Research and then describe it according to the properties and states of matter.

Math Link

Density and Volume
You have 80 mL of water. How many hollow cubes that are 2 cm on a side could you fill?

Focus on Skills

Inquiry Skill: Infer

When scientists see something interesting they make careful records of what they find. They then think about what they have seen and try to **infer** what it means about the world as a whole. Understanding why something happens allows you to draw conclusions about how objects react or what properties they may have.

Think about a material that is denser than water. It has more mass per milliliter than water—there is more matter in the same amount of space. Will a material denser than water float? By watching what happens to objects made of these materials, you can infer what keeps objects afloat in general.

▶ Learn It

When you **infer**, you form an opinion after analyzing recorded data. It's easier to analyze data if you organize the information on a chart or in a graph. That way you can quickly see differences among data and make inferences. Most metal objects, such as a spoon or nail, sink quickly. This is because they are denser than water. There are large metal boats, however, that regularly carry heavy cargo across oceans. How can they stay afloat? We will make several model boats to help us infer an answer. The boats will be made out of a material denser than water—the metal aluminum.

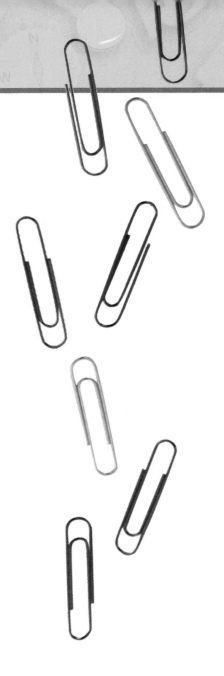

▶ **Try It**

Materials aluminum foil, paper clips, tank of water.

1 Take a sheet of aluminum foil. Use the foil to make a boat. Experiment with different designs. Draw a picture of the boat in a chart.

2 Float the boat in a large pan of water. Place the paper clips into the boat and record what happens. How many paper clips can the boat hold before it completely sinks? Try to **infer** why the boat is sinking.

▶ **Apply It**

1 Record the data and results from two other students in your chart.

2 Now it is time to analyze your data. Do you notice any pattern between the design of the boat and the number of paper clips?

3 As a class, design a boat that would carry the most paper clips possible. Use a final piece of aluminum foil to make the boat and record how many paper clips it can hold. Did this boat hold more paper clips than the others?

4 Think about all the models you have seen. Did the ones that held more paper clips have anything in common? What was happening as more paper clips were added to the boat? Use your observations to **infer** what makes an object float. Communicate your opinions by writing down your conclusions.

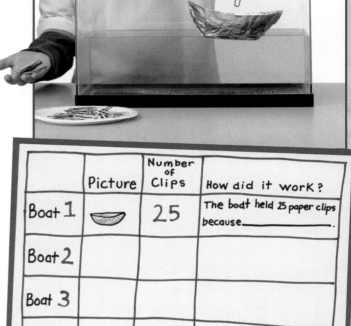

	Picture	Number of Clips	How did it work?
Boat 1		25	The boat held 25 paper clips because_____.
Boat 2			
Boat 3			
Boat 4			

Elements

Look and Wonder

Colored lights can be made by passing electricity through certain gases. These gases are examples of elements. How could you determine which element is in each tube?

Explore

How can you know what's "inside" matter?

Purpose

You will examine 4 sealed boxes to determine what is inside them.

Procedure

1. **Observe** Examine the 4 boxes, but do not open them. You can lift them, shake them, listen to the noises they make, feel the way they shift when you move them, and so on. Don't forget to use the magnet and equal pan balance to learn more about what is inside the boxes. Record your observations.

2. **Infer** Try to determine what is in each box.

Draw Conclusions

3. **Communicate** Describe what you think is in each box.

4. What evidence did you use to make your decisions?

5. When everyone in the class is finished, open the boxes and reveal what is inside. Were your inferences correct? Now that you know what is in each box, explain any wrong guesses you made.

Explore More

What if you were the one to fill the boxes before this experiment? What kind of items could you choose to make the experiment easier? To make it harder? Choose a few items that would fit inside the boxes. Now design a series of tests that would prove that those specific items were inside the boxes.

Materials

- **4 sealed, opaque boxes of different sizes, shapes, and colors**
- **magnet**
- **equal pan balance with set of masses**

Step **1**

Step **1**

Main Idea

All matter is made of elements.

Vocabulary

element, p. 490

metal, p. 491

atom, p. 491

nucleus, p. 492

proton, p. 492

neutron, p. 492

electron, p. 492

molecule, p. 493

LOG ON e-Glossary
at www.macmillanmh.com

Reading Skill ✓

Main Idea and Details

Main Idea	Details

Matter is made of elements just like these models are made of the same building blocks.

What is matter made of?

The toy models in the photograph stand for different things. If you took the models apart, though, you would get the same basic building blocks. If you mix the blocks together you cannot tell which model a block came from. In a similar way, all matter is made of the same set of building blocks: the chemical elements (EL•uh•muhnts). An **element** is a material that cannot be broken down into anything simpler by chemical reactions.

The ancient Greek philosopher Aristotle believed all matter is made of the elements earth, air, water, and fire. Modern scientists know that Aristotle's elements are not true elements. Fire is not matter. Air and earth are made up of many different materials, not just one. Water can be broken down into simpler substances: hydrogen and oxygen. Hydrogen and oxygen, however, cannot be broken down into simpler substances by using chemical reactions. This tells us that hydrogen and oxygen are elements.

Today, scientists know of just over 112 elements with different properties. Three important properties of elements are: state of matter at room temperature, the way they combine with other elements, and whether they are metals, nonmetals, or metalloids.

Most elements are solid, some are gases, and a few are liquid at room temperature. Some elements are more likely to combine with other elements to form new substances. These elements are more

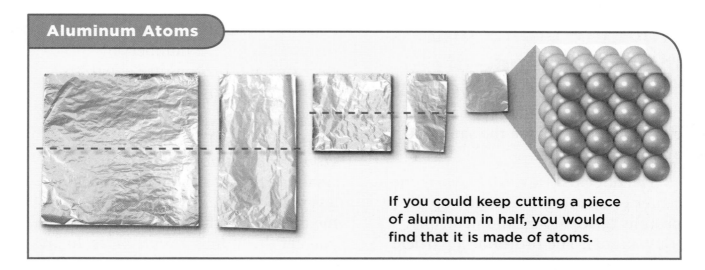

Aluminum Atoms

If you could keep cutting a piece of aluminum in half, you would find that it is made of atoms.

chemically reactive (KEM•i•kuhl•ee ree•AK•tiv). Magnesium, for example, is very reactive and is used in fireworks.

Metals (MET•uhlz) are elements that share common properties like shiny luster, conductivity, and flexibility. Nonmetals are elements that are dull, poor conductors, and brittle. Elements with properties in between are called *metalloids* (MET•uh•loydz).

Smaller and Smaller

If you cut a piece of an element in half, will it still be an element? Yes, the two halves have the same properties as the original element. What if you kept cutting it in half, again and again? Eventually you would have the smallest piece of element possible. John Dalton proposed in 1803 that elements are made of tiny particles. He believed that these particles could not be cut into smaller pieces. Today, we know that Dalton's particles do exist—we call them atoms (AT•uhmz). An **atom** is the smallest unit of an element that retains the properties of that element.

Magnesium adds brightness to fireworks.

 Quick Check

Main Idea and Details What do we mean when we say that matter is made of basic building blocks?

Critical Thinking When two elements combine to form a new substance, is the new substance an element? Why or why not?

What are atoms and molecules made of?

Atoms are made up of even smaller particles. These particles are not elements, but they are the same for every type of atom.

The **nucleus** (NEW•klee•uhs) is the center of an atom. It is made up of protons (PROH•tonz) and neutrons (NEW•tronz). The **proton** is a particle with one unit of positive electric charge. The number of protons in an atom is called the *atomic number* (uh•TOM•ik) and determines which element it is. The **neutron** is a particle with no electric charge—it is neutral.

Atoms also contain **electrons** (i•LEK•tronz), which are smaller particles with one unit of negative electric charge each. Electrons move around in the space outside the nucleus. Most of an atom is empty space. Usually, the numbers of electrons and protons are equal, so atoms have no overall charge.

Protons and neutrons have about the same mass. This mass is called an *atomic mass unit (amu)*. Electrons are smaller and have about 1,800 times less mass than 1 amu. If you add up the mass of all the particles in an atom, you get the atom's *atomic mass*. In an oxygen atom, there are 8 protons, 8 neutrons (usually), and 8 electrons, so its atomic mass is about 16 amu.

In 1913, Niels Bohr pictured an atom's electrons moving around the nucleus like planets moving around a star. Today, we know that the real picture is more complex. The electrons around the nucleus act in many ways like a cloud of electric charge.

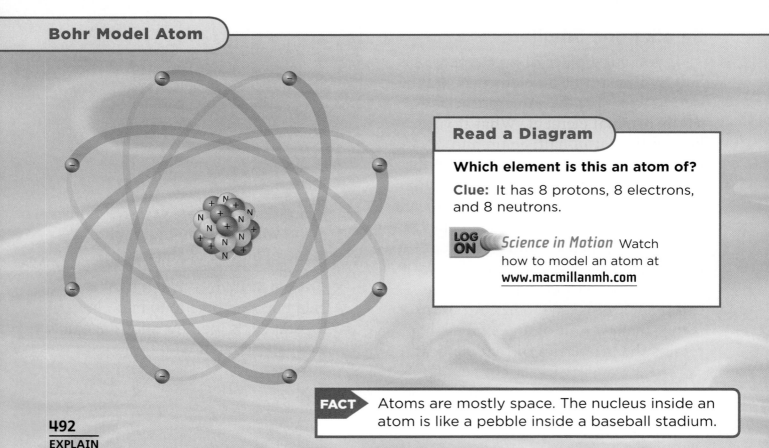

Bohr Model Atom

Read a Diagram

Which element is this an atom of?

Clue: It has 8 protons, 8 electrons, and 8 neutrons.

LOG ON *Science in Motion* Watch how to model an atom at **www.macmillanmh.com**

FACT Atoms are mostly space. The nucleus inside an atom is like a pebble inside a baseball stadium.

Molecules

When you snap toy bricks together to make something, many parts act as one part. The same thing happens for atoms when they form molecules (MOL•uh•kyewlz). **Molecules** are particles with more than one atom joined together. Most of the atoms in the world exist as part of a molecule, not on their own. Objects in the world are just many molecules grouped together.

When a molecule forms from elements, atoms link together through their electrons. This causes molecules to have different properties from their elements. With about 112 elements, the number of different kinds of molecules that can be made is nearly infinite. Molecules allow those few elements to combine into different materials and provide all the variety around you.

Scientists describe molecules by combining letters and numbers into a *chemical formula*. The letters tell which type of elements are in the molecule. The numbers are known as *subscripts*, and they describe the amount of each element. The oxygen we breathe is a molecule formed by two oxygen atoms and its chemical formula is O_2.

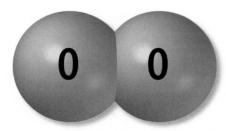

The oxygen molecule is made of two oxygen atoms that are joined together.

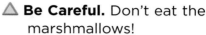

Quick Lab

Inside Atoms and Molecules

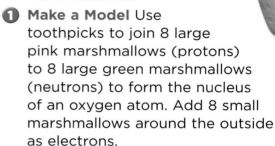

⚠ **Be Careful.** Don't eat the marshmallows!

❶ **Make a Model** Use toothpicks to join 8 large pink marshmallows (protons) to 8 large green marshmallows (neutrons) to form the nucleus of an oxygen atom. Add 8 small marshmallows around the outside as electrons.

❷ Make another oxygen atom or share with another student. Use 2 pipe cleaners to join the atoms by 2 electrons. This is an oxygen molecule (O_2).

❸ How do the shapes of your model atoms compare to the diagrams in this book?

❹ **Communicate** Draw pictures of your atoms and molecule that show their actual shapes better.

❺ In a molecule, electrons move and are sometimes traded between atoms. How could you represent this in your model?

✔ Quick Check

Main Idea and Details How are atoms and molecules different?

Critical Thinking Do you think that molecules are mostly space? Why or why not?

How are elements grouped?

Each chemical element has a name and a symbol. The symbols for most elements are one or two letters. The first letter is always a capital. Second letters are never capitals. Recently discovered elements have temporary symbols with three letters.

Symbols may look like an element's English name, like C for carbon. Many come from ancient names, such as Au for gold, whose Latin name is *aurum*.

The Periodic Table of Elements

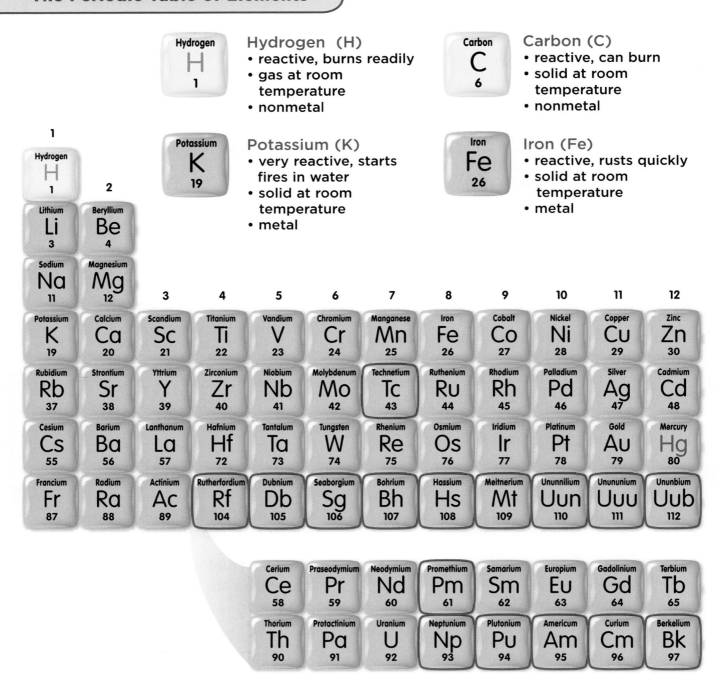

Hydrogen (H)
- reactive, burns readily
- gas at room temperature
- nonmetal

Carbon (C)
- reactive, can burn
- solid at room temperature
- nonmetal

Potassium (K)
- very reactive, starts fires in water
- solid at room temperature
- metal

Iron (Fe)
- reactive, rusts quickly
- solid at room temperature
- metal

The element symbols are the same letters used in chemical formulas.

In 1869, Dmitri Mendeleev made element index cards. He ordered the cards from lightest to heaviest. This led him to a discovery—the properties of the elements repeat in cycles! He made the cycles of elements into rows in a table. Each column contains elements with similar chemical properties. Mendeleev's table is called a *periodic table* (peer•ee•OD•ik) because the properties repeat in cycles (periods).

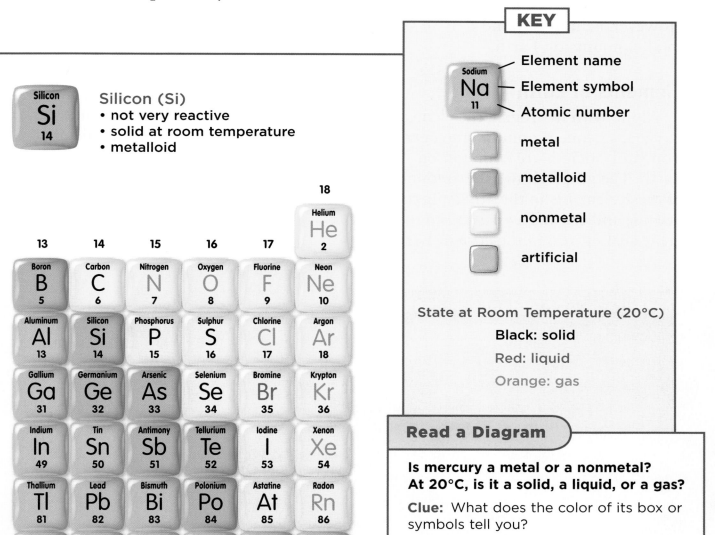

Silicon (Si)
• not very reactive
• solid at room temperature
• metalloid

KEY

Element name
Element symbol
Atomic number

metal
metalloid
nonmetal
artificial

State at Room Temperature (20°C)

Black: solid
Red: liquid
Orange: gas

Read a Diagram

Is mercury a metal or a nonmetal? At 20°C, is it a solid, a liquid, or a gas?

Clue: What does the color of its box or symbols tell you?

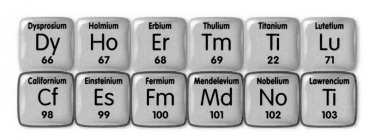

✓ *Quick Check*

Main Idea and Details What is an element's symbol?

Critical Thinking Why do you think that elements 58–71 and 90–103 are placed where they are?

Which are the most common elements?

In space, the most common elements by far are hydrogen and helium. These 2 elements make up 98 percent of the mass of the universe. On Earth, hydrogen is common in water. However, helium is found only in very small amounts on Earth.

Elements of Earth

Along with hydrogen, the elements oxygen, silicon, aluminum, nitrogen, iron, and calcium are common on Earth. The graph shows the amounts of these elements in the atmosphere, oceans, and crust. There also is a great deal of iron in the core of Earth.

Scientists think the inner core is solid iron with a layer of liquid iron surrounding it.

Just like all matter, plants and animals are made of elements. The amounts of elements are shown in the graph. Much of the oxygen and hydrogen come from water. In fact, about 60 percent of animal body weight is water! Most animal bodies are made from carbon, oxygen, hydrogen, nitrogen, phosphorus, and a dash of chlorine and sulfur. Bones and teeth contain most of the calcium shown in the graph.

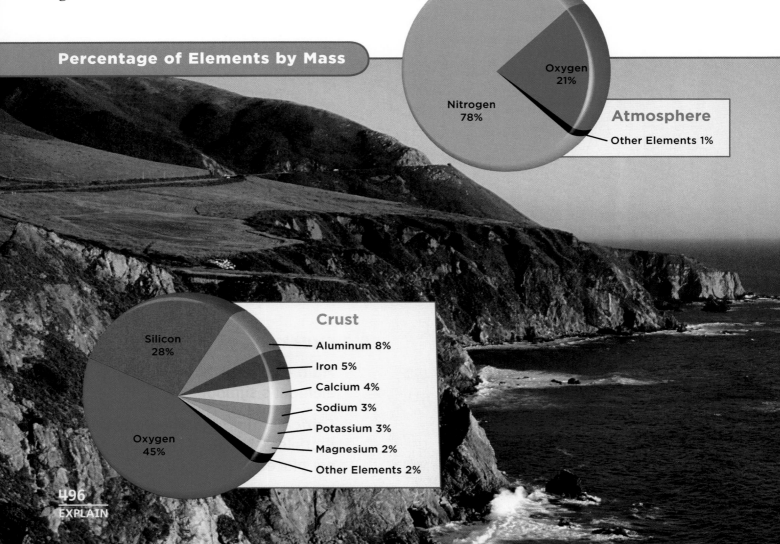

Percentage of Elements by Mass

Atmosphere
- Nitrogen 78%
- Oxygen 21%
- Other Elements 1%

Crust
- Oxygen 45%
- Silicon 28%
- Aluminum 8%
- Iron 5%
- Calcium 4%
- Sodium 3%
- Potassium 3%
- Magnesium 2%
- Other Elements 2%

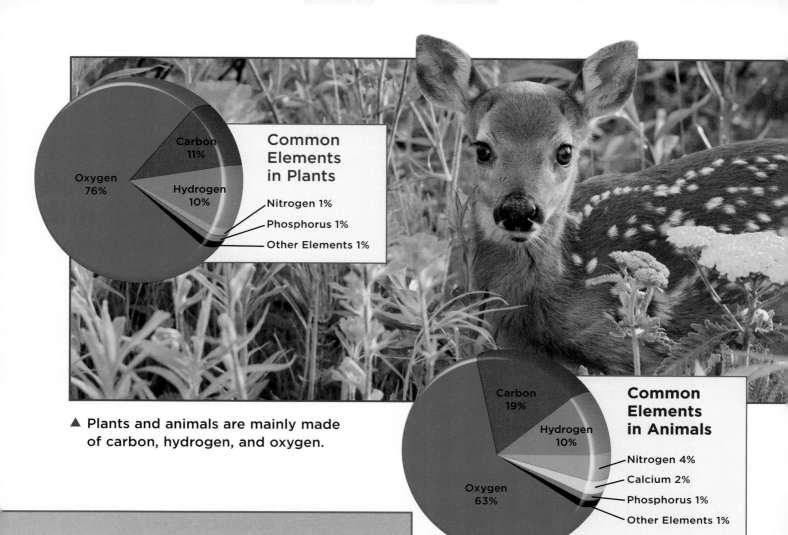

Common Elements in Plants

Oxygen 76%
Carbon 11%
Hydrogen 10%
Nitrogen 1%
Phosphorus 1%
Other Elements 1%

▲ Plants and animals are mainly made of carbon, hydrogen, and oxygen.

Common Elements in Animals

Carbon 19%
Hydrogen 10%
Oxygen 63%
Nitrogen 4%
Calcium 2%
Phosphorus 1%
Other Elements 1%

Heavier elements tend to collect in the crust while lighter elements reside in the oceans and atmosphere.

Ocean

Oxygen 85%
Hydrogen 11%
Other Elements 4%

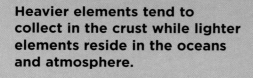

Quick Check

Main Idea and Details Why are oxygen and hydrogen so common in animals and on Earth?

Critical Thinking Why do you think Earth's crust has a more diverse set of elements than the ocean or atmosphere?

How do we examine elements?

Atoms are incredibly small. A single hydrogen atom is only 0.0000000001 meters across—that's around 1 millionth the width of a human hair! Even large molecules are too small to be seen with an ordinary light microscope. Special electron microscopes use electrons instead of light particles to examine a sample. They can show single atoms, or the arrangement of atoms. Another special microscope, called the scanning tunneling microscope, also shows single atoms. This instrument has a needle tip that moves over a surface. The needle is so sensitive that it moves up and down over each atom in the surface. The up-and-down motions are turned into an image like the one shown below.

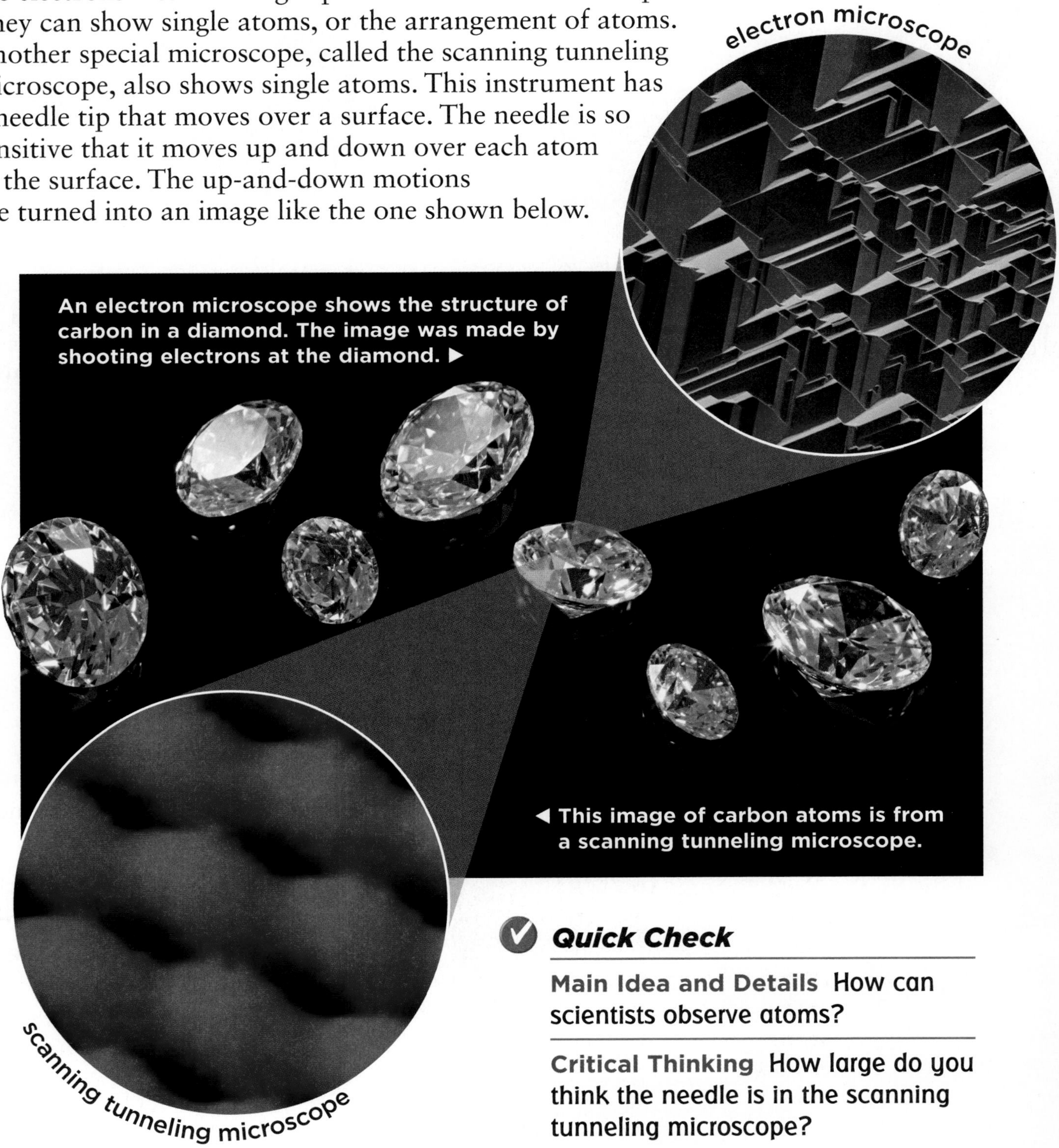

electron microscope

An electron microscope shows the structure of carbon in a diamond. The image was made by shooting electrons at the diamond. ▶

◀ This image of carbon atoms is from a scanning tunneling microscope.

scanning tunneling microscope

✔ Quick Check

Main Idea and Details How can scientists observe atoms?

Critical Thinking How large do you think the needle is in the scanning tunneling microscope?

Lesson Review

Visual Summary

Matter is made of **elements**.

Each element is made of one type of **atom**.

The properties of elements allow them to be organized into a **periodic table**.

Make a FOLDABLES™ Study Guide

Make a Three-Tab Book. Use the titles shown. Summarize the topic under each tab.

Think, Talk, and Write

❶ **Main Idea** How could you tell if a sample of matter is an element?

❷ **Vocabulary** The smallest particle of an element is a(n) _____.

❸ **Main Idea and Details** What are atoms like on the inside?

Main Idea	Details

❹ **Critical Thinking** Does a scanning tunneling microscope let you see atoms like you might see tiny beads? Why or why not?

❺ **Test Prep** Which element is MOST LIKELY to conduct heat and electricity?

A nitrogen
B aluminum
C helium
D oxygen

❻ **Test Prep** Which element is MOST LIKELY to be dull?

A carbon
B aluminum
C mercury
D sodium

Writing Link

Explanatory Writing

Pretend that John Dalton is still alive. Write a letter to him explaining what parts of his ideas about atoms were right and what parts had to be changed.

Math Link

Finding Oxygen

The mass of a sample of air is made of 23.2% oxygen. To obtain 46.4 kilograms of pure oxygen, how many kilograms of air would be needed?

Element Discovery

When Dmitri Mendeleev laid out element cards to create the periodic table in 1869, he found gaps in the order. He suspected there were elements that were not yet discovered. Mendeleev predicted they would be discovered and the gaps would eventually be filled.

Helium
He
2

1868-1895

Helium—Joseph Lockyer discovers helium in 1868 by studying the Sun's spectrum during a solar eclipse. It is named after the Greek god of the Sun. In 1895, helium is found on Earth in uranium minerals.

Hydrogen
H
1

1766

Hydrogen—Henry Cavendish isolates an element he calls "flammable air." The element is renamed for the Greek words meaning "water forming" when another scientist discovers that water is made of hydrogen and oxygen.

Oxygen
O
8

1772-1774

Oxygen—Joseph Priestley and Carl Wilhelm Scheele independently discover a new kind of "air." The gas takes its name from Greek words meaning "acid former." When oxygen combines with other elements, the compounds are usually acidic.

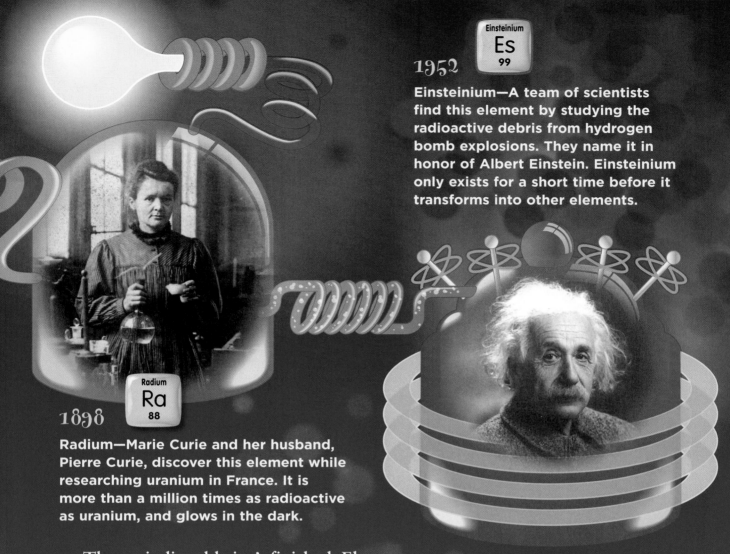

1952

Es Einsteinium 99

Einsteinium—A team of scientists find this element by studying the radioactive debris from hydrogen bomb explosions. They name it in honor of Albert Einstein. Einsteinium only exists for a short time before it transforms into other elements.

Ra Radium 88

1898

Radium—Marie Curie and her husband, Pierre Curie, discover this element while researching uranium in France. It is more than a million times as radioactive as uranium, and glows in the dark.

The periodic table isn't finished. Elements are still being added to it. In the past 75 years, over 25 new elements have been added to the table. That's about one element every three years! If you found a new element, what would you name it?

 Write About It
Classify

1. Which elements were discovered as gases?
2. Which elements have names that describe their properties? How are the other elements named?

LOG ON **e-Journal** Research and write about it online at **www.macmillanmh.com**

Classify

▶ Arrange ideas or objects into groups that have something in common.

▶ List the properties that objects in that group share.

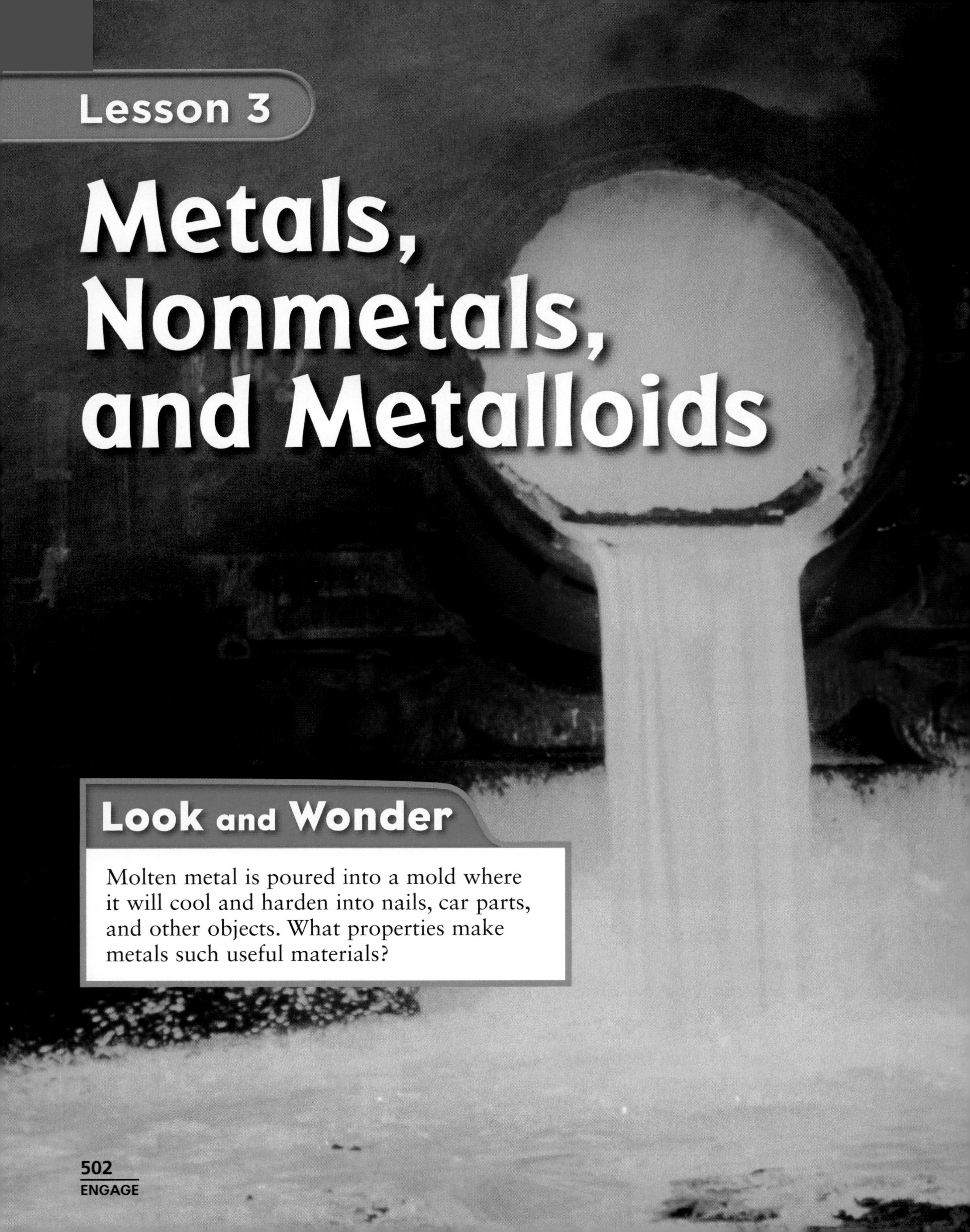

Metals, Nonmetals, and Metalloids

Look and Wonder

Molten metal is poured into a mold where it will cool and harden into nails, car parts, and other objects. What properties make metals such useful materials?

How can you tell if it is metal?

Purpose

In this activity, you will observe, compare, and contrast metal and nonmetal objects. You will describe each object as a strong or weak example of several important properties.

Procedure

❶ Prepare a table to record your observations. Label it as shown.

❷ **Experiment** Test for thermal conductivity: Place each object half in the sun or half under a lamp. Which materials seem hot to the touch on the unlit half?

❸ Test for shiny luster: Look at the aluminum foil and sheet of paper. Which reflects more light?

❹ △ **Be Careful.** Wear goggles. Test for flexibility: Bend a tie with wire in half. Bend a toothpick in the same way. Which holds its new shape without breaking?

- **plastic, metal, and glass rods**
- **aluminum foil**
- **paper**
- **safety goggles**
- **ties with steel wire**
- **wood toothpicks**

Step ❶

Properties of Metals and Nonmetals		
Property	Strong Example	Weak Example
Thermal Conductivity		
Shiny Luster		
Flexibility		

Draw Conclusions

❺ **Classify** Use your observations to decide if objects were strong or weak examples of the properties you tested.

❻ **Communicate** Based on your observations, summarize the properties of metals and nonmetals.

Explore More

Are the properties of all metals the same? Are some stronger examples of some properties, but not others? Plan and conduct an experiment to find out.

Step ❷

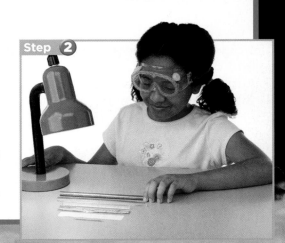

▶ Main Idea

Elements can be grouped into metals, nonmetals, and metalloids, based on their properties.

▶ Vocabulary

malleability, p. 504
ductility, p. 505
corrosion, p. 505
semiconductor, p. 509

LOG ON e-Glossary
at www.macmillanmh.com

▶ Reading Skill ✔

Compare and Contrast

Different   Alike   Different

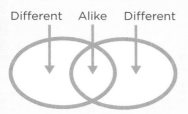

Gold is both malleable and ductile. A small gold nugget can be shaped into elaborate objects.

What are metals?

The chemical elements have many properties that are different from element to element. There are some properties, however, that are very similar among elements. Scientists divide the elements into three groups based on similar properties. The groups are metals, nonmetals, and metalloids.

Of all the elements, about 75 percent are classified as metals and they are on the left side of the periodic table. Metals share a common set of properties: They shine when polished, conduct heat and electricity well, and can be shaped without breaking. The more an element displays these properties, the more metallic it is. So, some metals are more metallic than others.

A shiny surface is one of the easiest ways to identify a metal. Most metals are shiny and many can be polished to be reflective. Metal surfaces will typically reflect not only light, but some heat as well.

Most metals conduct heat very well. If you look at the cookware in your kitchen, you'll see that most of it is made of metal. A metal cooking pan helps spread heat evenly under the food so that it can cook. Metals are also good conductors of electricity, especially copper, gold, and silver.

Metals are easy to shape because they are malleable. **Malleability** (mal•lee•uh•BIL•i•tee) is the ability to be bent, flattened, or hammered without breaking. Gold is one of the most malleable metals.

Copper is shiny, conducts heat and electricity well, and can be easily shaped.

A single gram of gold can be flattened into 1 square meter! Metals are also ductile. **Ductility** (duk•TIL•uh•tee) is the ability to be pulled into thin wires without breaking. Copper is often drawn out into wires for conducting electricity.

Chromium is the hardest metal, while cesium is the softest metal. You can compare the hardness of metals using the Mohs hardness scale. Mercury is the only metal that is a liquid at room temperature.

The ways metals tend to react with other elements—the metals' chemical properties—vary quite a bit. Some metals, like gold, are very unreactive. Other metals, especially the metals in the first column of the periodic table, are very reactive.

When left outside, many metals will corrode. **Corrosion** (kuh•ROH•zhuhn) is when metals combine with nonmetals from the environment. Iron corrodes by rusting, which causes the corroded iron to flake away. The most reactive metals corrode the fastest. Sodium and potassium, for instance, must be stored under oil to keep them from rapidly combining with oxygen in the air.

 Quick Check

Compare and Contrast In what ways are all metals similar? In what ways are they different?

Critical Thinking Do you think that harder metals are more or less malleable than softer metals? Why?

FACT The Statue of Liberty is not painted. Its green color comes from corroded copper.

How do we use metals?

Some periods of human development are named after the metal that was most commonly used at that time. For example, the Iron Age marks the time when humans first produced iron tools. Metals in these times were used primarily as tools but also for jewelry and even medicine. Today, metals are still some of the most important materials in our lives. Think of all the places you see metals—buildings, bridges, utensils, and more!

Some metals are used because they are strong and flexible. Iron is a good example. When combined with other metals iron can become very strong and flexible. Using this form of iron, skyscrapers can be built hundreds of feet tall. Not only does the metal support the weight of the building, it also allows the building to sway slightly in the wind. If skyscrapers didn't bend, they would be blown over!

One of the most versatile metals we use today is aluminum. Aluminum is often used in mirrors because it is inexpensive and can be polished to be reflective. Aluminum foil wrapped around food will trap the heat inside by reflecting it. Just like copper, aluminum can be used to conduct energy cheaply. Both metals are used in electrical wiring, water heaters, and radiators. Aluminum is easily coated with a thin layer of oxygen. This helps prevent corrosion.

Building with Metal

Read a Photo

What properties of metals are being used in this building?

Clue: Look at the part of the building that is being constructed with metal.

You may have heard of someone getting metal placed in their body as part of surgery. Artificial teeth, hips, and even hearts can be made of metal. Sometimes a doctor will put a metal pin inside a broken bone. The metal pin supports the bone as it heals. Some doctors use metal staples to hold large wounds closed.

Whenever metal is used in surgery, doctors must be sure the metal will not react with elements in the body. Gold and certain kinds of silver and titanium are all safe because they are unreactive in the human body.

Other metals are useful because of their reactivity. Batteries use the reactivity of metals to release electrons and generate electricity. Cadmium, nickel, zinc, mercury, lead, and lithium are all used in batteries.

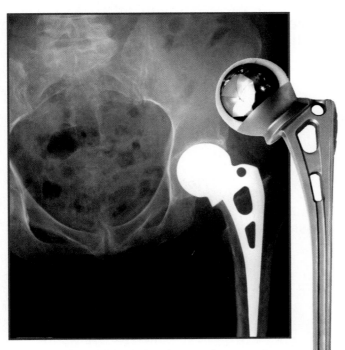

This artificial hip joint is used to replace damaged hips inside the human body.

Quick Lab

Hardness vs. Malleability

1. ⚠ **Be Careful.** Wear goggles to protect your eyes. Bend one end of a paper clip 90° and then bend it back to its original position. Try the same action with a copper wire.

2. **Predict** How many times can you repeat step 1 before the paper clip breaks? Before the copper wire breaks? Find out how many bends are required to break each.

3. Will the copper wire scratch the paper clip, or will the paper clip scratch the copper wire? Record the results when you try to scratch each metal with the other.

4. **Infer** Which metal was harder? Which was more malleable? Explain your reasoning.

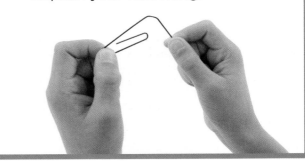

✓ Quick Check

Compare and Contrast In what ways are the uses of copper and aluminum similar? In what ways are they different?

Critical Thinking How could you use a metal that was unreactive, not too hard, and conducted heat well?

Which elements are nonmetals and metalloids?

Have you ever wondered why the wooden, plastic, or ceramic handle of a cooking pot stays cool even if the pot is very hot? Wood, plastic, and ceramic are poor conductors of heat. They also are poor conductors of electricity. These properties are signs of nonmetals. Most nonmetals are located on the right side of the periodic table.

Nonmetals

Nonmetals have properties that are basically opposite to those of metals. Besides being poor conductors of heat and electricity, nonmetals lack luster. They also generally break or crush rather than bend. Many are gases at room temperature. Still others are solids with many colors and forms. Only one nonmetal is liquid at room temperature: bromine.

The most reactive nonmetals are fluorine, chlorine, bromine, and iodine. The column in the periodic table right after fluorine's column contains unreactive, gaseous elements: helium, neon, argon, krypton, xenon, and radon. These elements rarely react with others. For this reason, they are called the inert or noble gases.

Metalloids

Elements get more metallic to the left across any row in the periodic table. Elements also get more metallic toward the bottom of any column. In the middle, the properties switch from nonmetallic to metallic. The elements at the middle points are called the metalloids. They are boron, silicon, germanium, arsenic, antimony,

chlorine (gas)

bromine (liquid)

Nonmetals exist in all three states of matter at room temperature.

iodine (solid)

Carbon
C
6

◀ Carbon is a nonmetal.

Aluminum
Al
13

Silicon
Si
14

◀ Silicon is a metalloid. It has properties between those of nonmetals and metals.

▲ Aluminum is a metal.

Germanium
Ge
32

◀ Germanium is also a metalloid.

Tin
Sn
50

◀ Tin is a metal.

metallicity →

← metallicity

Read a Diagram

Is tin or carbon more metallic?

Clue: Look at the luster of the samples or their position on the periodic table.

tellurium, and polonium. Silicon is the second most abundant element in Earth's crust. It makes up about 26 percent of the mass of the crust. The other metalloids are much rarer.

Metalloids have properties that are between metals and nonmetals. Metalloids look like metals, only they are not as shiny. They also are not as malleable or ductile—like nonmetals.

Metalloids conduct electricity better than nonmetals, but not as well as metals. For this reason, metalloids are called semiconductors (sem•ee•kuhn•DUK•tuhrz). A **semiconductor** is a material that has properties in between conductors and insulators.

Metalloids vary in their chemical reactivity. Some are very reactive with metals and not with nonmetals. Others act in just the opposite manner.

✔ Quick Check

Compare and Contrast How are the noble gases like the nonmetals in fluorine's column of the periodic table? How are they different?

Critical Thinking How could you demonstrate that a metalloid was neither a metal nor a nonmetal?

How do we use nonmetals and metalloids?

Nonmetals are excellent insulators of heat and electricity. Air is mostly made of the nonmetals nitrogen and oxygen and can insulate heat very well. Your winter coat works by trapping air to keep you warm. Nonmetals in plastics insulate electrical cords and keep you from getting shocked.

Nonmetals vary in their reactivity. Chlorine (Cl_2) has a high reactivity that makes it deadly to small living things. It is often added to drinking water and pool water to kill bacteria. Have you ever noticed the sharp smell at a swimming pool? That's the chlorine!

Argon is very unreactive. Even after being electrified or heated for many hours it will not corrode metal. This allows argon to be used in long-lasting and colorful electric lights.

Metalloids can be used like both metals and nonmetals. Fibers of pure boron are lightweight and very strong, like some metals. Boron is used to strengthen modern aerospace

Silicon and other metalloids are used to create this computer chip.

structures. Antimony is unreactive and a good insulator, like some nonmetals. Antimony is used in homes and businesses as a way of making things flameproof.

Silicon and other metalloids are used to make computer chips that use the properties of semiconductors. Computer chips are the heart of modern electronic devices. They allow computers to do math, draw pictures, or even translate languages.

 ## Quick Check

Compare and Contrast What are the similarities and differences between the uses of metalloids and the uses of nonmetals?

Critical Thinking How might you use an unreactive nonmetal gas?

Chlorine kills bacteria in pools and keeps them safe to play in.

Lesson Review

Visual Summary

About three-fourths of the elements are **metals**.

The properties of metals and **nonmetals** are generally the opposite of one another.

Metalloids have properties that are a blend of the properties of metals and nonmetals.

Make a FOLDABLES Study Guide

Make a Trifold Book. Use the titles shown. Summarize what you have learned about each topic in the column.

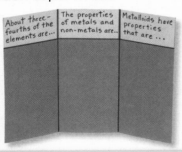
About three-fourths of the elements are... The properties of metals and non-metals are... Metalloids have properties that are ...

Think, Talk, and Write

① **Main Idea** How could you tell if a substance was a metal?

② **Vocabulary** The ability to be bent, flattened, or hammered without breaking is _____.

③ **Compare and Contrast** How are the chemical properties of metals and nonmetals similar and different?

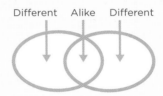
Different   Alike   Different

④ **Critical Thinking** How could you use mercury to make a switch to turn on a light when the trunk of a car is opened?

⑤ **Test Prep** Which material is a metalloid?
- **A** brass
- **B** iron
- **C** boron
- **D** liquid nitrogen

⑥ **Test Prep** Which substance is often used to kill bacteria?
- **A** calcium oxide
- **B** sodium
- **C** chlorine
- **D** nitrogen

 Math Link

How Much Money Could You Get?
A company produces 4 computer chips per gram of silicon. Computer chips sell for $160 each. How much money could you make by using 100 grams of silicon?

 Social Studies Link

How Could Metal Change History?
The Stone Age in a culture is the time when people do not use metal tools. How would learning to use metal affect the development of a culture?

Be a Scientist

battery

battery holder

alligator clips

wire

miniature bulb

bulb holder

copper, iron, wood, and graphite samples

Structured Inquiry

How can you compare the electrical conductivity of metals and nonmetals?

Form a Hypothesis

Are some materials better conductors than others? What happens if you use a poor conductor in an electrical circuit? Will the brightness of a light bulb in the circuit change? Write your answer as a hypothesis in the form "If the electrical conductivity in an electrical circuit decreases, then the brightness of the light bulb . . ."

Test Your Hypothesis

1. Place the battery in the battery holder. Connect one alligator clip and wire to one end of the battery holder. Connect another alligator clip and wire to the other end of the battery holder.

2. Connect an alligator clip from the battery to the miniature bulb in the bulb holder. Use a third alligator clip and wire to attach the bulb holder to one end of the copper strip. The copper strip is the material you will test and change—it is your independent variable.

3. **Experiment** Connect the second wire from the battery to the other end of the copper to close the circuit. Observe how brightly the light bulb glows—this is your dependent variable.

4. **Observe** Repeat the test for the other materials. Observe and record your results.

5. **Classify** Rank the four materials' conductivities in order from highest to lowest.

Step 1

Step 3

Draw Conclusions

6 **Infer** Did the light bulb act as a conductivity tester? Why do you think copper is used to make wire?

7 Do your results support your hypothesis? Explain.

Guided Inquiry

How does combining materials affect conductivity?

Form a Hypothesis

You know how the electrical conductivity of metals and nonmetals compare. How will that property change if different materials are combined? Write your answer as a hypothesis in the form "If a good conductor is combined with a poor conductor, then the new object's conductivity will be . . ."

Test Your Hypothesis

Design and perform an experiment to determine how conductivity changes when materials are combined. You may want to use steel (a combination of iron, other metals, and carbon) or a pencil (a combination of graphite and wood). Write out the resources you need and the steps you follow. Remember to describe your variables and record your results.

Draw Conclusions

Did your experiment support your hypothesis? Why or why not?

Open Inquiry

Do materials conduct heat energy as well as they conduct electrical energy? Design an experiment to answer your question. Organize your experiment to test only one independent variable, or one item being changed. Write your experiment so that someone else can complete the experiment by following your instructions.

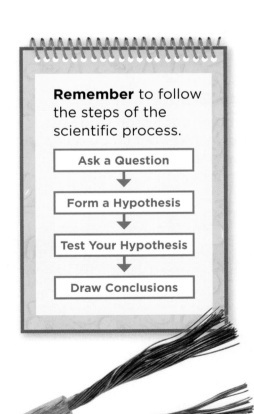

Remember to follow the steps of the scientific process.

Ask a Question

↓

Form a Hypothesis

↓

Test Your Hypothesis

↓

Draw Conclusions

Visual Summary

Lesson 1 Matter can be described by many different properties.

Lesson 2 All matter is made of elements.

Lesson 3 Based on their properties, elements can be grouped into metals, nonmetals, and metalloids.

Make a FOLDABLES™ Study Guide

Assemble your lesson study guide as shown. Use your study guide to review what you have learned in this chapter.

Fill in the blank with the best term from the list.

atom, p. 491

density, p. 482

element, p. 490

malleability, p. 504

mass, p. 480

metal, p. 491

semiconductor, p. 509

volume, p. 481

1. The ability to be bent, flattened, or hammered without breaking is called _____.

2. You divide an object's mass by its volume to calculate its _____.

3. The amount of space matter takes up is its _____.

4. The amount of matter in an object is its _____.

5. A material that cannot be broken down into anything simpler by chemical means is a(n) _____.

6. Shiny luster, conductivity, and flexibility are all properties of a(n) _____.

7. The smallest unit of an element that still has the properties of that element is a(n) _____.

8. A metalloid has properties between those of conductors and insulators, so it is a(n) _____.

Answer each of the following in complete sentences.

9. **Main Idea and Details** How can two items of the same shape and size have different densities?

10. **Classify** What type of structure does the picture show? What is it made of?

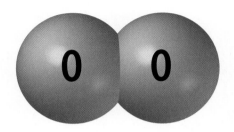

11. **Infer** You test an object made out of an unknown element. You find that the object does not conduct electricity, looks dull, and breaks easily. How would you classify the element that the object is made of?

12. **Critical Thinking** Why would it be dangerous to hold a metal pole in a thunderstorm?

13. **Writing a Fictional Story** Write a short story in which John Dalton goes back in time to meet Aristotle and they compare their ideas about elements.

14. What properties can help us classify matter?

Know Your Elements

Research an element in the periodic table.

What to Do

1. Select an element from the periodic table.

2. Research the properties of your element. How many protons and electrons do its atoms have? What are its boiling and melting points? Is it a metal, a nonmetal, or a metalloid?

3. Research your element's history and uses. When was it discovered, by whom, and how? Where can your element be found? Is the element part of any important technology?

4. Use your research to create a pamphlet about your element. Include photos, drawings, charts, or other graphics.

1. **What property of metal is shown in the image below?**

A ductility

B malleability

C corrosion

D buoyancy

CHAPTER 10

Physical and Chemical Changes

The Big Idea

What causes matter to change?

Key Vocabulary

sublimation
the process of changing directly from a solid to a gas without first becoming a liquid (p. 521)

mixture
a physical combination of two or more substances that are blended together without forming new substances (p. 530)

solution
a mixture of substances that are blended so completely that the mixture looks the same everywhere (p. 532)

compound
a substance that is formed by the chemical combination of two or more elements and that acts like a single substance (p. 542)

chemical change
a change of matter that occurs when atoms link together in a new way, creating a new substance (p. 544)

ion
an electrically charged atom or molecule with unequal numbers of protons and electrons (p. 554)

Changes of State

Look and Wonder

If all the ice in the world melted, the oceans would rise by more than 65 meters (215 feet)! This iceberg is melting in Paraiso Bay, Antarctica. What happens to ice while it is melting?

What happens when ice melts?

Make a Prediction

If you warm ice cubes, they will melt. What happens to the temperature of a cup of ice cubes and water as the ice melts? Write a prediction in the form "If a cup of ice and water is steadily warmed, then the temperature of the ice water will . . ."

Test Your Prediction

1 **Measure** Fill a cup halfway with cool water and add four ice cubes.

2 Record the mass of the cup of ice water. Do you think the mass will change as the water warms?

3 **Observe** Swirl the ice and water gently for 15 seconds (don't splash!). Then record the temperature of the mixture. Next, place the cup under the heat source.

4 Repeat step 3 every 3 minutes until you have 5 readings after the ice has fully melted.

5 Record the mass of the cup of water again.

Draw Conclusions

6 Use your data to make a graph of the temperature of water versus time.

7 **Interpret Data** Describe the temperature and mass of the ice water as the ice melted.

8 **Communicate** Did your observations support your prediction? Write a report that describes whether or not your prediction was correct.

Explore More

How would the temperature of water change with time as it freezes? Write a hypothesis and design an experiment to test it. Conduct your experiment and report on your findings.

Materials

- plastic or paper cup
- cool water
- ice cubes
- balance
- watch or clock
- thermometer
- heat source (lamp or sunlight)

Step 2

Step 3

Read and Learn

▶ **Main Idea**

Matter can change state when heat is added or removed.

▶ **Vocabulary**

physical change, p. 520
sublimation, p. 521
melting point, p. 522
boiling point, p. 522
freezing point, p. 523
thermal expansion, p. 524
thermal contraction, p. 524

e-Glossary
at www.macmillanmh.com

▶ **Reading Skill** ✔

Fact and Opinion

Fact	Opinion

How can matter change state?

What changes when you sharpen a pencil? Does it stay a pencil? Yes. You only change how the pencil is shaped, not the type of elements inside. A **physical change** alters the form of an object, without changing what type of matter it is. Sharpening your pencil is an example of a physical change.

Remember that there are three common states of matter: solid, liquid, and gas. The states of matter are physical properties. If you put a chip of ice on the kitchen counter, it will melt into a puddle. Overnight, the puddle is likely to disappear as the water turns into a gas. These are both physical changes. What makes water go through changes of state like these?

As you know, particles in objects move around. In solids, particles only vibrate. In liquids, particles vibrate as they move past one another. In gases, particles move quickly and far from one another. The average movement of particles in an object is measured by its temperature. Changes in temperature occur when an object gains or loses heat.

Changes of State

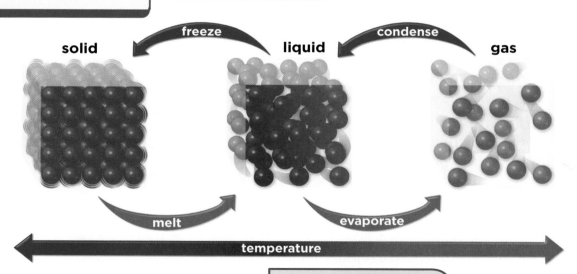

As heat is absorbed, particles move faster and become less organized.

Read a Diagram

Does a gas gain heat when it condenses?

Clue: Look at the movement of particles.

Heat is energy that flows between objects with different temperatures. If a solid gains enough heat, its particles start moving too fast to stay together. The solid then changes into a liquid. Add more heat and the liquid boils into a gas. What if you take away heat? Particles slow down and group together. Gases lose heat and condense into liquids. Liquids lose heat and freeze into solids.

Most matter can change state—it only takes a change in heat. Oxygen molecules are usually a gas, but at very low temperatures they condense into a liquid.

Some solid substances skip the liquid state of matter when turning to gas. **Sublimation** (suhb•luh•MAY•shuhn) is the change of state directly from solid to gas. A material called dry ice can sublimate at room temperature. Water can also sublime. The frost in a freezer comes from gaseous water sublimating out of uncovered food or ice cubes. The gaseous water from the food condenses on the cold freezer walls to make the frost.

In most materials, particles draw closer together as they lose heat. So a material is usually denser as a solid than as a liquid. Water, however, is different. Just before freezing, water molecules move farther apart and line up in a solid arrangement. For this reason, ice is less dense than liquid water.

 Quick Check

Fact and Opinion Frost can form in freezers so it is better to buy a frost-free freezer. Which part of this statement is fact and which part is opinion?

Critical Thinking How could snow disappear without making a puddle?

▲ This liquid oxygen is just starting to boil at a frigid –183°C.

▲ Dry ice, made of carbon and oxygen, sublimes at room temperature.

Water is denser as a liquid than as a solid.

When does matter change states?

We put ice in drinks to keep them cold. However, ice does not keep drinks cool just because it is cold. It keeps drinks cool mainly by melting! How can melting ice keep things cool? Melting ice absorbs heat from the drink. This heat changes the state of the ice instead of raising its temperature.

The temperature of water does not rise while it is melting or boiling. How can the water absorb heat and not get hotter? Usually, adding heat causes particles to gain in temperature. During a change of state, heat is spent breaking particles apart from one another instead. Once the particles are separated they can move around and increase in temperature again.

The diagram below is called a heating curve for water. It shows what happens to water when heat is steadily supplied, starting with ice. Do you see that water melts at 0°C (32°F) and boils at 100°C (212°F)? The temperature at which a substance melts is called its **melting point**. The melting point of water is 0°C. Similarly, the temperature at which a substance boils is called its **boiling point**. The boiling point of water is 100°C. Neither the melting point nor the boiling point of any substance depends on mass.

Heating Curve for Water

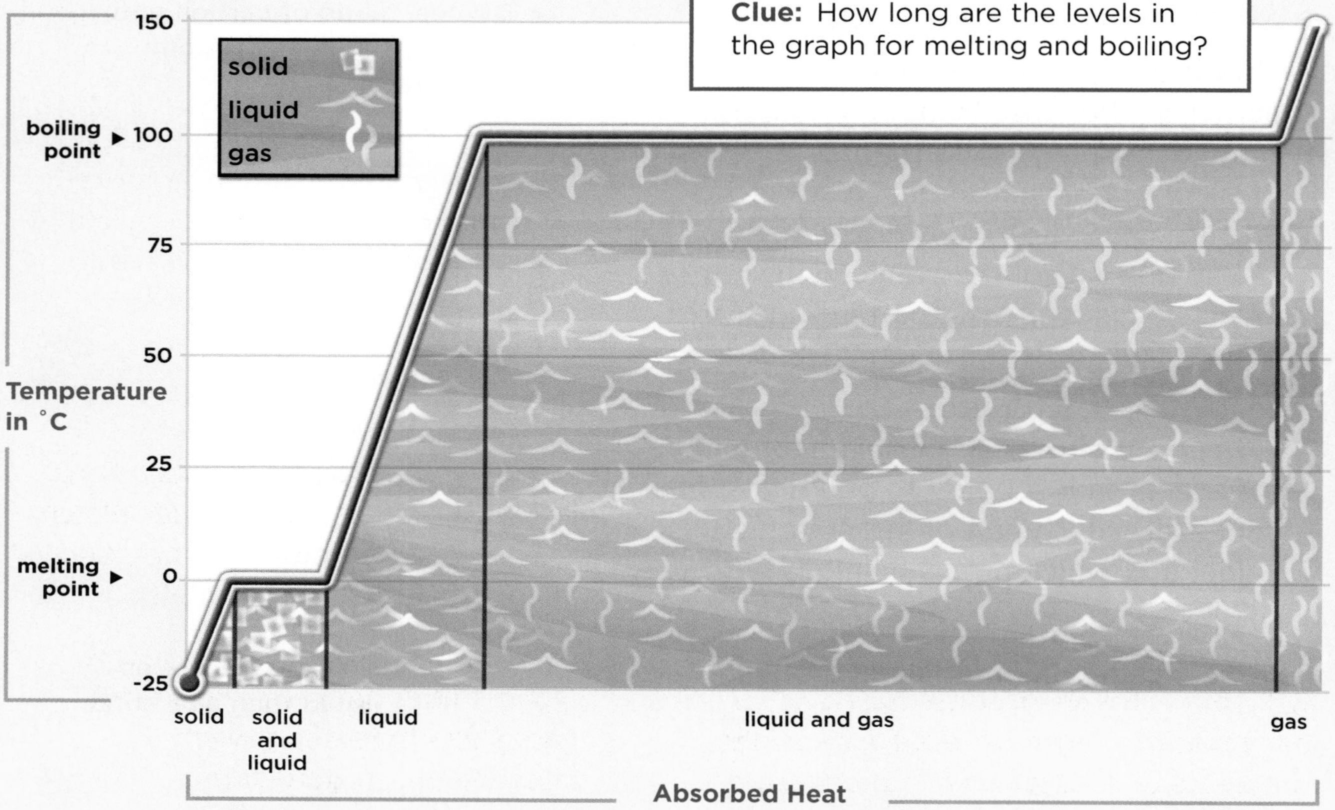

Read a Diagram

Which takes more heat—melting a sample of water or boiling it?

Clue: How long are the levels in the graph for melting and boiling?

solid
liquid
gas

boiling point ▶ 100

150

75

50

Temperature in °C

25

melting point ▶ 0

-25

solid | solid and liquid | liquid | liquid and gas | gas

Absorbed Heat

Changes of State for Some Common Materials

Name	Melting Point	Boiling Point
copper	1,083°C	2,567°C
nitrogen	-210°C	-196°C
water	0°C	100°C
table salt	801°C	1,465°C
iron	1,538°C	2,861°C

What happens as you remove heat from water? You follow a cooling curve. It is just the heating curve in reverse. The water condenses at the boiling point and freezes at the melting point. We call the temperature at which a substance freezes its **freezing point**.

Every pure substance melts and boils at specific temperatures. A material with high melting and boiling points has particles that are attracted strongly to one another. A material with low melting and boiling points has particles that are weakly attracted to one another. Many metals will not melt until they reach very high temperatures. Nonmetal gases will not condense unless it is very cold.

If you get out of a swimming pool on a windy day, you start to dry as the liquid water on your skin turns to gas. How could water turn to gas on your skin when body temperature is much less than the boiling point of 100°C?

The temperature in a liquid is the average energy of all the particles within it. However, some particles have slightly more or less energy than the average. Particles with more energy

≡ Quick Lab

Changing Balloons

1. **Predict** What will happen to the volume of a balloon filled with warm air as it is cooled? Record your prediction.

2. Blow up a balloon. Air from your lungs is warm. Tie off the balloon and measure it around with a piece of string.

3. Submerge the balloon in a pail of ice water for a few minutes. Measure it again with the piece of string. Record your observations.

4. **Infer** How does the motion of molecules explain what you observed? Write out your ideas.

can escape the surface of the liquid and become a gas. This process is what we call evaporation. Evaporation is the slow change from liquid to gas at temperatures well below the boiling point. It is always happening. Liquids cool as they evaporate because their highest energy particles are leaving.

✔ Quick Check

Fact and Opinion Your friend says ice in a soda keeps the drink cool but makes it taste bad. Which part of that statement is fact, and which part opinion? Explain.

Critical Thinking Some people take steam baths to relax. Why does steam feel hot when it condenses on your skin?

What are expansion and contraction?

The volume of an object changes as it gains heat. Particles in the object start to move around faster and push on each other. This causes the object to grow. **Thermal expansion** (THUR•muhl eks•PAN•shuhn) is the increase in an object's volume due to a change in heat.

When an object loses heat, its particles slow down and the object shrinks. **Thermal contraction** (kuhn•TRAK•shuhn) is the decrease in an object's volume due to a change in heat. The word *thermal*, meaning "warmth," tells you that the expansion or contraction is due to heat. Gases expand and contract more than liquids. Liquids expand and contract more than solids.

Many building materials will expand and contract as the seasons change. So, builders leave space

This expansion joint keeps the bridge from breaking or buckling on hot days.

between sections of material. These spaces are called expansion joints.

Ice is less dense than water. So water expands, instead of contracts, when it freezes. Water can slip into small cracks as a liquid, then widen those cracks when it freezes. Builders make sure rain does not collect anywhere that it might cause damage as ice.

Thermometers use the thermal expansion and contraction of alcohol to show temperature. As the temperature near a thermometer increases, the alcohol inside expands. Markings along the thermometer tell you the temperature indicated by the volume of alcohol.

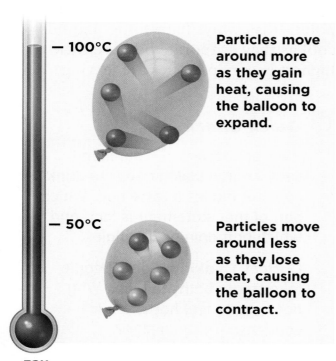

— 100°C

Particles move around more as they gain heat, causing the balloon to expand.

— 50°C

Particles move around less as they lose heat, causing the balloon to contract.

 Quick Check

Fact and Opinion Do you agree that expansion and contraction only cause problems? Why or why not?

Critical Thinking What would happen if there weren't any spaces in between sections of a sidewalk?

Lesson Review

Visual Summary

As heat is added to or removed from matter, the matter may go through **changes of state**.

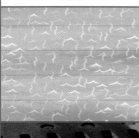

Each pure substance has its own **melting point** and **boiling point**.

Changes in heat cause objects to **expand** or **contract**.

Make a FOLDABLES™ Study Guide

Make a Layered-Look Book. Use the titles shown. Summarize what you learned in each topic.

Changes of State

Adding and removing heat...

Melting and boiling points...

Expansion and contraction...

Think, Talk, and Write

1. **Main Idea** What would happen to a sample of water at 30°C if it is heated until the temperature reaches 120°C?

2. **Vocabulary** The temperature at which a substance melts is called its _____.

3. **Fact and Opinion** Is it a bad idea to put a glass bottle full of water in a freezer? Support your opinion with facts.

Fact	Opinion

4. **Critical Thinking** Why wouldn't turning up the heat on the stove make boiling water cook things faster?

5. **Test Prep** In general, the state of matter with the most energy is
   - **A** solid.
   - **B** liquid.
   - **C** gas.
   - **D** none of the above.

6. **Test Prep** What usually happens when an object's temperature is rising?
   - **A** expansion
   - **B** contraction
   - **C** condensation
   - **D** freezing

Writing Link

Descriptive Writing
Pretend you live in a world where water contracts upon freezing. Ice would be less dense than liquid water. Write about how your world would be different from our normal world.

Math Link

Boiling
At the boiling point of water It takes about 2260 J of heat to completely boil a gram of liquid water into a gas. How much heat will it take to completely boil 5.5 g of water at the boiling point?

Focus on Skills

Inquiry Skill: Use Variables

Particles in hot liquid move faster than those in cold. Since hot water has more energy to get rid of before it freezes, it shouldn't freeze as fast as cold water.

That's what many people thought. But scientists wanted to know for sure, so they did a series of experiments and recorded their observations. Those experiments changed only one thing at a time. That way the scientists knew what caused the outcome they observed. What they changed is called the independent **variable**. They learned that sometimes hot water freezes faster than cold water— this is called the Mpemba effect.

▶ Learn It

When you use variables, you change one thing at at time to see how it affects the outcome of the experiment. The thing you change is the independent **variable**. The outcome is the dependent variable. The way the dependent variable changes depends on the way the independent variable changes.

For this experiment the independent variable is the starting temperature of the water. The time it takes the water to freeze is the dependent variable. You will change the starting temperature of the water and record how this affects the time it takes the water to freeze.

▶ Try It

Materials	hot water, cool water, plastic cups, thermometer, graduated cylinder, labels, freezer

1. Make a chart like the one shown to record your data.

2. Fill a cup with 120 mL of hot water. Label it *Hot Water*. Fill a cup with 120 mL of cold water. Label it *Cold Water*. Fill a cup with 80 mL cold and 40 mL hot water. Label it *Cool Water*. Fill a cup with 80 mL hot and 40 mL cold water. Label it *Warm Water*.

3. Record the temperature of each cup of water on your table. This is the independent **variable**.

4. Place all the cups in a freezer at the same time. The cups should be close together and on the same level.

5. Check the freezer every 10 minutes. Record when the water in each cup starts to freeze. Record when the water in each cup is completely frozen. These are both dependent variables.

▶ Apply It

1. Which water froze first: cold, cool, warm, or hot water? Repeat the experiment to confirm your findings.

2. Scientists changed the independent **variable** to learn about the Mpemba effect. What did you learn from your results? Do you agree that the Mpemba effect is real?

3. What do you think would happen if you used really icy or even hotter water? Are you still changing the same independent **variable**? Try it and record data about the investigation. Use that data to help you develop an opinion about how water freezes.

Time to Freeze

Temperature	Starts Freezing	Ends Freezing
Hot water (70°C)		
Warm water (52°C)		
Cool water (35°C)		
Cold water (18°C)		

Mixtures

Look and Wonder

How many different-colored paints do you see? Even the portions of paint that seem to be just one color are often many colors blended together. What do you think helps substances mix together?

How can you speed up mixing?

Materials

- **sugar cubes**
- **cold and hot water**
- **plastic cups**
- **stopwatch**
- **spoon**

Make a Prediction

Which do you think will speed up the mixing of sugar in water more: crushing the sugar, stirring the water, or heating the water? Record your prediction.

Test Your Prediction

1 Prepare a table to record your observations. Label it as shown.

2 **Experiment** Take 1 whole sugar cube and place it in $\frac{1}{2}$ cup of cold water. Record the time it takes to completely dissolve. This is the control mixing time.

3 Repeat step 2 using 1 crushed cube, then with 1 cube while stirring the water, then with 1 cube in $\frac{1}{2}$ cup of hot water.

Draw Conclusions

4 **Interpret Data** Look at your table. Which method had the shortest mixing time? Was it close to or very different from the control?

5 **Infer** How do you think you could create the shortest mixing time possible? Write a report and justify your answer.

Explore More

Do you think there are any other methods that could decrease the mixing time? Design an experiment that could provide information and test your prediction. Carry out the experiment and record your results.

Step 1

	Water	Temperature	Sugar	Time to Dissolve
Control	½ cup	Cold	Whole	
Test 1	½ cup	Cold	Crushed	
Test 2	½ cup	Cold	Stirred	
Test 3	½ cup	Hot	Whole	

Step 2

Read and Learn

▶ Main Idea

Mixtures are physical combinations of different kinds of matter.

▶ Vocabulary

mixture, p. 530

colloid, p. 531

solution, p. 532

solute, p. 532

solvent, p. 532

alloy, p. 532

solubility, p. 533

distillation, p. 535

LOG ON ⊝ -Glossary
at **www.macmillanmh.com**

▶ Reading Skill ✓

Infer

Clues	What I Know	What I Infer

▶ Technology

Explore Mixtures with Team Earth

Trail mix is a heterogeneous mixture.

What are mixtures?

Pretend you are making trail mix for you and your friends. You could use cereal, pretzels, crackers, and other ingredients stirred together. Has stirring the ingredients together made new and different substances? No, the pieces of food have just been physically combined. We know this because each piece of food retains its properties. The pretzels are still salty, the crackers are still crunchy, and so on.

Trail mix is an example of a mixture (MIKS•chuhr). A **mixture** is a physical combination of substances. These substances remain the same even though they are close together. You can separate mixtures back into their original substances. For instance, you could pick out all the pretzels from your trail mix.

Mixtures like trail mix that have different parts you can plainly see are said to be *heterogeneous* (het•uhr•uh•JEE•nee•uhs). Not all heterogeneous mixtures look "speckled" or "chunky" to your eye. They may look smooth or creamy. If you look at them under a microscope, however, you can see the different parts clearly. This kind of heterogeneous mixture is called a *suspension* (suh•SPEN•shuhn).

Over time, one or more parts of a suspension will settle to the bottom like mud in a stream. Stirring or shaking the mixture, however, will make it look smooth again. This happens with mud at the bottom of a stream when you step in the water. The mud rises up and darkens the water by reforming a suspension.

You may also have suspensions around your house. Do you ever take medicine or use a food product, like orange juice, that says "shake well before using"? If so, it is probably a suspension.

▲ Most of the clay has settled out of the muddy water.

▲ The appearance of muddy water under a microscope shows that it is a heterogeneous mixture.

Read a Photo

How can you tell that the muddy water is not a colloid?

Clue: Look at the difference between the first and second photos.

What if the parts of a heterogeneous mixture do not settle eventually? That type of mixture is called a colloid (KOL•oid). A **colloid** is a mixture like a suspension, except that its parts do not settle. The suspended particles are just small enough not to layer out. At the same time, they are large enough to make the mixture look cloudy or creamy. Smoke, mayonnaise, and foam are all examples of colloids.

 Quick Check

Infer Dust settles out of air onto furniture and shelves. What kind of mixture is dusty air? Why?

Critical Thinking What do you think would happen if you mixed two suspensions?

531

What are solutions?

When you mix sugar into water and stir, the sugar dissolves and seems to disappear. You can taste the sugar, but can't see it in the liquid. Even under a microscope, the sugar water looks the same everywhere—it's a *homogeneous* (hoh•muh•JEE•nee•uhs) mixture. A mixture like sugar water is called a solution (suh•LEW•shuhn). A **solution** is a mixture with parts that blend so that it looks the same everywhere, even under a microscope.

The part of a solution in the smaller amount and that is dissolved is called the **solute** (SOL•yewt). The part of the solution in the larger amount that dissolves the other substance is called the **solvent** (SOL•vuhnt). In sugar water, for example, the sugar is the solute and the water is the solvent. Solutions can be made with solids, liquids, and gases. Usually gases form solutions easier than liquids, which form solutions easier than solids. An **alloy** (AL•oy) is a solution of a metal and another solid (often another metal).

Many common household products are solutions, such as window cleaner, bleach, vinegar, and beverages. Some of these are very dangerous when mixed. Mixing bleach and ammonia cleaners, for instance, produces deadly gases.

Humid air is an important solution of water vapor in air. When it condenses, it forms clouds.

Seltzer (carbonated water) is a solution of carbon dioxide gas in liquid water.

Most alloys are solutions of one type of metal in another. ▶

FACT Solutions can be formed from every state of matter.

Solution Limits

When there is just a little sugar in water, it is called a dilute sugar solution. This water is not very sweet. Adding more sugar makes the solution more concentrated and also sweeter. Could you dissolve more and more sugar and concentrate the solution without limit? No, after a certain amount, additional sugar will not dissolve. You could stir and stir, but the added sugar crystals would just settle undissolved to the bottom.

The maximum amount of a solute that can dissolve in a solvent is called the **solubility** (sol•yuh•BIL•i•tee). The solubility of table sugar in water at room temperature is 2.1 g of sugar per 1 g of water. Table salt, in contrast, has a solubility of 0.4 g of salt per 1 g of water at room temperature.

Solubility often depends on temperature. Many substances become more soluble at higher temperatures, but not all. For example gases such as oxygen usually become less soluble in warmer water. The solubility of table salt in water hardly changes between temperatures of 0°C and 100°C.

Solubility can be given for many solvents besides water. Water, though, is often called the universal solvent because it can dissolve many things.

✓ Quick Check

Infer Why might fish not get enough oxygen in hot water?

Critical Thinking How could you increase the amount of solute that will dissolve in a solvent?

≡ Quick Lab

Temperature in Solutions

1 **Predict** Do you think you could dissolve more sugar in hot water or cold water? Why? Write down your reasons.

2 **Observe** Fill a cup with cold water. A level spoonful of sugar is about 28 g. Record how many grams of sugar will dissolve in the water as you stir. Repeat with hot water.

3 Which water dissolved the most sugar? How could you tell?

4 Was your prediction correct? Write out your findings.

Solutions are formed when solvents, like water, dissolve solutes by pulling their particles apart.

How can you take mixtures apart?

Making a mixture requires a physical change. Physical changes are also needed to take a mixture apart. Parts of a mixture with different properties will act differently when changed in the same way. So you can use a physical change to push, pull, lift, or otherwise separate one part of a mixture from another. Density, solubility, particle size, magnetism, melting points, and boiling points are all good properties to use when separating mixtures.

Dissolving sugar in water separates it from sand. Filtering and evaporating the water recovers the sugar.

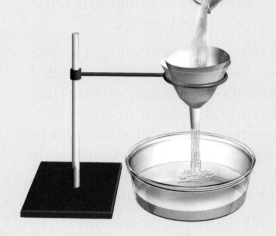

Sand particles cannot pass through the pores in the filter, but water particles easily can.

Iron is attracted to a magnet, while sand is nonmagnetic.

In water, the low-density sawdust floats while the higher-density sand sinks.

Read a Diagram

How would you separate a mixture of sand, sawdust, iron filings, and sugar?

Clue: Each diagram shows one separation method

 Science in Motion Watch how mixtures are separated at **www.macmillanmh.com**

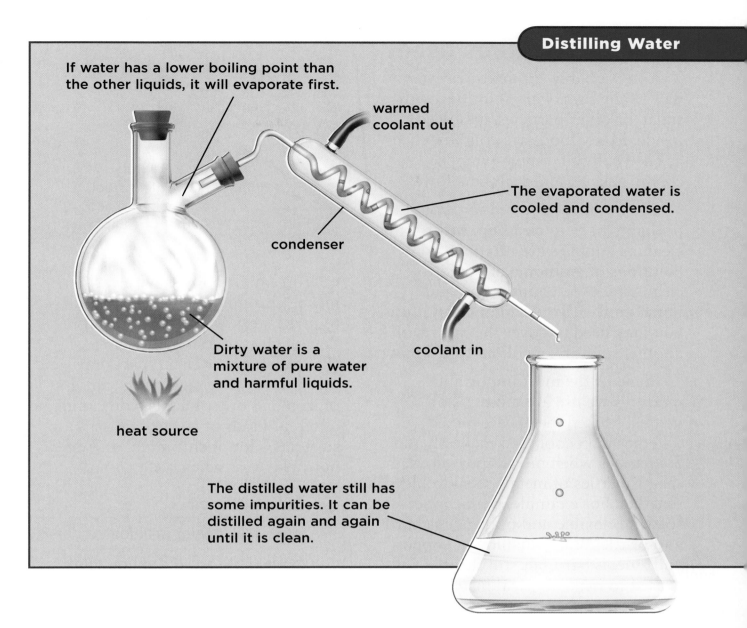

If water has a lower boiling point than the other liquids, it will evaporate first.

warmed coolant out

The evaporated water is cooled and condensed.

condenser

Dirty water is a mixture of pure water and harmful liquids.

coolant in

heat source

The distilled water still has some impurities. It can be distilled again and again until it is clean.

Separating Liquids

What if you need to separate two liquids? If the two liquids had different boiling points you could use distillation (dis•tuh•LAY•shun). **Distillation** is the process of separating liquids by using evaporation and condensation.

You can also separate liquids, or the particles in liquids, using *chromatography* (kroh•muh•TOG•ruh•fee). In this process, a liquid travels up special chromatography paper and carries small substances with it. The substances in the liquid travel at different speeds up the paper. Bands of color appear on the paper as the substances separate.

 Quick Check

Infer Two liquids boil at about the same temperature. Would it be easy to separate them by distillation? Why or why not?

Critical Thinking How could you separate clean air from smoke?

How are mixtures used?

You probably eat mixtures every day. Many drinks are solutions with sugar, or other particles, dissolved inside. Many foods are colloids, such as cheese, low-fat milk, whipped cream, gelatin, and marshmallows.

After you get done eating, you'll probably have to clean up. Most cleaning supplies are mixtures, too! Solutions of ammonia are used to clean windows, countertops, and much more. If you cannot clean it up, you may need to cover the mess with another mixture: a colloid called paint.

Some of the most important mixtures we use are mixtures of metals. Melted metals are mixed together then cooled to make an alloy. Sometimes when mixed into alloys, the properties of metals seem to blend together. For example, pure copper is soft and flexible and pure zinc is hard and brittle. Brass, an alloy of copper and zinc, is hard but still flexible and is used to make musical instruments.

Stainless steel is a strong alloy that does not rust.

Iron can be alloyed with carbon, nickel, chromium, and other metals to make steel. Steel is strong and flexible. It is used in nails, cars, silverware, paper clips, spaceships, and much more. By changing the amount of iron and other metals, you can make different kinds of steel. Stainless steel has a lot of chromium and does not easily rust when wet.

✅ Quick Check

Infer Will low-fat milk form layers?

Critical Thinking Do you think chromium would corrode easily? Why or why not?

Gelatin dessert is a colloid, while fruit salad is a heterogeneous mixture.

Lesson Review

Visual Summary

Mixtures are physical combinations whose parts keep their properties.

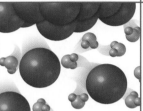

Mixtures can be heterogeneous combinations, **colloids**, or **solutions**.

Mixtures can be **separated** by physical changes.

Make a ![FOLDABLES™] Study Guide

Make a Three-Tab Book. Use the titles shown. Summarize what you learned about each topic under the tab.

> Properties of mixtures...
>
> Types of mixtures...
>
> Separating mixtures...

Think, Talk, and Write

1 **Main Idea** What would tell you if a sample of matter is a mixture?

2 **Vocabulary** A solution of metals is called a(n) _____.

3 **Make Inferences** Why would mixtures of gases rarely form colloids?

Clues	What I Know	What I Infer

4 **Critical Thinking** A solution of sugar in water has a solubility limit. Do you think a heterogeneous mixture of sugar and water has a limit? Why or why not?

5 **Test Prep** Which mixture is MOST LIKELY a solution?
 A muddy water
 B cranberry juice
 C potting soil
 D milk

6 **Test Prep** Which would make a sugar and water solution more dilute?
 A Let water evaporate away.
 B Add sugar and stir.
 C Add water and stir.
 D Heat to boil away water.

Writing Link

Descriptive Writing
Pollutants in water include pieces of trash, suspensions of sewage, and dissolved chemicals. Describe how you could remove these pollutants from the water.

Math Link

Making a Soil Mix
Potting soil is $\frac{1}{2}$ loam, $\frac{1}{4}$ peat, $\frac{1}{8}$ sand, and $\frac{1}{8}$ gypsum by mass. To make 16 kg of potting soil, how many kg of each ingredient do you need?

Be a Scientist

Materials

sand

gravel

2 bowls

spoon

iron filings

sieve

tweezers

bar magnet

How can you separate mixtures?

Form a Hypothesis

Are all mixtures made in the same way? Will different methods of separation work equally well on the same mixture? Write your answer as a hypothesis in the form "If the method for separating a mixture changes, then . . ."

Test Your Hypothesis

1. Take a cup of sand and gravel and pour it into a bowl. Add a spoonful of iron filings and mix them into the sand and gravel.

2. **Experiment** For about one minute, use a sieve to try and separate the mixture into another bowl. Record how well the mixture separated—the dependent variable for this experiment.

3. Remix the ingredients. For about one minute use the tweezers to try and separate the mixture. Record your results.

4. Repeat step 3 using a magnet.

Draw Conclusions

5. **Use Variables** What was the independent variable for this experiment? Were there controlled variables?

6. **Interpret Data** Rank the separation methods from least to most effective. Make sure to give reasons for your ranking.

7. Did your results support your hypothesis? Write a report explaining why or why not.

Step 2

Step 3

Step 4

How can water separate a mixture?

Form a Hypothesis

You have seen how a mixture's properties affect how you can separate a mixture. Adding water to a mixture changes the properties of that mixture. How would this change the way you could separate a mixture of salt, sand, and sawdust? Write your answer as a hypothesis in the form "If water is added to a mixture of salt, sand, and sawdust, then the best method to separate the mixture will . . ."

Test Your Hypothesis

Briefly try to separate a mixture of salt, sand, and sawdust using only a filter. Next, design a procedure that uses both water and a filter to completely separate the mixture into three piles: sand, salt, and sawdust. Write out the resources and steps you would follow. Record your variables, results, and observations as you follow your plan.

Draw Conclusions

Did your experiment support your hypothesis? Why or why not?

For example, how can you separate black ink into different-colored inks? Design an experiment using chromatography paper to answer your question. Your experiment must be written so that someone else can complete the experiment by following your instructions.

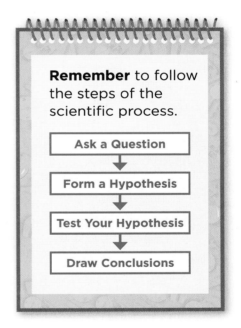

Remember to follow the steps of the scientific process.

Ask a Question

↓

Form a Hypothesis

↓

Test Your Hypothesis

↓

Draw Conclusions

Salt is separated from sea water using evaporation.

Compounds and Chemical Changes

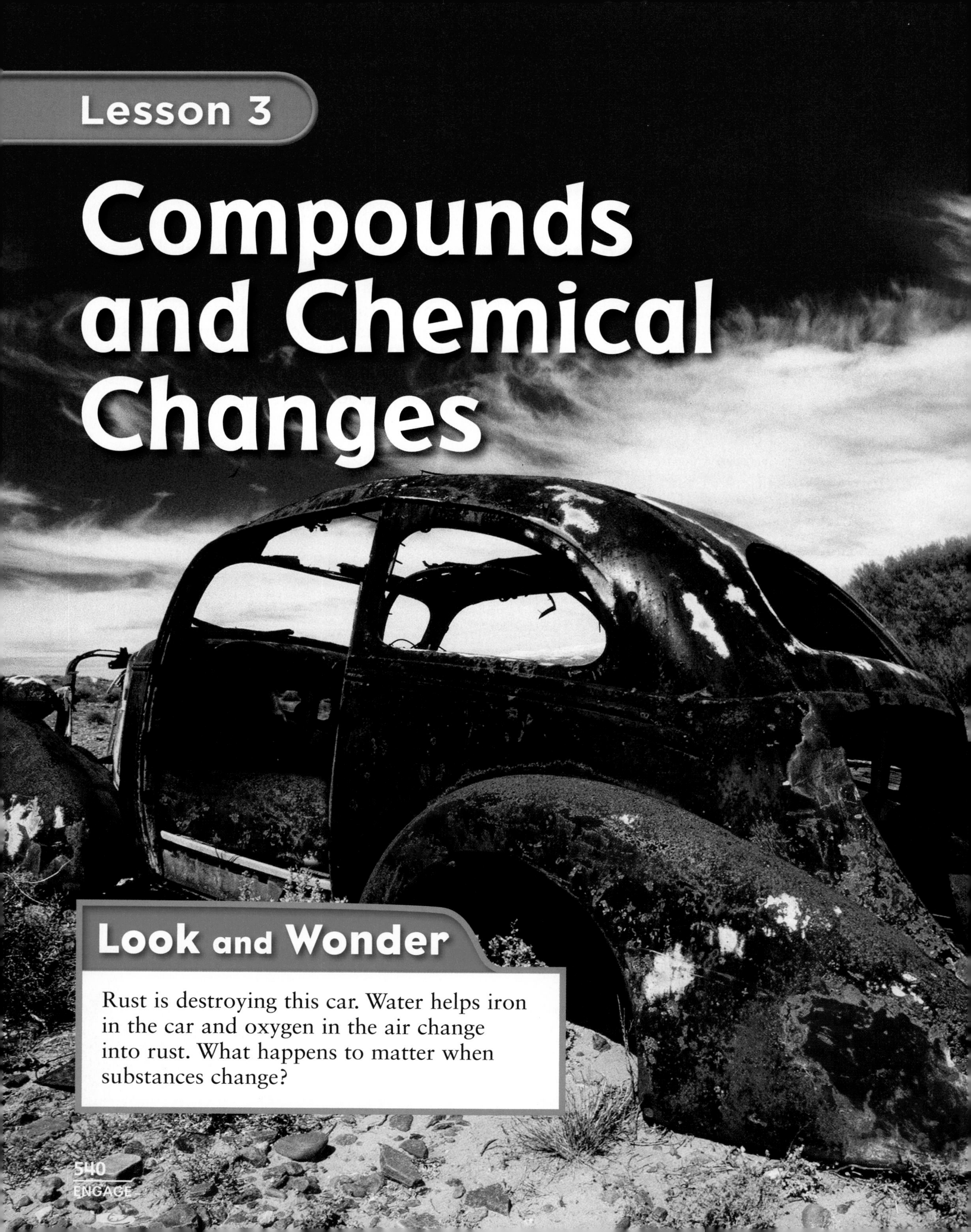

Look and Wonder

Rust is destroying this car. Water helps iron in the car and oxygen in the air change into rust. What happens to matter when substances change?

Does mass change in a chemical change?

Materials

Form a Hypothesis

Does the total mass of matter change when one substance turns into another? Think about chemical changes you have observed: an egg being cooked or wood burning in a fireplace. Write your answer as a hypothesis in the form "If a chemical reaction occurs, then the total mass . . ."

- safety goggles
- washing-soda solution (sodium carbonate)
- sealable bag
- Epsom-salt solution (hydrated magnesium sulfate)
- small paper cup
- equal pan balance

Test Your Hypothesis

1. ⚠ **Be Careful.** Wear safety goggles! Pour 40 mL of washing-soda solution into a bag. Place 40 mL of Epsom-salt solution in a paper cup. Put the cup inside the bag so that it rests upright. Seal the bag.

2. **Measure** Place the bag on a balance. Don't mix the solutions! Record the mass—it is the dependent variable for this experiment.

3. **Observe** Without opening the bag, pour the solution in the cup into the solution in the bag to cause a chemical change.

4. Record the mass of the bag and its contents.

Draw Conclusions

5. What is the independent variable in this experiment? Were there other variables you controlled?

6. **Interpret Data** How did mass change during the chemical reaction?

7. Does the data support your hypothesis? If not, how would you change your hypothesis?

Step 2

Explore More

Do you think that volume is conserved in a chemical change? Plan an experiment that would provide information to support your conclusion.

What are compounds?

You've probably sprinkled table salt on food to change its taste. Did you know table salt can be made from a poisonous gas and a metal that explodes in water? Two elements, sodium metal and yellowish chlorine gas, are combined. In a flash of heat and light, the sodium and the chlorine gas are changed. Only bits of table salt, also called sodium chloride, remain. Why are the properties of the sodium chloride so different from the properties of the sodium metal and the chlorine gas?

Sodium chloride forms from sodium and chlorine atoms. The joining of the atoms is what gives the sodium chloride new and different properties. Sodium chloride is an example of a **compound** (KOM•pound). A compound is formed by the combination of two or more elements. Compounds have properties different from their elements.

Forming a Compound

▼ Sodium is a soft, reactive metal that explodes on contact with water.

◀ Chlorine is a poisonous, yellow-green gas. Putting sodium in chlorine causes a fiery reaction.

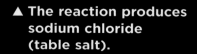

▲ The reaction produces sodium chloride (table salt).

Names and Formulas

Have you ever left a metal spoon out in the rain and come back later to find it covered with rust? Rust is a compound formed by the combination of iron in the spoon and oxygen in the air.

All compounds have chemical names and many have common names as well. The chemical name indicates which elements come together to make the compound. The chemical name for rust is iron oxide.

Chemical names use the names of elements. Often, the second element has the last part of its name changed slightly. You can see this change in *iron ox-ide* and *sodium chlor-ide.*

rust: Fe_2O_3

▲ There are 2 iron atoms and 3 oxygen atoms in the rust compound.

Sometimes we use prefixes to indicate the number of atoms in a compound. For example, we use *mon-*, meaning "one"; *di-* or *bi-*, meaning "two"; and *tri-*, meaning "three." Carbon dioxide is a gas with 1 carbon atom for every 2 oxygen atoms.

Compounds have chemical formulas just like the elements you've learned about before. Compounds, however, have more than one elemental symbol because they are made from more than one kind of element. For example, the chemical formula for iron oxide is Fe_2O_3. This formula tells us that 2 iron atoms combine with 3 oxygen atoms to form iron oxide.

fructose: $C_6H_{12}O_6$

▲ To form fructose, or fruit sugar, 6 carbon atoms, 12 hydrogen atoms, and 6 oxygen atoms combine.

✓ Quick Check

Draw Conclusions What is the minimum number of atoms in a compound? Why?

Critical Thinking How are chemical names and formulas related?

What are chemical changes?

Suppose you spilled vinegar with its strong, sharp odor. What a smell! How could you get rid of it? If you mixed it with water you could still taste it and smell it. If you froze it, it would still be vinegar, just in a solid form. As long as you are only making physical changes, the vinegar will always be vinegar.

To remove the vinegar, you have to change it chemically. A **chemical change** occurs when atoms link together in new ways to create a substance different from the original substances. It is also known as a *chemical reaction* (ree•AK•shuhn).

If you stir baking soda into vinegar, something happens. The baking soda changes and bubbles form. The bubbles are a new substance. Also, a new

white solid is left behind as the liquid evaporates away. The new substances show that this is a chemical change! The new substances have different properties than the vinegar and baking soda. The bubbles are a gas, and the white solid does not react with vinegar.

Atoms in the baking soda and vinegar link together in new ways. The new groupings form sodium acetate, water, and carbon dioxide. The bubbles from the reaction are carbon dioxide. The white solid that evaporation leaves behind is sodium acetate.

Chemical Equations

In studying math, you probably have seen equations like $2 + 6 = 8$ or $3 + 7 = 6 + 4$. Chemists write equations for chemical changes that are similar to math equations. For example, hydrogen gas and oxygen gas combine to form water. A chemical equation keeps track of which substances are used, and in what ratio. The number in front of the chemical formula tells how many of those molecules are used.

The chemicals on the left side of a chemical equation are called **reactants** (ree•AK•tuhnts). The chemicals on the right side of the equation are called **products** (PROD•ukts). For the water-forming reaction, the reactants are hydrogen and oxygen and the product is water. The equation reads: "2 hydrogen molecules plus 1 oxygen

Bubbles of carbon dioxide form when baking soda reacts with vinegar.

vinegar:
$HC_2H_3O_2$

baking soda:
$NaHCO_3$

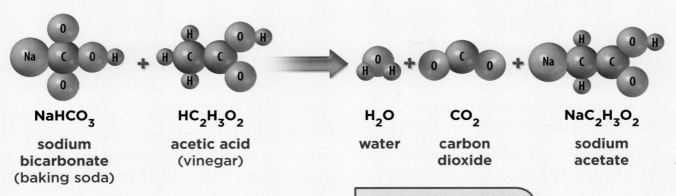

$NaHCO_3$	$HC_2H_3O_2$	H_2O	CO_2	$NaC_2H_3O_2$
sodium bicarbonate (baking soda)	acetic acid (vinegar)	water	carbon dioxide	sodium acetate

Read a Diagram

Is mass conserved in this reaction?

Clue: Count the number of each type of atom on each side of the arrow.

molecule produce 2 water molecules." The reactants and products of a chemical reaction may be in different states of matter.

In a math equation, the left side must equal the right side. How does the left side of a chemical equation "equal" the right side? The total mass of the reactants equals the total mass of the products in a reaction. This is known as the *law of conservation of mass*. In other words, the total number of each type of atom must be the same in the reactants and the products.

Math equations make sense whether you read them right to left or left to right. Chemical equations are similar. Most chemical changes are reversible, or can be undone. When a chemical reaction is reversed, the products break apart or combine to form the original reactants. Water can be broken down back into hydrogen and oxygen gas.

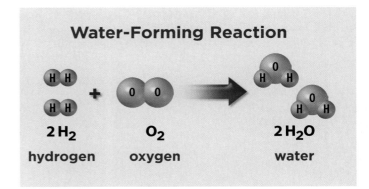

Water-Forming Reaction

$2H_2$	O_2	$2H_2O$
hydrogen	oxygen	water

✓ Quick Check

Draw Conclusions What are the reactants and products of the vinegar and baking soda reaction?

Critical Thinking If 32 atoms of hydrogen react completely with 16 atoms of oxygen, how many molecules of water will be made? Why?

FACT Like physical changes, many chemical changes can be reversed.

How can you spot a chemical change?

Chemical changes produce new substances. These substances have different properties than the previous ones. Often, you can see, hear, or smell the formation of new substances as a chemical change occurs.

A change in color is one signal of a chemical change. When bleach whitens a piece of cloth, it does so by chemically changing the dyes on the cloth or the cloth itself.

Chemical changes form layers on metals, and often dull their color. Rust, for example, is reddish, while the iron from which it forms is shiny. In fact, whenever metals corrode it is due to a chemical change. When corrosion of a metal causes a color change it is called *tarnish* (TAHR•nish).

When you place an antacid tablet in water, a chemical reaction occurs and bubbles start to form. The appearance of bubbles is another signal that a chemical change has occurred. Do you recall what happens when baking soda and vinegar are mixed? Bubbles of carbon dioxide show that a chemical change takes place there too!

Signs of a Chemical Change

changes color

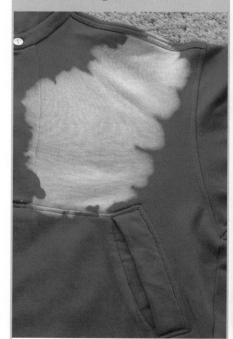

Bleach removes the color from clothing by means of a chemical change.

forms tarnish

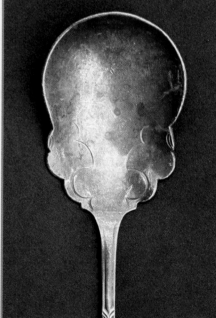

The silver in this spoon reacts with oxygen or sulfur to form tarnish.

releases gas

As an antacid reacts with water, bubbles of carbon dioxide gas form.

Chemical changes can produce more than just gases. A **precipitate** (pri•SIP•i•tit) is a solid formed from the chemical reaction of two solutions. When a sink isn't cleaned, it may develop a sticky white layer. This "soap scum" is a precipitate formed from solutions of soap and water.

Some chemical changes involve the release of light and heat. A burning candle, for instance, produces a hot flame. The heat and light come from the combination of atoms in the candle and in the wick with oxygen from the air. If a chemical reaction releases energy, then reversing that chemical reaction will absorb energy.

Quick Lab

Chemical Cents

1. Pennies are coated in copper, which corrodes easily. Find a dull and tarnished penny.

2. **Observe** Put the penny in a cup of salt and vinegar. Record your observations.

3. Are there any signs of a chemical reaction? Take the penny out and let it dry in the air. Do any more chemical reactions occur? How do you know?

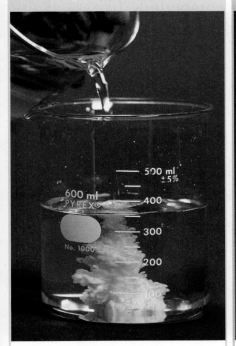

forms a precipitate

When two solutions form a precipitate, a chemical change has occurred.

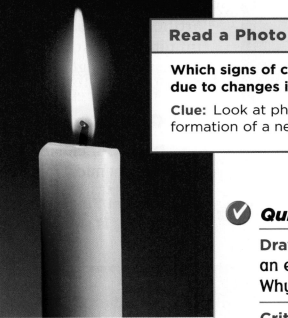

releases energy

A release of energy as light or heat can indicate a chemical change.

Read a Photo

Which signs of chemical changes are due to changes in states of matter?

Clue: Look at photos that show the formation of a new solid, liquid, or gas.

✓ Quick Check

Draw Conclusions Is frying an egg a chemical change? Why or why not?

Critical Thinking What signs tell you that a log burning in a fire is a chemical change?

How can you use chemical changes?

Plants and animals use two important chemical reactions to create food and release energy: photosynthesis and respiration. Photosynthesis requires the energy of the Sun to help create simple sugars in plants. Respiration reverses that chemical reaction and releases energy. Your body uses that energy to fuel its cells.

Machines also can use the energy from chemical reactions. A space shuttle needs huge amounts of energy to fly into space. In its main engines, hydrogen and oxygen molecules chemically react and release hot gases to push the shuttle upward. The product from this chemical reaction is safe for the environment—it's just water!

Chemical reactions also are the only way to form compounds. Some compounds, like fossil fuels, are formed in nature. Others are artificial, such as plastics. There is almost an infinite variety of useful compounds that can be made through chemical changes.

✓ Quick Check

Draw Conclusions How are compounds and chemical reactions related?

Critical Thinking Where do you think energy is stored during photosynthesis?

The space shuttle uses a chemical reaction of hydrogen and oxygen to launch into space.

Lesson Review

Visual Summary

Compounds have different properties than the two or more elements that form them.

Chemical changes occur when atoms link together in new ways.

Signs of a chemical change include **precipitates,** bubbles, and changes in color, light, or heat.

Make a FOLDABLES™ Study Guide

Make a Three-Tab Book. Use the titles shown. Summarize what you have learned about each topic under the tabs.

Compounds...

Chemical changes...

Signs of a chemical Change...

Think, Talk, and Write

1. **Main Idea** What happens to the atoms in compounds when they undergo a chemical change?

2. **Vocabulary** Substances on the left side of a chemical equation are called _____.

3. **Draw Conclusions** What would happen if you removed one of the reactants during a chemical reaction?

Text Clues	Conclusions

4. **Critical Thinking** Where does a burning candle go over time?

5. **Test Prep** Which is an example of a chemical reaction?
   - **A** ice melting
   - **B** salt stirred into water
   - **C** wood burning
   - **D** rain falling

6. **Test Prep** Which compound could be tarnish on a metal?
   - **A** CO_2
   - **B** $C_6H_{12}O_6$
   - **C** HCO_3
   - **D** Al_2O_3

Math Link

How Much Will You Get?
For every 4 grams of hydrogen burned, 36 grams of water are made. How many kilograms of water are made if 100 kg of hydrogen are burned?

Art Link

Chemical Changes and Paintings
A painting may be protected by a layer of varnish. When that varnish is worn away, chemical reactions can occur between the air and the painting. What effects would this have on the painting?

Writing in Science

The Case of the Mystery Compounds

Expository Writing

A good exposition

▶ **develops the main idea with facts and supporting details**

▶ **summarizes information from a variety of sources**

▶ **uses transition words to connect ideas**

▶ **draws a conclusion based on the facts and information**

Scientists can discover what a compound is by using chemical reactions. They use a set of chemicals whose chemical properties are well known. First, they combine those known chemicals with the unknown compound one at a time. Then they observe the reactions. The unknown substance will react with some of the known chemicals but not with others.

Scientists carefully record their observations. They then compare the chemical properties of the unknown compound with the chemical properties of known compounds. If two compounds have the same chemical properties, they are the same compound. The mystery is solved! This process of discovering the compound's identity is called *qualitative analysis*.

Write About It

Expository Writing Do research and write a report about how scientists can test water for pollutants and dangerous chemical compounds. Which chemical reactions do they use to perform the tests? Give the steps of the process in order.

 LOG ON e-Journal Research and write about it online at **www.macmillanmh.com**

Math in Science

Reacting with Water (H_2O)

Your body needs water, but other substances find it very destructive. When sodium metal (Na) is placed in water, it reacts to form hydrogen gas (H_2), which catches fire or even explodes! Sulfuric acid (H_2SO_4) in water will boil and spit. Hydrogen peroxide (H_2O_2), used to treat cuts and scrapes, can split into water and oxygen gas (O_2), causing bubbles to form rapidly.

Chemical formulas are used to represent the atoms and elements in a compound or element. The number in front of the formula tells you how many molecules there are. The subscript, the small number behind a letter, tells you how many atoms of an element there are. To find out how many atoms are in a formula, you multiply the subscripts by the number in front of the molecule.

Sodium in water produces a bright yellow flame.

Multiplying Whole Numbers

To find the number of atoms

▶ multiply the **number** in front of the molecule by the **subscript**

$2H_2$ $2 \times 2 = 4$ atoms of hydrogen

▶ If there is no number in front of the molecule, then use the number 1

O_2 $1 \times 2 = 2$ atoms of oxygen

▶ If there is no number behind the element, then use the number 1

$3S$ $3 \times 1 = 3$ atoms of sulfur

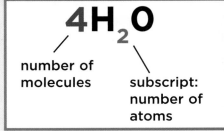

$$4H_2O$$

number of molecules

subscript: number of atoms

Solve It

1. How many atoms of hydrogen and oxygen are in 2 molecules of water ($2H_2O$)?

2. How many atoms of hydrogen and oxygen are in 3 molecules of hydrogen peroxide ($3H_2O_2$)?

3. How many atoms of hydrogen, sulfur, and oxygen are in 1 molecule of sulfuric acid (H_2SO_4)?

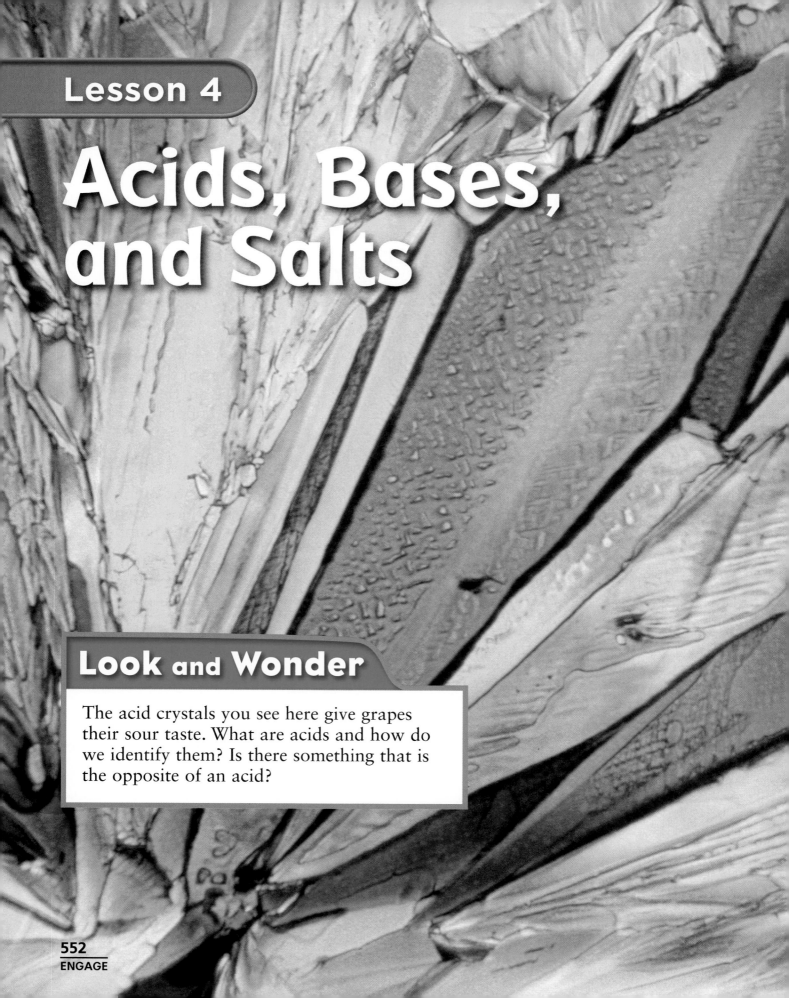

Acids, Bases, and Salts

Look and Wonder

The acid crystals you see here give grapes their sour taste. What are acids and how do we identify them? Is there something that is the opposite of an acid?

Which are acids and which are bases?

Purpose

Acids turn blue litmus paper red and have no effect on red litmus paper. Bases turn red litmus paper blue and have no effect on blue litmus paper. You will examine household solutions to see if they are acids or bases.

Procedure

1. **Predict** Create a table like the one shown. Use your sight and sense of smell to predict whether each household solution is an acid or base.

2. △ **Be Careful.** Wear goggles, gloves, and an apron! Record what happens as you dip a strip of each color of litmus paper into sample #1.

3. **Classify** Repeat step 2 with other solutions your teacher has provided and record your results.

4. Now test each solution with a small strip of pH paper. It tells you how strong an acid or base is. Use the color scale provided to find and record the pH.

Draw Conclusions

5. Organize the samples you tested into acids and bases. What does pH tell you about acids and bases?

6. **Communicate** Do all the acids or all the bases have anything in common? Do any seem more acidic or more basic than others? Why do you think so? Write down your reasons in a report.

Explore More

Some people claim that cola drinks are very acidic and will even dissolve iron nails left in them overnight! Design and perform an activity to test the acidity of colas. Are they really strong acids?

Materials

- goggles
- gloves
- apron
- red and blue litmus papers
- samples of household solutions
- pH paper

Step ❶

	Predict: Acid or Base	Effect on Red Litmus	Effect on Blue Litmus	pH strip	Result: Acid or Base
Sample #1					
Sample #2					
Sample #3					

Step ❸

▶ Main Idea

Acids and bases are chemical compounds that can be classified by their properties.

▶ Vocabulary

acid, p. 554

ion, p. 554

base, p. 555

acidity, p. 556

alkalinity, p. 556

neutralization, p. 558

electrolyte, p. 558

e-Glossary
at **www.macmillanmh.com**

▶ Reading Skill ✓

Summarize

Summary

What are acids and bases?

Have you ever bitten into a slice of lemon? It tastes sour! The sour taste comes from an acid (AS•id) in the lemon juice called citric acid. The word *acid* comes from an ancient word meaning "sour."

A sour taste is one property of acids. There are other properties of acids. An **acid**

- tastes sour and burns if you touch it
- turns blue litmus to red
- reacts with metals to make hydrogen gas

Acid compounds contain hydrogen. An acid gives off hydrogen ions (EYE•uhnz), $H+$, in water. **Ions** are atoms or molecules that have gained or lost electrons. Hydrogen ions have lost an electron, and have a positive charge. Water molecules will join with hydrogen ions to form *hydronium ions* (hye•DROH•nee•uhm), H_3O+.

The acid shown in the diagram, hydrochloric acid, is in our stomachs. This acid helps us digest food. Fortunately, our stomachs are coated with a substance that keeps the acid from digesting our stomachs as well. Hydrochloric acid is also used to clean steel and make plastics. Acids like this are dangerous and should never be touched or tasted.

Lemons taste sour because they contain an acid.

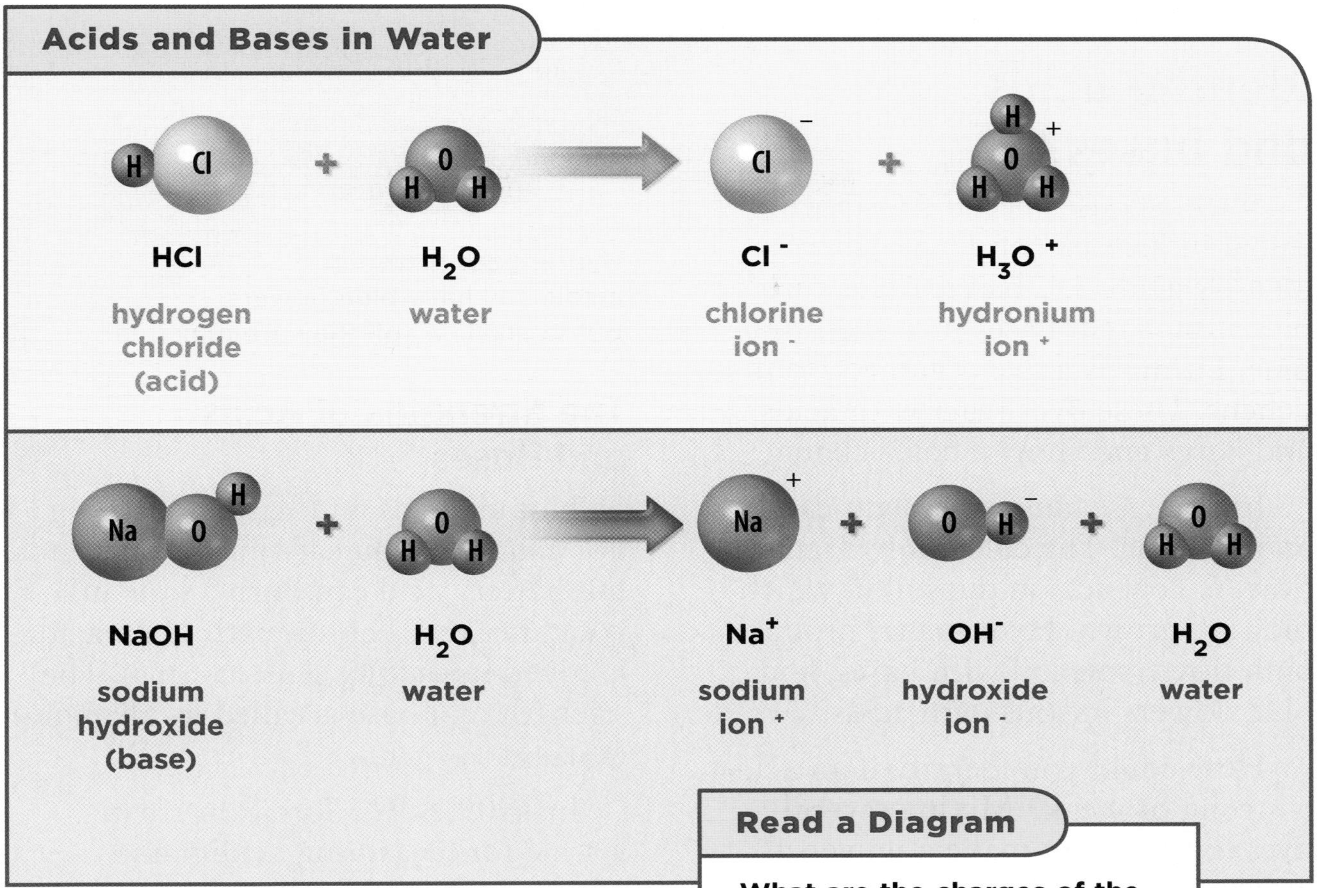

Acids and Bases in Water

HCl
hydrogen
chloride
(acid)

+

H₂O
water

→

Cl⁻
chlorine
ion⁻

+

H₃O⁺
hydronium
ion⁺

NaOH
sodium
hydroxide
(base)

+

H₂O
water

→

Na⁺
sodium
ion⁺

+

OH⁻
hydroxide
ion⁻

+

H₂O
water

Read a Diagram

What are the charges of the ions of sodium and chlorine?

Clue: Look at the plus and minus signs next to their chemical formulas.

Soap, drain cleaner, and ammonia all have something in common—they are bases (BAY•siz). A **base**

- tastes bitter
- feels soapy
- turns red litmus to blue

Base compounds often contain oxygen and hydrogen pairs, OH. Bases make *hydroxide ions* (hye•DROK•side), OH-, when in water. Hydroxide ions have gained an electron and have a negative charge.

Strong bases can dissolve hair and foods. They can be used to clear clogged drains in your home. Ammonia is used to make fertilizers. Sodium hydroxide, called lye, is used to make cloth, soaps, and some plastics.

 Quick Check

Summarize Show some properties of acids and bases in a table format.

Critical Thinking Why does zinc bubble when placed in vinegar?

How can indicators identify acids and bases?

Scientists use special substances called *indicators* (IN•di•kay•tuhrz) to identify acids and bases. Acids turn blue litmus red. Bases turn red litmus blue. Litmus is a dye obtained from lichens. These dyes react with acids and bases and show a color change.

In one case, an entire plant can be an indicator! The color of hydrangea flowers depends on the soil in which they are grown. Hydrangeas produce pink flowers in soil with bases, and blue flowers in soils with acids.

How could you identify if an acid is strong or weak? Mixing several dyes together can make a universal indicator. This indicator will turn different colors depending on the strength of an acid or a base.

Hydrangeas grown in acidic soil have blue flowers, but in alkaline soil they are pink.

The Strengths of Acids and Bases

Not all acids and bases are strong. For example, you eat vinegar on salads, but battery acid can burn a hole in your sneaker! The strength of an acid is called its **acidity** (uh•SID•i•tee). The strength of a base is called its **alkalinity** (al•kuh•LIN•i•tee).

In 1909, S.P.L. Sorenson created a scale for measuring acidity and alkalinity. The *pH scale* stands for p*otential for* H*ydrogen*. Low pH numbers indicate strong acids. High pH numbers indicate strong bases.

▼ Universal indicators change color with the acidity or alkalinity of substances.

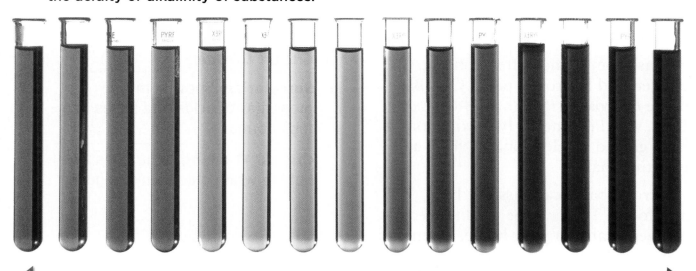

acid   base

The pH of most solutions is between pH 1 and pH 14. A pH of 1 means that there are many hydronium ions in a solution. The solution is very acidic. A pH of 14 means that there are many hydroxide ions in a solution. The solution is very alkaline. A pH of 7 means that the solution is neutral.

 Quick Check

Summarize How does the pH scale show acidity and alkalinity?

Critical Thinking Why might red cabbage juice be a different color in vinegar than in baking soda?

≡Quick Lab

Indicating Ink

1 Dip a cotton swab in baking soda solution. Use it to write a message to a partner on a piece of paper.

2 Allow the paper to dry. Then switch papers with your partner.

3 **Observe** Is the message invisible? Use another swab to "paint" the paper with grape juice. Record your observations.

4 **Infer** Is the grape juice an acid-base indicator? Why or why not?

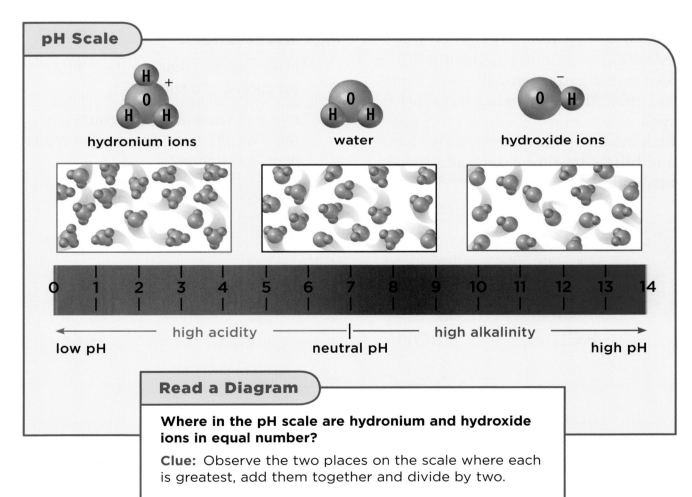

pH Scale

hydronium ions   water   hydroxide ions

0 1 2 3 4 5 6 7 8 9 10 11 12 13 14

◄——— high acidity ——— | ——— high alkalinity ———►

low pH   neutral pH   high pH

Read a Diagram

Where in the pH scale are hydronium and hydroxide ions in equal number?

Clue: Observe the two places on the scale where each is greatest, add them together and divide by two.

What are salts?

Hydrochloric acid is dangerous. Sodium hydroxide base is also dangerous. Yet if these two are mixed, ordinary table salt and water form!

This reaction is an example of an important pattern: *Mixing an acid with a base produces a salt plus water*. In this reaction, the acid and base cancel each other's properties. **Neutralization** (new•truhl•i•ZAY•shuhn) is when acids and bases react to form salt and water.

You already know one salt: table salt, or sodium chloride (NaCl). There are many other salts. Potassium chlorate is used in matches and explosives! A salt is any compound made of positive ions and negative ions. The ions form a crystal. Many salts are a metal (+) with a nonmetal (-). Salts usually have a high melting point. They are also hard and brittle. In some cases, salts dissolve readily in water.

In table salt (sodium chloride), ions form crystals even at the microscopic level.

Acids, bases, and salts give off ions when in water. Ions in a solution conduct electricity. If a substance forms ions in water it is called an **electrolyte** (i•LEK•truh•lite). Acids, bases, and salts are all electrolytes.

✓ Quick Check

Summarize What are the typical properties of a salt?

Critical Thinking How could you tell if a substance dissolved in water is an electrolyte?

Formation of Salt

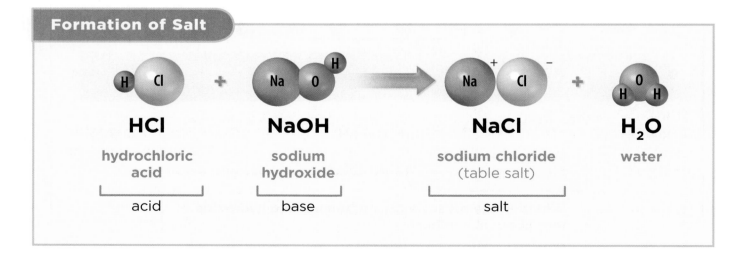

HCl	NaOH	NaCl	H₂O
hydrochloric acid	sodium hydroxide	sodium chloride (table salt)	water
acid	base	salt	

FACT There are many kinds of salt besides table salt.

Lesson Review

Visual Summary

Acids and **bases** can be classified by their chemical properties.

Indicators can show if a material is acidic or alkaline.

Salts form from the neutralization of an acid and a base.

Make a FOLDABLES™ Study Guide

Make a Folded Table. Use the titles shown. Summarize what you have learned about each topic in the columns provided.

Acids and bases...	Indicators...	Salts...

Think, Talk, and Write

1 Main Idea How can you tell acids and bases apart?

2 Vocabulary The strength of a base is called its _____.

3 Summarize What substances in your kitchen are acids and bases?

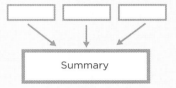

Summary

4 Critical Thinking How would pH paper help you to predict the color of hydrangea flowers?

5 Test Prep Which acid is in our stomachs??

  A sulfuric acid, H_2SO_4
  B nitric acid, HNO_3
  C hydrochloric acid, HCl
  D phosphoric acid, H_3PO_4

6 Test Prep Which is NOT a property of a base?

  A turns blue litmus red
  B dissolves hair
  C dissolves fats and oils
  D tastes bitter

 Math Link

Neutralizing Water

A swimming pool holds 115,000 L of water. pH tests show that the water is too acidic by 0.1 pH units. To raise the pH by 0.1 unit, you must add 4 g of soda ash (a base) per 1,000 L of water. How much soda ash must you add?

 Social Studies Link

Pollution Laws

Burning coal to make electricity creates acidic air pollution. The pollution can be cleaned up, but doing so is expensive. We can pass laws that require pollution controls on power plants. What are the pros and cons of such laws?

Meet CHRISTINA ELSON

Christina Elson is a scientist at the American Museum of Natural History. She studies how salt was used by the ancient Aztec culture. During the twelfth through sixteenth centuries, the Aztecs lived in the area that is now Mexico. This area was very rich in salt, which is a natural mineral resource that is mined from the ground.

Christina studies a region in Mexico where salt was taken from deposits near a dried lake bed. The Aztecs turned these deposits into salt. First, they collected the salty soils by digging them out of the ground. Then they filtered water through the soils to dissolve out the salts into big pots. The final step required boiling the salt solution so the water evaporated away. The salt remained behind in the form of crystals.

Christina Elson

Aztecs used salt for much more than cooking. In one Aztec town, Christina found thousands of ceramic fragments. These were pieces of clay pots that were used to transport salt for trade.

She also found that salt was used to dye cloth. Colorfully dyed cotton cloth was valuable because it was desired by the Aztec nobles. The cloth was dyed with pigment in a hot, watery dye-bath. When salt was added to the dye-bath, it helped the pigment "stick" to the cloth. The salt combined with the color pigment to make a compound that could not be dissolved in water.

Salt was important to many other ancient cultures, and continues to be important today. Salt can be used to preserve food so it can be stored for a long time without refrigeration. It can be used to prepare and preserve animal skins for clothing. It can also be used to make soap. Salt's value stems from its usefulness, durability, and portability.

▼ **Christina Elson records data at an archeological "dig" site in Mexico.**

Write About It
Infer

1. How did the Aztecs change a mineral resource into a finished product?
2. What would happen to the colors in Aztec cloth when washed if salt were not part of the dye-bath?

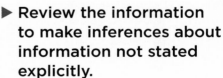

LOG ON **e-Journal** Research and write about it online at **www.macmillanmh.com**

Infer

▶ Review the information to make inferences about information not stated explicitly.

▶ List the details that support the inferences you make.

AMERICAN MUSEUM OF NATURAL HISTORY

Visual Summary

Lesson 1 Matter can change state when heat is added or removed.

Lesson 2 Mixtures are physical combinations of different kinds of matter.

Lesson 3 Compounds form when chemical changes cause atoms to link together in new ways.

Lesson 4 Acids and bases are chemical compounds that can be classified by their properties.

Make a FOLDABLES™ Study Guide

Assemble your lesson study guide as shown. Don't forget to include your Lesson 4 study guide in the back!

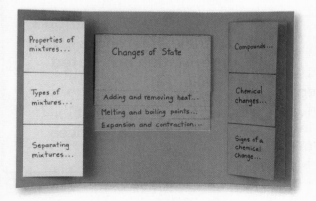

Fill in each blank with the best term from the list.

acid, p.554

alloy, p.532

electrolyte, p.558

product, p.544

reactant, p.544

solution, p.532

sublimation, p.521

thermal expansion, p.524

1. Litmus paper will turn from blue to red when it comes in contact with a(n) _____.

2. A chemical on the left side of a chemical equation is a(n) _____.

3. When the parts of a mixture blend together so that it looks the same everywhere it is a(n) _____.

4. The change of state directly from solid to gas is called _____.

5. A substance that dissolves to form ions is a(n) _____.

6. When the particles of an object start to move faster and cause it to grow, it is called _____.

7. A solution of at least one metal and another solid is a(n) _____.

8. The result of a chemical reaction is a(n) _____.

Answer each of the following in complete sentences.

9. **Compare and Contrast** How is dry ice different after it goes through sublimation? How is it the same?

10. **Summarize** What signal shows that a chemical change is occurring in the picture below? What are some other signals of chemical changes?

11. **Use Variables** You are doing an experiment to see how to mix sugar and water into a solution more quickly. What is one variable you might change in your experiment? What would be the control test?

12. **Critical Thinking** If you tasted a new food that was very bitter, would you guess that it was an acid or a base? Explain.

13. **Expository Writing** Explain how this chemical equation demonstrates the law of conservation of mass:
    $2 H_2 + O_2 = 2 H_2O$.

14. How can matter be changed?

Fizzy Evidence

Your goal is to determine whether lemon juice or apple juice is more acidic when they react with baking soda—a weak base—to form water and a salt.

What to Do

1. Add a $\frac{1}{4}$ teaspoon of baking soda to each of two small cups.

2. Add a tablespoon of lemon juice to one cup and repeat with the other using clear apple juice.

3. Evaporate the liquid to recover the salt that formed.

Analyze Your Results

▶ What evidence supports the idea that a chemical reaction occurred in the glasses?

▶ What evidence suggests that lemon juice contains a stronger acid?

1. **The chemical reaction below shows the formation of a(n) _____.**

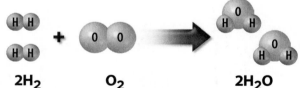

$2H_2$	O_2	$2H_2O$
hydrogen	oxygen	water

   A mixture.

   B compound.

   C acid.

   D salt.

Food Science Technician

For many students the most interesting and fun part of science class is doing lab experiments. Are you one of these students? Then you like learning new things from working in a lab. Do you also have a special interest in food? For example, you might be curious about how to keep foods from spoiling. If so, then you might enjoy a career as a food science technician. As an assistant to the food scientist in charge, you would operate laboratory instruments. You might do tests on chemicals that are added to foods to make them taste better or keep them fresh. These tests are needed to make sure the color, texture, and nutrients in foods meet government standards. The best way to get started in this career is to get an associate's degree in a two-year college program.

▲ Food science technicians study the chemistry of foods.

▼ Green chemists keep the benefits of chemistry from harming the environment.

Green Chemist

Chemists develop many useful materials in the laboratory. These include dyes, plastics, medicines, food products, and building materials. In the past, developing these materials also meant creating harmful chemical wastes. These wastes then polluted the environment. However, a movement called Green Chemistry is now helping to eliminate pollution. Green chemists work to create reactions that produce safe, useful products without harmful side effects. To become a green chemist, you will need a college degree in chemistry followed by graduate studies.

Forces and Energy

Rockets travel at speeds near 11 kilometers per second (7 miles per second) to leave Earth's atmosphere.

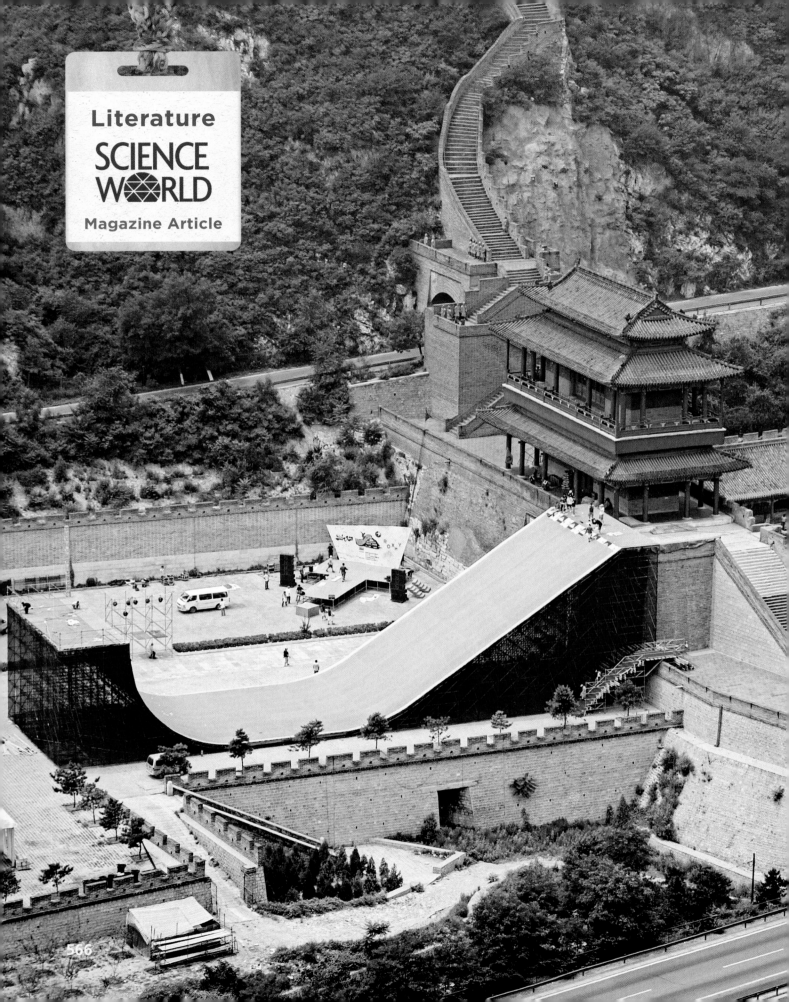

This daredevil even managed a spin while airborne.

Danny Way standing on the Great Wall of China.

from *Science World*

The Great Jump in China

During a hot day in July, Danny Way rolled into the record books. He became the first person to leap across the Great Wall of China . . . on a skateboard!

Way began his wall-jumping roll from the top of a towering J-shaped ramp. "The higher up he is, the more gravitational potential energy (stored energy due to height) he'll have," says Louis Bloomfield, a physicist at the University of Virginia. When Way descended, that stored energy converted into energy of motion, or kinetic energy, and gave him maximum speed.

Why the need for speed? When Way blasted off the ramp and up into the air, gravity pulled him downward. With more upward speed he could fight gravity's tug longer to gain big air.

Way also needed forward speed to cross the 21-meter-wide (70-foot-wide) wall. That's why his takeoff ramp sloped gradually up toward the wall. The gentle angle shot him both upward and forward onto a ramp on the other side.

Write About It

Response to Literature This article describes how an athlete used a ramp to jump over a large object. If you were a professional athlete, what other kinds of devices might you use? Write a fictional narrative describing one of these devices and its uses.

 LOG ON e-Journal Write about it online at www.macmillanmh.com

Using Forces

The Big Idea

How do forces move objects?

Sailing in Oslofjord, Norway

Key Vocabulary

velocity
the speed and direction of a moving object (p. 575)

momentum
the mass of an object multiplied by its velocity (p. 578)

force
any push or pull by one object on another (p. 584)

work
the use of force to move an object a certain distance (p. 598)

energy
the ability to perform work or change an object (p. 600)

simple machine
a machine that changes the direction, distance, or strength of one force (p. 608)

Lesson 1

Motion

Look and Wonder

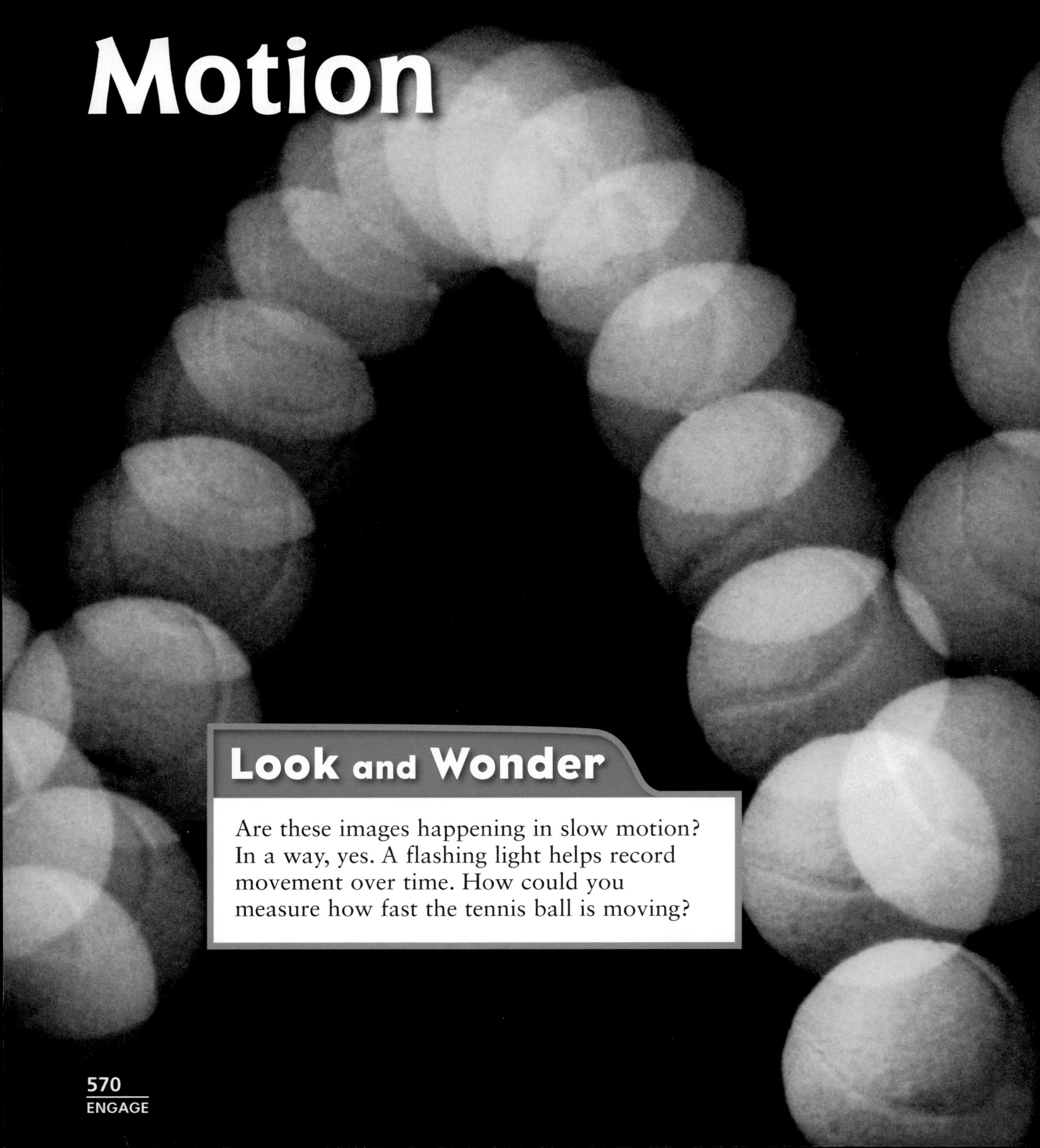

Are these images happening in slow motion? In a way, yes. A flashing light helps record movement over time. How could you measure how fast the tennis ball is moving?

How is speed measured?

Form a Hypothesis

How do you think speed depends on the distance an object travels? Write your answer as a hypothesis in the form "If the distance a marble travels increases, then…"

Test Your Hypothesis

1. Make a marble launcher out of an index card. Use the pattern provided. Place the launcher on a long, flat, and smooth surface.

2. Place a piece of tape in front of the launcher—this is your starting point. Use a meterstick to place a piece of tape one meter from your starting point. This is your "finish line" and your independent variable.

3. **Measure** Roll a marble down the launcher. Use a stopwatch to time it as it travels from the starting point to the finish line. Repeat twice more and calculate an average time—this is your dependent variable.

4. Repeat step 3 for finish lines at 2 m and 3 m.

Step 1

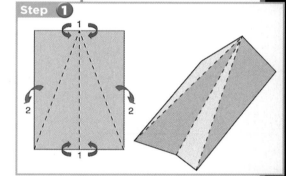

Draw Conclusions

5. **Use Numbers** Divide each distance by its average time. This value is the average speed of the marble over that distance.

6. **Communicate** Did the launcher roll the marble at the same speed every time? Write a report describing the motion of the marble out of the launcher.

Step 3

Explore More

What would be the marble's speed if it traveled on a curved path? Would it move faster or slower than on a straight path? Write a hypothesis and design an experiment to test it.

Main Idea

Motion occurs when an object moves from one position to another.

Vocabulary

position, p. 572

motion, p. 572

frame of reference, p. 573

speed, p. 574

velocity, p. 575

acceleration, p. 576

momentum, p. 578

LOG ON e-Glossary
at www.macmillanmh.com

Reading Skill ✔

Main Idea and Details

Main Idea	Details

Positions on a Grid

What is motion?

Where are you? Are you in a state, in a city, in a classroom? Are you a certain number of steps from a door? Is the door to the left or right? To answer these questions you need to know your position (puh•ZISH•uhn). A **position** is the location of an object. It answers the question, "Where is the object?"

Positions of objects can be described by a grid. In a grid, you describe a position using points on each axis, or axes. When an object changes its position on the grid, you can draw an arrow between the old position and the new. This arrow represents the motion (MOH•shuhn) of the object. **Motion** is a change in position over time.

A motion has two parts: distance and direction. Distance is the length of the arrow on the grid, and can be measured with a ruler. We use units such as meters (m), kilometers (km), feet (ft), or miles (mi) to describe distance.

Direction is where the arrow is pointing. On a map, we use words like north, east, south, and west. To measure direction you can use a compass or protractor, and units of degrees.

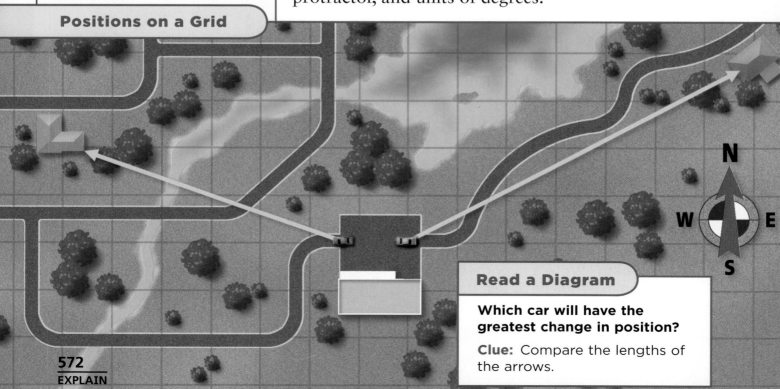

Read a Diagram

Which car will have the greatest change in position?

Clue: Compare the lengths of the arrows.

Frame of Reference

Suppose your friend tells you he is north. Do you think you know where he is? You need to ask him, "North of what?" Positions and motions only make sense if you have a frame of reference (fraym uv REF•uhr•uhns). A **frame of reference** is a group of objects from which you can measure a position or motion. Your classroom and the objects inside are a frame of reference. If your friend told you he was moving three meters north of his desk, then you could find him easily.

Almost anything can be a frame of reference: A baseball field, a fish tank, or the solar system. It is easiest to describe positions and motions when the frame of reference is a grid. This is why maps often have grids placed on top of them.

Frames of reference can move. For example, the inside of a moving car is a frame of reference. If you move in a car, other passengers inside that car see you moving normally. In your frame of reference, the car's motion does not seem to affect you at all.

Different frames of reference see things differently, however. To anyone outside the car, you appear to be moving very fast. Why? They see the motion of the car added to your motion. The same thing occurs when you look out the car window. The ground seems to be moving very fast, even though you know it's really not moving at all. You add the motion of the car to all the objects outside your frame of reference.

In the frame of reference of a car, outside objects seem to be moving quickly.

In the frame of reference of the ground, the car is moving quickly.

 Quick Check

Main Idea and Details How can you measure the distance an object has moved?

Critical Thinking How could you be moving in one frame of reference, but not in another?

What is speed?

You're at the starting line of a 100-meter dash. The race starts, and you spring from the starting blocks. What is your goal? To cover the 100-meter distance in the shortest time possible! Whoever travels fastest wins the race.

"Fastest" in a race means running with the greatest speed. **Speed** is how fast an object's position changes over time. To calculate speed, you divide the distance traveled by the time spent traveling. Units of speed are units of distance per unit of time, such as meters per second (m/s), kilometers per hour (km/h), or miles per hour (mph).

The speed of a moving object can change. A runner in a longer race, for example, might go fast at first, slow down in the middle, and then go fast again at the end.

We determine the runner's average speed by dividing the total distance by the total time. Over short distances, like 100 meters, the fastest human can run at an average speed of about 10 m/s (22 mph). Over longer distances, like 50 km, the fastest human can run at an average speed of about 5.6 m/s (12.5 mph).

Calculating Speed

data: distance = 100 m, time = 10 s

speed = distance ÷ time
= 100 m ÷ 10 s
= 10 m/s

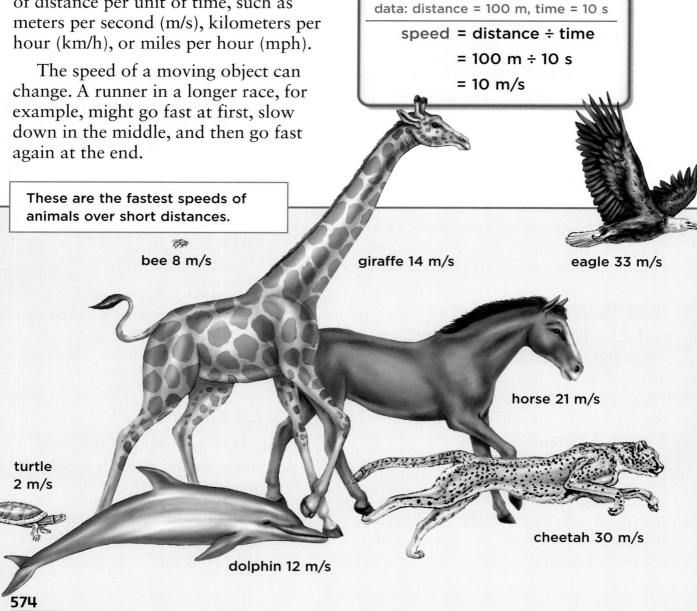

These are the fastest speeds of animals over short distances.

bee 8 m/s

giraffe 14 m/s

eagle 33 m/s

horse 21 m/s

turtle 2 m/s

cheetah 30 m/s

dolphin 12 m/s

Speed with Direction

If you were a pilot flying a plane, you would want to know how fast your plane could fly and how far the trip would be. With this data, you could calculate how long your trip would take. You would also need to know in which direction to fly, or you would miss your destination. **Velocity** (vuh•LOS•i•tee) is the measurement that combines both the speed and direction of a moving object. As a pilot you would want to know the velocity of the plane as you traveled.

Airplane Velocities

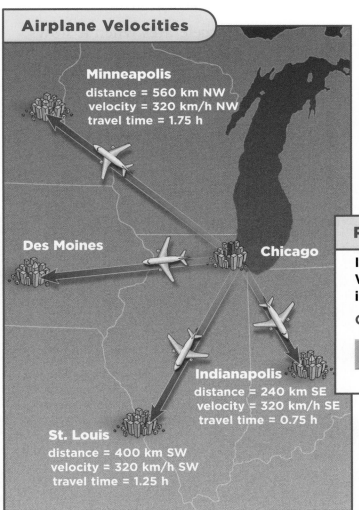

Minneapolis
distance = 560 km NW
velocity = 320 km/h NW
travel time = 1.75 h

Des Moines

Chicago

Indianapolis
distance = 240 km SE
velocity = 320 km/h SE
travel time = 0.75 h

St. Louis
distance = 400 km SW
velocity = 320 km/h SW
travel time = 1.25 h

≡ Quick Lab

Stepping Speed

1. Organize into teams of three: a quick-stepper, a timer, and a measurer.

2. **Measure** At the command "Go!", the stepper takes two quick steps, covering as much distance as possible. The timer records the time and the measurer records the distance. The stepper then repeats this process for three, four, and five quick steps.

3. Repeat the set of measurements two more times with everyone changing roles.

4. Make a line graph of your data for distance and time. Distance should go on the vertical axis and time on the horizontal axis.

5. **Interpret Data** Does distance change steadily with time in your graph? Why or why not?

Read a Diagram

It's 448 km from Chicago to Des Moines. What velocity would you need to fly there in 1.4 hours?

Clue: Be sure to state a direction of travel.

 Science in Motion Watch the planes move at **www.macmillanmh.com**

✔ Quick Check

Main Idea and Details How do you calculate average speed?

Critical Thinking What is the difference between speed and velocity? Give an example.

What is acceleration?

Suppose you are in a race car at the starting line facing north. The light changes from red to green and the driver steps on the gas. When you reach 180 m/s (403 mph), the driver lets off on the gas and the car travels at a constant speed. On your watch, you note that it took 6 seconds for the car to go from rest (0 m/s) to 180 m/s.

When the position of an object changes, it is in motion and it has a velocity. When the velocity of an object changes, it is accelerating. **Acceleration** (ak•sel•uh•RAY•shuhn) is the change in velocity over time for an object. Units of acceleration are units of velocity divided by units of time: meters per second per second ((m/s)/s). Just like motions and velocity, accelerations have a direction. So you might say the race car accelerates at 30 (m/s)/s north when the driver steps on the gas.

> ### Calculating Acceleration
> data: change in speed = 180 m/s, time = 6 s
>
> **acceleration = change in speed ÷ time**
>
> = 180 m/s ÷ 6 s
>
> = 30 (m/s)/s

In the example of the race car the value of the acceleration is 30 (m/s)/s. What does "(m/s)/s" really mean? Each time a second passes, the car gains 30 m/s of speed. After 6 seconds, the car has reached a final speed of 180 m/s. After the driver lets off on the gas, the car is traveling at a constant velocity, so it is no longer accelerating.

A car also accelerates when it slows down. An example would be a car stopping at a red light. Acceleration for decreasing speed is given as a negative number. For instance, a stopping car might accelerate at −30 (m/s)/s. We could also say that the car *decelerates* (dee•SEL•uh•rayts) at 30 (m/s)/s.

0:00   0 m/s

The driver steps on the gas and the car begins accelerating at 30 (m/s)/s.

0:02   60 m/s

0:06   0 m/s

After 6 more seconds, the car has decelerated to a stop.

0:04   60 m/s

Changing Directions

Pretend you are traveling forward in a canoe. You have a forward velocity. You can decelerate by paddling backward. If you paddle backward enough, you start moving backward. Your acceleration changes the direction of your velocity! Remember the definition of acceleration. A change in velocity does not just mean a change in speed—it can also mean a change in direction.

You accelerate whenever you change direction. When you travel around a curve, the direction of your velocity changes even if your speed does not change.

Velocity and acceleration can both be represented by arrows. By adding the two arrows, you can find what the velocity will be after the acceleration.

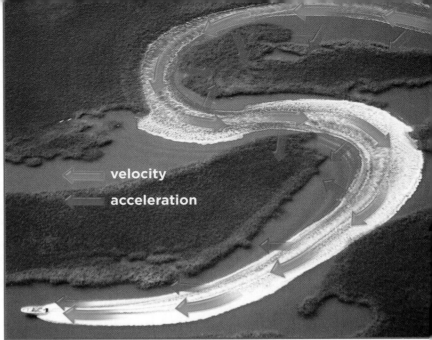

velocity

acceleration

✓ Quick Check

Main Idea and Details A car is accelerating from rest at a rate of 2 (m/s)/s. How fast will it be going after 4 seconds?

Critical Thinking What is the direction of acceleration when traveling in a curve?

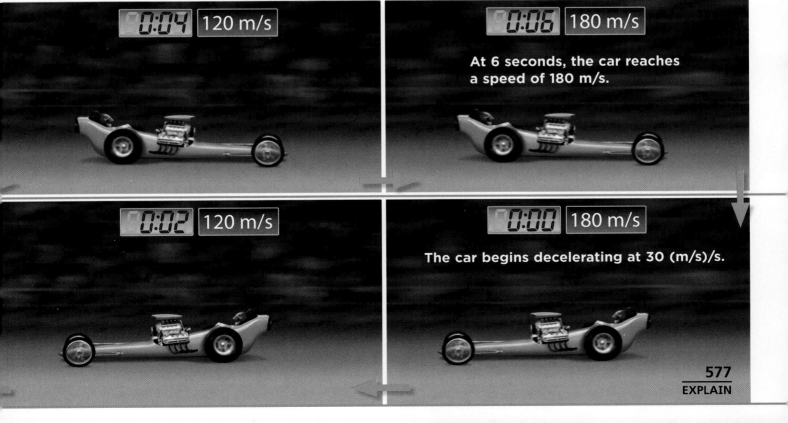

0:04 120 m/s

0:06 180 m/s

At 6 seconds, the car reaches a speed of 180 m/s.

0:02 120 m/s

0:00 180 m/s

The car begins decelerating at 30 (m/s)/s.

What is momentum?

Have you ever gone bowling? In this game, there are many pins at the end of a lane. A player knocks pins over with a large bowling ball. How could you knock over the most pins? You could use a heavier ball, you could roll your ball faster, or you could aim it in a different direction.

When you change mass or velocity, you also change momentum (moh·MEN·tuhm). **Momentum** is the product of mass multiplied by velocity. The more momentum an object has, the easier it is for that object to move other objects. Units of momentum are equal to units of mass times units of velocity, most often kilogram meters per second (kg m/s) or gram meters per second (g m/s).

A fast-moving and heavy bowling ball can knock over many lighter bowling pins.

> ### Calculating Momentum
> data: mass = 4 kg,
> velocity = 5 m/s up the lane
>
> momentum = mass x velocity
>
> = 4 kg x 5 m/s
>
> = 20 kg m/s

When you want to change an object's velocity, you have to overcome its *inertia* (i·NUR·shuh). Inertia is the tendency of any object to resist a change in motion or of a moving object to keep moving in a straight line. The more mass an object has, the more inertia it will have. The more inertia an object has, the harder it is to change its momentum. A very heavy bowling ball is hard to get rolling because of its inertia. Once rolling, it also has a lot of momentum. When it hits the pins, the bowling ball's momentum will overcome the inertia of the pins and knock them over.

 Quick Check

Main Idea and Details Who would be harder to stop, a professional hockey player skating at 4 m/s, or a fifth-grader skating at 4 m/s? Why?

Critical Thinking What changes the momentum of an object?

Lesson Review

Visual Summary

Motion is a change of an object's position over time.

Velocity is the speed and direction of a moving object.

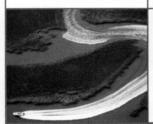

Acceleration is an obejct's change in velocity over time.

Make a FOLDABLES™ Study Guide

Make a Three-Tab Book. Use the titles shown. On the inside of each tab summarize how that topic describes how objects move.

Motion is...

Velocity is...

Acceleration is...

Think, Talk, and Write

1. **Main Idea** Is speed or velocity more helpful to a pilot? Why?

2. **Vocabulary** The property of a moving object that is equal to its mass times its velocity is its _____.

3. **Main Idea and Details** How could an object accelerate while still traveling at a constant speed?

Main Idea	Details

4. **Critical Thinking** Earth's surface is moving at about 1,000 mph as Earth spins on its axis. How can you be going so fast and not sense it?

5. **Test Prep** Which unit would properly label an acceleration?
   - **A** m
   - **B** m/s
   - **C** (m/s)/s
   - **D** kg m/s

6. **Test Prep** What describes how objects tend to resist changes in motion?
   - **A** distance
   - **B** speed
   - **C** acceleration
   - **D** inertia

Math Link

Stopping Safely

A jogger is moving at 5 m/s as she approaches a busy street. She needs to stop in 2 seconds in order to stay safe. What average deceleration must she have in order to stop in time?

Social Studies Link

Reconstructing Accidents

You look over the scene of a car accident. How would knowing about speed, acceleration, and momentum help you to piece together what happened? Give examples.

The Positions of EARTH and the SUN

Look at the sky, and you'll see the universe in motion. The Sun and Moon move in patterns, and the stars change with the seasons. Long ago, people thought Earth was the center of the universe and that everything revolved around it. After all, the Sun does seem to move across the sky. Today we know that Earth's rotation makes it appear this way. We see the Sun move because we are in Earth's frame of reference. How did people discover that Earth revolves around the Sun?

384–322 BCE Aristotle

This Greek philosopher thought that Earth was the center of the universe. His model had stars and planets attached to hollow spheres, or shells, that moved around Earth.

1473–1543 CE Copernicus

This Polish astronomer challenged Ptolemy's view. He proposed that the Sun was at the center of the solar system and that Earth and other planets revolved around it. He claimed that Earth's rotation and revolution around the Sun explained how the stars and planets appeared to move. His idea was not accepted for many years.

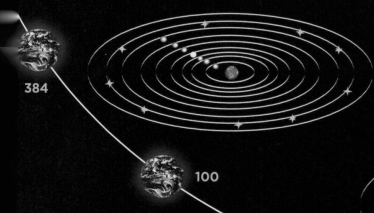

384

100

100–178 CE Ptolemy

This Greek astronomer followed Aristotle's Earth-centered model. He made careful observations of the positions of planets and stars. Then he used geometry to accurately predict the way the Sun, Moon, and planets

Today, with the help of new technology, astrophysicists like Margaret Geller continue to develop our understanding of the universe. She started the field of precision cosmology with the first 3-D map of the universe. **Today**

1879–1955 CE Einstein

By the time this German physicist was born, it was common knowledge that Earth revolves around the Sun. He used physics and mathematics to explain how gravity puts objects in motion. His theories helped astrophysicists answer important questions about the movement of planets, stars, galaxies, and the universe.

1879

1564–1642 CE Galileo

This Italian physicist and astronomer built a telescope and discovered Jupiter's moons and Saturn's rings. His observations supported Copernicus's theory. The view that the Sun is the center of the solar system became more widely accepted.

Main Idea and Details

▶ Look for the central point of a selection to find the main idea.

▶ Details are important parts of the selection that support the main idea.

Write About It

Main Idea and Details

1. Think about the selection you just read. Look for the main topic or central idea of the selection.
2. Write the main idea of the selection and give one detail that supports the main idea.

 **-Journal** Research and write about it online at **www.macmillanmh.com**

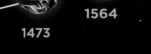

1564

1473

AMERICAN
MUSEUM

Forces and Motion

Look and Wonder

This skydiver falling over the Hawaiian Islands could reach speeds of more than 183 km/h (115 mph) before opening a parachute. Why might some skydivers fall faster than others?

Do heavier objects fall faster?

Form a Hypothesis

In the late 1500s, Galileo argued that weight should not affect how fast something falls. Do you agree? Write a hypothesis in the form "If the mass of an object increases, then . . ."

Test Your Hypothesis

1 **Observe** Use a balance and standard masses to determine the mass of each object. Order the objects from lightest to heaviest and write down your list.

2 **Experiment** Hold two of the objects at the same height in front of you. Drop them at exactly the same time. Record which object hits first or if they hit at the same time. Repeat two more times to verify your result.

3 Repeat step 2 until you've tested all possible pairs of objects.

Draw Conclusions

4 **Interpret Data** Was your hypothesis correct? Write a brief explanation of your answer.

5 **Infer** In your experiment, the objects were falling through air. There is no air on the Moon. How would the falling rate of a tennis ball and a cotton ball compare on the Moon? Why?

Explore More

How would the results of this experiment change if you dropped objects with the same mass, but different densities? Write out a hypothesis. Then use balloons at different levels of inflation to test your hypothesis. Write a summary of your results.

Materials

- equal pan balance
- golf ball
- tennis ball
- cotton ball

Step 1

Step 2

Read and Learn

Main Idea

Forces are pushes, pulls, or lifts that may cause changes in motion.

Vocabulary

force, p. 584
friction, p. 587
balanced force, p. 588
unbalanced force, p. 588
action force, p. 592
reaction force, p. 592

e-Glossary
at www.macmillanmh.com

Reading Skill ✔

Problem and Solution

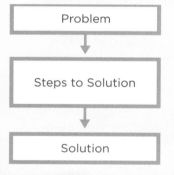

What are forces?

Have you ever been in a tug-of-war? You push with your feet against the ground and you pull as hard as you can. Pushes, pulls, and lifts are all known as forces. A **force** is any push or pull from one object to another. Units of force are the newton (N) and the pound (lb). When we draw diagrams of forces, we often use arrows to represent the direction and strength of the force.

Many forces occur as one object touches another, such as the pull of a tow truck on a broken car. Other forces, though, may act without objects touching. Think about a compass needle. It swings to point north because it is pulled by the magnetic force of Earth. There is nothing actually touching the needle, but it still feels a force.

You know about buoyancy, which is a lifting force caused by a difference in densities. Buoyancy lifts lighter substances out of denser substances.

An airplane has special names for its forces. The engines push or pull the plane forward. This is known as *thrust*. As the plane moves forward, air moves around the wings and creates a force that raises the plane into the air. This force is called *lift*.

The greater pulling force wins a tug-of-war.

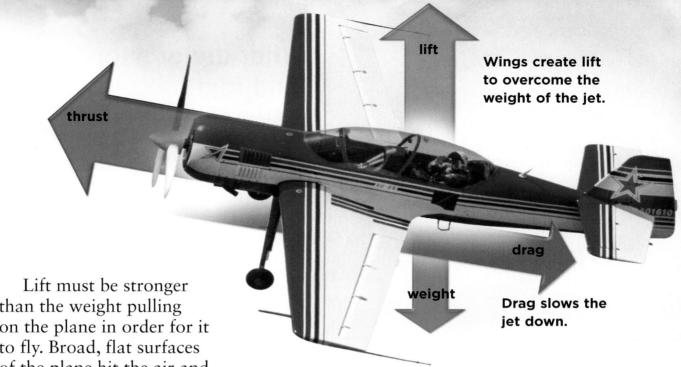

lift

Wings create lift to overcome the weight of the jet.

thrust

drag

weight

Drag slows the jet down.

Lift must be stronger than the weight pulling on the plane in order for it to fly. Broad, flat surfaces of the plane hit the air and pull back on the rest of the plane. This pull, called drag, slows down the plane.

You use forces in many different ways. Forces can crush, stretch, or twist objects and deform them. For example, you can crush an aluminum can if you squeeze it hard with your hands. The harder a substance is, the more force it takes to change its shape.

A force that acts over a short period of time can still have a large effect.

Most often, though, we use forces to move objects. A force can cause an object to start moving, speed up, change direction, slow down, or stop. Do you notice anything about all of these motions? They all involve accelerations. Forces accelerate objects when they affect their motion.

Some forces, such as a bat hitting a ball, act for a very short time. You know that the bat still accelerates objects, however, because the ball flies away quickly. Other forces act continuously. A cyclist pedaling steadily and a balloon rising slowly are examples of continuous forces.

 Quick Check

Problem and Solution How could you make an airplane rise faster into the air?

Critical Thinking Give an example of a force that changes both an object's shape and motion.

drag

gravity

The drag force on a parachute is almost as strong as gravity.

What are gravity and friction?

Has anyone ever said to you, "What goes up, must come down"? If so, he or she may have been talking about gravity, the force that attracts all matter together. If you throw a ball upward, the gravity between it and Earth will make it fall back to the ground. Without gravity, the ball would fly off Earth.

Sir Isaac Newton, after whom the unit of force is named, researched gravity in the 1600s. He believed that every object in the universe was pulling on every other object. This theory is called Newton's Law of Universal Gravitation. Newton said that gravity depends on the masses of the objects and the distance between them. Increasing the mass increases the force, and increasing the distance decreases the force.

Gravity pulls all objects together, big or small. The gravity between small objects, though, is weak. Two bowling balls an inch apart will never roll together because of gravity. Their masses are too small. The huge masses of moons, planets, and stars, however, make their gravity powerful. The force of gravity between Earth and the Moon is 200 billion billion newtons!

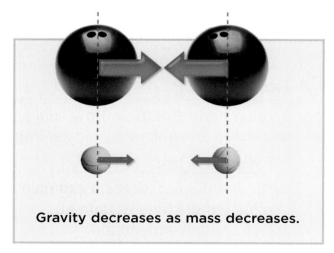

Gravity decreases as mass decreases.

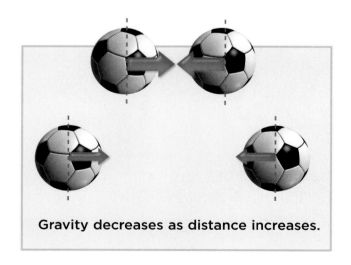

Gravity decreases as distance increases.

Friction

You've probably slid down a slide at a playground. For the slide to be fast the surface must be very slippery. Friction (FRIK•shuhn) makes it difficult to slide on rough surfaces. **Friction** is a force that opposes the motion of one object moving past another.

Friction depends on the surfaces of two objects and how hard the objects are pushed together. Smooth surfaces usually have less friction than rough surfaces. Friction increases when surfaces are pressed together with greater force. Friction also increases with the weight of an object.

Have you ever rubbed your hands together to keep them warm? Friction between your hands slows them down and also generates heat. Heat is created whenever there is friction.

Air Resistance

When an object moves through air, the air hits the object and slows it down. This air resistance increases with velocity while friction usually does not. Liquids also exert resistance. This is why water can slow down a boat.

Drag forces are a result of air resistance. Broad flat surfaces have the most drag. This is why a feather falls slower than a pencil. Without air, the two would fall at the same rate. Drag is affected by the movement of liquids and gases. This is why it is harder to paddle upstream in a boat, or to fly into the wind in an airplane.

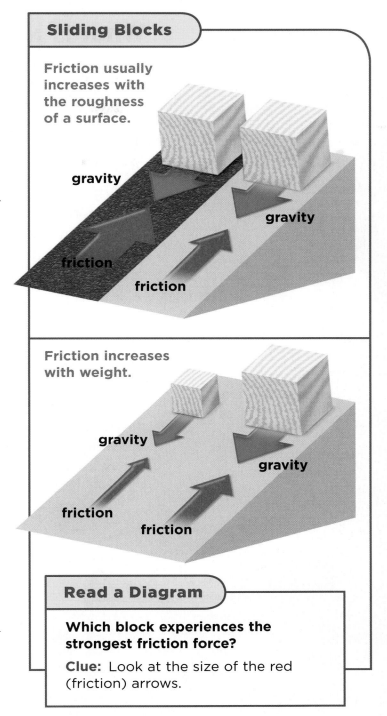

Sliding Blocks

Friction usually increases with the roughness of a surface.

gravity

gravity

friction

friction

Friction increases with weight.

gravity

gravity

friction

friction

Read a Diagram

Which block experiences the strongest friction force?

Clue: Look at the size of the red (friction) arrows.

 Quick Check

Problem and Solution How could you increase the traction (friction) of car tires on an icy road?

Critical Thinking What would the world be like without friction?

What is Newton's First Law?

Suppose you want to hang a picture on the wall. The force of gravity is pulling the picture down, but you do not want the picture to fall down. What can you do? Use a piece of string to hold up the picture. The force of the string pulling up on the picture is equal to the force of gravity pulling it down, but in the opposite direction.

When forces act on an object without changing its motion, they are called **balanced forces**. Balanced forces often point in opposite directions, and they always add up to zero. There may be more than one pair of balanced forces acting on an object.

The forces acting on stationary objects are always balanced. However, balanced forces can act on moving objects, too. Think of a bus moving at a constant speed down a straight road. The force of the engine pushing the bus forward is balanced by the force of drag, and the friction of the tires. Although the bus is moving, its velocity is not changing, so the forces acting on it are balanced. As long as those forces remain balanced, the bus will continue at the same speed and travel in a straight line.

Most car trips, however, do not follow a straight line. Eventually the driver will have to change direction, slow down, or speed up. When the car speeds up the forward force is greater than the friction force and the car accelerates. A force that causes an object to change its motion is called an **unbalanced force**.

Sir Isaac Newton studied balanced and unbalanced forces. He then wrote his first law of motion.

rope

gravity

The forces acting on the piñata are balanced, so it will not fall.

SCHOOL BUS

◀ The forces on the bus are balanced so it will continue to travel at a constant velocity.

engine

drag and friction

FACT ▶ Moving objects will not stop until acted upon by an unbalanced force.

> ### NEWTON'S FIRST LAW
> *An object at rest tends to stay at rest, and an object in constant motion tends to stay in motion, unless acted upon by an unbalanced force.*

Newton's first law is sometimes called the law of inertia. This is because the law describes inertia: Objects do not change their motion unless forced to.

If there were no forces, such as friction or drag, an object in motion could travel forever in a straight line. Objects in space, like the *Voyager* space probe, are doing just that. On Earth, however, friction and drag often act as unbalanced forces to bring objects to rest.

Objects in space like the Voyager space probe may travel forever in a straight line.

≡ Quick Lab

Unbalanced Balloon Force

1. Pass thread or string through two short lengths of soda straw. Then stretch the string tightly between two chairs.

2. Inflate a balloon. Hold the neck closed while you tape it to the straws.

3. **Observe** Let go of the balloon's neck and record what happens.

4. **Infer** Did an unbalanced force act on the balloon? Explain.

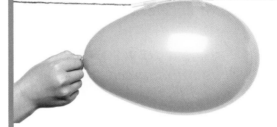

5. How would the balloon's motion change if you inflate the balloon more than before? Write down your prediction and experiment to test it. Report what you discover.

✔ *Quick Check*

Problem and Solution How could you keep a balloon from rising or falling in air?

Critical Thinking How does Newton's first law explain how seat belts prevent injuries in car accidents?

What is Newton's Second Law?

Two boats in a race are acted upon by unbalanced forces. Which boat will win the race? You know both boats will accelerate because of the unbalanced forces. It would help, however, to know how much they will accelerate.

Suppose you are rowing in one of the boats. To start going faster, you row harder. Acceleration increases as your force increases as long as your mass is not also changing.

What if you carried things in your boat? You could row just as hard, but you would go slower. Acceleration decreases as your mass increases as long as your force is not also changing.

Sir Isaac Newton studied these same results and created his second law of motion. Instead of acceleration, he talked about things in terms of momentum. Later, scientists reworded the law in terms of acceleration.

> NEWTON'S SECOND LAW
> *The unbalanced force on an object is equal to the mass of the object multiplied by its acceleration: F = m x a.*

Newton's second law tells you that a total force of 1 newton (F) will accelerate a 1 kg object (m) at 1 (m/s)/s (a). Notice that a newton is equal to a kilogram times a meter per second per second (kg (m/s)/s). The direction of the acceleration is always in the same direction as the total force.

Using more rowers increases the force and increases acceleration.

Having more passengers increases the mass and decreases acceleration.

You can also rearrange the law to say a = F / m. Then you could calculate the acceleration of an object. You would just need to know the total force and the object's mass.

The total force is the sum of all the forces acting on an object. When those forces are balanced, the total force is zero and the object has zero acceleration—just like a hanging picture. Do you see how Newton's second law agrees with his first law? When the forces acting on an object are unbalanced, the total force is not zero. You then use Newton's second law of motion to tell you exactly how the object will accelerate—just like the boats in the race.

 Quick Check

Problem and Solution What could you do to a race car to make it accelerate faster?

Critical Thinking What will happen to the acceleration of an object if you double both the mass and the unbalanced force acting on it?

Calculating Acceleration

data: total force = 80 N forward, mass of boat = 160 kg

$$F = m \times a$$
$$a = F/m$$
$$a = 80 \text{ N} / 160 \text{ kg}$$
$$a = 0.5 \text{ (m/s)/s}$$

Accelerating Boat

rowing force
50 N

water resistance
10 N

rowing force
50 N

water resistance
10 N

total force
80 N

Read a Photo

What would be the boat's acceleration if the total force was doubled?

Clue: Look at the calculations.

What is Newton's Third Law?

You are ice skating with a friend. You give her a small push to help her go faster. As you push her forward you find yourself moving backward. Why are you moving? Wasn't your friend the object being pushed?

Actually, you received a push, too. When one object pushes on a second object, the second pushes back on the first with the same strength. Commonly, the push of the first object on the second is called the **action force**. The push of the second object back on the first is called the **reaction force**.

Newton summarized this idea in his third law of motion.

> ### NEWTON'S THIRD LAW
> *All forces occur in pairs, and these two forces are equal in strength and opposite in direction.*

When one skater pushes or pulls on the other, they each feel an equal but opposite force acting on them.

This astronaut feels "weightless" because there is nothing in space to create a reaction force.

More commonly people say, "For every action there is an opposite but equal reaction." It is important to remember that action and reaction forces are not balanced forces. This is because action and reaction forces always act on separate objects.

When you sit in a chair, your weight is pushing down on the chair. A reaction force from the chair pushes back up on you. This reaction force is what you feel as your weight. When you are falling (or when astronauts are in space) the force of gravity is still acting upon you. You do not feel gravity, however, because there is nothing providing a reaction force. This is why you feel "weightless" when you are in free fall.

✓ Quick Check

Problem and Solution How could a toy rocket accelerate after take off?

Critical Thinking What are the action and reaction forces acting on your body when you walk?

Lesson Review

Visual Summary

Forces are pushes, pulls, or lifts.

Forces acting on an object may be **balanced** or **unbalanced**.

Unbalanced forces cause acceleration according to **F = m x a,** Newton's second law.

Make a FOLDABLES Study Guide

Make a Trifold Book. Use the titles shown. Fill in each box with the type of information described by the column heading.

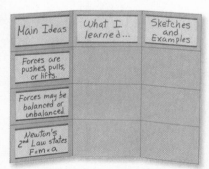

Think, Talk, and Write

1. **Main Idea** What forces are at work on a floating object? Are they balanced?

2. **Vocabulary** One force that opposes motion is _____.

3. **Problem and Solution** How could you decrease the drag on an airplane?

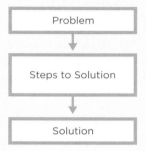

4. **Critical Thinking** How could working under water help astronauts practice being in space?

5. **Test Prep** If an unbalanced force acting on an object is increased, the object will
   A accelerate more.
   B accelerate less.
   C remain at constant speed.
   D remain still.

6. **Test Prep** Which is a unit of force?
   A gram
   B (m/s)/s
   C Newton
   D m/s

Writing Link

Narrative Writing
What would the forces feel like in a rocket launch? Write a narrative story about the forces an astronaut would feel when a rocket flies into space.

Math Link

Upward Forces
The total force on a hot air balloon is 600 newtons. Its mass is 300 kg. How fast will it accelerate?

Inquiry Skill: Measure

If an unbalanced force acts on an object, that object will accelerate. As the force changes, how will the acceleration change? Will an increasing force increase the final velocity of an object? Scientists often ask themselves questions they don't know the answer to. To find the answers, they **measure** and observe the things around them.

▶ Learn It

When you observe you use one or more of your senses to identify or learn about an object. When you **measure** you find the size, distance, time, velocity, mass, weight, or temperature.

It is important to record measurements and observations you make during your experiment. Once you have enough information, you can make predictions about what might happen if you changed a variable in the experiment.

	Time 1	Time 2	Time 3	Average Time
One Rubber Band				
Two Rubber Bands				

▶ **Try It**

Materials masking tape, meterstick, 2 wood blocks with thumb-tacks, rubber bands, safety goggles, model car, stopwatch

Newton's second law of motion ($F = m \times a$) tells you how unbalanced forces, mass, and acceleration relate.

You can launch a model car with rubber bands and then record its final speed. How will the car's speed depend on the number of rubber bands used? Write a hypothesis in the form "If the number of rubber bands used to launch a model car increases, then . . ."

1. Place a 10 cm strip of masking tape on the floor. Hold the wood blocks on either side of the tape. Stretch a rubber band between the thumbtacks.

2. Tape a "finish line" on the floor 1.5 m (59 in.) from the rubber band.

3. ⚠ **Be Careful.** Wear goggles. Pull a model car back against the rubber band a few centimeters. Mark this starting point with a small piece of tape and use it for the rest of the experiment.

4. **Measure** Launch the car and use a stopwatch to time how long it takes for the car to travel 1.5 m. Repeat this procedure two more times and find the average time. Record this on a table like the one shown.

5. Repeat step 4 with 2 rubber bands.

▶ **Apply It**

Now use the data you **measured** and recorded to answer questions.

1. Divide the distance (1.5 m) by the average time for 1 rubber band to get the car's speed. Repeat for the trials with two rubber bands.

2. According to the speeds you found, was your hypothesis correct? Explain.

3. Can you predict how the car's speed would change if you were to tape a second car on top of it? How do Newton's second law and your experiments support this prediction?

Work and Energy

Pleasureland Amusement Park, UK

Look and Wonder

Passengers on a roller coaster may feel forces twice as strong as gravity. Where does the energy come from to take the roller coaster through the ride?

What happens to energy?

Form a Hypothesis

The energy an object has changes as it is pulled by gravity. What will happen if you let a marble roll in a bicycle tire? Write a hypothesis in the form "If the height a marble is released from is increased, then…"

Materials

- **section of old bicycle tire (or piece of split garden hose)**
- **masking tape**
- **marble**

Test Your Hypothesis

❶ You will need to work in a group. One member of your group should hold the section of tire firmly on a tabletop. Use a piece of masking tape to mark a starting point on one side of the tire.

❷ **Observe** Release the marble at the starting point and let it roll in the tire. Observe what the marble does until it comes to a stop. The marble's actions are your dependent variable. Repeat several times to verify your observations.

❸ Repeat steps 1 and 2 for two more starting points. They should be at different heights. The height of the marble is your independent variable.

Draw Conclusions

❹ **Interpret Data** According to your observations, was your hypothesis correct? Explain.

❺ **Infer** When did the marble move the fastest? Did it have more or less energy there than when it began? How do you know?

Step ❷

Explore More

Why did the marble eventually stop? What effect did the texture of the inside of the bicycle tire have? Write a hypothesis and design an experiment to test it.

Read and Learn

▶ Main Idea

Moving objects and causing change requires work and energy.

▶ Vocabulary

work, p.598

energy, p.600

potential energy, p.600

kinetic energy, p.600

law of conservation of energy, p.602

e-Glossary
at **www.macmillanmh.com**

▶ Reading Skill ✓

Infer

Clues	What I Know	What I Infer

What is work?

Placing boxes on a shelf can be tiring work. You have to lift the boxes from the ground up to the shelf. Lighter boxes require less force to move, so it is less work to put them on a shelf. The lowest shelf is closest to the ground, so it is less work to place boxes there than on higher shelves. But what do we actually mean when we say *work*?

Work is the measurement of the energy used to perform a task. When work is done on an object, the amount of energy it has changes. Work is equal to the force used times the distance the force was applied. If the force and the distance are in the same direction, the work is positive. If the force and the distance are in opposite directions, the force is negative. Lifting a box is positive work, lowering a box is negative work.

The units of work are the units of force times the unit of distance: newton-meters (N m). If you lift a box that weighs 10 newtons onto a shelf that is 1 meter high, you are performing 10 N m of work. Newton-meters are also known as joules (J).

Lifting Boxes

work = force x distance

force

distance

Read a Diagram

Which boxes took the most work to place on the shelves?

Clue: Look at the size of the boxes and the height of the shelves.

There are many things that seem like work but are not. For instance, do you think it is work to hold a ball over your head? Lifting it there is definitely work, but just holding it there is not. Why? A force must be applied over a distance in order to qualify as work. When you lift the ball, you are applying a force over a distance. When you are holding the ball you are still applying a force, but the ball is not moving, so the distance equals zero.

You may notice that each example of work involves an unbalanced force. Unbalanced forces cause acceleration and motion. Motion is a necessary part of work, so you will always see work when there is an unbalanced force.

Consider a toy car that you and your friend are pushing on from opposite directions. If you both push with the same force, the car will not move.

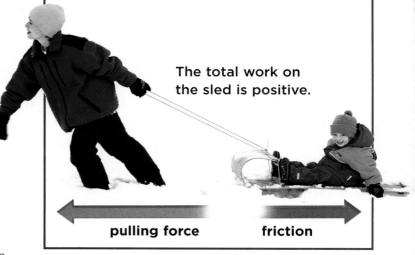

The total work on the sled is positive.

pulling force   friction

Work is not being done. What if you push harder than your friend? The car starts to move and work is done.

Your work on the car is positive because it is in the direction of motion. Your friend's work is negative because it is against the direction of motion. The total work is the sum of the positive and negative work. When we talk about work we may talk about the total work or the work of individual forces, like you and your friend.

Friction often performs negative work on objects as you move them. When a car is traveling at a constant velocity, the engine, friction, and air resistance may all be performing work on the car, but the total work is zero.

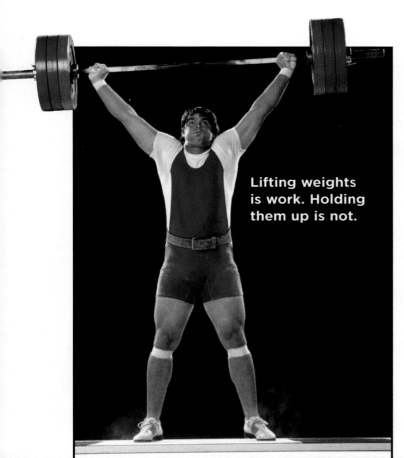

Lifting weights is work. Holding them up is not.

 Quick Check

Infer How does friction affect the work it takes to push a box across the floor?

Critical Thinking You lift a box and then start to walk at a constant velocity. When are you working?

 FACT Not everything that makes you tired is work.

What is energy?

When you feel tired you may say, "I don't have any energy." **Energy** is the ability to perform work or to change an object. The units of energy are the same as the units of work—joules (J). When you are low on energy, you probably cannot do much work.

Objects can also have energy. When you stretch a spring, it pulls back on your hand. If you release it, the spring does work by pulling itself back to its original length.

When the spring is stretched, it has energy, but it is not moving. It has the potential to do work. **Potential energy** (puh•TEN•shuhl) is energy that is stored in the position, or structure, of an object. When you release the spring, it moves. **Kinetic energy** (ki•NET•ik) is the energy of a moving object. Vibrating like a spring is called *periodic* (peer•ee•OD•ik) motion. In periodic motion, energy changes back and forth from potential to kinetic.

When you do positive work, you add energy to an object. If you throw a ball, you increase its velocity and add to its kinetic energy. If you lift a ball, you increase the distance gravity can pull it and add to its potential energy. If you drop a ball, gravity does work and changes the ball's potential energy to kinetic energy.

When you release a spring, potential energy becomes kinetic energy.

Using Energy

dropping

throwing

lifting

Read a Photo

Which image shows an increase in potential energy?

Clue: Where is the ball the highest?

Forms of Energy

There are many forms that potential and kinetic energy can take. There is potential energy in the links between atoms and molecules. This is chemical energy. Nuclear energy is potential energy stored in the links between the protons and neutrons in an atom. Magnetic energy is another form of potential energy. It acts like gravity, pulling objects together, but it can also push some objects apart. Electrical energy can be potential energy when particles of different electrical charges are attracted to one another.

Kinetic energy can take many forms as well. Heat is the kinetic energy in the vibrations of particles. Electricity is related to the kinetic energy of electrons. Sound is the kinetic energy of particles as they move in waves. Light is also kinetic energy that moves in the form of waves.

All forms of energy have one thing in common—they can perform work! Some forms of energy change the structure of objects rather than move them but this is still work. You know that heat can melt or boil substances. You also know that chemical reactions change one type of substance into another. These are examples of work.

✔ Quick Check

Infer Which can do more work, a joule of heat energy or a joule of sound energy?

Critical Thinking What potential and kinetic energies exist when you dive into water?

≡ Quick Lab

Measuring Used Energy

1 Tie a loop of string around a book. Hook the string to a spring scale.

2 **Measure** Slide the book across a table by pulling on the spring scale. Keep the force reading on the spring scale constant. Record your results.

3 Hang the book from the spring scale. Record the book's weight.

4 Which is more work, sliding the book for 1 meter or lifting the book a distance of 1 meter? Explain.

5 **Infer** If you lift the book up to a certain height, it gains potential energy. When you slide the book a given distance, it is neither lifted nor left with kinetic energy. Where does the energy from the work go when you slide the book?

Magnets do work by pulling objects together.

How can energy change?

Scientists believe that energy cannot be created or destroyed, it can only change form. This theory has been observed so many times that it is called the **law of conservation of energy** (kon•suhr•VAY•shuhn). A roller coaster, for example, cannot gain kinetic energy without losing potential energy.

You might think that a roller coaster destroys energy. After all, the roller coaster steadily slows down. The "lost" energy, however, is not destroyed. It has become heat and sound through the work of friction.

In fact, whenever energy is used to do work, that energy changes. The kinetic energy of water does work by moving the arms of a water turbine. The arms of the turbine do work and generate electricity. Electricity does work in an oven by moving particles around and changing into heat. Heat does work on a loaf of bread and

The kinetic energy of falling water is changed into electricity at a hydroelectric dam.

changes to chemical energy. Chemical energy in the bread does work and changes into kinetic energy in your muscles. Your muscles can do work when they build another turbine!

Energy can sometimes do undesirable work. This often happens when there is friction. Friction inside the turbine, for instance, causes some kinetic energy to be changed into heat, not electricity. Although this energy is not lost, it is much less useful, and may be harmful to the turbine.

Kinetic energy is changed into heat energy through the work of friction.

 Quick Check

Infer A dropped ball doesn't bounce back to its starting height. How does this situation fit the law of conservation of energy?

Critical Thinking How might the heat energy in an oven perform both desirable and undesirable work?

Lesson Review

Visual Summary

	Work is done when a force moves an object over a distance.
	Energy is needed to do work or to cause changes in matter.
	One **form of energy** can be changed into another.

Make a  Study Guide

Make a Folded Table. Use the titles shown. In each box summarize how that topic can help you describe how work is done.

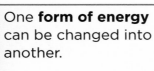

Think, Talk, and Write

1. **Main Idea** When is work done when you throw and catch a ball?

2. **Vocabulary** Energy that comes from the position of an object or its organization is _____.

3. **Infer** When might kinetic energy turn into sound energy?

Clues	What I Know	What I Infer

4. **Critical Thinking** A pendulum is a weight swinging back and forth on a cord. What energy changes occur as a pendulum swings?

5. **Test Prep** If a car accelerates on a level road, it gains
   A chemical energy.
   B kinetic energy.
   C light energy.
   D potential energy.

6. **Test Prep** Which is a unit of work or energy?
   A joule
   B watt
   C newton
   D horsepower

 Writing Link

Descriptive Writing
Solar energy is energy from the Sun. Research solar energy and describe how it can be used by people to do work.

 Math Link

Elevator Work
An elevator lifts a 200-newton object 10 meters while overcoming friction. Friction provided 1,000 joules of negative work. How much work did the elevator do?

Be a Scientist

Materials

wax paper

aluminum foil

clear plastic wrap

masking tape

cardboard

books

ruler

wooden block

Structured Inquiry

What affects potential and kinetic energy?

Form a Hypothesis

Potential energy is the amount of energy stored in an object. Kinetic energy is the energy that an object has due to its motion. Gravity converts potential energy into kinetic energy as an object falls. Friction can decrease the kinetic energy an object has.

Picture sliding a block down a smooth ramp. How would friction affect this block as it slid? Write your answer as a hypothesis in the form "If friction is increased, then the amount of potential energy that becomes kinetic energy . . ."

Test Your Hypothesis

1. **Observe** Examine the wax paper, aluminum foil, and plastic wrap. Which do you think would create more friction? Why?

2. Tape a piece of wax paper to one side of a piece of cardboard. The material on the ramp is your independent variable.

3. Use four books to create a ramp with the wax paper on the cardboard facing up.

4. **Measure** Record the height of the books. Using tape, mark where the cardboard rests on the table. These are variables you want to keep the same every time.

5. **Experiment** Place the wooden block at the top of the ramp and let it slide down. Record how far the block slid. Repeat two more times and take an average. This is your dependent variable.

6. Repeat the experiment with aluminum foil and plastic wrap.

Step 4

Step 5

Draw Conclusions

7 Was your hypothesis correct? Explain why or why not.

8 **Infer** What material caused the ball to lose the most kinetic energy? Where do you think this energy went?

How does gravity affect potential energy?

Form a Hypothesis

How can potential energy be changed by gravity? Write your answer as a hypothesis in the form "If the height a ball is dropped from increases, then the potential energy of the ball . . ."

Test Your Hypothesis

You know that gravity changes the potential energy of falling objects into kinetic energy. Design an experiment to investigate how distance from the ground will affect a ball's potential energy. Write out the materials you need and the steps you will follow. Record your results and observations.

Draw Conclusions

Did your results support your hypothesis? Why or why not? How did you measure the amount of potential energy of the ball? What became of the potential energy during the experiment?

What can you learn about kinetic energy? For example, what other types of forces affect kinetic energy? Use research materials to answer your question. Your experiment must be written so that another group can complete the experiment by following your instructions.

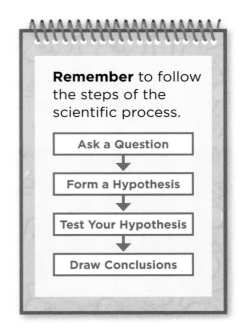

Remember to follow the steps of the scientific process.

Ask a Question

↓

Form a Hypothesis

↓

Test Your Hypothesis

↓

Draw Conclusions

605

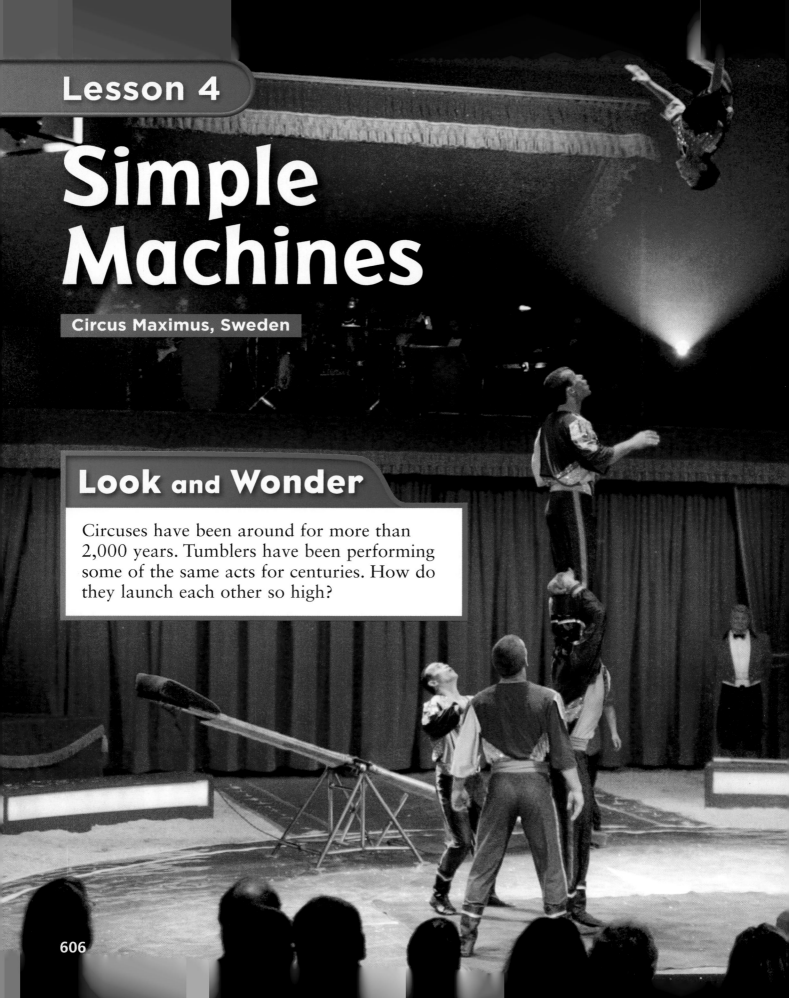

Simple Machines

Circus Maximus, Sweden

Look and Wonder

Circuses have been around for more than 2,000 years. Tumblers have been performing some of the same acts for centuries. How do they launch each other so high?

What makes work easier?

Make a Prediction

Will pulling a toy car up a ramp be more work than lifting it straight up? Write a prediction stating whether you think that pulling the car up the ramp will be more or less work than pulling it straight up.

- **toy car**
- **spring scale**
- **books**
- **ruler**

Test Your Prediction

1. Hang the toy car from the spring scale and read its weight in newtons. Record your result.

2. Use four books to build a ramp as shown. Measure the height of the ramp using the ruler. Pull the car up the ramp at a steady speed. Read the force required in newtons. Measure the distance the car traveled along the ramp using a ruler. Record your results.

3. Repeat your measurements to verify your results.

Draw Conclusions

4. **Use Numbers** Calculate the work it takes to lift the car to the height of the ramp and the work it takes to pull the car up the ramp. Remember work = force x distance. Were your predictions correct? Explain.

5. **Communicate** To get something you want, you usually have to pay for it. What "price" do you pay when you use a ramp to help lift something?

6. **Infer** Are there any additional forces acting on the car when you use the ramp? How might these forces affect the work you did?

Step **1**

Explore More

How would changing the angle of the ramp affect the pulling force? Make a prediction and design an experiment to test it. Perform the experiment to check to see if your prediction was correct.

Step **2**

▶ Main Idea

Machines change the forces and distances used to do work.

▶ Vocabulary

simple machine, p. 608
effort, p. 608
load, p. 608
fulcrum, p. 610
compound machine, p. 616
efficiency, p. 616

LOG ON ℮-Glossary
at www.macmillanmh.com

▶ Reading Skill ✓

Classify

What are simple machines?

It's time to paint a wall. You try to open a can of paint, but the lid is stuck. You can't apply enough force to open it. Fortunately, you can use a screwdriver to increase the force you apply and open the lid. A screwdriver is an example of a machine.

A **simple machine** (SIM•puhl muh•SHEEN) is a machine that takes one force and changes its direction, distance, or strength. The force you apply to a machine is called the **effort**. The force a machine supplies is called the *output*. The object moved by the output is called the **load**.

The part of the simple machine that receives the effort is called the *effort arm*. The part of the machine that delivers the output is called the *resistance arm*. The ratio of the lengths of the effort arm and resistance arm is a machine's MA or *mechanical advantage* (muh•KAN•i•kuhl ad•VAN•tij). You can find the strength of the output by multiplying the effort by the MA.

When the effort arm is longer than the resistance arm, the output will be greater than the effort. So a simple machine can increase the force we supply. This happens with the screwdriver and paint can.

Parts of a Machine

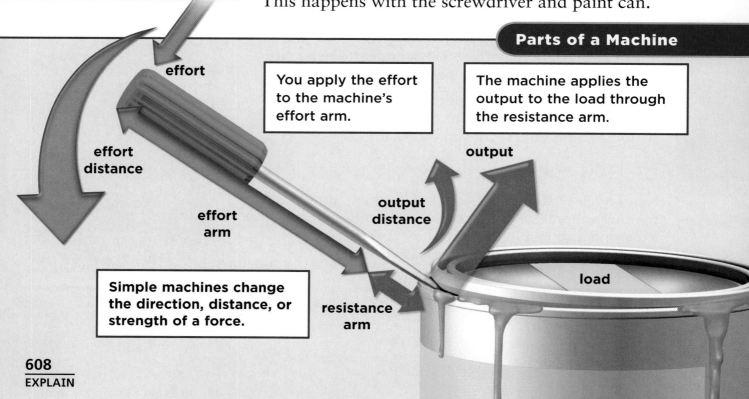

effort

You apply the effort to the machine's effort arm.

The machine applies the output to the load through the resistance arm.

effort distance

output

effort arm

output distance

Simple machines change the direction, distance, or strength of a force.

resistance arm

load

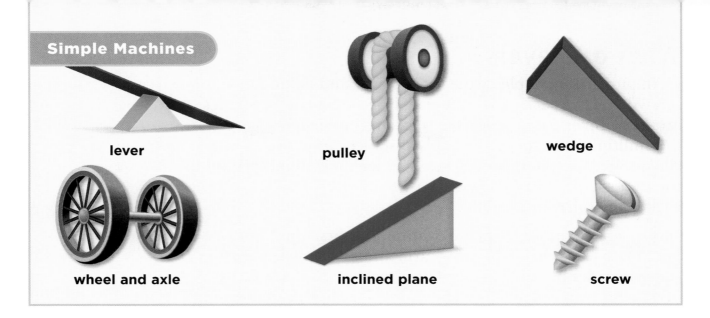

Simple Machines

lever

pulley

wedge

wheel and axle

inclined plane

screw

If the effort arm is shorter than the resistance arm, then the output distance will be greater than the effort distance. A simple machine can increase the distance or speed of a force. A broom is an example of this type of simple machine.

Do simple machines change the amount of work you do? Remember that work = force x distance. When a simple machine multiplies your effort, it also divides the distance over which that effort is used. When a simple machine multiplies the distance a force is applied, it also divides the strength of that force. So, simple machines can increase the force or distance, but the work stays the same.

If you can't get free work out of simple machines, why use them? Sometimes it is easier for you to apply a small force over a long distance, and let the machine apply a strong force over a small distance. By letting the user choose the ratio of force to distance, simple machines can decrease the amount of time it takes to do work.

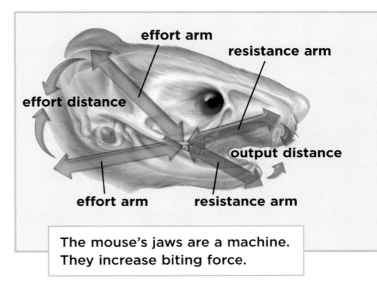

The mouse's jaws are a machine. They increase biting force.

There are also living examples of simple machines. Many animals have jaws that act like simple machines. Rodents can bite through hard wood because they have powerful jaws.

 Quick Check

Classify What is the load moved by an animal's jaw?

Critical Thinking If a machine multiplies your effort, how must the effort distance and the output distance compare?

FACT Simple machines can be found in nature.

What are levers?

In our first example of the screwdriver and paint can, the screwdriver was acting as a lever. A lever is a bar that rotates around a pivot point called the **fulcrum** (FUL•kruhm). Levers can multiply an effort or multiply distance and speed. Some also redirect effort. Levers are easy to make. A stick lying on top of a rock is a lever. The stick is the rotating bar, and the rock is the fulcrum.

Crowbars and seesaws are examples of first-class levers. First-class levers have the effort arm and resistance arm on opposite sides of the fulcrum. The effort and the output are in opposite directions. The ratio of the forces depends on the ratio of the lengths of the arms.

You can pick up and move heavy loads if you use a wheelbarrow. A wheelbarrow is a second-class lever. Second-class levers have both arms on the same side of the fulcrum,

Classes of Levers

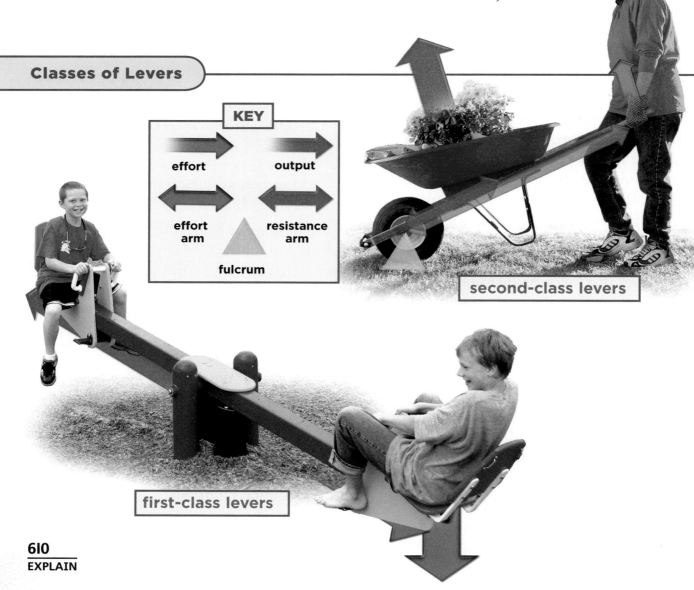

KEY

effort

output

effort arm

resistance arm

fulcrum

second-class levers

first-class levers

and the effort arm is longer. The output is greater than the effort and both are in the same direction.

When you go fishing, you're using a third-class lever. Both its arms are on the same side of the fulcrum, and the resistance arm is longer. A third-class lever has an output less than the effort. Its output distance, though, is greater than the effort distance. A fishing rod turns a short movement of your hand into a longer, faster movement at the tip of the rod. This helps you to cast.

Two levers can be attached at their fulcrum. Tweezers are two third-class levers attached at the fulcrum. This is called a double-action third-class lever.

Read a Photo

Which types of levers change the direction of a force?

Clue: Compare the directions of the effort and the output.

Quick Lab

Levers and Effort

1. Hang a meterstick by its center so that it balances.

2. Use a paper-clip hook to hang a weight 25 cm from the middle of the meterstick. Position a second hook on the other side, 25 cm from the middle. Attach a spring scale to this hook and measure the downward force it takes to hold the bar in a level position. Record your reading.

3. Repeat with the spring scale positioned at 15 cm and 35 cm. Record your readings.

4. **Interpret Data** In each case, the resistance arm was 25 cm long. How was the length of the effort arm (the spring-scale arm) related to the force needed to keep the meterstick level? Explain.

third-class levers

✔ Quick Check

Classify What class of levers are pliers? Why?

Critical Thinking If the effort arm in a lever is half as long as the resistance arm, what is the ratio of the speeds of the arms?

Which machines are like levers?

The wheel is another simple machine that is easy to make. Even a rounded rock can roll like a wheel. It is when you put an axle, or rod, on a wheel, however, that the machine becomes more useful. The wheel and axle is a rigid rod through the center of a wheel that can multiply force, speed, or distance like a lever. An axle acts like the fulcrum of a lever, and the wheel acts as the two arms of a lever.

Gears are a type of wheel and axle. The effort and resistance arms are the radii of the different gears.

effort arm

resistance arm

axle (fulcrum)

A pulley is a grooved wheel with a rope running along the groove. The wheel in a pulley acts as a lever, but we talk about the effort and output distances in terms of the rope. The length of the rope you move while applying the effort is the effort distance. The distance the load travels is the output distance. The mechanical advantage of a pulley is the ratio of the effort distance to the output distance.

A simple pulley just redirects effort. If a pulley does not move as the load is lifted, it is called a fixed pulley.

A winch uses a wheel and axle to wind up a length of rope.

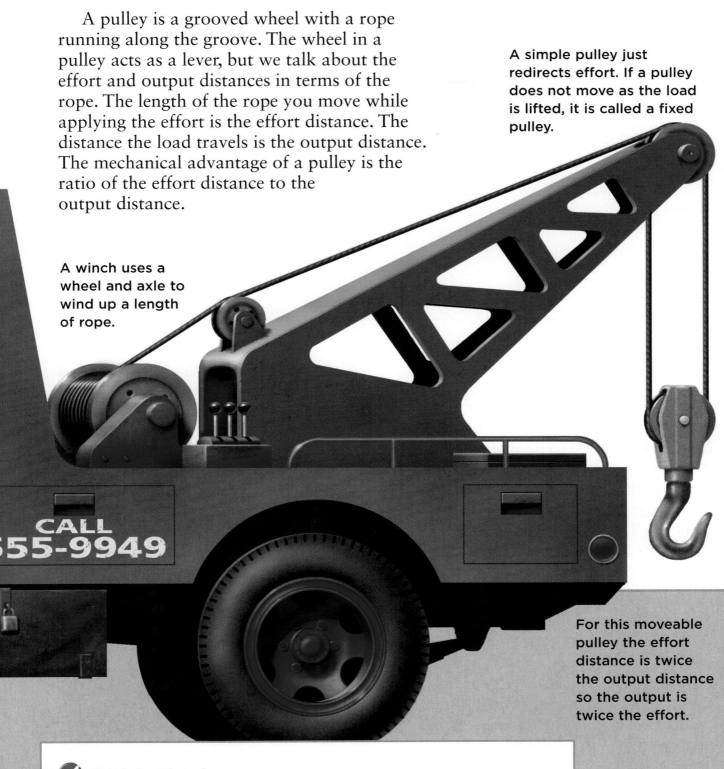

For this moveable pulley the effort distance is twice the output distance so the output is twice the effort.

✅ **Quick Check**

Classify Is a doorknob a pulley or a wheel and axle? Why?

Critical Thinking How could a pulley multiply the distance but not the force of the effort?

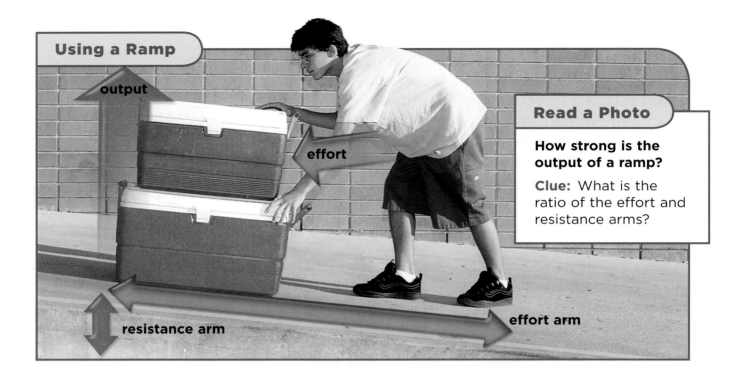

output

effort

resistance arm

effort arm

How strong is the output of a ramp?

Clue: What is the ratio of the effort and resistance arms?

What are inclined planes?

Have you ever climbed a tall hill or a mountain? The steeper your path, the more force it takes to climb. Many paths wind back and forth to create a gradual rise that is easier to walk up. The same trick is seen at the entrance to many buildings. Ramps are sometimes used in place of stairs.

A ramp is a type of simple machine called an inclined plane. As with all simple machines, comparing the effort distance to the output distance tells us how much the machine will multiply effort. The shorter the effort distance, the faster the load can be raised, but it takes more effort to do so.

Ramps are often used with wheels and axles. A rolling object easily gains kinetic energy when it travels down a ramp. A skater can use this energy to perform jumps and other stunts.

This winding road is an inclined plane.

Wedges and Screws

If an inclined plane is pushed into an object, the effort of pushing it in will cause a spreading force and split the two halves of the load. This is what happens when you chop firewood. When an inclined plane is used to separate two objects it is called a wedge. Wedges may have one or two sloping sides. They may also be driven under an object to lift it. In this way a wedge acts very much like a ramp.

Scissors, knives, and other cutting tools take advantage of the wedge. The sloped sides of the blade create the spreading force that results in cutting.

A screw is an inclined plane wrapped around a cylinder. Like a wedge, a screw moves into material. However, a wedge is hammered or driven, while a screw must be turned.

The inclined plane of a screw can be "steep" or "shallow." In other words, the threads can be far apart or close together. The farther apart the threads, the faster a screw moves into a material for each turn. However, the farther apart its threads, the more effort it takes to turn the screw.

Friction acts along the entire length of the threads of a screw. The more threads, the greater the friction force that acts on the screw.

✔ Quick Check

Classify Is a boat propeller a type of inclined plane? Explain.

Critical Thinking How would a screw act if there were no friction?

The spreading force of the wedge splits this log.

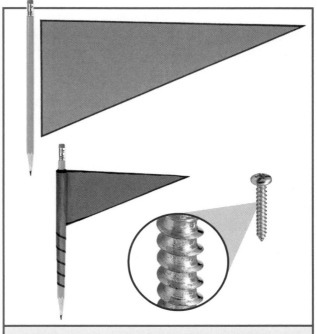

Wrapping an inclined plane around a cylinder makes a screw.

What are compound machines?

When two or more simple machines are combined they form a **compound machine**. The tow truck you saw earlier, for example, has several simple machines. There were wheels and axles, and pulleys. Some tow trucks also have levers and wedges to help lift a car. Threaded screws hold many of the parts together on the truck.

Some compound machines use one simple machine repeatedly. An elevator uses several pulleys to raise and lower it. In many elevators, the effort is provided by an electric winch. Heavy weights called *counterbalances* are attached to the effort end of the cord to provide additional force.

In a machine, the rubbing of surfaces converts some of the mechanical energy into heat. In turn, this reduces the output work that the machine can do. The more work a machine does for a given input of energy, the more efficient it is.

Efficiency (i•FISH•uhn•see) is the ratio of input energy to output work, usually written as a percentage. Electric motors like the winch on an elevator are fairly efficient. They output about 85 percent of the energy they use. Automobiles, however, are fairly inefficient. A typical car outputs only 17 percent of the energy in its fuel!

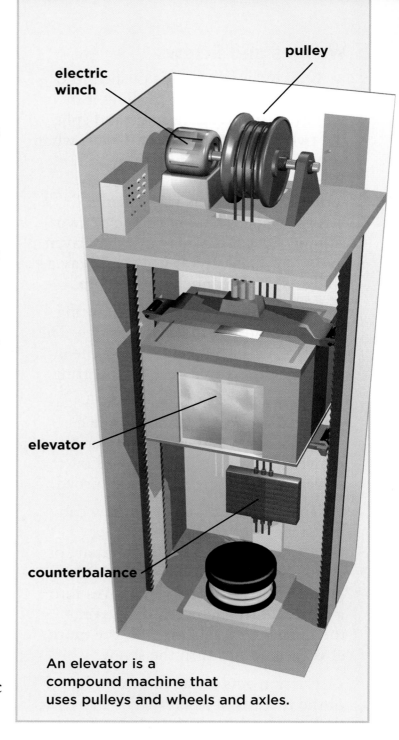

electric winch

pulley

elevator

counterbalance

An elevator is a compound machine that uses pulleys and wheels and axles.

 Quick Check

Classify Are scissors a compound machine? Why or why not?

Critical Thinking How could lubrication help a machine's efficiency?

Lesson Review

Visual Summary

Simple machines can change the direction, strength, or distance of a force.

There are six simple machines: **lever, pulley, wheel and axle, inclined plane, wedge,** and **screw**.

Compound machines are made up of two or more simple machines.

Make a FOLDABLES Study Guide

Make a Layered-Look Book. Use the titles shown. Under each tab summarize the main idea of each topic.

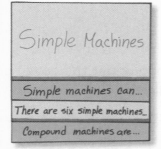

Simple Machines
Simple machines can...
There are six simple machines...
Compound machines are...

Think, Talk, and Write

1. **Main Idea** How could several pulleys help you lift a very heavy object?

2. **Vocabulary** The pivot point of a lever is called its _____.

3. **Classify** Write a list of at least three parts in a car and state what kind of simple machine each is.

4. **Critical Thinking** Why do animals benefit from jaws that are levers?

5. **Test Prep** A lever that has its fulcrum between the effort and the output is a
   - **A** first-class lever.
   - **B** second-class lever.
   - **C** third-class lever.
   - **D** compound machine.

6. **Test Prep** Which is an inclined plane wrapped around a cylinder?
   - **A** wedge
   - **B** screw
   - **C** wheel and axle
   - **D** pulley

 Writing Link

Narrative Writing

What would life be like without any machines? Write a story about a day when all of the machines in the world have disappeared.

 Math Link

Calculating Efficiency

A hybrid gas-electric car gets 45 miles to the gallon. It is 90% efficient. How many miles to the gallon would the car get if it was 100% efficient?

Writing in Science

A HUMANE MOUSETRAP

Explanatory Writing

A good explanation

▶ **develops the main idea with facts and supporting details**

▶ **lists what happens in an organized and logical way**

▶ **uses time-order words to make the description clear**

A mouse in the house can be a problem. But most mousetraps cause mice to die a painful death. Imagine a mousetrap that doesn't hurt or kill. Many people have applied their brainpower to developing a humane mousetrap. Here's how one basic type works.

The main parts are a base tray and cover. A bent stick or other rod acts to hold the cover up just enough so that the mouse can squeeze in.

First, bend the stick at a 90 degree angle about one quarter of the way up. This angle acts as the fulcrum of a lever. Next, use a toothpick to hold the angle in place. Loosely tape the ends of the toothpick to the sides of the stick. (You will form a right triangle.) Finally, attach your bait to the end of the long part of the stick. Cheese works just great!

The stick acts as a lever. When the mouse touches the bait, even a small effort will move it. Once the stick moves, the cover falls onto the tray. The mouse is trapped under the cover, but it isn't hurt. You can now take the mouse out of the house!

This mousetrap uses a lever attached to bait.

Write About It

Explanatory Writing Do research online about bird feeders that keep squirrels from stealing birdseed. Write an explanation of how this kind of bird feeder works by using simple machines. Provide steps for making this device. You can also invent your own.

LOG ON **e-Journal** Research and write about it online at **www.macmillanmh.com**

Math in Science

Measuring Mechanical Advantage

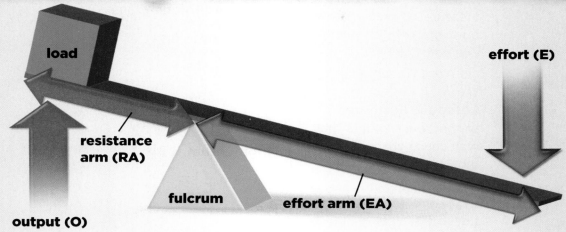

load

effort (E)

resistance arm (RA)

output (O)

fulcrum

effort arm (EA)

In a first-class lever the ratio of the output force to the effort force is called its *mechanical advantage*. You can calculate mechanical advantage by dividing output, the force applied to the load, by your effort.

$$MA = O/E$$

What if you do not know either the output or the effort? Use the following formulas to find either one:

$$O = EA \times E/RA$$
$$E = RA \times O/EA$$

 Solve It

1. For 400 N of effort, a lever applies 4,000 N of output to its load. What is the lever's mechanical advantage?

2. A lever has a 2 m long resistance arm and a 4 m long effort arm. What would be its output for an effort of 50 N?

3. A lever has a 9 m long resistance arm and a 3 m long effort arm. What effort was applied if its output was 600 N?

Multiply Fractions

To multiply a fraction by a whole number

▶ write the whole number as a fraction by placing it over the denominator 1

▶ then multiply numerators and denominators

▶ then reduce the fraction

If RA = 5 m, EA = 10 m, and E = 70 N, you can find the output force.

O= EA x E/RA

O= 10/1 x 70/5

O= 700/5

O= 140

Visual Summary

Lesson 1 Motion can be measured by seeing how quickly an object changes position.

Lesson 2 Forces are pushes, pulls, or lifts that may cause changes in motion.

Lesson 3 Moving objects and causing change requires work and energy.

Lesson 4 Machines change the forces and distances used to do work.

Make a FOLDABLES™ Study Guide

Assemble your lesson study guide as shown. Don't forget to include your Lesson 3 study guide in the back.

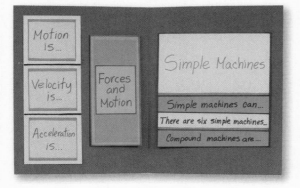

Fill in each blank with the best term from the list.

balanced forces, p. 588

efficiency, p. 616

effort, p. 608

force, p. 584

kinetic energy, p. 600

momentum, p. 578

motion, p. 572

work, p. 598

1. A push, pull, or lift from one object to another is a(n) _____ .

2. A change in the position of an object over time is _____.

3. The measurement of the energy used to perform a task is called _____.

4. The force applied to a machine is known as _____.

5. The ratio of input energy to output work is called _____.

6. When you drop a ball, gravity changes potential to _____.

7. The measurement of the mass times the velocity of an object is called its _____.

8. An object's motion will not change if you only apply _____ .

Answer each of the following in complete sentences.

9. **Main Idea and Details** Friction is a force that opposes the motion of one object moving past another. Provide two details that help explain how friction works.

10. **Infer** Describe what is happening in this image. How would it look different to someone standing in the background?

11. **Use Numbers** Calculate how much work is done when an 80-pound student climbs a 3-meter-tall ladder. One pound = 4.5 newtons of force. Give your answer in joules.

12. **Critical Thinking** Suppose you are designing a model car for a race. What features would it need to help it travel as fast as possible?

13. **Explanatory Writing** Write a detailed caption for a display on the law of conservation of energy. Use a roller coaster as an example.

14. How do forces move objects? What are three kinds of forces that affect an object's motion?

Mad Machines

Design a machine that people can use to help move objects in the kitchen.

What to Do

1. Think of a problem that people sometimes encounter in the kitchen.

2. Copy the table below and fill it in to review the uses of simple machines. Which of these machines could help deal with the problem?

Simple Machine	What It Does
lever	
wheel and axle	
pulley	
inclined plane	
wedge	
screw	

3. Design an invention using one or more of these. Make a diagram or model of your invention.

Analyze Your Results

▶ Write a paragraph for a catalog that explains the function of the invention and its usefulness.

1. **A fast-moving and heavy ball could knock over many lighter objects. This illustrates the principle of**

   A deceleration.

   B friction.

   C momentum.

   D balanced forces.

Using Energy

What form does energy have?

Key Vocabulary

thermal conductivity
the ability of a material to transfer heat (p. 630)

sound wave
a series of rarefactions and compressions traveling through a substance (p. 639)

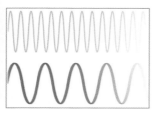

frequency
the number of times an object vibrates per second (p. 642)

wavelength
the distance from one peak to the next on a wave (p. 652)

electricity
the movement of electrons (p. 666)

electromagnet
an electric circuit that produces a magnetic field (p. 680)

More Vocabulary

Heat

Los Padres National Forest, California

Look and Wonder

The flames in a forest fire can be over 50 meters (164 feet) tall and reach temperatures greater than 1,000°C (1,832°F)! Do you think all the objects in the fire give off the same amount of heat?

Which can give you more heat?

Form a Hypothesis

If you mix the same amount of water or oil into ice water, which one will warm the ice water more? Write your answer as a hypothesis in the form "If the same amount of room-temperature water and oil is added to ice water, then…"

Test Your Hypothesis

1. Pour 100 mL of ice water (but don't include any ice) into 2 cups. Pour 100 mL of room-temperature water and cooking oil into 2 different cups. Record the temperature of each.

2. **Experiment** Mix a cup of ice water into the cup of room-temperature water and stir for two minutes. Record its temperature. Repeat this process for the cooking oil.

3. **Use Numbers** Subtract the starting temperature of the ice water from the final temperature of each mixture. This gives you the temperature change of the ice water for each experiment.

Draw Conclusions

4. **Interpret Data** How do the temperature changes compare? Was your hypothesis correct? Explain.

5. **Infer** Based on your answers in step 4, is heat the same thing as temperature? Explain.

Explore More

Which would cool faster starting at the same high temperature, 100 mL of cooking oil or 100 mL of water? Write a hypothesis. Then design an experiment to test it.

Materials

- ice
- water
- graduated cylinder
- plastic cups
- cooking (corn) oil
- thermometer

Step 1

Step 2

▶ Main Idea

Heat flows between objects when they have different temperatures.

▶ Vocabulary

heat, p. 626
temperature, p. 626
conduction, p. 628
convection, p. 628
radiation, p. 628
thermal conductivity, p. 630

LOG ON ⊖-**Glossary**
at www.macmillanmh.com

▶ Reading Skill ✔

Draw Conclusions

Text Clues	Conclusions

What is heat?

Have you ever taken a sip of hot cocoa and burned your tongue? Ouch! Energy in the hot drink flows into your tissues and damages them. The energy that causes such a burn is heat. **Heat** is thermal energy that flows between objects due to a difference in temperature. Heat is measured in joules (J). Your tongue is cooler than the drink, so heat flows from the drink into your tongue.

Heat moves from an object with a high temperature to an object with a lower temperature. But what is temperature? **Temperature** is a measurement of the average kinetic energy of particles in an object. All the particles in an object are vibrating with kinetic energy. Objects with a higher temperature have particles that are vibrating faster. Objects with a lower temperature have particles that do not vibrate as much.

When a hot object touches a cold object, the particles in each object bump into each other. During these collisions, the particles from the hot object pass on some of their energy to the particles in the cold object. The cold object is now warmer, and the

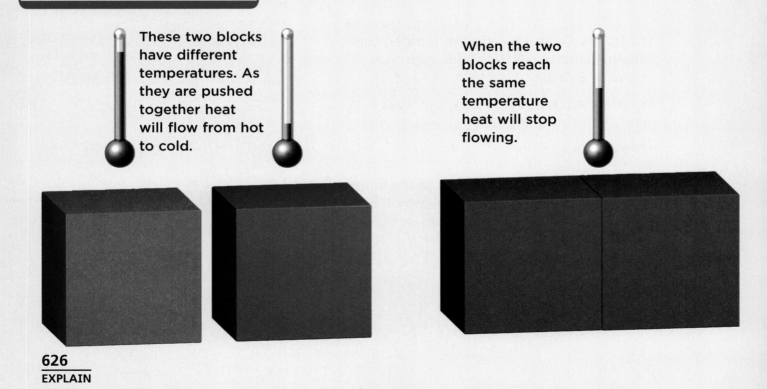

These two blocks have different temperatures. As they are pushed together heat will flow from hot to cold.

When the two blocks reach the same temperature heat will stop flowing.

The soup has a higher temperature, but the lake's greater mass releases more heat than the soup.

hot object is now cooler. When heat moves from one object to another, the temperature of each object changes. Heat will continue to flow until both objects are at the same temperature.

When two things rub together, they can get warm. This is because friction between objects changes kinetic energy into thermal energy. How does this happen? Particles in the objects bump into one another as they are rubbed together. Each bump causes the particles to vibrate a little more, and the temperatures of the objects rise.

We often measure temperature by using a thermometer. When a thermometer touches an object, heat will flow either into or out of the thermometer. When the thermometer and the object are at the same temperature heat will stop flowing. The thermometer then gives the temperature of the object in units of degrees Celsius (°C) or degrees

Fahrenheit (°F). Many thermometers work by measuring the way liquids expand or contract with temperature.

We often use the word *thermal* when discussing heat and temperature. Keep in mind, though, that temperature and heat are not the same thing. Temperature measures the average kinetic energy of particles. Heat is the total amount of thermal energy an object releases. A small bowl of steaming soup has a higher temperature than a cool lake. The lake, though, could release more heat because it has many more particles.

 Quick Check

Draw Conclusions Two objects have been touching for hours. What can you conclude about their temperatures? Explain.

Critical Thinking How could a typical liquid thermometer measure its own temperature?

FACT ▶ Heat and temperature are not the same.

How does heat travel?

You're stirring some hot cocoa. Suddenly, you notice that the spoon is too hot to hold! The cocoa warmed the spoon, and the spoon is warming your fingers! How did the heat get from the cocoa to your fingers?

Particle vibrations have spread from the cocoa to the spoon and then to your fingers. **Conduction** (kuhn•DUK•shuhn) is the passing of heat through a material while the material itself stays in place. It occurs between objects that are touching.

Inside the cocoa, heat spreads as warm and cool parts of the cocoa move around. **Convection** (kuhn•VEK•shuhn) is the flow of thermal energy through a liquid or gas, caused by hot parts rising and cool parts sinking. As hot and cool parts of the liquid move, they cause rotating currents. The currents spread thermal energy throughout the material. Remember that convection causes many of the weather patterns of Earth.

Earth's surface is warmed by radiation from the Sun. **Radiation** (ray•dee•AY•shuhn) is the transfer of energy through *electromagnetic rays* (i•lek•troh•mag•NET•ik rayz). These rays include visible light, X rays, and radio waves. Matter is not needed to transmit radiation. Electromagnetic rays can travel through empty space. Otherwise, heat from the Sun could not warm Earth!

Transmitting Heat

In conduction, heat transfers directly from the stove to the pan to the eggs.

In convection, hotter water rises as colder water falls.

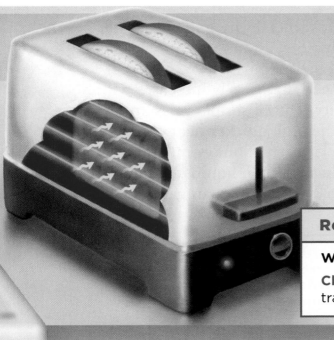

An infrared camera took this image of cars radiating heat in a parking lot.

In radiation, electromagnetic rays carry energy from the hot wires to the toast.

Infrared Rays

Hot objects radiate heat. The electromagnetic rays they produce are called *infrared rays* (in•fruh•RED). They are called infra "red" rays because they are close in color to red visible light. You cannot see infrared rays.

You can feel infrared rays, however, when you stand in the sun. Cooks use infrared rays when they broil food in an oven. A red-hot coil gives off infrared radiation to cook foods quickly. Infrared rays heat the surface of an object more than its center.

Some snakes and other animals have special sensory organs to "see" infrared rays. Scientists have built special instruments and cameras so that you can do the same. Computers can artificially color the images that these cameras take. You can use infrared cameras to take pictures of objects even in the middle of the night.

✔ Quick Check

Draw Conclusions How is heat transferred in a pot of water on a stove top?

Critical Thinking How could a large, hot surface move the air above it?

Read a Diagram

Which method moves both heat and matter?

Clue: Look at the arrows that show the transfer of energy.

What is thermal conductivity?

When heat travels, it does not always travel at the same speed. Radiation will carry thermal energy at the speed of light (300,000,000 m/s)! Convection currents can be measured like wind speed, and are rarely faster than 56 m/s (125 mph). Conduction typically carries thermal energy much slower than either of the other two methods. Why?

In conduction, particles bump into other particles and give them energy. Thermal energy can only move from one side of an object to another by moving through all the particles in between. This exchange of thermal energy takes time.

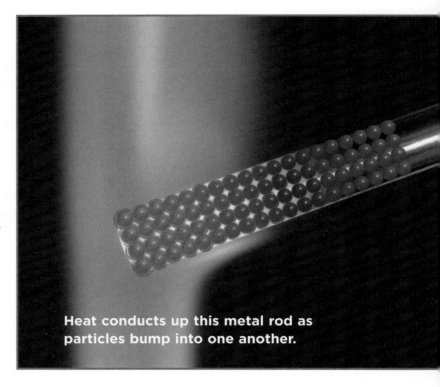

Heat conducts up this metal rod as particles bump into one another.

Thermal conductivity (kon·duk·TIV·i·tee) is the ability of a material to transfer heat. If a material conducts heat easily, we say it is a good *thermal conductor*. If a material conducts heat poorly, we say it is a good *thermal insulator*.

Most metals are good thermal conductors, and most nonmetals are good thermal insulators. Thermal conductivity usually increases with density. The closer particles are together, the quicker heat can move through them. Solids are better conductors than liquids. Liquids are better conductors than gases.

Conductivity

material	how many times better than air it conducts heat
oak wood	6
water	23
brick	25
glass	42
stainless steel	534
aluminum	8,300
copper	15,300
silver	16,300
diamond	35,000 or more

Read a Table

Which material from the table would be the best thermal insulator?

Clue: It has the lowest thermal conductivity.

Heat Capacity

Not all materials change temperature at the same rate. A gram of cooking oil warms more than a gram of water with the same amount of heat. We say that oil has a lower *capacity* (kuh•PAS•i•tee) for heat than water. Materials with low heat capacity change temperature quickly.

A material's heat capacity partly depends on how its particles hold together. Water molecules are attracted to one another. This is why water has a high heat capacity. Every substance has its own unique heat capacity.

 Quick Check

Draw Conclusions Animal fur traps air within it. Would it be a good insulator? Why or why not?

Critical Thinking Would you want a material with low or high heat capacity in a fireplace? Why?

Drinking cold water cools you down because it has a large heat capacity.

 *Quick Lab*

Thermal Differences

1. Fill one cup with water and the other with sand. Place a thermometer in each material. Record the temperatures.

2. **Predict** Which material do you think will heat more rapidly when placed under a lamp?

3. Arrange a lamp so that it shines evenly on both cups. Every minute for ten minutes, measure and record the temperatures.

4. Graph your data for each material as temperature versus time.

5. Was your prediction correct? How do you know?

6. **Infer** Which might cool faster, sand or water? Explain.

When is heat waste?

Besides keeping you warm, heat has many uses. Many chemical reactions require heat in order to occur. For example, the heat released from chemical reactions powers the engine in a fossil-fuel-burning car. Sometimes metals are heated to make them harder and more brittle. Infrared cameras are used in "night-vision" goggles to allow you to view objects without visible light.

Yet, heat can also be an unwanted waste product. Over time, heat generated by friction causes many machines to break. For instance, friction and heat wear down car parts and those parts must then be replaced. Also, friction often lowers the efficiency of machines. The energy that is lost to friction becomes thermal energy. This energy is hard to use to perform work. It is often considered to be wasted.

When energy changes form or performs work it will often produce waste heat. Electrical energy produces heat when electrons move through electrical wires. Muscles are heated when chemical reactions release energy from food. Although it can be very useful, heat is the most common waste product from work.

Water absorbs the heat caused by the friction of a saw cutting stone.

Cars will stop working if they cannot release their extra heat.

✔ Quick Check

Draw Conclusions How is energy wasted as heat during a bike ride?

Critical Thinking When would heat from friction not be wasted?

Lesson Review

Visual Summary

 **Heat** is thermal energy flowing from a warmer object to a cooler one.

 Heat travels by **conduction, convection,** and **radiation**.

 **Thermal conductivity** is the ability of a material to conduct heat.

Make a FOLDABLES™ Study Guide

Make a Three-Tab Book. Use the titles shown. Under each tab, summarize what you learned about that topic.

Heat is thermal energy flowing

Heat travels by...

Thermal conductivity is the ability...

Think, Talk, and Write

1. **Main Idea** A candle is placed on a steel block and begins to melt. How do you know whether the candle or the steel block is hotter?

2. **Vocabulary** The ability of a material to transfer heat is called _____.

3. **Draw Conclusions** An infrared image of a house shows that the roof is brighter than the rest of the house. What does this mean?

Text Clues	Conclusions

4. **Critical Thinking** Why is water a material used to cool hot machinery?

5. **Test Prep** The transfer of heat by electromagnetic rays is called
   A conduction.
   B convection.
   C radiation.
   D insulation.

6. **Test Prep** Which is most responsible for winds on Earth?
   A conduction
   B convection
   C radiation
   D insulation

Writing Link

Writing a Story
You live on a planet where no materials are good conductors of heat. Write a story about your life on this planet.

Math Link

Heating Gold
It takes 128 J of heat to raise the temperature of 1 kg of gold from 20°C to 21°C. A heat source supplies 1,536 J. How much gold can it heat from 20°C to 21°C?

Inquiry Skill: Form a Hypothesis

You know that heat flows from warmer to cooler objects until they both reach the same temperature. How is that temperature affected by the amount of each object?

Scientists use observations or theories to help them **form a hypothesis**. When you form a hypothesis, you make a testable statement about what you think is logically true.

▶ Learn It

A hypothesis is a statement about the effect of one variable on another. It should be based on observations or collected data. For example, when you drink hot chocolate, you might notice that it cools faster when you add ice to it. Based on this observation, you might **form a hypothesis** like "If increasingly colder substances are added to hot chocolate, then it will cool faster."

A hypothesis is tested by conducting an experiment. In this experiment, you will test how much hot water cools when room-temperature water is added. Think about observations you have made in the past involving temperature changes. Write a hypothesis in the form "If increasingly larger amounts of room-temperature water are added to hot water, then…"

Skill Builder

▶ **Try It**

Materials graduated cylinder, hot and room-temperature water, cups, thermometer, stopwatch

In this activity, you will observe how water temperatures change in order to test your **hypothesis**.

1 Use the graduated cylinder to pour 25 mL of room-temperature water into one cup and record its temperature on a chart like the one shown.

2 Pour 75 mL of hot water into a different cup and record its temperature on the chart.

3 Add the room-temperature water to the hot water and start the timer on the stopwatch. Place a thermometer in the water and observe its temperature after two minutes. Record the new temperature of the hot water.

4 Repeat steps 1 to 3 with 50 mL, 75 mL, and 100 mL of room-temperature water. You are changing this variable in order to test your hypothesis.

▶ **Apply It**

1 Subtract the final temperature of the hot water from its starting temperature for each trial. Record your results on your chart.

2 Use the data in your chart to form a graph. On the horizontal axis plot the amount of room-temperature water added to the hot water. On the vertical axis plot the change in temperature of the hot water.

3 Was your **hypothesis** correct? How do you know?

4 Did the results of the first three trials make it easier to understand what would happen the last time? Why or why not?

Water Added	Added Water's Temperature	Hot Water's Starting Temperature	Hot Water's Ending Temperature	Hot Water's Temperature Change
25 mL				
50 mL				
75 mL				
100 mL				

Sound

Look and Wonder

A cloud forms as this jet breaks the sound barrier and creates a sonic boom. What do you think you might feel if you were near a sonic boom?

What makes sound?

Form a Hypothesis

When you pluck the rubber band on the "instrument" shown, it makes sound. How will this sound depend on the way you pluck the rubber band? Write your answer as a hypothesis in the form "If the rubber band is plucked with increasing force, then the sound..."

Test Your Hypothesis

1. ⚠ **Be Careful.** Wear goggles. Make a rubber-band "instrument" as shown. Poke a small hole in the bottom of the cup with a toothpick. Tie one end of a cut rubber band to the toothpick. Thread the toothpick through the hole in the cup. Tie the stretched rubber band to the ruler and tape the ruler to the cup.

2. **Observe** Wrap one hand around the cup while you pluck the rubber band. What do you hear and feel? Record your observations.

3. Pluck the rubber band both gently and forcefully. Record how the sound is affected. Repeat your actions to verify your results.

Draw Conclusions

4. **Interpret Data** Based on your observations, was your hypothesis correct? Explain.

5. **Infer** How do you think your rubber-band "instrument" made sound? Use your observations from step 2 to help you.

Explore More

How will stretching the plucked rubber band affect whether the pitch is high or low? Write out your hypothesis. Then carry out experiments to test it.

Materials

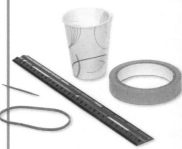

- goggles
- paper cup
- toothpick
- rubber band
- wood or plastic ruler
- masking tape

Step 1

Step 2

► Main Idea

Sounds are produced by vibrating objects.

► Vocabulary

sound wave, p. 639

medium, p. 639

vacuum, p. 640

absorption, p. 641

frequency, p. 642

pitch, p. 642

amplitude, p. 644

echolocation, p. 646

LOG ON e-Glossary
at **www.macmillanmh.com**

► Reading Skill ✔

Fact and Opinion

Fact	Opinion

How is sound produced?

Have you ever noticed the sound from a low-flying jet rattling the dishes in the kitchen? Perhaps you've noticed something similar when someone plays a stereo system too loudly. What causes objects to shake when there are loud sounds nearby?

When an object makes sound, it vibrates back and forth. The vibrations of a drum alternately squeeze air particles and then spread them out. This creates regions of air that have many particles, called *compressions* (kuhm•PRESH•uhnz), and regions of air that have few particles, called *rarefactions* (rayr•uh•FAK•shuhnz). The compressions and rarefactions move through the air, carrying sound energy. Each region of the air is only moved back and forth. Drums do not create a steady wind.

compression

rarefaction

Sound waves vibrate in the same direction that they travel.

Vibrations caused by the helicopter blades produce loud sound waves.

The density of the air, but not the air itself, moves.

peak

dip

The density of the air can be shown as a series of peaks and dips.

A series of rarefactions and compressions traveling through a substance is called a **sound wave**. The substance through which the wave travels is called the **medium** for the wave. As with all waves, sound waves carry energy. When they pass through a medium, the medium is not permanently moved. Energy, however, is permanently moved from one place to another.

Sound waves vibrate the medium in the same direction that the energy moves. They are called *longitudinal* (lon•ji•TEW•duh•nuhl) waves. We can also represent sound waves as a series of peaks and dips. The peaks represent the high density of air in compressions. The dips indicate the low density of air in rarefactions. Remember that air does not move up and down like the peaks and dips.

When sound waves hit an object, it can start vibrating. The object is moved by the energy of the wave. This is how sound from a loud airplane or stereo rattles dishes. You can literally feel the vibrations caused by such loud sounds.

✔ Quick Check

Fact and Opinion Should homes be built next to airports? Support your opinion with facts.

Critical Thinking Describe the density of air in a room when music is played.

How does sound travel?

Could you hear sounds in space? No, space is a **vacuum** (VAK•yew•uhm), a region that contains few or no particles. Since space has few particles, there is no medium for sound to travel through. You could not hear a radio, even if it were playing next to you!

Sound can travel through solids, liquids, and gases. In fact, sound tends to travel with the greatest speeds in solids and the slowest speeds in gases. In steel, for example, sound travels at almost 6,000 m/s. In air, though, sound only travels at about 343 m/s.

These differences in the speed of sound are due to how far apart the particles are. The collision of particles is what carries sound energy. In a solid, the particles are close together. They collide quickly and move sound rapidly. In gases, particles are far apart. Collisions are less frequent, so sound travels more slowly.

The temperature of the medium also affects the speed of sound. In warmer air, particles move faster. As a result, they collide more often. With more collisions, the particles in warmer air transmit sound faster.

FACT Sound cannot travel through outer space.

Water is a good medium for sounds like dolphin songs.

Changing How Sound Travels

Have you ever been in a sound-proof room? Such rooms are often used by musicians. When a sound wave hits soft, thick, or uneven materials, the energy of the wave is absorbed. **Absorption** (ab•SAWRP•shuhn) is the transfer of energy when a wave disappears into a surface. Absorbed sound waves become kinetic or thermal energy on that surface.

Have you ever heard an echo? When sound waves hit a flat, firm surface much of their energy bounces back. Echoes are sound waves that have reflected back at the speaker. Reflection is the bouncing of a wave off a surface. Whenever a sound wave reflects off a surface, at least some of it is absorbed. Echoes are never as loud as the original sound wave.

The walls of this room are made to absorb sound.

Quick Lab

Sound Carriers

1. **Predict** Will you be able to hear the sound from a radio better through air, water, or wood?

2. Put a radio on a wooden table. Put your ear on the other side of the table and listen. Now lift your head. Record your observations.

3. Fill a plastic bag with water. Hold the bag against your ear. Then hold the radio against the other side of the bag. How loud is the radio? Move your ear away from the bag. How loud is the radio now? Record your observations.

4. Rate wood, air, and water as sound mediums from worst to best.

5. **Infer** Foam is less dense than wood or water, but more dense than air. How well do you think foam will carry sound?

✓ Quick Check

Fact and Opinion A friend says echoes are scary because they sound softer than regular voices. Which part of this statement is fact, and which is opinion?

Critical Thinking How would putting your ear to the floor let you hear a sound sooner than you would hear it in the air?

What is pitch?

How high can you sing? How low can you sing? What is changing when you go from singing a high note to a low note? The sound wave that reaches your ears is different. The series of peaks and dips in the wave get closer together the higher you sing. **Frequency** (FREE•kwuhn•see) is the number of times an object vibrates per second. Its units are cycles per second (1/s) or hertz (Hz). High notes have a greater frequency than low notes.

Musical notes like the ones you sing are defined by their pitch. **Pitch** is how high or low a sound is, and is related to frequency. In music, pitch is often given a letter name of "C," "D," "E," "F," "G," "A," or "B." The series repeats itself so that the eighth note is "C" again. A series of eight notes is called an *octave* (OK•tiv).

If the first "A" you sing is at a frequency of 55 Hz, the "A" in the next octave will be 110 Hz. You will hear that the second A is twice as high in pitch as the first. The third "A" will in turn sound three times as high in pitch as the first "A." Its frequency, however, will not be 165 Hz, but 220 Hz! The frequency doubles for every octave.

Pitch and frequency are two different ways to describe sound waves. Pitch is the way our ears perceive frequency. It is closely related to the number of peaks in a sound wave, but is not the same thing as frequency.

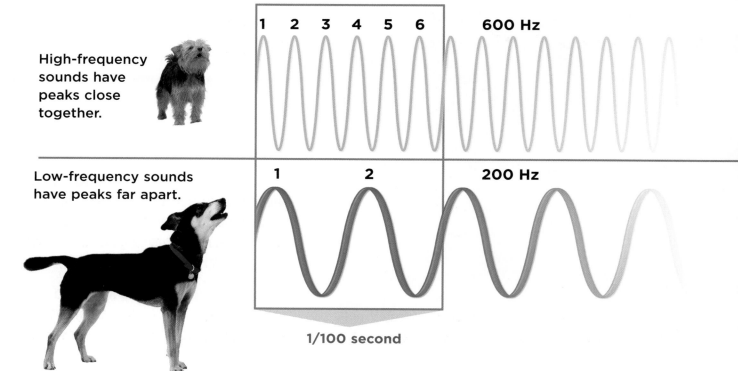

High-frequency sounds have peaks close together.

1 2 3 4 5 6   600 Hz

Low-frequency sounds have peaks far apart.

1   2   200 Hz

1/100 second

Read a Photo

Would a train's whistle sound higher or lower in pitch than normal?

Clue: Is the train moving toward or away from you?

Changing Pitch

To make a sound higher in pitch, you increase the number of times it vibrates each second. On a string instrument, shortening the string increases pitch. On a wind instrument, shortening the tube increases pitch. A shorter tube produces a higher pitch because the air inside vibrates faster.

You can increase the frequency of a sound wave by moving toward it. How? Frequency is just the number of peaks of a wave per second. If you move toward a wave, you will hear the peaks quicker than if you were standing still. If you move away from a wave, the peaks arrive at your ear more slowly and the pitch is lower.

A change in frequency due to moving away or toward a wave is called the *Doppler effect* (DOP•luhr i•FEKT). Any movement can cause a Doppler effect, but only fast speeds will change a pitch enough to be heard by your ears.

The pitch of a trombone changes with the length of its tubes.

✔️ *Quick Check*

Fact and Opinion A classmate says that higher notes are irritating because they vibrate your ear faster. Which part of that statement is fact and which is opinion?

Critical Thinking How do you think you change the pitch of your voice?

What is volume?

Pretend you are in a room when someone turns up the volume on a radio too much. Is it easy to hear other noises? What makes a sound so loud?

The height of a sound wave is called its *amplitude* (AM•pli•tewd). The **amplitude** is how dense the air is in the compressions or rarefactions compared to normal air. The loudness, or volume, of a sound depends on the amplitude of the sound's waves.

Scientists measure the volume of sounds with decibels (dB). A 20 dB noise has 10 times more energy than a 10 dB noise. A 30 dB noise has 100 times more energy than a 10 dB noise.

Our ears hear things in a different way. A 30 dB noise sounds twice as loud as a 20 dB noise, and four times as loud as a 10 dB noise. Sounds above 85 decibels damage your hearing. Wear earplugs if you are near loud sounds!

Volume of Sounds

decibel level	sound
180 dB	rocket engine at 30 m (98 feet)
130 dB	threshold of pain, train horn at 10 m (33 feet)
120 dB	rock concert
110 dB	chainsaw at 1 m (3.3 feet)
100 dB	jackhammer at 2 m (6.6 feet)
85 dB	threshold of damaging hearing
80 dB	vacuum cleaner at 1 m
60 dB	normal conversation
50 dB	rainfall
30 dB	theater (without talking)
10 dB	human breathing at 3 m (10 feet)
0 dB	threshold of human hearing (with healthy ears)

Read a Table

Could the sound from a rocket engine 30 m away cause pain in your ears?

Clue: Compare the volume for the rocket engine and the threshold of pain.

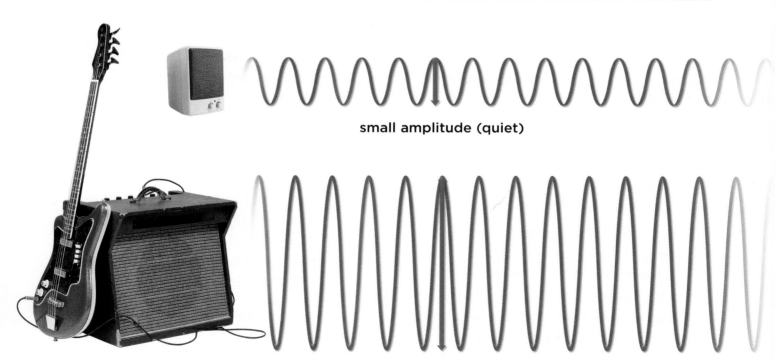

small amplitude (quiet)

large amplitude (loud)

As the sound wave from the bell travels, it becomes quieter.

Changing Volume

You can make sounds louder by using more energy. For example, you can pluck a string harder, use more air in your voice, or hit a drum with more force. The extra energy increases the density of the particles in the compressions. Also, the rarefactions will be less dense than before.

Changing the medium of a sound wave will also change its amplitude. A wave in a dense material will have a smaller amplitude than in air. The wave, however, will have the same amount of energy. Even though the amplitude is smaller there are more particles moving in the wave.

The volume of a sound will be smaller the farther you travel from its origin. Why? Think of ripples in a pond. At their center, the ripples are high, but as they expand outward they get smaller. The same amount of energy in the wave is being spread out over a larger and larger area. So as you move away from the origin of a sound, the energy in the wave at any point gets smaller. Less energy means less volume, and you hear the difference.

 Quick Check

Fact and Opinion Listening to loud music damages your hearing. Is that statement fact or opinion?

Critical Thinking You hear the sound of a drum as 45 dB, then 55 dB, then 65 dB. How might this be happening?

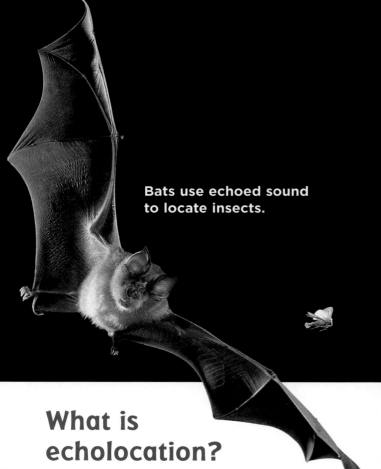

Bats use echoed sound to locate insects.

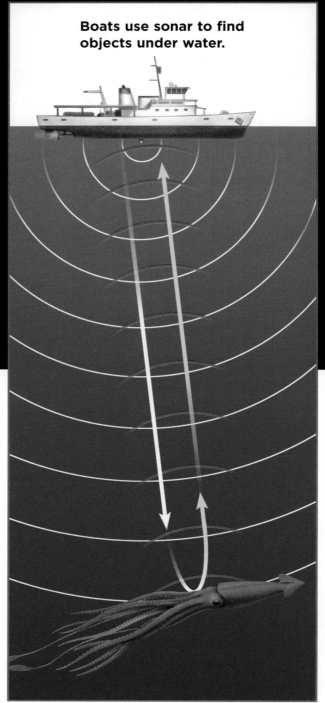

Boats use sonar to find objects under water.

What is echolocation?

Echoes can be put to good use. Bats, for example, make sounds that echo off of their prey. The returning echoes tell the bat where the prey is located. Finding food or other objects in this manner is known as **echolocation**. Whales and dolphins, too, use echolocation to orient themselves and to find food.

Scientists have developed a system called *sonar* that works like echolocation does for animals. Sonar stands for "sound navigation and ranging." It is used under water to find objects. The sonar system sends out sound waves that reflect off of objects. It then detects the reflected sound waves. The return time and direction of the sonar echoes are used to calculate the location of the object.

 Quick Check

Fact and Opinion Dolphins are cuter than whales when using echolocation. Is this fact or opinion?

Critical Thinking Could sonar work on land? Why or why not?

646

Lesson Review

Visual Summary

Vibrating objects produce **sound waves** in a **medium**.

Sounds may be **transmitted, absorbed,** or **reflected** by materials or objects.

As the **frequency** of a sound wave increases, the **pitch** becomes higher.

Make a FOLDABLES™ Study Guide

Make a Folded Chart. Use the title shown. Fill in the chart with information you learned about each topic.

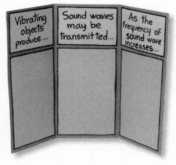

Vibrating objects produce ...

Sound waves may be transmitted ...

As the frequency of sound wave increases ...

Think, Talk, and Write

❶ **Main Idea** How could a stereo playing loudly rattle dishes?

❷ **Vocabulary** The number of compressions that pass by in a given time is called a sound wave's _____.

❸ **Fact and Opinion** Should you wear earplugs while using a vacuum cleaner? Support your opinion with facts.

Fact	Opinion

❹ **Critical Thinking** Is there more energy in a 30 dB or 40 dB sound wave? Why?

❺ **Test Prep** At what volume do sounds start damaging hearing?

  A 10 decibels
  B 65 decibels
  C 85 decibels
  D 150 decibels

❻ **Test Prep** An echo is an example of a sound wave being

  A transmitted.
  B absorbed.
  C reflected.
  D surfed.

 Math Link

Finding Frequencies
A particular "C" note has a frequency of 33 Hz. What will be the frequency of a "C" note 5 octaves higher than that note?

 Music Link

Why would guitars have a top made of thin wood? Why would they have a big, hollow body? Write out your ideas.

Be a Scientist

Materials

scissors

10 straws

ruler

masking tape

Structured Inquiry

How can you change a sound?

Form a Hypothesis

Increasing or decreasing the number of vibrations in a second changes the pitch of a sound. For example, on a guitar, the highest notes are played when the strings vibrate the fastest. For instruments that have tubes, the length of each tube determines how quickly air inside it vibrates.

How does the length of tubes affect the pitch of sounds they make? Write your answer as a hypothesis in the form "If the tube of a wind instrument is shortened, then the pitch… "

Test Your Hypothesis

1. **Make a Model** Use scissors to cut a straw to a length of 15 cm.

2. Cut the next straw to be 1 cm shorter than the last one. Repeat this procedure until all of the straws are cut. The last straw should be 6 cm long.

3. Lay the straws on the table. Place a piece of tape over all of the straws.

4. **Experiment** Hold the instrument to your mouth and blow across the straws to create sound.

Step 2

Step 4

Draw Conclusions

5 **Observe** What do the shortest and longest pipes sound like? Was your hypothesis correct? Why or why not?

6 **Infer** Would the 12 cm straw sound identical to the 6 cm straw if it was cut in half? Why or why not?

How are pitch and tension related?

Form a Hypothesis

How do you think tension in a rubber band would affect the sound it makes? Write your answer as a hypothesis in the form "If the tension in a rubber band is increased, then the pitch of the sound will…"

Test Your Hypothesis

⚠ **Be Careful.** Wear goggles. Design an experiment to investigate the effect that tension in a rubber band has on the pitch of its sound. Write out the materials you need and the steps you will follow. Record your results and observations.

Draw Conclusions

Did your results support your hypothesis? Why or why not?

What other variables might affect the pitch of sounds? For example, how is sound affected by different mediums? Determine the materials needed for your investigation. Your experiment must be written so that another group can complete the experiment by following your instructions.

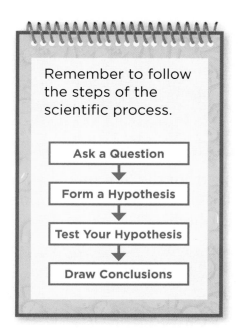

Remember to follow the steps of the scientific process.

Ask a Question

↓

Form a Hypothesis

↓

Test Your Hypothesis

↓

Draw Conclusions

Light

Garibaldi Provincial Park, British Columbia, Canada

Look and Wonder

Light from the Sun hits Earth at an angle. What kind of path do you think it follows to get here?

What path does the light follow?

Materials

- masking tape
- flat mirror
- 2 pencils
- 2 erasers
- protractor

Form a Hypothesis

When you look in a mirror, you see light that travels to the mirror, bounces off, and travels to your eye. How does the angle of the light hitting the mirror compare to the angle of the light bouncing to your eye? Write your answer as a hypothesis in the form "If the angle at which light strikes a mirror decreases, then . . . "

Test Your Hypothesis

1. Using two pieces of tape, form a large letter *T*. Place the mirror upright on the top of the *T*. Stick each pencil, point down, into an eraser so that they can stand up on their own.

2. **Experiment** Place a pencil on the left side of the *T*. Place your head on the right side. Move your head until the pencil appears to be in the center of the mirror at the top of the *T*. Now place the second pencil so that it completely blocks your view of the first pencil in the mirror.

Step 1

3. **Measure** Move the mirror and place a protractor on the top of the *T*. Find the angle between the top left of the *T* and the first pencil. This is your independent variable. Find the angle between the top right of the *T* and the second pencil. This is your dependent variable.

4. Repeat steps 2 and 3 three more times, moving the first pencil farther from the *T* each time.

Draw Conclusions

5. **Interpret Data** Look at the angles you measured. Was your hypothesis correct? Why or why not?

Explore More

What would happen if one pencil was close to the mirror while another was far away? Would the angles change? Write a hypothesis and carry out an experiment to test it.

Step 2

▶ Main Idea

Light travels as waves, but can also be described as particles.

▶ Vocabulary

wavelength, p.652
photon, p.653
translucent, p.654
image, p.656
refraction, p.657
prism, p.658
spectrum, p.658
electromagnetism, p.660

 e-Glossary
at www.macmillanmh.com

▶ Reading Skill ✓

Summarize

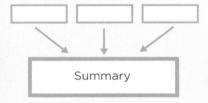

Summary

What is light?

Light from the Sun travels over 90 million miles to Earth in only $8\frac{1}{3}$ minutes! Light is made of vibrating electric and magnetic energy. This energy travels as a wave—it has both frequency and amplitude. Light waves vibrate in the direction perpendicular to the direction of their motion. They are called *transverse* (trans•VURS) waves.

Light does not depend on compressions or rarefactions. In fact, light waves can travel with or without a medium. In a vacuum, light travels very fast—about 300,000 km/s (186,000 mile/s). Light travels slightly slower through mediums like air, water, or glass. The speed of light is so fast that some scientists think that nothing travels faster. The speed of light may be the speed limit of our universe.

We often wish to know the wavelength (WAYV•lengkth) of light. Wavelength is the distance between one peak and the next in a wave. When you multiply the wavelength of a wave by its frequency, you get the speed of that wave.

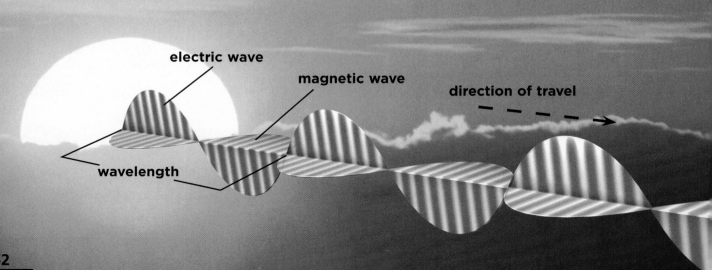

Light is a wave made from electric and magnetic energy.

electric wave

magnetic wave

direction of travel

wavelength

Light Is Also a Particle

Although light is a wave of energy, it is also a particle. How can something be both a wave and a particle? Scientists were confused about that question for a very long time. They performed many experiments and found that light has properties of both waves and particles, so they had to conclude it was both.

Light is like a particle in several ways. It travels in straight lines called *light rays*. Light does not have mass like a particle, but it does have momentum like a particle. When light hits an object, it acts just like a tiny particle. Light can even change the direction of atoms and other small particles. When light hits camera film, it produces little dots instead of forming an image all at once. Over time, those points will eventually add up and form the original image.

Particles of light are called photons (FOH•tonz). A **photon** is a tiny bundle of energy by which light travels. The energy of a single photon is very small: a photon of red light has only about 0.0000000000000000003 J of energy! Each photon also acts like a wave with a frequency. If a photon has a higher frequency, it also has more energy.

 Quick Check

Summarize What properties of particles does light have?

Critical Thinking How could you find the wavelength of light if you knew its speed and frequency?

FACT Light is both a wave and a particle.

Photons hit a piece of film individually. When enough of them have hit, the image the camera took appears.

How does light make shadows?

When light strikes an object's surface, photons can bounce off at random angles. This is called *scattering* (SKAT•uhr•ing) light. We see objects because light has scattered off them and entered our eyes.

Sometimes when light hits an object, a photon is absorbed. When photons are absorbed by objects, those objects gain energy. This energy is usually transformed into heat. Darker objects absorb more light than lighter objects.

Light may also pass through objects. Objects that allow most light through are called *transparent* (trans•PAYR•uhnt). Objects that blur light as it passes through are called **translucent** (trans•LEW•suhnt). If an object allows little to no light through, it is called *opaque* (oh•PAYK).

Whether an object is opaque, translucent, or transparent depends on its material, its thickness, and the color of light. Thicker objects have more particles to absorb photons so they are more likely to be opaque. Some objects will be opaque, transparent, or translucent in only one color of light.

Opaque and translucent objects block light. The area behind these objects is darker—they have a shadow! Shadows are just the absence of light.

| Opaque objects let little to no light pass through. | Translucent objects blur light that passes through. | Transparent objects allow almost all light through. |

You can find the size and shape of a shadow by tracing light rays.

When an object is between a light source and another object, it will cast a shadow on the other object. Light sources can be natural (like the Sun) or artificial (like a flashlight).

You cast shadows on the ground when the Sun shines. Have you ever seen how long your shadow is at sunrise? The Sun is low in the sky. Light from the Sun travels toward you at a small angle. At such a small angle, it takes a long distance for the light to hit the ground behind you. As the Sun rises, the angle of the sunlight increases and your shadow becomes shorter.

Shadows depends on the angle and distance between a light source and an object, and between the object and the place where the shadow is cast. Drawing light rays helps you trace the outline of a shadow. The closer a light source is to an object, the larger the shadow an object will cast.

 Quick Check

Summarize What are the ways in which light interacts with matter?

Critical Thinking What makes a white shirt cooler than a dark one on a sunny day?

How does light bounce and bend?

When you look into a mirror, you see an image. An **image** is a "picture" of the light source that light makes bouncing off a shiny surface. Light reflects off a mirror like sound echoes off a cliff. The image in a mirror is clear because most of the light wave reflects the same way off the mirror's smooth surface. Reflection is just the organized scattering of a wave.

When light hits a mirror, it obeys the *law of reflection*: the angle of an incoming light ray equals the angle of the reflected light ray. An image in a flat mirror appears to be behind the mirror. The distance to the image is equal to the distance the light traveled from the object to the mirror.

This girl's image is enlarged and reflected by a curved mirror.

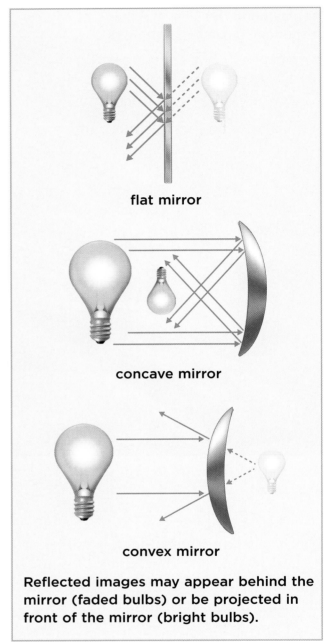

flat mirror

concave mirror

convex mirror

Reflected images may appear behind the mirror (faded bulbs) or be projected in front of the mirror (bright bulbs).

Mirrors can also be made with curved surfaces. If they curve in, they are *concave* (kon•KAYV). If they curve out, they are *convex* (kon•VEKS). Curved mirrors can form many kinds of images. They may be upright or upside down. They may also be enlarged or reduced. Convex mirrors always produce images that are upright and reduced.

Light Can Bend

When you place an object in a glass of water, it appears to bend. Yet, if you pull the object out it is still straight. How can this be? It is the light from the object that is bending!

When light changes mediums it also changes speed. When waves change speed, they refract. **Refraction** (ri•FRAK•shuhn) is the bending of waves as they pass from one substance into another. Refraction is not very noticeable with sound waves. With light waves, however, you see it easily.

Refraction causes this pencil to appear bent.

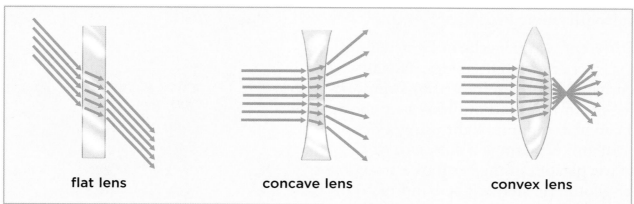

flat lens   concave lens   convex lens

Rays entering a denser medium bend to make a steeper angle with the surface. Rays leaving a denser medium bend in the opposite direction.

Lenses use refraction to shape images. Convex lenses work like concave mirrors and concave lenses work like convex mirrors.

Lenses are used in eyeglasses to make objects appear in focus. We also use lenses in cameras and telescopes to change the size of the image we see. The size and location of the image depend on where the object and lenses are in relation to each other.

Eyeglasses focus light to help your vision.

Quick Check

Summarize What properties do images have if they are formed by concave lenses or convex mirrors?

Critical Thinking How is bouncing a basketball to a friend a model of how light reflects from a surface?

Why do we see colors?

When sunlight hits raindrops in the sky, a rainbow appears. Where do these colors come from? Actually, the colors are already in the sunlight that produces the rainbow.

Our eyes see light waves with different wavelengths as different colors. Visible light waves with longer wavelengths look red. Visible light waves with shorter wavelengths look violet. All the colors in between have wavelengths in the middle of those two. White light, like the kind from the Sun, is actually just a collection of many different wavelengths mixed together.

Different wavelengths of light will reflect and refract at different angles. This is why when white light is refracted by raindrops in the sky, it is spread out into a rainbow. You can also separate light using a prism (PRIZ•uhm). A **prism** is a cut piece of clear glass (or plastic) in the form of a triangle or other geometric shape. The band of color in a rainbow, or from light passing through a prism, is called a **spectrum** (SPEK•truhm).

Opaque objects are the color of light that they scatter.

Translucent objects are the color of light that passes through them.

Creating a Spectrum

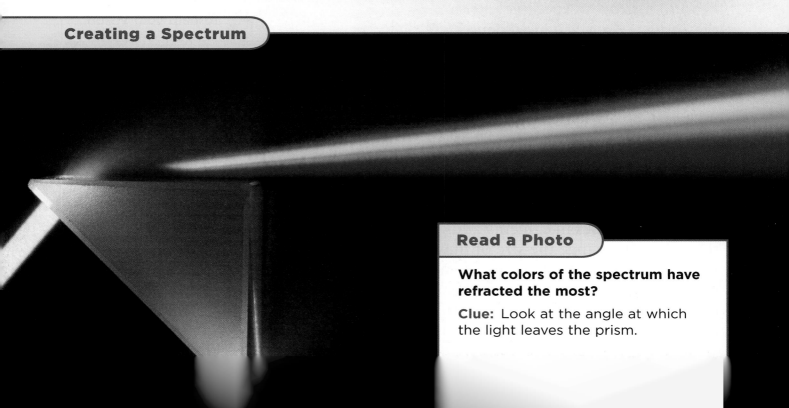

Read a Photo

What colors of the spectrum have refracted the most?

Clue: Look at the angle at which the light leaves the prism.

Overlapping Colors

Whether an object scatters, absorbs, or transmits light may depend on the wavelength of the light. When light hits an opaque object it is scattered and/or absorbed. Opaque objects appear the color of light that they scatter. They absorb all other colors of light.

When light hits a translucent object, some colors are absorbed and others pass through. Translucent objects appear the color of the light that passes through them. They absorb all other colors of light.

The picture on a color television is made up of red, green, and blue dots of light. Why are these colors used? We can create any color of light by mixing red, green, and blue light in the right

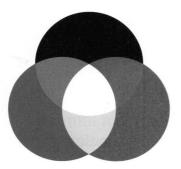

When equal parts of red, green, and blue light rays are mixed, they form white light.

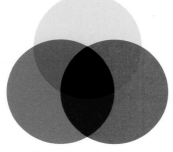

When equal parts of magenta, cyan, and yellow materials are mixed, they absorb all light and appear black.

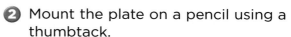

Quick Lab

Mixing Colors

1. Divide a paper plate into six sections. Color two sections red, two sections blue, and two sections green.

2. Mount the plate on a pencil using a thumbtack.

3. **Observe** Roll the pencil between your palms to spin the wheel. What color do you see? Why?

amounts. For this reason, red, green, and blue are called the primary colors of light. If mixed equally, red, green, and blue light produce white light.

Magenta, cyan, and yellow are often used to create color by scattering. For instance, we want a part of a picture on a page to look blue when white light strikes it. We could mix equal amounts of magenta and cyan paint. Magenta scatters only red and blue. Cyan scatters only blue and green. When the two are mixed, magenta absorbs cyan's green, and cyan absorbs magenta's red. Together they only scatter blue, the color they share.

 Quick Check

Summarize What colors are made by mixing red, green, and blue light two at a time in equal amounts?

Critical Thinking What would happen if you shined yellow light on a blue opaque object?

larger
wavelength

AM

radio
waves

FM

TV

radar

infrared

visible

light

ultra-
violet

X rays

gamma
rays

higher
frequency

Is all light visible?

The way in which electric and magnetic forces interact is called **electromagnetism** (i•lek•troh•MAG•ni•tiz•uhm). You know that light is made of electric and magnetic waves that can move through space. Light is just a form of electromagnetic radiation.

Scientists know of many forms of electromagnetic radiation besides visible light. All travel at the speed of light and can move through a vacuum. They differ, however, in wavelength and energy. Together, they make up the electromagnetic spectrum.

What single source could produce all forms of electromagnetic radiation? If you said the Sun, you're right! Most of the radiation from the Sun is infrared, visible and ultraviolet light. Solar flares, however, give off all forms of electromagnetic radiation when they erupt.

 Quick Check

Summarize What forms of electromagnetic radiation does the Sun give off?

Critical Thinking Why would exposure to X rays be more dangerous than radio waves?

Read a Diagram

Do radio or gamma-ray photons have more energy?

Clue: Higher frequency photons have more energy.

Lesson Review

Visual Summary

Light travels as **electromagnetic** waves, but can also be thought of as particles called **photons**.

Light **reflects** off surfaces and **refracts** when entering a new material.

The color of light depends on its **wavelength**.

Make a FOLDABLES™ Study Guide

Make a Trifold Table. Use the titles shown. Summarize what you have learned in the boxes provided.

Think, Talk, and Write

1 **Main Idea** What makes light able to move through empty space?

2 **Vocabulary** A material or object that blocks light completely is _____.

3 **Summarize** How does light act like a wave?

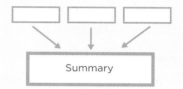

4 **Critical Thinking** How does light change when it enters a new medium?

5 **Test Prep** The law of reflections states that incoming and outgoing angles are
  A always the same.
  B never the same.
  C always large.
  D always small.

6 **Test Prep** Which kind of light has a wavelength shorter than green light?
  A red light
  B radio waves
  C X rays
  D yellow light

Math Link

Solar Energy
Some infrared photons have half the frequency of blue-light photons. How many blue-light photons would it take to equal the energy in 5,000 of these infrared photons?

Art Link

Photography
Colored transparent windows, called filters, can be placed over the lens of a camera. How could these colors be used? Write down your ideas.

How We Use Lasers

Expository Writing

A good exposition

▶ **develops the main idea with facts and supporting details**

▶ **summarizes information from a variety of sources**

▶ **uses transition words to connect ideas**

▶ **draws a conclusion based on the facts and information**

Lasers can be used to align objects or measure distances. ▼

Lasers play a role in almost all aspects of our lives. Laser shows amaze audiences. Lasers read compact discs (CDs). Laser light carries television signals along optical fibers.

Lasers play an important role in business and industry, too. Bar code scanners help stores keep track of purchases. Laser beams can focus a huge amount of energy on a single spot. They can cut metal very precisely. Jewelers use lasers to cut diamonds because they are so precise.

Lasers have changed the face of medicine. A laser can cut out diseased parts without hurting healthy areas of the body. It can destroy diseased cells by heating them. It can join cells together for healing. Most delicate eye surgery is done with lasers. Dentists use lasers to remove decay from teeth, bond teeth, and whiten teeth.

Engineers and surveyors use lasers to measure distances more accurately than ever before. Lasers have been used to measure the distance from Earth to the Moon.

Write About It

Expository Writing Find out more about one of the uses of lasers. Write an expository essay giving important information about this use. Support your main idea with facts and details. Reach a conclusion at the end.

 e-Journal Research and write about it online at **www.macmillanmh.com**

Math in Science

Graphing Wavelengths of Light

Have you ever looked at a rainbow and wondered about the colors? Why do they always appear in the same order? It's because the colors show up in order of wavelength, with the longest wavelength on the outside. Use the information in the table to find out the order of the colors in a rainbow.

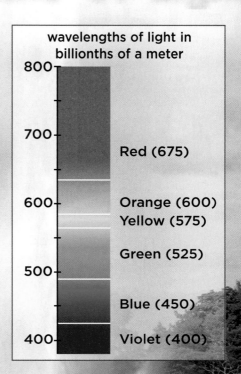

wavelengths of light in billionths of a meter

800
700 — Red (675)
600 — Orange (600)
Yellow (575)
Green (525)
500 —
Blue (450)
400 — Violet (400)

Make a Bar Graph

To make a bar graph using data

▶ have each axis represent one variable

▶ if an axis has numbers, use even increments (for example: 350, 400, 450, 500...) and label the units

▶ use your data to draw a bar of the correct height for each point on the horizontal axis

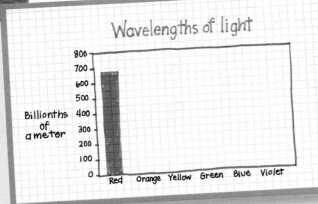

Wavelengths of light

Solve It

1. Which color has the longest wavelength? What is it?

2. What is the difference in wavelength between yellow and orange light?

3. Make a bar graph using the colors and wavelengths listed in the chart.

663
EXTEND

Electricity

Theater of Electricity, Boston Museum of Science

Look and Wonder

A Van de Graaff generator can create giant arcs of electrons. How could you control this much energy?

Which bulbs does each switch control?

Purpose

A bulb will light if there is an unbroken path through it from one end of a battery to the other. You will examine several different electrical paths with switches. You will then predict which light bulbs will be lit when a switch is opened or closed.

Procedure

1. Assemble the electric circuit as shown in the diagram, and leave all the switches open.

2. **Predict** Examine the top switch. When it closes, which bulbs will have an unbroken path from one end of a battery to another? Which bulbs will light when the switch closes? Record your prediction.

3. **Experiment** Close the top switch and record your observations. Then open the switch.

4. Repeat steps 2 and 3 for the other switches.

Draw Conclusions

5. **Interpret Data** Look at the observations you wrote down. How many of your predictions were correct? For any that were incorrect, explain what was wrong in your thinking.

Explore More

Which switch should be closed to provide the most light from a single bulb? What if you could close more than one switch? Design a procedure to test which closed switches produce the most light. Follow your procedure and record your results.

Materials

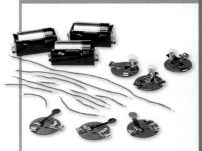

- **three switches**
- **three 1.5-volt light bulbs and stands**
- **three 1.5-volt batteries and stands**
- **insulated wire with stripped leads**

Step 1

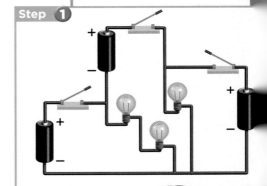

Step 2

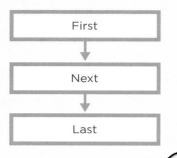
What is static electricity?

Have you ever shocked yourself by touching a doorknob on a cold, dry day? A spark jumps to your finger! Lightning during a thunderstorm is a larger version of this spark. Both are examples of electricity. **Electricity** is the movement of electrons. Its energy is measured in joules (J). We use units called volts (V) to measure how strongly electrons will move.

You may remember that in an atom, there are protons and electrons. Protons have a positive charge (+) and electrons have a negative charge (-). Particles with opposite charges are attracted to each other. Particles with the same charge are repelled from each other. When two objects rub against each other, electrons are sometimes knocked off one object and onto the other. This causes static electricity. **Static electricity** is the buildup of charged particles.

Static Electricity

A buildup of electrons on a shoe will discharge back to the carpet they came from.

Read a Diagram

Does the shoe have an overall charge?

Clue: Count the number of protons and electrons.

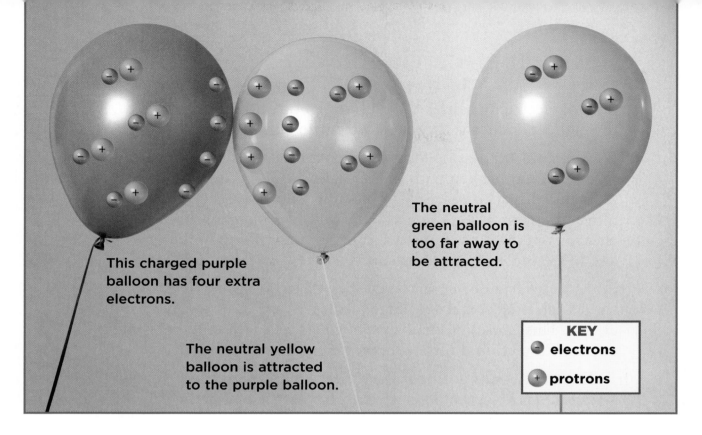

This charged purple balloon has four extra electrons.

The neutral yellow balloon is attracted to the purple balloon.

The neutral green balloon is too far away to be attracted.

KEY
● electrons
⊕ protrons

The attractive force between electrons and protons is strong. Static electricity causes electrons to jump through the air toward nearby protons. A spark is formed. The electrons have been discharged and returned to the protons. The objects are now neutral. A neutral object has equal numbers of protons and electrons.

Two oppositely charged objects will stick together; this is called *static cling*. It happens often when clothes rub together in a tumble dryer. Charged objects can also attract neutral objects. How? When a charged object nears a neutral object, it pulls on one type of charge and pushes on the other. Then the neutral object will act like it is slightly charged on one side and attract the charged object.

When static electricity forms on a good electrical conductor, like a metal, the charges can move freely.

Like charges will push on each other and spread themselves out. When static electricity forms on an electrical insulator, charges cannot move freely.

Earth is a large neutral conductor. You can protect objects from static electricity (including lightning) by grounding them to Earth with a wire. **Grounding** occurs when a conductor shares its excess charge with a much larger conductor. Instead of building up static electricity, grounded objects pass their charge onto Earth. The charges then can spread out so that they are barely noticed.

 Quick Check

Sequence What happens when a balloon with excess electrons is brought closer to a wall?

Critical Thinking What would happen if two oppositely charged conductors touched?

How can electricity flow?

When you're in the dark, a flashlight comes in handy. A flow of electricity causes the bulb to light. A flow of electricity through a conductor is called an **electric current**.

A **circuit** (SUR•kit) is formed when an electric current passes through an unbroken path of conductors. Often the path of a circuit consists of wires. Circuits must also have a device to move electrons along the path. Such devices increase the volts of electrons in the circuit and are called *voltage sources*. Batteries are an example of a voltage source.

A *switch* is a device that can open or close the path. When the switch is closed, the voltage pushes on the electrons in the circuit. This causes electrons to move. Protons feel a force in the opposite direction. Protons, however, are not free to move.

Electricity does not flow the same way through every part of a circuit. An object in an electrical circuit that resists the flow of electrons is called a **resistor** (ri•ZIS•tuhr). Resistance is measured in units called *ohms* (Ω). Electrons lose energy when moving through a resistor. This energy can be transformed into heat or light. A light bulb is a resistor.

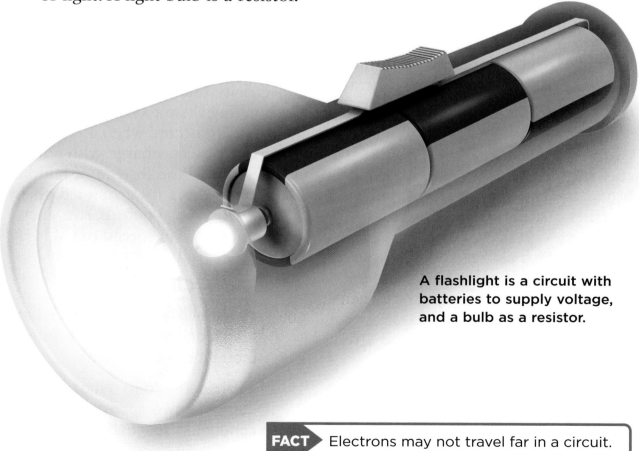

A flashlight is a circuit with batteries to supply voltage, and a bulb as a resistor.

FACT ▷ Electrons may not travel far in a circuit.

Electric current in a circuit travels fast—almost at the speed of light. Electrons, however, travel just a few millimeters per second. Why? Electrons only need to move far enough to push another electron. That electron then pushes another and so on.

The amount of electric charge moving in a circuit is measured in units called *amperes* or *amps* (A). There are about six billion billion electrons moving every second in one amp of current. Even currents as small as 0.05 A can seriously hurt you, so be careful!

 Quick Check

Sequence How does energy change form in a flashlight?

Critical Thinking How is the resistance of a resistor like friction?

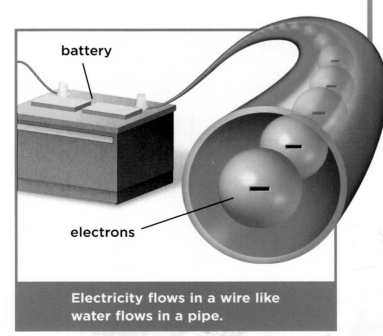

battery

electrons

Electricity flows in a wire like water flows in a pipe.

≡Quick Lab

Measuring Electric Current

1 Build a flashlight circuit using a battery, a switch, and a light bulb.

2 **Observe** Close the switch and record your findings.

3 Open the circuit and add another battery. Make sure the positive end of one battery touches the negative end of the other.

4 Close the switch again. Is the light bulb the same brightness as before? Why?

5 **Infer** When was there more electricity flowing through the circuit? How do you know?

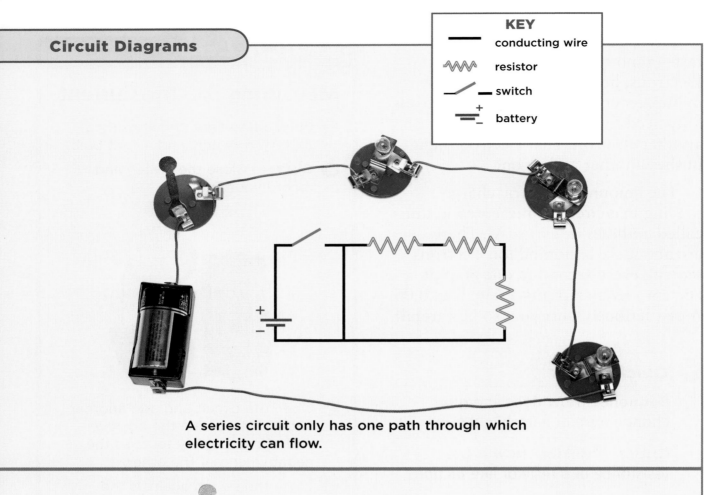

A series circuit only has one path through which electricity can flow.

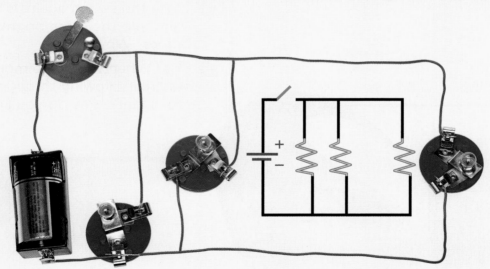

A parallel circuit has more than one path through which electricity can flow.

Read a Diagram

Which bulbs will be brightest when the switches are closed?

Clue: Which paths have the least resistance?

What kinds of circuits are there?

Look at the circuits in the photograph. The illustrations next to them are called *circuit diagrams*. Can you match the parts of the photo to the parts of the diagram?

If there is only one conductive path it is called a *series circuit* (SEER•eez). In this type of circuit, the resistance increases with each resistor added. Electricity travels through all the resistors one after another. As resistors are added, the energy each resistor receives is decreased.

Did you ever have a string of holiday lights where if one bulb was taken out, none of the other bulbs would light? Light strings like this are a series circuit. If appliances in our homes were wired in series circuits, it would cause problems. Turning off one appliance would turn off all of the others!

Circuits in your house are in parallel (PAR•uh•lel). A *parallel circuit* has more than one conductive path. Since there is more than one path the overall resistance of the circuit is smaller, and more current will flow.

Electricity flows through all paths in a parallel circuit at the same time. The smaller the resistance of the path, the more current flows through it. What happens if one of the paths is broken? The current flows through the remaining paths.

If a conductor accidentally forms a path in a circuit, it can short that circuit. A *short circuit* is a path with little or no resistance that connects the two ends of an electrical source. The tiny resistance in short circuits causes large currents to flow across them. These large currents can damage appliances, or start fires. Frayed wires are a common cause of short circuits.

This frayed wire is dangerous— it may cause a short circuit.

 Quick Check

Sequence What happens to the brightness of bulbs each time one is added into a series circuit?

Critical Thinking How would the electric current compare for two identical bulbs in a series circuit versus in a parallel circuit?

▲ Breakers protect circuits from too much current.

GFI outlets are often used in bathrooms.

How can you use electricity safely?

Plugging too many appliances into a power strip dangerously heats wires. Each time an appliance is plugged in, another branch is added to the parallel circuit. This increases the current. Adding too many appliances leads to currents large enough to start fires.

To protect against large currents, homes have fuses or breakers. A *fuse* is a wire that breaks if too much current flows through it. A *breaker* is a switch that opens when it detects too much current. Homes have separate fuses or breakers for different circuits.

Delicate electronics, like computers, are often plugged into surge protectors. Surge protectors prevent sudden spikes in current from entering electronics and damaging them.

In bathrooms and kitchens, outlets have small buttons saying "test" and "reset." These are part of a *ground fault interrupter* (in•tuh•RUP•tuhr) (GFI). A GFI is sensitive to changes in current. It will turn an outlet off if a short circuit forms. It will also turn the outlet off if electricity starts to flow through water.

The electrical energy that comes to your home through power lines is dangerous. Never reach up into power lines to get a toy that is stuck there. If you touch two power lines at the same time, or one power line and the ground, it can be deadly.

▼ Never go near fallen power lines.

 Quick Check

Sequence How might an electrical fire start?

Critical Thinking How is a fuse like a switch? How is it different?

Lesson Review

Visual Summary

Static electricity is a buildup of electric charge.

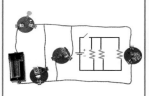

Electric current is a flow of electricity through a conductor.

Electricity flows in either **series** or **parallel circuits**.

Make a FOLDABLES™ Study Guide

Make a Three-Tab Book. Use the titles shown. Summarize what you have learned about each topic under the tab.

Static electricity is...

Electric current is...

Electricity flows in either...

Think, Talk, and Write

1 **Main Idea** Why will a comb rubbed with wool pick up bits of paper?

2 **Vocabulary** When a conductor shares its excess charge with a much larger conductor it is called _____.

3 **Sequence** What happens as objects rub together and form sparks?

First

↓

Next

↓

Last

4 **Critical Thinking** Do electrons from a battery reach a bulb before it lights?

5 **Test Prep** Adding branches to a parallel circuit
 A increases current.
 B decreases current.
 C keeps current the same.
 D reverses the direction of current.

6 **Test Prep** Which protects a home from high currents?
 A on/off switches
 B resistors
 C circuit breakers
 D electrical sources

Math Link

Lighting with Lightning
A small lightning bolt produces about 500,000,000 J of energy. A light bulb uses 100 J/s. How many hours could the lightning bolt keep the bulb lit?

History Link

Discovering Electricity
Benjamin Franklin performed many important experiments with electricity. Research and summarize those experiments.

Building a Better Battery

You find them in cars, in flashlights, and in radios. They're batteries: devices that store chemical energy and provide electricity. Over the years, batteries have changed from the inside out. They all have a basic structure in common, however. Positive and negative electrodes attach to wires, and an electrolyte provides the medium through which current can flow.

1748 CE
Benjamin Franklin
experiments with charged glass plates. He calls his device a *battery* because the shock it produces feels like a beating.

1859 CE

French physicist Raymond Gaston Planté takes two lead electrodes, separates them with rubber, and places them in a solution of sulfuric acid and water. This *wet-cell* battery uses its liquids as electrolytes. The battery is rechargeable, and cars today still use lead-acid batteries.

1800 CE
Alexander Volta,
an Italian nobleman, conducts experiments with battery designs. He uses copper and zinc discs separated by cloth soaked in salt water. This "cell" produces a steady current of electricity. Many such cells stacked together form a *voltaic pile battery.*

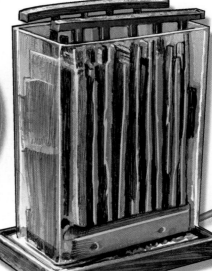

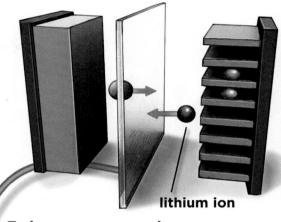

lithium ion

1959 CE

American researcher Lewis Urry uses powdered zinc and manganese dioxide to build a battery that provides more energy and lasts longer. Newer versions of this "alkaline battery" are used today to power devices like flashlights and remote controls.

Today, many companies make lithium-ion batteries. The lithium ions travel from one electrode to the other. Electrons move along with the ions and produce electricity.

Today's batteries are getting smaller and longer-lasting. They are also used in many new devices. Laptop computers need batteries that are lightweight and powerful. Lithium-ion batteries provide laptops with the energy they need for hours and can then be recharged.

Draw Conclusions

▶ Use information in the text and background knowledge.

▶ Support your conclusions with information found in the text.

 Write About It
Draw Conclusions

1. What makes batteries useful?
2. What is the electrolyte in a lead-acid battery?

LOG ON e-Journal Research and write about it online at www.macmillanmh.com

AMERICAN MUSEUM OF NATURAL HISTORY

Magnetism

Look and Wonder

The aurora borealis (or "Northern Lights") shines brightly in Finland. It is shaped by Earth's magnetism applying forces on solar particles. What other ways might magnetism apply forces?

How do magnets apply forces?

Make a Prediction

Magnets push and pull on other magnets. Where on a bar magnet do you think the strongest forces are felt? Write down your prediction.

Test Your Prediction

1 Observe Lay a sealed bag containing iron filings over a bar magnet. Do the iron filings form a pattern? Draw a sketch.

2 Experiment Hang one magnet from a meterstick. Take another magnet and move it toward the hanging magnet. Watch how it moves. Record your observations. Repeat for each side of the bar magnet.

3 Place a compass at 0 cm of a meterstick lying flat on a table. Align the meterstick west-east. Move a bar magnet from the 100 cm mark toward the compass. Record at what distance the compass first starts to move. Repeat for each side of the bar magnet.

Draw Conclusions

4 Interpret Data Look at all of your observations. Which support your prediction and which disprove your prediction? Explain. Was your prediction correct? Why or why not?

Explore More

Suppose you put two bar magnets in a line, the north pole of one touching the south pole of the other. Where do you think this double magnet would be strongest? Design an experiment to test your prediction and report on how accurate it was.

Materials

- bag
- iron filings
- 2 bar magnets
- string
- meterstick
- books
- compass

Step **2**

Step **3**

▶ **Main Idea**

Magnets have north and south poles that apply forces to other magnets and magnetic materials.

▶ **Vocabulary**

magnetism, p.678
magnetic field, p.679
electromagnet, p.680
generator, p.682
alternating current, p.682
magnetic levitation, p.684

e-Glossary
at www.macmillanmh.com

▶ **Reading Skill** ✔

Compare and Contrast

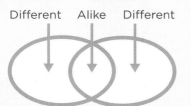

Different   Alike   Different

What is magnetism?

You're hiking in a forest. The trees are so thick that you can't see the Sun. How could you tell which way to go? Pull out your compass! Its needle tells you which direction is north. You're on your way!

How does a compass point north? The needle inside a compass is a magnet. **Magnetism** (MAG•ni•tiz•uhm) is the ability of an object to push or pull on another object that has the magnetic property. Magnets will also apply forces to certain metals like iron or nickel.

Magnets have two poles: north (N) and south (S). Like poles repel one another; different poles attract. You may notice that this is similar to what happens with electric charges. Magnetic poles always exist in north-south pairs. If you cut a magnet in half, each half will form a new magnet with two poles.

Are the pole names familiar? Earth has a North and South pole. Is Earth magnetic? Yes! Compass needles point toward Earth's North Pole. The geographic North Pole and magnetic north pole are in slightly different places, however.

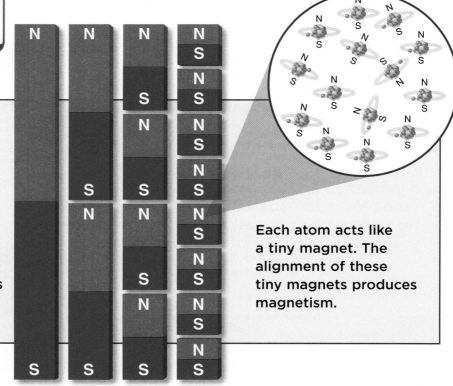

Cut a magnet in half and you will form new magnets with two poles.

Each atom acts like a tiny magnet. The alignment of these tiny magnets produces magnetism.

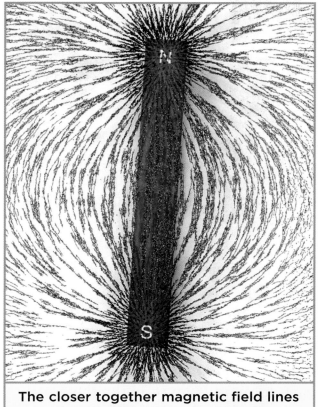

The closer together magnetic field lines are, the stronger the magnetic force.

Earth's magnetic field is like the field of a large permanent bar magnet.

Forming Magnets

Atoms also act like magnets. This magnetism comes from the properties and movement of electrons. In most materials, the north and south poles of atoms point in random directions. The forces from the random poles cancel each other out.

If the poles of many of the atoms line up in the same direction, a *permanent magnet* is formed. The forces from the aligned poles of the atoms add up and give a magnet its strength. The bar magnets you have used are permanent magnets.

Iron, nickel, cobalt, and a few other metals are attracted to magnets. Their atoms can line up to match the alignment of magnets. They will then act like weak magnets.

When small pieces of these metals are sprinkled over a magnet, they form lines. These lines are the directions of the magnetic forces around a magnet, and are called the **magnetic field**. The closer together these lines appear, the stronger the magnetic forces are in that area. The magnetic forces around Earth are similar to those of a bar magnet.

 Quick Check

Compare and Contrast How is Earth like or unlike a bar magnet?

Critical Thinking How could you make a piece of iron into a permanent magnet?

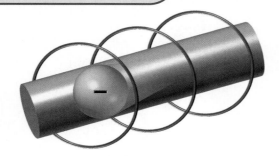

Electrons moving in a wire produce a magnetic field.

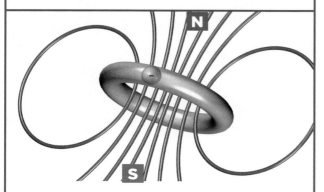

A loop of electric current will have north and south magnetic poles.

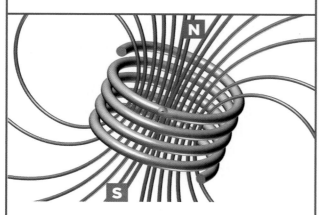

The magnetic field of a coil is like that of a bar magnet.

Read a Diagram

Which electromagnet has the strongest magnetic field?

Clue: Look at the magnetic field lines.

What are electromagnets?

What do doorbells, television sets, and electric motors have in common? They all use electromagnets. An **electromagnet** (i•LEK•troh•mag•nit) is an electric circuit that produces a magnetic field. The moving electrons in electricity generate magnetic fields. When the current stops, the magnetic field disappears.

The simplest electromagnet is a straight wire. The magnetic field circles around the wire when current is flowing. When you wrap a wire into a loop, you increase the strength of the magnetic field. Many loops together can make a coil. The magnetism from each loop adds up to make the coil a stronger electromagnet. The shape of its magnetic field is like a bar magnet's.

Placing a rod of iron in a coil will magnetize the iron. This adds to the strength of the electromagnet's magnetic field. Increasing the current also strengthens the field.

An iron rod in an electromagnet's coil is pulled toward the center of the coil. If you try to pull it out, it will spring back. This action is used in a number of devices, such as doorbells.

A *voice coil* operates audio speakers. The voice coil sits in a permanent magnetic field. Current changes in the coil alter its magnetic field. This causes the forces of the permanent magnet to move it back and forth. The coil is connected to a cone of paper or metal. The coil's vibrations make the cone move back and forth, creating sound waves in the air.

If several voice coils were placed in a circle, changes in electric current would cause them to rotate back and forth. Something very similar happens in electric motors. An axle is attached to many coils which are between two permanent magnets. Forces between the permanent magnets and the coils acting as electromagnets cause the coils to rotate. Electric motors are used in many devices, from ceiling fans to cars.

✓ Quick Check

Compare and Contrast How are electromagnets and permanent magnets alike and different?

Critical Thinking How could a coil of wire and an iron rod be turned into a doorbell?

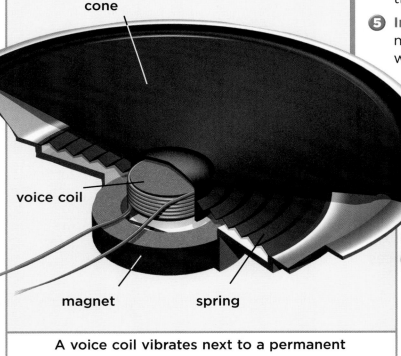

cone

voice coil

magnet   spring

A voice coil vibrates next to a permanent magnet to create sound in a speaker.

☰ *Quick Lab*

Building an Electromagnet

1. Coil a length of insulated wire around a pencil 25 times. Remove the pencil.

2. **Observe** Place a compass right under the wire coil. Turn the coil so that it is crosswise to the compass needle. Touch the ends of the wire to a battery. Write down what you observe.

3. Hold the ends of the wire to the battery and try to pick up small steel paper clips with the coil. What's the largest chain of paper clips you can lift?

4. Repeat steps 2 and 3 with a nail inserted into the coil. Then repeat the test with a longer coil.

5. **Interpret Data** How would you make the strongest electromagnet with the materials you used?

How can magnets produce electricity?

What would happen if you turned the axle of an electric motor by hand? You would be using the electric motor as a generator (JEN•uh•ray•tuhr). A **generator** is a device that creates electric current by spinning an electric coil between the poles of a magnet.

Energy is used to turn the axle of the generator. As the coil moves through the magnetic field, forces push on its electrons and create an electric current. Wires attached to the loop allow the current to flow as the loop rotates.

Whenever a loop moves past the pole of a magnet, the direction of the magnetic forces changes. This causes changes in the direction of the electric current. Electric current that rapidly changes directions is called **alternating current** (AWL•tuhr•nayt•ing).

In real generators, there are several coils of wire that spin past many magnets. U.S. generators produce alternating current that changes direction 120 times each second.

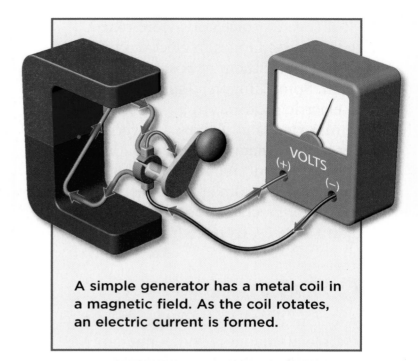

A simple generator has a metal coil in a magnetic field. As the coil rotates, an electric current is formed.

 Quick Check

Compare and Contrast What are the similarities and differences between electric motors and generators?

Critical Thinking What would happen in a generator if the permanent magnet rotated instead of the coils?

Devices called transformers use magnetism to lower the voltage to the 120 V used in homes.

Electric Generator

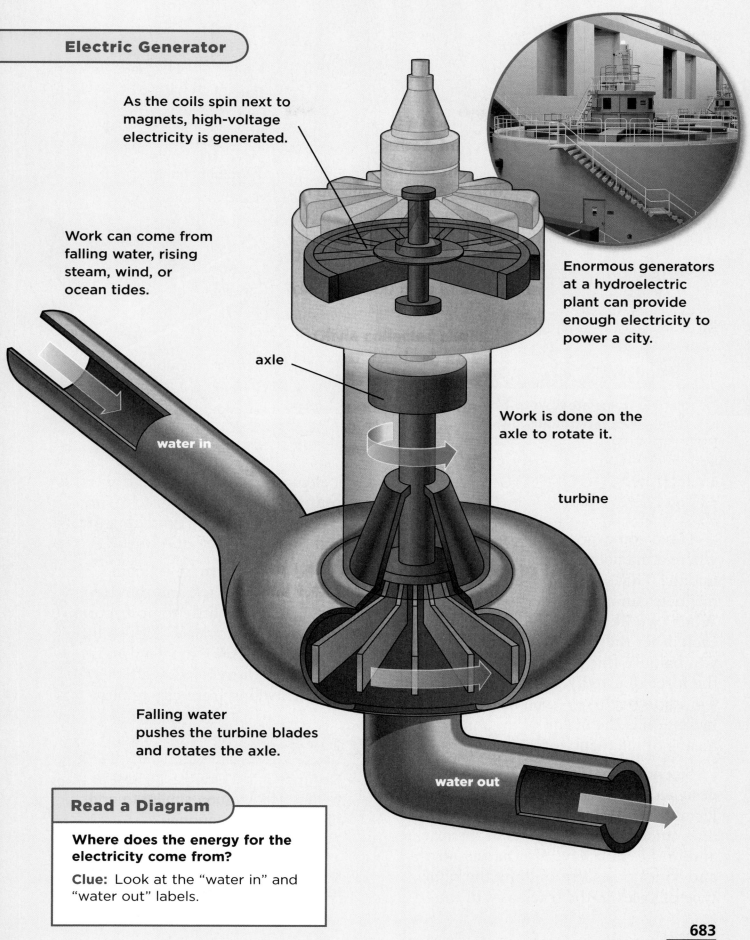

As the coils spin next to magnets, high-voltage electricity is generated.

Work can come from falling water, rising steam, wind, or ocean tides.

water in

axle

Enormous generators at a hydroelectric plant can provide enough electricity to power a city.

Work is done on the axle to rotate it.

turbine

Falling water pushes the turbine blades and rotates the axle.

water out

Read a Diagram

Where does the energy for the electricity come from?

Clue: Look at the "water in" and "water out" labels.

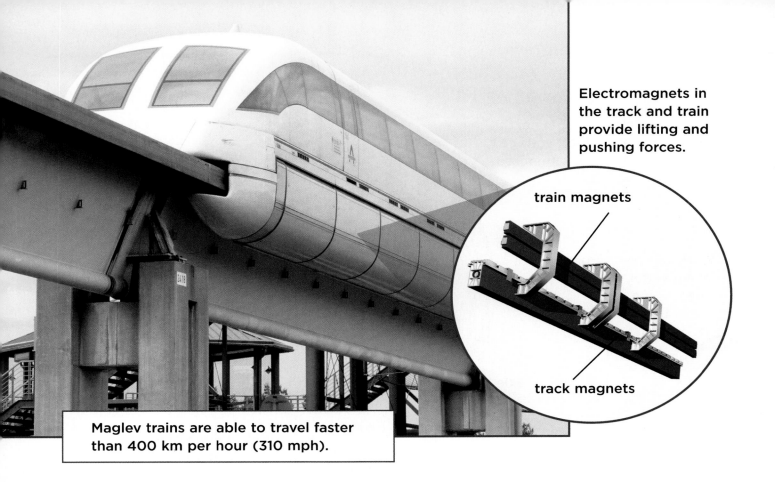

Electromagnets in the track and train provide lifting and pushing forces.

train magnets

track magnets

Maglev trains are able to travel faster than 400 km per hour (310 mph).

What is magnetic levitation?

Have you ever seen the magic trick where a magician floats someone in midair? That's just an illusion, but magnets can accomplish the real thing! When two like poles of magnets face each other they feel a pushing force. If you balance this pushing force against the force of gravity you have magnetic levitation (lev•i•TAY•shuhn). **Magnetic levitation** is the lifting of an object by means of magnetic forces.

Scientists and engineers have designed trains that use magnetic levitation (*maglev* for short) to travel on a track. Electromagnets in the track and in the train have alternating north and south poles. By aligning the right type of poles in the track and in the

train, electromagnets push the train a few centimeters above the track. The train moves forward by switching the poles in the track back and forth.

Maglev trains have no touching parts between the track and the train. This means there is little to no friction, though still air resistance. With so little energy lost to friction, maglev trains may be able to provide very efficient means of traveling from city to city.

 Quick Check

Compare and Contrast How is magnetic levitation similar to and different from buoyancy?

Critical Thinking How would electromagnetic poles be arranged in order to levitate a bar magnet?

Lesson Review

Visual Summary

Magnets have north and south poles and can apply forces on one another.

Electric currents create **electromagnets**.

Spinning a coil of wire in a magnetic field can **generate** electricity.

Make a FOLDABLES™ Study Guide

Make a Folded Chart. Use the titles shown. Summarize what you have learned in the boxes.

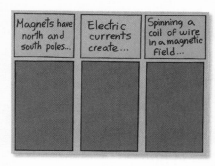

Think, Talk, and Write

1 **Main Idea** What happens when a bar magnet is cut in half?

2 **Vocabulary** Electric current that rapidly changes direction is called _____.

3 **Compare and Contrast** How are electric doorbells and speakers similar and different?

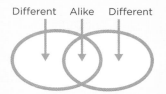

4 **Critical Thinking** How could heating a bar magnet reduce its magnetism?

5 **Test Prep** All of these increase the strength of an electromagnet EXCEPT

  A increasing the number of loops
  B adding an iron rod in the middle
  C increasing the resistance
  D increasing the electric current

6 **Test Prep** Which energy conversion happens in electric motors?

  A radiant to electrical
  B heat to mechanical
  C nuclear to electrical
  D electrical to kinetic

Math Link

Electromagnetic Forces

An electromagnetic coil can pick up 114 kg of iron. A strong bar magnet can pick up 33 kg of iron. What is the simplified ratio of their strengths?

Social Studies Link

Geography

Orienteering is a sport where you find your way to a location as fast as possible. Research and write a report on how magnetism is used in orienteering.

Be a Scientist

wire with leads

3-inch nail

battery

battery holder

compass

Structured Inquiry

How are electric current and electromagnets related?

Form a Hypothesis

A magnetic field is produced when a current is flowing in a circuit. An electromagnet can be produced in this way. Electromagnets produce a magnetic field similar to bar magnets. When the current stops, the magnetic field disappears.

Each electromagnet has a north and a south pole. A compass needle also has a north and south pole. The compass needle will point to the appropriate poles of other magnets. How do you think the direction of electric current affects the poles of the electromagnet? Write a hypothesis in the form "If the direction of electric current is reversed, then the poles on an electromagnet..."

Test Your Hypothesis

1. Coil the wire around the nail 35 times clockwise toward the flat end. Leave about 10 cm of straight wire at both ends.

2. Find the straight part of the wire near the flat side of the nail. Connect that end of the wire to the positive side of the battery.

3. Lay the compass on the flat end of the nail. Press the unconnected wire to the negative side of the battery. Record what happens.

4. Find the wire from the flat end of the nail. Disconnect it from the positive side of the battery and connect it to the negative side. Keep the compass on the flat end of the nail. Press the other end of the wire to the positive side of the battery. Record what happens.

Step ①

Step ②

Draw Conclusions

5 **Infer** Where did the compass point in step 3 and step 4? What do you think happened to the poles of the electromagnet?

6 **Communicate** Draw a picture of the electromagnet before and after the current was reversed. Mark to which side of the battery the wires were connected. Label the poles of the electromagnet as north or south.

How is an electromagnet affected by the direction of its coils?

Form a Hypothesis

Are the poles of an electromagnet only dependent on electric current? How does the direction in which a coil is wound affect an electromagnet? Write your answer as a hypothesis in the form "If the direction in which a coil is wound is reversed, then the poles of the electromagnet..."

Test Your Hypothesis

Design an experiment to investigate the effect changing the direction of coils will have on an electromagnet. Write out the materials you need and the steps you will follow. Record your observations.

Draw Conclusions

Did your results support your hypothesis? Explain.

What can you learn about electromagnets? For example, how are electromagnets used in electric motors? Determine the materials needed for your investigation. Your experiment should be written so that another group can complete it by following your instructions.

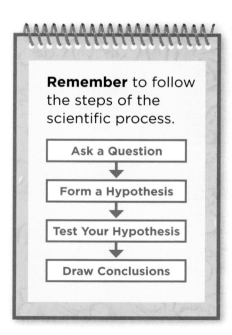

Remember to follow the steps of the scientific process.

Ask a Question
↓
Form a Hypothesis
↓
Test Your Hypothesis
↓
Draw Conclusions

Visual Summary

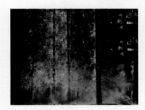

Lesson 1 Heat flows between objects when they have different temperatures.

Lesson 2 Sounds are produced by vibrating objects.

Lesson 3 Light travels as waves, but can also be described as particles.

Lesson 4 Electricity is a form of energy and can flow in a circuit.

Lesson 5 Magnets have north and south poles that apply forces to other magnets and magnetic materials.

Make a **FOLDABLES**™ Study Guide

Assemble your lesson study guide as shown. Don't forget to include your Lesson 5 study guide in the back.

Fill in each blank with the best term from the list.

amplitude, p. 644

circuit, p. 668

conduction, p. 628

electromagnet, p. 680

generator, p. 682

photon, p. 653

pitch, p. 642

refraction, p. 657

static electricity, p. 666

temperature, p. 626

1. A tiny bundle of energy by which light travels is called a(n) _____.

2. An unbroken path of conductors carrying an electric current is a(n) _____.

3. An electric circuit that produces a magnetic field is a(n) _____.

4. The average energy of molecules in an object is its _____.

5. A musical note will sound high or low depending on its _____.

6. When solid objects touch, heat can pass through _____.

7. The height of a wave is called its _____.

8. The bending of waves as they pass from one substance to another is called _____.

9. Lightning can result after a large buildup of _____.

10. A hydroelectric dam creates electricity when water powers its _____.

Answer each of the following in complete sentences.

11. Compare and Contrast What are the similarities and differences between heat and temperature?

12. Summarize How are the colors created in the rainbow below?

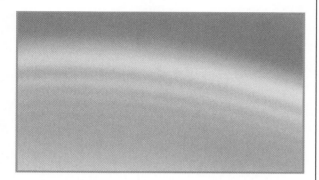

13. Form a Hypothesis Suppose that the lights in one room of your house went out but were still on in another room. Form a hypothesis to explain this. Tell how you might test your hypothesis.

14. Critical Thinking What would be the advantages and disadvantages of using magnetic levitation to run a public transportation system?

15. Expository Writing Write a paragraph explaining how echolocation works and provide an example.

16. What forms does energy take? Define each one and give an example of how we use it in our everyday lives.

Space Waves

Find out about the different kinds of waves that form the electromagnetic spectrum. These include:

radio waves	microwaves
infrared waves	visible waves
ultraviolet rays	X rays
gamma rays	cosmic rays

1. Use reference books or the Internet to research the characteristics of each.

2. Make a chart comparing and contrasting them. Your chart should compare wavelength, frequency, and two other variables, and provide an example of each kind of wave.

Analyze Your Results

▶ Write a paragraph about your results based on your chart.

Test Prep

1. Which of the following statements about reflection is false?

A Lenses in eyeglasses use reflection to make objects appear in focus.

B Incoming and reflected angles are equal.

C Lighter colors reflect more light than darker colors.

D Reflection is like an organized scattering of light.

Careers in Science

Mechanical Engineer

If you like to take things apart to find out how they work, you are in good company—that of mechanical engineers. They design, build, and operate many types of machines, such as those that run refrigerators, elevators, automobiles, and electric generators. Mechanical engineers also develop robots. Robotic machines do many kinds of jobs, from heavy manufacturing to delicate heart surgery. To qualify for a career in robotics, start developing your math, science, and computer skills while you are in school. In college and graduate school, you would specialize in mechanical engineering.

▼ Mechanical engineers build all types of machines—including some that go into space.

Auto Mechanic

Have you ever been in a car that broke down? If so, then it was probably an auto mechanic who came to the rescue. Auto mechanics do not only make repairs. They also service cars to keep them in good running condition. To make repairs, an auto mechanic uses hand tools, such as pliers and screwdrivers, power tools, such as pneumatic wrenches and welding equipment, and computers. Auto mechanics need to be good problem solvers, too. If you want to become an auto mechanic, the more you learn about automobiles, the better. Some schools offer training programs which allow you to earn a mechanic's certificate along with your high school diploma.

▲ An auto mechanic uses tools, including computers, to fix cars.

LOG ON e-Careers at www.macmillanmh.com

Reference

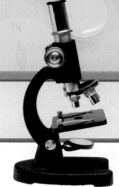

Science Handbook

Measurements

Units of Measurement

Table of Measurements		
International System of Units (SI)		**Customary Units**
Temperature Water freezes at 0°C (degrees Celsius) and boils at 100°C		**Temperature** Water freezes at 32°F (degrees Fahrenheit) and boils at 212°F
Length and Distance 1,000 meters (m) = 1 kilometer (km) 100 centimeters (cm) = 1 meter (m) 10 millimeters (mm) = 1 centimeter (cm)		**Length and Distance** 5,280 feet (ft) = 1 mile (mi) 3 feet (ft) = 1 yard (yd) 12 inches (in.) = 1 foot (ft)
Volume 1,000 milliliters (mL) = 1 liter (L) 1 cubic centimeter (cm^3) = 1 milliliter (mL)		**Volume** 4 quarts (qt) = 1 gallon (gal) 2 pints (pt) = 1 quart (qt) 2 cups (c) = 1 pint (pt) 8 fluid ounces (oz) = 1 cup (c)
Mass 1,000 grams (g) = 1 kilogram (kg)		**Mass (and Weight)** 2,000 pounds (lb) = 1 ton (T) 16 ounces (oz) = 1 pound (lb)
Weight 1 kilogram (kg) weighs 9.81 newtons (N)		

Making Measurements

Temperature

You use a thermometer to measure temperature. A thermometer is made of a thin tube with a liquid inside that is usually red in color. When the liquid inside the tube gets warmer, it expands and moves up the tube. When the liquid gets cooler, it contracts and moves down the tube.

1. Look at the thermometer shown here. It has two scales—a Fahrenheit scale and a Celsius scale.

2. What is the temperature on the thermometer? At what temperature does water freeze on each scale?

Length

1. Look at the ruler below. Each centimeter is divided into 10 millimeters. Estimate the length of the paper clip.

2. The length of the paper clip is about 3 centimeters plus 8 millimeters. You can write this length as 3.8 centimeters.

Try estimating the length of some objects found in your classroom. Then measure the length of the objects with a ruler. Compare your estimates with accurate measurements.

Time

You use timing devices to measure how long something takes to happen. Two timing devices are a clock with a second hand and a stopwatch. A clock with a second hand is accurate to one second. A stopwatch is accurate to parts of a second.

Measurements

Measuring Mass, Weight, and Volume

Mass

Mass is the amount of matter that makes up an object. You can use a balance to measure mass. To find the mass of an object, you balance or compare it with masses you know.

1 Place the balance on a level surface. Check that the two pans are empty, clean, and balanced with each other. The pointer should point to the middle mark. If it does not, move the slider to the right or left until the pans are balanced.

2 Gently place the object you want to measure in the left pan. The left pan will then move lower.

3 Now add masses to the right pan until both pans are balanced again. Add and get the total mass in the right pan. This total is the mass of the object in grams.

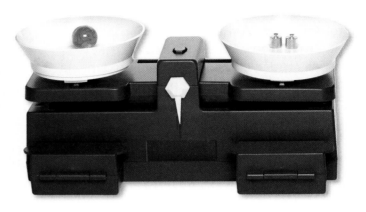

Weight

You use a spring scale to measure weight. Weight equals the amount of gravity pulling down on the mass of an object. Therefore weight is a force. Weight is measured in newtons (N).

1 To find the weight of your object, hold the spring scale by the top. Determine the weight of the empty plastic cup. Add the object to the cup.

2 Subtract the first measurement from the second, and the difference is the weight of the rock.

Volume

1 You can use a beaker or graduated cylinder to find the volume of a liquid.

2 You can also find the volume of a solid such as a rock. Add water to a beaker or graduated cylinder. Gently slide the object down into the beaker. To find the volume of the rock, subtract the starting volume of the liquid from the new volume. The difference is equal to the volume of the rock.

Collecting Data

Microscopes

1 Look at the photograph to learn the different parts of your microscope.

2 Always carry a microscope with both hands. Hold the arm of the microscope with one hand, and put your other hand beneath the base. Place the microscope on a flat surface.

3 Move the mirror so that it reflects light from the room up toward the stage. Never point the mirror directly at the Sun or a bright light. ⚠**Be careful.** Bright light can cause permanent eye damage.

4 Place a small piece of newspaper on a slide. Put the slide under the stage clips. Be sure that the area you are going to examine is over the hole in the stage.

5 Look through the eyepiece. Turn the focusing knob slowly until the newspaper comes into focus.

6 Draw what you see through the microscope.

7 Look at other objects through the microscope. Try a piece of leaf, a human hair, or a pencil mark.

Other Lenses

You use a hand lens to magnify an object, or make the object look larger. With a hand lens, you can see more detail than you can without the lens. Look at a few grains of salt with a hand lens and draw what you see. Binoculars are a tool that makes distant objects appear closer. In nature, scientists use binoculars to look at animals without disturbing them. These animals may be dangerous to approach or frightened at the approach of people. Cameras can act like binoculars or they can be used to see things up close. Cameras have the advantage of making a record of your observations. Cameras can make a record on film or as data on a computer chip.

Use Calculators

Sometimes after you make measurements, you have to analyze your data to see what they mean. This might involve doing calculations with your data. A handheld calculator helps you do calculations quickly and accurately, and can also be used to verify your own calculations.

Hints

- Make sure the calculator is on and previous calculations have been cleared.

- To add a series of numbers, press the + sign after you enter each number. After you have entered the last number, press the = sign to find the sum.

- If you make a mistake while putting numbers in, press the clear entry key. You can then enter the correct number.

- To subtract, enter the first number, then the – sign. Then enter the number you want to subtract. Then press the = sign for the difference.

- To multiply, enter the first number, then the ⊠ sign and enter the second number you want to multiply by. Then press the = sign for the product.

- To divide, enter the number you want to divide, press the ÷ sign and enter the number you want to divide by. Then press the = sign for the quotient.

- You can also find averages and percents with a calculator, and verify your own work.

Use Computers

A computer has many uses. You can write a paper on a computer. You can use programs to organize data and show your data in a graph or table. The Internet connects your computer to many other computers and databases around the world. You can send the paper you wrote to a friend in another state or another country. You can collect all kinds of information from sources near and far. Best of all, you can use a computer to explore, discover, and learn. You can also get information from computer disks that can hold large amounts of information. You can fit the information found in an entire encyclopedia set on one disk.

One class used computers to work on a science project. They were able to collect data from students in another state who were working on a similar project, and share their data with them. They were also able to use the Internet to write to local scientists and request information. The students collected and stored their data, moved paragraphs around, changed words, and made graphs. Then they were able to print their report to share their discoveries with others.

Organizing Data

Use Graphs

When you do an experiment in science, you collect information, or data. To find out what your data means, you can organize it into graphs. There are several different kinds of graphs. You can choose a type of graph that best organizes your data and makes it easier for you and for others to understand the data presented.

Bar Graphs

A bar graph uses bars to show information. For example, what if you performed an experiment to test the strength of a nail electromagnet and the number of coils of wire wrapped around it? This graph shows that increasing the number of coils increases the strength of the electromagnet.

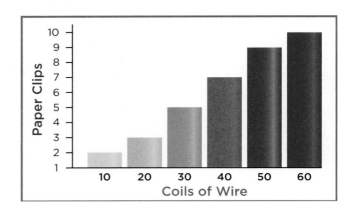

Circle Graphs

A circle graph is used to show how a complete set of data is divided into parts. This circle graph shows how water is used in the United States. In a circle graph, all the data must add up to 100.

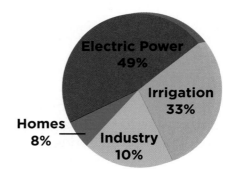

Line Graphs

A line graph shows information by connecting dots plotted on a graph. A line graph is often used to show changes that occur over time. For example, this line graph shows the relationship between temperature and time for a particular morning.

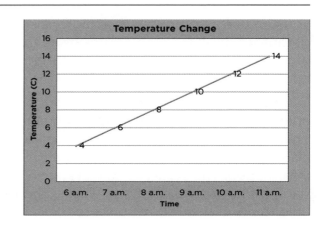

Use Tables and Maps

Tables

Tables help you organize data during experiments. Most tables have columns that run up and down, and rows that run horizontally. The columns and rows have headings that tell you what kind of data goes in each part of the table. The table here shows a record of the conductivity of several different kinds of substances.

Material	Thermal Conductivity
Aluminum	109.0
Copper	385.0
Wood	0.1
Packing foam	0.01

Idea Maps

This kind of map shows how ideas or concepts are connected to each other. Idea maps help you organize information about a topic. The idea map shown here connects different ideas about rocks.

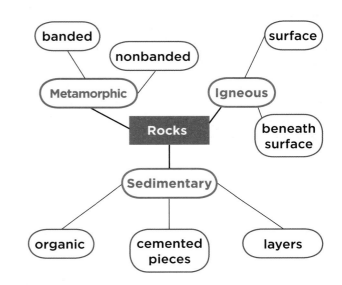

Maps

A map is a drawing that shows an area from above. Maps help you learn about a location. You are probably most familiar with road maps, which are often used to plan ways to travel from one place to another. Other kinds of maps show terrain. Hills and valleys, for example, can be shown on some types of maps. A good map also has a legend that shows the scale it was made to, and also a compass point that shows the direction of north and sometimes other directions as well.

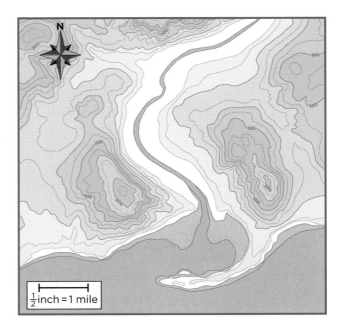

Human Body Systems

Organization of the Human Body

Like all organisms, humans are made up of cells. In fact, the human body is made of trillions of cells. These cells are organized into tissues, a group of similar cells that perform a specific function. The cardiac muscle in your heart is an example of tissue. Tissues, in turn, form organs. Your heart and lungs are examples of organs. Finally, organs work together as part of organ systems. For example, your heart and blood vessels are part of the circulatory system. The organ systems in the human body all function together to keep the body healthy.

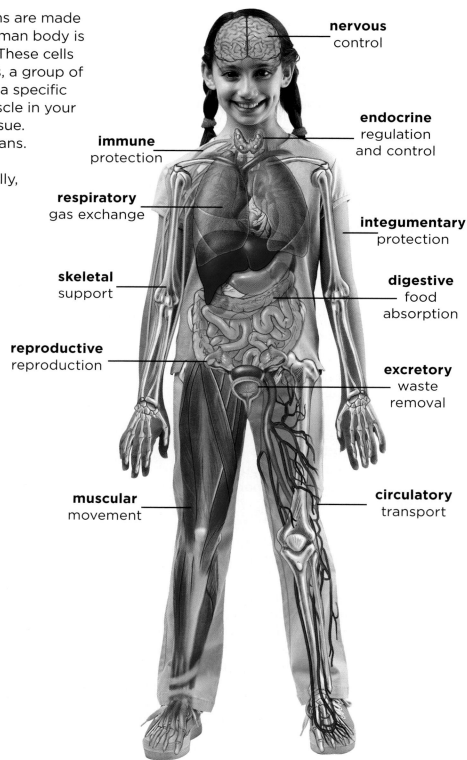

nervous
control

endocrine
regulation
and control

immune
protection

integumentary
protection

respiratory
gas exchange

digestive
food
absorption

skeletal
support

excretory
waste
removal

reproductive
reproduction

muscular
movement

circulatory
transport

The Skeletal and Muscular Systems

The body has a supporting frame called a skeleton, which is made up of bones. The skeleton gives the body its shape, protects organs in the body, and works with muscles to move the body.

Each of the 206 bones of the skeleton is the size and shape best fitted to do its job. For example, long and strong leg bones support the body's weight.

Three types of muscles make up the body—skeletal muscle, cardiac muscle, and smooth muscle. Cardiac muscles are found only in the heart. These muscles contract to pump blood throughout the body.

Smooth muscles make up internal organs such as the intestines, as well as blood vessels.

The muscles that are attached to and move bones are called skeletal muscles. Skeletal muscles pull bones to move them. Most muscles work in pairs to move bones.

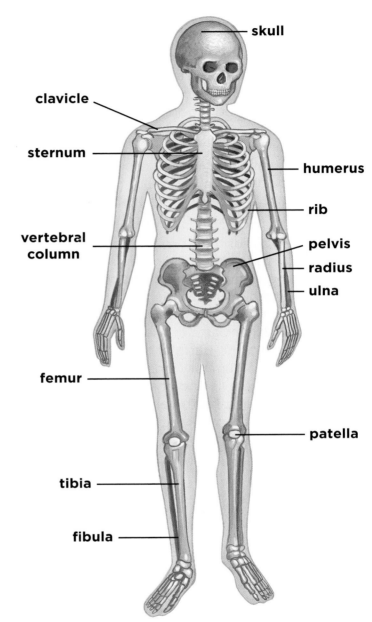

skull
clavicle
sternum
humerus
rib
vertebral column
pelvis
radius
ulna
femur
patella
tibia
fibula

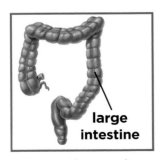

large intestine

smooth muscle

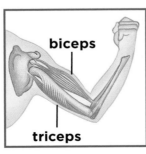

biceps
triceps

skeletal muscles

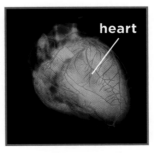

heart

cardiac muscle

Human Body Systems

The Circulatory and Respiratory Systems

The circulatory system consists of the heart, blood vessels, and blood. Circulation is the flow of blood through the body. Blood is a liquid that contains red blood cells, white blood cells, and platelets. Red blood cells carry oxygen and nutrients to cells. They also carry CO_2 and cellular wastes away from the cells. White blood cells work to fight germs that enter the body. Platelets are cell fragments that help make the blood clot.

The heart is a muscular organ about the size of a fist. Arteries carry blood away from the heart. Some arteries carry blood to the lungs, where red blood cells pick up oxygen. Other arteries carry oxygen-rich blood from the lungs to all other parts of the body. Veins carry blood from other parts of the body back to the heart. Blood in most veins carries the wastes released by cells and has little oxygen. Blood flows from arteries to veins through narrow vessels called capillaries.

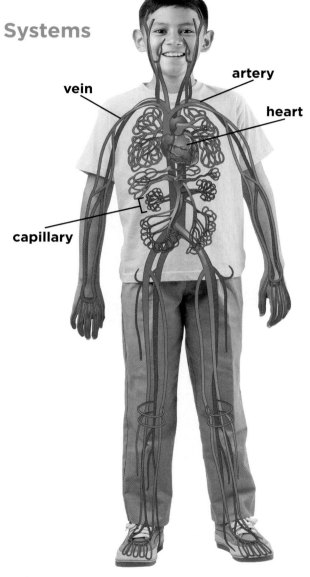

vein artery heart capillary

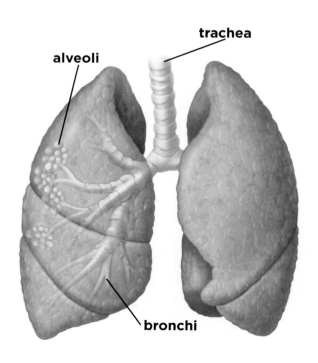

trachea alveoli bronchi

The process of getting and using oxygen in the body is called respiration. When a person inhales, air is pulled into the nose or mouth. The air travels down into the trachea. In the chest the trachea divides into two bronchial tubes. One bronchial tube branches into smaller tubes called bronchioles. At the end of each bronchiole are tiny air sacs called alveoli. The alveoli exchange carbon dioxide for oxygen.

The Digestive and Excretory Systems

Digestion is the process of breaking down food into simple substances the body can use. Digestion begins when a person chews food. Chewing breaks the food down into smaller pieces and moistens it with saliva. Food passes through the esophagus and into the stomach. The stomach mixes digestive juices with food before passing it on to the small intestine.

Digested food is absorbed in the small intestine. The walls of the small intestine are lined with villi, which are fingerlike projections. Digested food is absorbed through the surface of the villi. From the villi the blood transports nutrients to every part of the body. Water is absorbed from undigested food in the large intestine.

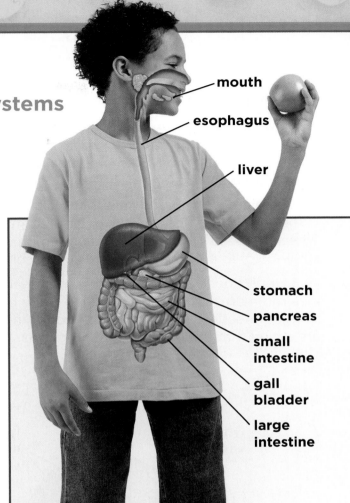

mouth
esophagus
liver
stomach
pancreas
small intestine
gall bladder
large intestine

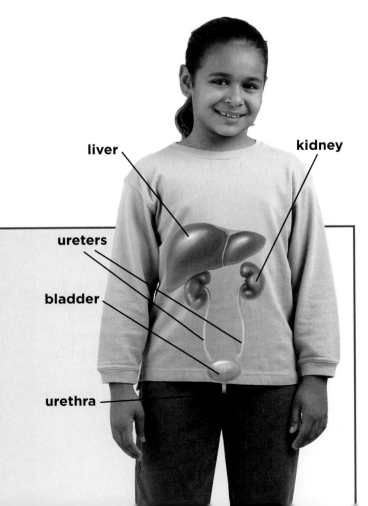

liver
kidney
ureters
bladder
urethra

Excretion is the process of removing waste products from the body. The liver filters nitrogen wastes from the blood and converts them into urea. Urea is then carried by the blood to the kidneys for excretion. Each kidney contains more than a million nephrons. Nephrons are structures in the kidneys that filter blood.

The skin takes part in excretion when a person sweats. Glands in the inner layer of skin produce sweat. Sweat is mostly water. There is also a tiny amount of urea and mineral salts in sweat.

Human Body Systems

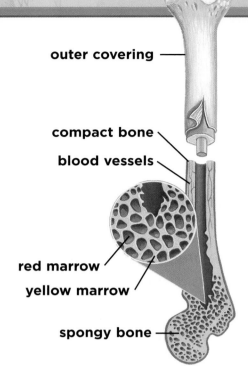

outer covering

compact bone

blood vessels

red marrow

yellow marrow

spongy bone

The Immune System

The immune system helps the body fight disease. A soft tissue known as red marrow fills the spaces in some bones. Red marrow makes new red blood cells, platelets that stop a cut from bleeding, and germ-fighting white blood cells.

There are white blood cells in the blood vessels and in the lymph vessels. Lymph vessels are similar to blood vessels. Instead of blood, they carry lymph. Lymph is a straw-colored fluid that surrounds body cells.

Lymph nodes filter out harmful materials in lymph. Like red marrow, they also produce white blood cells to fight infections. Swollen lymph nodes in the neck are a clue that the body is fighting germs.

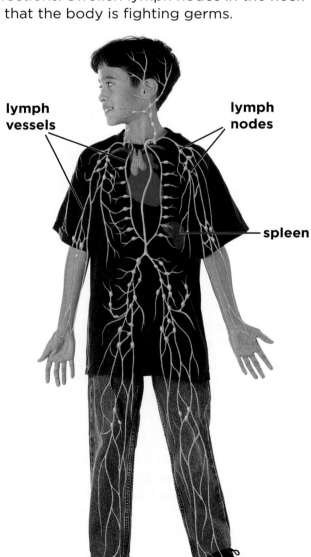

lymph vessels

lymph nodes

spleen

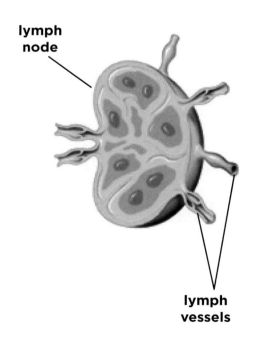

lymph node

lymph vessels

Communicable Diseases

A disease is anything that interferes with the normal functions of the body. Some diseases are caused by harmful materials in the environment. Many diseases, however, are caused by microscopic organisms and viruses and can be passed from person to person. This type of disease is called a communicable or infectious disease.

Disease-causing agents are called pathogens. Pathogens include many types of bacteria, as well as viruses. Diseases caused by pathogens are also called communicable diseases, because they can be passed from one person to another. Pathogens must enter the body before they can cause an illness. Once these invaders enter the body, the immune system works very hard to fight them off.

Human Infectious Diseases		
Disease	**Caused by**	**Organ System Affected**
common cold	virus	respiratory system
chicken pox	virus	skin
smallpox	virus	skin
polio	virus	nervous system
rabies	virus	nervous system
influenza	virus	respiratory system
measles	virus	skin
mumps	virus	digestive system and skin
tuberculosis	bacteria	respiratory system
tetanus	bacteria	muscular system
meningitis	bacteria or virus	nervous system
gastroenteritis	bacteria or virus	digestive and excretory system

Human Body Systems

The Nervous System

The nervous system has two parts. The brain and the spinal cord make up the central nervous system. All other nerves make up the outer, or peripheral, part of the nervous system.

The largest part of the human brain is the cerebrum. A deep groove separates the right half, or hemisphere, of the cerebrum from the left half. Both the right and left hemispheres of the cerebrum contain control centers for the senses. The cerebrum is the part of the brain where thought occurs.

The cerebellum lies below the cerebrum. It coordinates the skeletal muscles so they work smoothly together. It also helps in keeping balance.

The brain stem connects to the spinal cord. The lowest part of the brain stem is the medulla. It controls heartbeat, breathing, blood pressure, and the muscles in the digestive system.

The spinal cord is a thick band of nerves that carries messages to and from the brain. Nerves branch off from your spinal cord to all parts of your body. The spinal cord also controls reflexes. A reflex is a quick reaction that occurs without waiting for a message to and from the brain. For example, if you touch something hot, you pull your hand away without thinking about it.

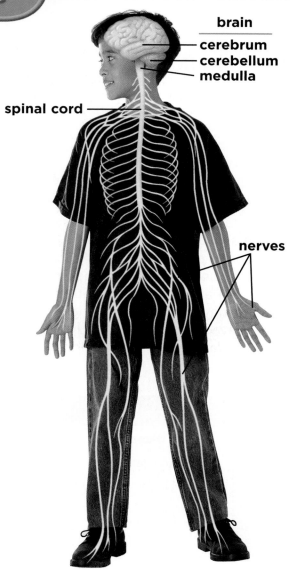

brain
cerebrum
cerebellum
medulla

spinal cord

nerves

Parts of a Neuron

The nerves in the nervous system are made up of nerve cells called neurons. Each neuron has three main parts—a cell body, dendrites, and an axon. Dendrites are branching nerve fibers that carry impulses, or electrical signals, toward the cell body. An axon is a nerve fiber that carries impulses away from the cell body.

When an impulse reaches the tip of an axon, it must cross a tiny gap to reach the next neuron. This gap between neurons is called a synapse.

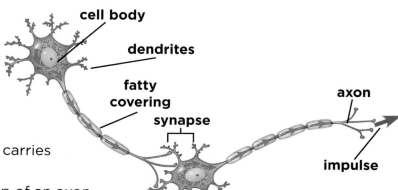

cell body

dendrites

fatty covering

synapse

axon

impulse

Stimulus and Response

The nervous system, the skeletal system, and the muscular system work together to help you adjust to your surroundings. Anything in the environment that requires your body to adjust is called a stimulus (plural: stimuli). A reaction to a stimulus is called a response.

As you learned, nerve cells are called neurons. There are three kinds of neurons: sensory, associative, and motor. Each kind does a different job to help your body respond to stimuli. Sensory neurons receive stimuli from your body and the environment. Associative neurons connect the sensory neurons to the motor neurons. Motor neurons carry signals from the central nervous system to the organs and glands.

In addition to responding to external stimuli, your body also responds to internal changes. Your body regulates its internal environment to maintain a stable condition for survival. This is called a steady-state condition.

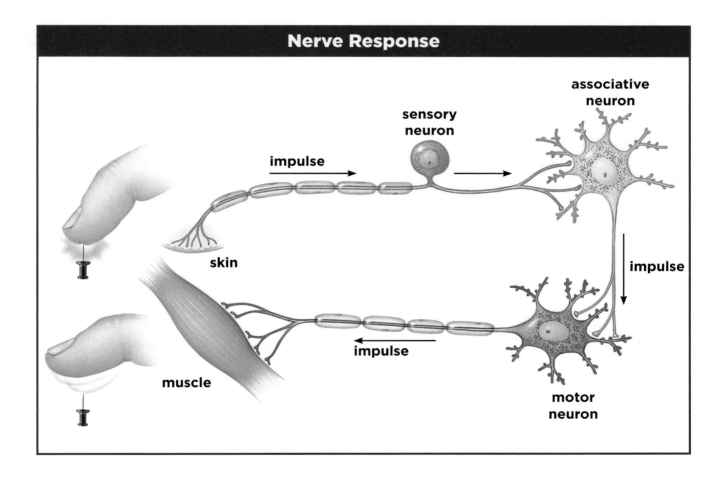

Nerve Response

impulse

skin

sensory neuron

associative neuron

impulse

muscle

impulse

motor neuron

Human Body Systems

The Senses

Sense of Sight

Light reflected from an object enters your eye and falls on the retina. Receptor cells change the light into electrical signals, or impulses. These impulses travel along the optic nerve to the vision center of the brain.

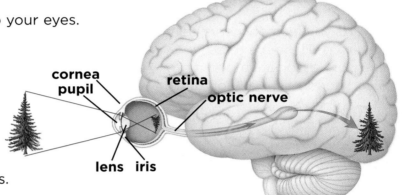

(1) Light reflects off the tree and into your eyes.

(2) The light passes through your cornea and the pupil in your iris.

(3) The lens bends the light so that it hits your retina.

(4) Receptor cells on your retina change the light into electrical signals.

(5) The impulses travel along neurons in your optic nerve to the seeing center of your brain.

Sense of Hearing

Sound waves enter your ear and cause the eardrum to vibrate. Receptor cells in your ear change the sound waves into impulses that travel along the auditory nerve to the hearing center of the brain.

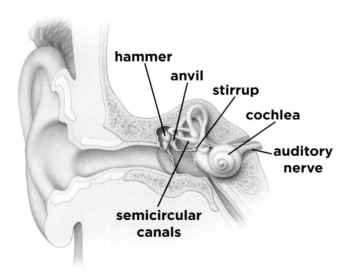

(1) Your outer ear collects sound waves.

(2) They travel down your ear canal.

(3) The eardrum vibrates.

(4) Three tiny ear bones vibrate.

(5) The cochlea vibrates.

(6) Receptor cells inside your cochlea change.

(7) The impulses travel along your auditory nerve to the brain's hearing center.

Sense of Smell

The sense of smell is really the ability to detect chemicals in the air. When you breathe, chemicals dissolve in mucus in the upper part of your nose or nasal cavity. When the chemicals come in contact with receptor cells, the cells send impulses along the olfactory nerve to the smelling center of the brain.

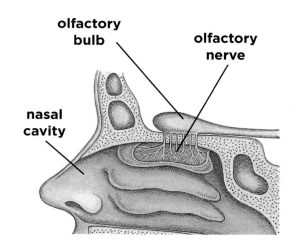

Sense of Taste

When you eat, chemicals in the food dissolve in saliva. Saliva carries the chemical to taste buds on the tongue. Inside each taste bud are receptors that can sense the four main tastes—sweet, sour, salty, and bitter. The receptors send impulses along a nerve to the taste center of the brain. The brain identifies the taste of the food, which is usually a combination of the four main taste categories.

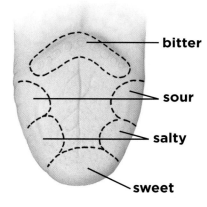

Sense of Touch

Receptor cells in the skin help a person tell hot from cold, wet from dry. These can also tell the light touch of a feather or the pressure of stepping on a stone. Each receptor cell sends impulses along sensory nerves to the spinal cord. The spinal cord then sends the impulses to the touch center of the brain.

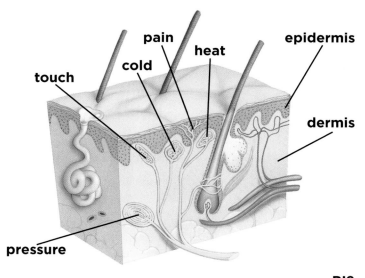

Human Body Systems

The Endocrine System

Hormones are chemicals that control body functions. An organ that produces hormones is called an endocrine gland.

The endocrine glands are scattered around the body. Each gland makes one or more hormones. Every hormone seeks out a target organ or organ system, the place in the body where the hormone acts. Changing levels of different hormones communicate important messages to target organs and organ systems.

The endocrine glands help to maintain a constant healthy condition in your body. These glands can turn the production of hormones on or off whenever your body produces too little or too much of a particular hormone.

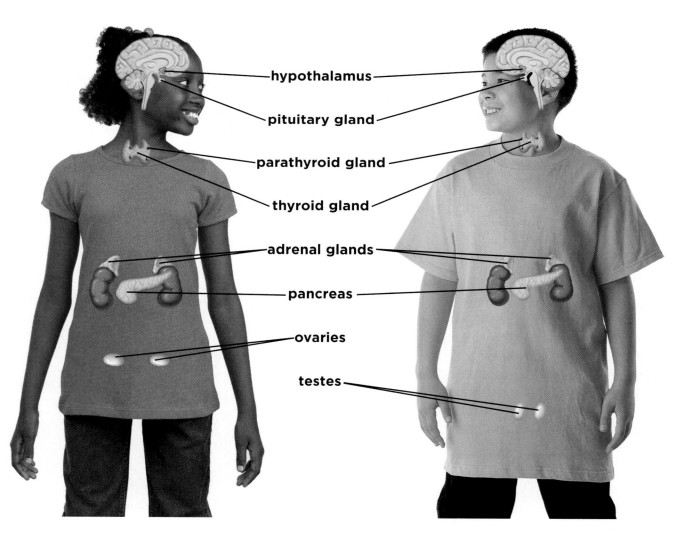

hypothalamus

pituitary gland

parathyroid gland

thyroid gland

adrenal glands

pancreas

ovaries

testes

by Dinah Zike

Folding Instructions

The following pages offer step-by-step instructions to make the Foldables study guides.

Half-Book

1. Fold a sheet of paper ($8\frac{1}{2}''$ x 11") in half.
2. This book can be folded vertically like a hot dog or …
3. … it can be folded horizontally like a hamburger

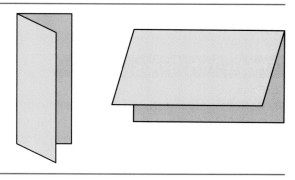

Folded Book

1. Make a Half-Book.
2. Fold in half again like a hamburger. This makes a ready-made cover and two small pages inside for recording information.

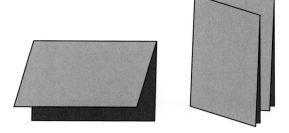

Pocket Book

1. Fold a sheet of paper ($8\frac{1}{2}''$ x 11") in half like a hamburger.
2. Open the folded paper and fold one of the long sides up two inches to form a pocket. Refold along the hamburger fold so that the newly formed pockets are on the inside.
3. Glue the outer edges of the two-inch fold with a small amount of glue.

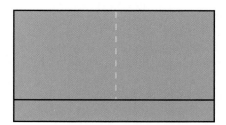

Shutter Fold

1. Begin as if you were going to make a hamburger, but instead of creasing the paper, pinch it to show the midpoint.
2. Fold the outer edges of the paper to meet at the pinch, or midpoint, forming a Shutter Fold.

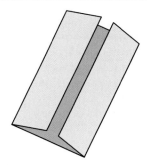

Trifold Book

1. Fold a sheet of paper ($8\frac{1}{2}$" x 11") into thirds.
2. Use this book as is, or cut into shapes.

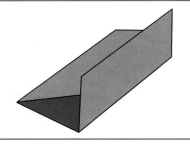

Three-Tab Book

1. Fold a sheet of paper like a hot dog.
2. With the paper horizontal and the fold of the hot dog up, fold the right side toward the center, trying to cover one half of the paper.
3. Fold the left side over the right side to make a book with three folds.
4. Open the folded book. Place one hand between the two thicknesses of paper and cut up the two valleys on one side only. This will create three tabs.

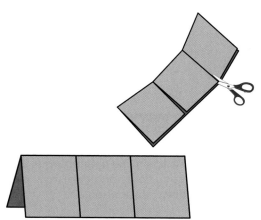

Layered-Look Book

1. Stack two sheets of paper ($8\frac{1}{2}$" x 11") so that the back sheet is one inch higher than the front sheet.
2. Bring the bottoms of both sheets upward and align the edges so that all of the layers or tabs are the same distance apart.
3. When all the tabs are an equal distance apart, fold the papers and crease well.
4. Open the papers and glue them together along the valley, or inner center fold, or staple them along the mountain.

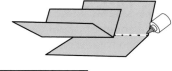

Folded Table or Chart

1. Fold the number of vertical columns needed to make the table or chart.
2. Fold the horizontal rows needed to make the table or chart.
3. Label the rows and columns.

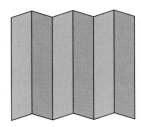
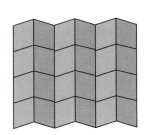

Glossary

Use this glossary to learn how to pronounce and understand the meanings of Science Words used in this book. Page numbers tell you where to find words in the book.

A

absolute age (ab′sə·lüt′ āj) A rock's age in years. (p. 328)

absorption (ab·sôrp′shən) The transfer of energy when a wave disappears into a surface. (p. 641)

acceleration (ak·sel′ə·rā′shən) Change in velocity with respect to time. (p. 576)

acid (as′id) A substance that tastes sour and turns blue litmus paper red. (p. 554)

acidity (ə·sid′i·tē) The strength of an acid. (p. 556)

action force (ak′shən fôrs) The initial push or pull of one object on another object. (p. 592)

adaptation (ad′əp·tā′shən) A characteristic that helps an organism to survive in its environment. (p. 168)

air mass (âr mas) A large region of the atmosphere in which the air has similar properties. (p. 384)

air pressure (âr presh′ər) The force put on a given area by the weight of the air above it. (p. 367)

alkalinity (al′kə·lin′i·tē) The strength of a base. (p. 556)

alloy (al′oi) A solution of a metal and at least one other solid which is often also a metal. (p. 532)

alternating current (ôl′tər·nāt·ing kûr′ənt) Electric current that changes directions many times per second. (p. 682)

alternative energy source (ôl·tûr′nə·tiv en′ər·jē sôrs) A source of energy other than the burning of a fossil fuel. (p. 332)

amplitude (am′pli·tüd′) The height of a sound wave, which determines its volume. (p. 644)

angiosperm (an′jē·ə·spûrm′) A seed plant that produces flowers. (p. 49)

aquifer (ak′wə·fər) An underground layer of rock or soil filled with water. (p. 343)

asexual reproduction (ā·sek′shü·əl rē′prə·duk′shən) The production of new organisms from only one parent. (p. 91)

Pronunciation Key
The following symbols are used throughout the Macmillan/McGraw-Hill Science Glossaries.

a	**a**t	e	**e**nd	o	h**o**t	u	**u**p	hw	**wh**ite	ə	**a**bout
ā	**a**pe	ē	m**e**	ō	**o**ld	ū	**u**se	ng	so**ng**		tak**e**n
ä	f**a**r	i	**i**t	ôr	f**or**k	ü	r**u**le	th	**th**in		penc**i**l
â	c**a**re	ī	**i**ce	oi	**oi**l	u̇	p**u**ll	th	**th**is		lem**o**n
ô	l**a**w	îr	p**ier**ce	ou	**ou**t	ûr	t**ur**n	zh	mea**s**ure		circ**u**s

′ = primary accent; shows which syllable takes the main stress, such as **kil** in **kilogram** (kil′ ə gram′).

′ = secondary accent; shows which syllables take lighter stresses, such as **gram** in **kilogram**.

---

asteroid (as'tə·roid') One of many small, rocky objects between Mars and Jupiter. (p. 450)

asymmetrical (ā'si·met'ri·kəl) Cannot be divided into mirror images. (p. 62)

atmosphere (at'məs·fîr') The layers of gases that surround Earth. (p. 246)

atom (at'əm) The smallest unit of an element that has the properties of that element. (p. 491)

balanced forces (bal'ənst fôrs'əz) Forces that act together on an object without changing its motion. (p. 588)

barometer (bə·rom'i·tər) A device for measuring air pressure. (p. 374)

base (bās) A substance that tastes bitter and turns red litmus paper blue. (p. 555)

benthos (ben'thos) Organisms that live on the bottom in aquatic ecosystems. (p. 220)

bilateral symmetry (bī·lat'ər·əl sim'ə·trē) A body plan in which an organism can be divided along only one plane of its body to produce two mirror images. (p. 63)

biome (bī'ōm) One of Earth's large ecosystems, with its own kind of climate, soil, plants, and animals. (p. 208)

black hole (blak hōl) An object whose gravity is so strong that light cannot escape it. (p. 460)

blizzard (bliz'ərd) A snowstorm with 35-mile-per-hour winds and enough snowfall that you can only see up to one quarter of a mile. (p. 396)

boiling point (boil'ing point) The temperature at which a substance changes state from a liquid to a gas. (p. 522)

buoyancy (boi'ən·sē) The upward push of a liquid or gas on an object placed in it. (p. 483)

cambium (kam'bē·əm) The layer in plants that separates the xylem from the phloem. (p. 53)

camouflage (kam'ə·fläzh') An adaptation in which an animal protects itself against predators by blending in with the environment. (p. 173)

carbon cycle (kär'bən sī'kəl) The continuous exchange of carbon dioxide and oxygen among living things. (p. 186)

carrier (kar'ē·ər) An individual who has inherited a factor for a trait but does not show the trait physically. (p. 128)

carrying capacity (kar'ē·ing kə·pas'i·tē) The maximum population size that an ecosystem can support. (p. 157)

cell (sel) The smallest unit of living matter. (p. 22)

cellular respiration (sel'yə·lər res'pə·rā'shun) The process in which energy is released from food (sugar) inside a cell. (p. 56)

chemical change (kəm'i·kəl chānj) A change of matter that occurs when atoms link together in a new way, creating a new substance different from the original substances. (p. 544)

chlorophyll (klôr'ə·fil') A green chemical in plant cells that allows plants to use the Sun's energy to make food. (p. 27)

cinder-cone volcano (sin'dər cōn vol·kā'nō) A steep-sided mountain that forms from explosive eruptions of thick lava. (p. 264)

circuit (sûr'kit) A loop formed when electric current passes through an unbroken path of conductors. (p. 668)

circulatory system (sûr′kyə·lə·tôr′ē sis′təm) The organ system that consists of the heart and blood vessels and that moves blood through the body. (p. 78)

classification (klas′ə·fi·kā′shən) The science of organizing categories for living things. (p. 34)

climate (klī′mit) The average weather pattern of a region. (p. 408)

climax community (klī′maks kə·mū′ni·tē) The final stage of succession in an area, unless a major change happens. (p. 201)

colloid (kol′oid) A type of mixture in which the particles of one material are scattered through another and block the passage of light without settling out. (p. 531)

comet (kom′it) A mixture of ice, frozen gases, rock, and dust left over from the formation of the solar system. (p. 450)

commensalism (kə·men′sə·liz·əm) A relationship between two kinds of organisms that benefits one without harming the other. (p. 161)

community (kə·mū′ni·tē) All the living things in an ecosystem. (p. 143)

complete metamorphosis (kəm·plēt′ met′ə·môr′fə·sis) A series of four distinct growth stages. (p. 114)

composite volcano (kəm·poz′it vol·kā′nō) A mountain formed from alternating eruptions of lava and ash. (p. 265)

compost (kom′pōst) A mixture of dead plant material that can be used as fertilizer. (p. 190)

compound (kom′pound) A substance that is formed by the chemical combination of two or more elements. (p. 542)

compound machine (kom′pound mə·shēn′) A combination of two or more simple machines. (p. 616)

condensation (kon′den·sā′shən) The changing of a gas into a liquid. (p. 184)

conduction (kən·duk′shən) The passing of heat through a material while the material itself stays in place. (p. 628)

conifer (kon′ə·fər) Any of a group of gymnosperms that produces seeds in cones and has needlelike leaves. (p. 108)

conservation (kon′sər·vā′shən) Saving, protecting, or using natural resources wisely. (p. 320)

constellation (kon′stə·lā′shən) Any of the patterns formed by groups of stars in the night sky. To people in the past, these patterns looked like pictures of animals or people. (p. 462)

Little Dipper

Polaris

Big Dipper

convection (kən·vek′shən) The flow of heat through a liquid or gas, caused by hot parts rising and cooler parts sinking. (p. 628)

corrosion (kə·rō′zhən) The gradual wearing away of a metal by combining with nonmetals in its environment. (p. 505)

crust (krust) The rocky surface that makes up the top of the lithosphere and includes the continents and the ocean floor. (p. 246)

current (kûr′ənt) An ocean movement; a large stream of water that flows in the ocean. (p. 410)

cyclone (sī′klōn) Any storm with a low-pressure center that causes a circular pattern of winds to form. (p. 401)

D

deciduous forest (di·sij′ü·əs   fôr′ist) A forest biome with many kinds of trees that lose their leaves each autumn. (p. 213)

density (den′si·tē) The amount of matter in a certain volume of a substance; found by dividing the mass of an object by its volume. (p. 482)

deposition (dep′ə·zish′ən) The process of dropping off pieces of eroded rock. (p. 288)

desert (dez′ərt) A sandy or rocky biome, with little precipitation and little plant life. (p. 209)

dicot (dī′kot) An angiosperm with two cotyledons in each seed. (p. 107)

digestive system (di·jes′tiv   sis′təm) The organ system that breaks down food so it can be absorbed into the bloodstream. (p. 76)

distillation (dis′tə·lā′shən) The process of separating the parts of a mixture by evaporation and condensation. (p. 535)

dominant trait (dom′ə·nənt   trāt) The form of an inherited trait that masks the other form of the same trait. (p. 126)

ductility (duk·til′i·tē) The ability to be pulled into thin wires without breaking. (p. 505)

E

earthquake (ûrth′kwāk′) A sudden shaking of Earth's crust. (p. 272)

echolocation (ek′ō·lō·kā′shən) Finding an object by using reflected sound. (p. 646)

ecosystem (ēk′ō·sis′təm) All the living and nonliving things in an environment, including their interactions with each other. (p. 142)

efficiency (i·fish′ən·sē) The measure of how much useful work a machine puts out compared to the amount of work put into it. (p. 616)

effort (ef′ərt) The force applied to a machine. (p. 608)

electric current (i·lek′trik   kûr′ənt) A flow of electricity through a conductor. (p. 668)

electricity (i·lek·tris′i·tē) The movement of electrons. (p. 666)

electrolyte (i·lek′trə·līt′) A substance that forms ions when dissolved and that conducts electricity. (p. 558)

electromagnet (i·lek′trō·mag′nit) An electric circuit that produces a magnetic field. (p. 680)

electromagnetism (i·lek′trō·mag′ni·tiz′əm) The way electric and magnetic forces interact. (p. 660)

electron (i·lek′tron) A particle in the space outside the nucleus of an atom that carries one unit of negative electric charge. (p. 492)

element (el′ə·mənt) A pure substance that cannot be broken down into any simpler substances through chemical reactions. (p. 490)

El Niño (el   nēn′yō) A change in weather conditions caused by the sinking of the cold current in the Pacific Ocean. Higher tides, heavy rains, and storms occur along the coasts of North and South America. (p. 412)

embryo (em′brē·ō) A developing organism that results from fertilization. (p. 106)

endangered species (en·dān′jərd   spē′shēz) A species that is in danger of becoming extinct. (p. 199)

endocrine system (en′də·krin   sis′təm) The set of ductless organs that secrete hormones directly into the bloodstream to regulate life processes. (p. 80)

energy (en'ər·jē) The ability to perform work or change an object. (p. 600)

energy pyramid (en'ər·jē   pir'ə·mid') A diagram that shows the amount of energy available at each level of an ecosystem. (p. 148)

epicenter (ep'i·sen'tər) The point on Earth's surface directly above the focus of an earthquake. (p. 273)

era (îr'ə) Long stretches of time used to measure Earth's geological history. (p. 329)

erosion (i·rō'zhən) The process of carrying away soil or pieces of rocks. (p. 286)

estuary (es'chü·er·ē') The boundary where a freshwater ecosystem meets a saltwater ecosystem. (p. 226)

evaporation (i·vap'ə·rā'shən) The change of a liquid into a gas below the boiling point. (p. 184)

excretory system (ek'skri·tôr'ē sis'təm) The organ system that removes cellular wastes from the body. (p. 77)

external fertilization (ek·stûr'nəl fûr'tə·lə·zā'shən) The process in which sperm and egg cells come together outside the female's body. (p. 116)

extinct species (ek·stingkt' spē'shēz) A species that has died out completely. (p. 198)

fault (fôlt) A deep crack in Earth's crust. (p. 256)

fertilization (fûr'tə·lə·zā'shən) The joining of a sperm cell with an egg cell to make one new cell, a fertilized egg. (p. 90)

floodplain (flud'plān) Land near a river that is likely to be under water during a flood. (p. 290)

focus (fō'kəs) The point where an earthquake starts. (p. 273)

food chain (füd   chān) The path that energy and nutrients follow in an ecosystem. (p. 144)

food web (füd   web) The overlapping food chains in an ecosystem. (p. 146)

force (fôrs) Any push or pull by one object on another. (p. 584)

fossil (fos'əl) Any remains or imprint of living things from the past. (p. 326)

fossil fuel (fos'əl   fū'əl) A fuel formed from the decay of ancient forms of life. (p. 327)

frame of reference (frām   uv   ref'ər·əns) A group of objects from which a position or motion is measured. (p. 573)

freezing point (frēz'ing   point) The temperature at which a substance changes state from a liquid to a solid. (p. 523)

frequency (frē'kwən·sē) The number of times an object vibrates per second. (p. 642)

friction (frik'shən) A force that opposes the motion of one object moving past another. (p. 587)

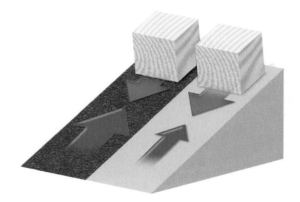

front (frunt) The boundary between two air masses. (p. 384)

fulcrum (fúl'krəm) The pivot point of a lever. (p. 610)

galaxy (gal′ək·sē) A collection of billions of stars. (p. 465)

gene (jēn) A portion of a chromosome that contains chemical instructions for inherited traits. (p. 126)

generator (jen′ə·rā′tər) A device that creates electric current by spinning an electric coil between the poles of a magnet. (p. 682)

geologist (jə·ol′ə·jist) A scientist who uses rocks to learn about the history and structure of Earth. (p. 252)

germination (jer′mə·nā′shən) The development of a seed into a new plant. (p. 106)

glacier (glā′shər) A large sheet of ice and snow that moves slowly over land. (p. 287)

global wind (glō′bəl wind) A wind that blows steadily in a predictable direction over long distances. (p. 370)

grassland (gras′land′) A biome where grasses, not trees, are the main plant life. (p. 214)

gravity (grav′i·tē) The force of attraction between any two objects due to their mass. (p. 422)

grounding (ground′ing) Connecting an object to Earth with a conducting wire to prevent the buildup of static electricity. (p. 667)

groundwater (ground′ wô′tər) Precipitation that seeps into the ground and is stored in tiny holes, or pores, in soil and rocks. (p. 184)

gymnosperm (jim′nə·spûrm′) A seed plant that does not produce flowers. (p. 49)

habitat (hab′i·tat) The place where a plant or animal lives and grows. (p. 158)

hardness (härd′nəs) How well a mineral resists scratching. (p. 303)

heat (hēt) Thermal energy that flows between objects due to a difference in temperature. (p. 626)

heredity (hə·red′i·tē) The passing down of inherited traits from one generation to the next. (p. 124)

hot spot (hot spot) A stationary location in Earth's mantle where magma melts through a tectonic plate. (p. 266)

humidity (hū·mid′i·tē) The amount of water vapor in the air. (p. 369)

humus (hū′məs) Decayed plant or animal material in soil. (p. 317)

hurricane (hûr′i·kān′) A very large, swirling storm with very low pressure at the center, and wind speeds greater than 119 km/h. (p. 400)

hydrosphere (hī′drə·sfîr′) Earth's water, whether found on continents or in oceans, including the fresh water in ice, lakes, rivers, and underground. (p. 246)

igneous rock (ig′nē·əs rok) A rock formed when magma or lava cools and hardens. (p. 306)

image (im′ij) A "picture" of the light source that light rays make in bouncing off a polished, shiny surface. (p. 656)

incomplete metamorphosis (in′kəm·plēt′ met·ə·môr′phə·sis) A series of three growth stages that occur gradually. (p. 115)

inertia (i·nûr′shə) The tendency of a moving object to keep moving in a straight line or of any object to resist a change in motion. (p. 423)

inherited trait (in·her′i·təd trāt) A characteristic that is passed from parent to offspring. (p. 124)

inner core (in′ər kôr) A solid layer of iron and nickel inside Earth. (p. 246)

insolation (in′sə·lā′shən) The amount of the Sun's energy that reaches Earth. (p. 364)

instinct (in′stingkt′) An inherited behavior, one that is not learned but is done automatically. (p. 124)

internal fertilization (in·tûr′nəl fûr′tə·lə·zā′shən) The joining of sperm and egg cells inside a female's body. (p. 117)

intertidal zone (in′tər·tī′dəl zōn) The shallowest part of the ocean ecosystem, where the ocean floor is covered and uncovered as the tide goes in and out. (p. 224)

invertebrate (in·vûr′tə·brit) An animal that does not have a backbone. (p. 36)

ion (ī′ən) An electrically charged atom or molecule with unequal numbers of protons and electrons. (p. 554)

island arc (ī′lənd ärk) A string of volcanic islands made from melted rock rising up from beneath the sea floor. (p. 266)

island chain (ī′lənd chān) A formation of islands aligned above a volcanic hot spot. (p. 266)

kinetic energy (ki·net′ik en′ər·jē) The energy of a moving object. (p. 600)

kingdom (king′dəm) The largest group into which an organism can be classified. (p. 34)

landform (land′fôrm′) A physical feature on Earth's surface. (p. 240)

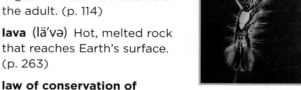

larva (lär′və) An immature stage in complete metamorphosis where the organism does not resemble the adult. (p. 114)

lava (lä′və) Hot, melted rock that reaches Earth's surface. (p. 263)

law of conservation of energy (lô uv kon′sər·vā′shən uv en′ər·jē) Energy may change form but it cannot be created or destroyed. (p. 602)

light-year (līt′yîr′) The distance light travels in a year. (p. 463)

limiting factor (lim′it·ing fak′tər) Anything that controls the growth or survival of a population. (p. 156)

load (lōd) The object being moved by a machine. (p. 608)

lunar eclipse (lü′nər i·klips′) A situation that occurs when the Sun, Earth, and Moon are in a straight line and Earth's shadow falls across the Moon. (p. 435)

luster (lus′tər) The way a mineral reflects light from its surface. (p. 302)

M

magma (mag′mə) Hot, melted rock below Earth's surface. (p. 254)

magnetic field (mag·net′ik fēld) A region of magnetic force around a magnet, represented by lines. (p. 679)

magnetic levitation (mag·net′ik lev′i·tā′shən) The lifting of an object by means of magnetic forces. (p. 684)

magnetism (mag′ni·tiz′əm) The ability of an object to push or pull on another object that has the magnetic property. (p. 678)

magnitude (mag′ni·tüd′) The amount of energy released by an earthquake. (p. 276)

malleability (mal′ē·ə·bil·i·tē) The ability to be bent, flattened, hammered, or pressed into new shapes without breaking. (p. 504)

mantle (man′təl) A nearly melted layer of hot rock below Earth's crust. (p. 246)

marsupial (mär·sü′pē·əl) A mammal in which the female has a pouch where offspring develop after birth. (p. 68)

mass (mas) A measure of the amount of matter in an object. (p. 480)

matter (mat′ər) Anything that has mass and takes up space. (p. 481)

meander (mē an′dər) A bend or S-shaped curve in a river. (p. 288)

medium (mē′·dē·əm) Substance through which a wave travels. (p. 639)

melting point (melt′ing point) The particular temperature at which a substance changes state from a solid to a liquid. (p. 522)

metal (met′əl) Any of a group of elements that conducts heat and electricity, has a shiny luster, and is flexible. (p. 491)

metamorphic rock (met′ə·môr′fik rok) A rock that forms from another kind of rock under heat and pressure. (p. 306)

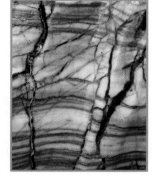

metamorphosis (met′ə·môr′fə·sis) A series of distinct growth stages that are different from one another. (p. 114)

meteor (mē′tē·ər) A chunk of rock from space that travels through Earth's atmosphere. (p. 451)

mimicry (mim′i·krē) An adaptation in which an animal is protected against predators by its resemblance to another, unpleasant animal. (p. 174)

mineral (min′ər əl) A solid, nonliving material of Earth's crust with a distinct composition. (p. 302)

mixture (miks′chər) A physical combination of two or more substances that are blended together without forming new substances. (p. 530)

molecule (mol′ə·kūl′) A particle that contains more than one atom joined together. (p. 493)

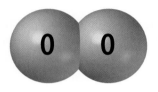

momentum (mō·men′təm) The mass of an object multiplied by its velocity. (p. 578)

monocot (mon′ə·kot′) An angiosperm with one cotyledon in each seed. (p. 107)

monotreme (mon′ə·trēm′) A mammal that lays eggs. (p. 68)

moon (mün) A natural object that orbits a planet. (p. 448)

motion (mō′shən) A change in an object's position over time. (p. 572)

multicellular (mul′ti·sel′yə·lər) Many-celled organism. (p. 23)

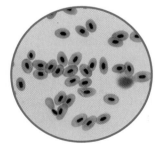

muscular system (mus′kyə·lər sis′təm) The body system made up of muscles that move bones. (p. 74)

mutualism (mū′chü·ə·liz′əm) A relationship between two kinds of organisms that benefits both. (p. 160)

N

nebula (neb'ye·lə) A huge cloud of gas and dust in space that is the first stage of star formation. (p. 458)

nekton (nek'ton) Organisms that swim through the water in aquatic ecosystems. (p. 220)

nervous system (nûrv'əs sis'təm) The set of organs that uses information from the senses to control other body systems. (p. 80)

neutralization (nü'trəl·ī·zā'shən) The chemical change of an acid and a base into a salt and water. (p. 558)

neutron (nü'tron) A particle in the nucleus of an atom that has no electric charge. (p. 492)

niche (nich) The role of an organism in an ecosystem. (p. 158)

nitrogen cycle (nī'trə jən sī'kəl) The continuous trapping of nitrogen gas into compounds in the soil and its return to the air. (p. 188)

nonrenewable resource (non'ri·nü'ə·bəl rē'sôrs') A resource that cannot be replaced within a short period of time or at all. (p. 331)

nonvascular (non·vas'kye·lər) Containing no plant tissue through which water and food move. (p. 38)

nucleus (nü'klē·əs) The center of an atom that has most of its mass. (p. 492)

nymph (nimf) A stage of metamorphosis where the organism is similar to an adult form, but is smaller. (p. 115)

 O

orbit (ôr'bit) The path one object travels around another object. (p. 423)

organ (ôr'gən) A group of tissues working together to do a certain job. (p. 28)

organ system (ôr'gən sis'təm) A group of organs that work together to do a certain job. (p. 28)

organism (ôr'gə·niz·əm) Any living thing that can carry out its life on its own. (p. 22)

outer core (ou'tər kôr) A liquid layer of iron and nickel below Earth's mantle. (p. 246)

ozone (ō'zōn) A form of oxygen gas that makes a layer in the atmosphere that screens out much of the Sun's ultraviolet rays. (p. 348)

 P

parasitism (par'ə·sī·tiz'əm) A relationship in which one organism lives in or on another organism and benefits from that relationship while the host organism is harmed by it. (p. 162)

pedigree (ped'i·grē) A chart used to trace the history of traits in a family. (p. 128)

phase (fāz) The appearance of the shape of the Moon at a particular time. (p. 433)

phloem (flō'em) The tissue through which food from the leaves moves throughout the rest of a plant. (p. 53)

photon (fō′ton) A tiny bundle of energy through which light travels. (p. 653)

photosynthesis (fō′tə sin′thə sis) The food-making process in green plants that uses sunlight. (p. 54)

physical change (fiz′i kəl chānj) A change of matter in size, shape, or state that does not change the type of matter. (p. 520)

pioneer community (pī′ə·nîr′ cə·mü′ni·tē) The first community living in a lifeless area. (p. 200)

pioneer species (pī′ə·nîr′ spē′shēz) The first species living in an otherwise lifeless area. (p. 200)

pitch (pich) How high or low a sound is. (p. 642)

placental mammal (pləsen′təl mam′əl) A mammal whose young develops within the mother. (p. 68)

planet (plan′it) A large object that orbits a star and does not produce its own light. (p. 444)

plankton (plangk′tən) Organisms that float on the water in aquatic ecosystems and are unable to swim. (p. 220)

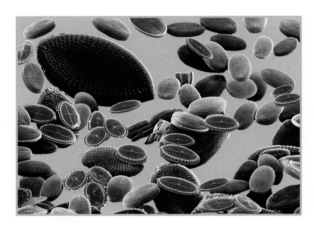

plate tectonics (plāt tek·ton′iks) A scientific theory that Earth's crust is made of moving plates. (p. 254)

pollen (pol′ən) Powdery grains in a flower that contain its male sex cells. (p. 104)

pollination (pol′ə·nā′shən) The transfer of a pollen grain to the egg-producing part of a plant. (p. 104)

pollution (pə·lü′shən) The addition of harmful substances to the environment. (p. 319)

population (pop′yə·lā′shən) All the members of one species in an area. (p. 143)

position (pə·zish′ən) The location of an object. (p. 572)

potential energy (pə·ten′shəl en′ər·jē) Energy stored in the position or structure of an object. (p. 600)

precipitate (pri·sip′i·tit) A solid formed by the chemical reaction of two solutions. (p. 547)

precipitation (pri sip′i tā′shən) Any form of water that falls from the atmosphere and reaches the ground. (p. 184)

predator (pred′ə tər) An animal that hunts other animals for food. (p. 147)

prey (prā) A living thing that is hunted for food. (p. 147)

primary succession (prī′mer·ē sək·sesh′ən) The beginning of a community where few, if any, living things exist, or where earlier communities were wiped out. (p. 200)

prism (priz′əm) A cut piece of clear glass (or plastic) with two opposite sides in the shape of a triangle or other geometric shape. (p. 658)

product (prod′ukt) A substance at the end of a chemical reaction. (p. 544)

protective coloration (prə·tek′tiv kul′ə·rā′shən) A type of camouflage in which the color of an animal blends in with its background, protecting the animal against predators. (p. 173)

protective resemblance (prə·tek′tiv ri·zem′bləns)
A type of camouflage in which the color and shape of an animal blends in with its background, protecting it against predators. (p. 173)

proton (prō′ton) A particle in the nucleus of an atom that carries one unit of positive electric charge. (p. 492)

pupa (pū′pə)
A non-feeding stage in complete metamorphosis in which a hard, caselike cocoon surrounds the organism. (p. 114)

radial symmetry (rā′dē·əl sim′ə·trē) A body plan in which all body parts of an organism are arranged around a central point. (p. 62)

radiation (rā′dē·ā′shən) The transfer of heat through electromagnetic rays. (p. 628)

rain shadow (rān shad′ō) The dry area on the leeward side of a mountain. (p. 411)

reactant (rē·ak′tənt) An original substance at the beginning of a chemical reaction. (p. 544)

reaction force (rē·ak′shən fôrs) The push or pull of a second object back on the object that started the push or pull. (p. 592)

recessive trait (ri·ses′iv trāt) The form of a trait that is hidden, or masked, in the hybrid generation. (p. 126)

refraction (ri·frak′shən) The bending of waves as they pass from one substance into another. (p. 657)

relative age (rel′ə·tiv āj) The age of one rock or fossil as compared to another. (p. 328)

relief map (ri·lēf′ map) A map that shows the elevation of an area using colors or shading. (p. 244)

renewable resource (ri·nü′ə·bəl rē′sôrs′) A resource that can be replanted or replaced naturally in a short period of time. (p. 331)

reservoir (rez′ər·vwär′) A storage area for fresh water. (p. 343)

resistor (ri·zis′tər) An object in an electrical circuit that resists the flow of electrons. (p. 668)

respiratory system (res′pər ə tôr′ē sis′təm) The organ system that brings oxygen to body cells and removes waste gas. (p. 78)

revolution (rev′ə·lü′shən) One complete trip of one object around another object. (p. 424)

rock cycle (rok sī′kəl) A never-ending process in which rocks change from one kind into another. (p. 306)

rotation (rō·tā′shən) A complete spin on an axis. (p. 426)

runner (run′ər) A thin plant stem that puts down roots and gives rise to new plants. (p. 93)

runoff (run′ôf′) Precipitation that flows across the land's surface or falls into rivers and streams. (p. 184)

satellite (sat′ə·līt) A natural or artificial object in space that circles around another object. (p. 448)

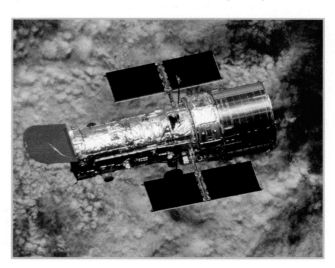

secondary succession (sek'ən der'ē sək·se s h 'ən) The beginning of a new community where an earlier community already exists. (p. 202)

sediment (sed'ə·mənt) Particles of soil or rock that may be eroded and deposited. (p. 288)

sedimentary rock (sed'ə·mən'tə·rē rok) A rock made of compacted and cemented materials. (p. 306)

seed coat (sēd kōt) The outer covering of a seed. (p. 106)

semiconductor (sem'ē· kən·duk'tər) A material that has properties in between conductors and insulators. (p. 509)

sexual reproduction (sek'shü·əl rē'prə·duk'shən) The production of a new organism from a female sex cell and a male sex cell. (p. 90)

shield volcano (shēld vol·kā'nō) A wide, gently sloped cone that forms from flowing lava. (p. 264)

simple machine (sim'pəl mə·shēn') A machine that changes the direction, distance, or strength of one force. (p. 608)

skeletal system (skel'i·təl sis'təm) The organ system made up of bones, tendons, and ligaments. (p. 74)

smog (smog) A type of air pollution formed by particles produced by burning fossil fuels. (p. 348)

soil (soil) A mixture of bits of rock and bits of once-living plants and animals. (p. 316)

soil horizon (soil hə·rī'zən) Any of the layers of soil from the surface to the bedrock. (p. 317)

solar eclipse (sō'lər i·klips') A blocking of the Sun's light that happens when Earth passes through the Moon's shadow. (p. 434)

solubility (sol'yə·bil'i·tē) The maximum amount of a substance that can be dissolved by another substance. (p. 533)

solute (sol'ūt) A substance that is dissolved by another substance to form a solution. (p. 532)

solution (sə·lü'shən) A mixture of substances that are blended so completely that the mixture looks the same everywhere. (p. 532)

solvent (sol'vənt) A substance that dissolves one or more other substances to form a solution. (p. 532)

sound wave (sound wāv) A series of rarefactions and compressions traveling through a substance. (p. 639)

species (spē'shēz) A group of similar organisms in a genus that can reproduce more of their own kind. (p. 34)

spectrum (spek'trəm) A band of colors produced when light goes through a prism. (p. 658)

speed (spēd) How fast an object's position changes with time at any given moment. (p. 574)

star (stär) An object in space that produces its own energy, including heat and light. (p. 458)

static electricity (stat'ik i·lek·tris'i·tē) The buildup of charged particles. (p. 666)

storm surge (stôrm sûrj) A bulge of water in the ocean caused by hurricane waves and winds. (p. 401)

sublimation (sub·lə·mā'shən) The process of changing directly from a solid to a gas without first becoming a liquid. (p. 521)

succession (sək·sesh'ən) The process of one ecosystem changing into a new and different ecosystem. (p. 200)

supernova (sü'pər·nō'və) A star that has produced more energy than gravity can hold together and explodes. (p. 460)

symbiosis (sim'bī·ō'sis) A relationship between two kinds of organisms over time. (p. 160)

taiga (tī′gə) A cool forest biome of conifers in the upper northern hemisphere. (p. 211)

telescope (tel′ə·skōp′)
An instrument that makes distant objects appear closer and larger. (p. 442)

temperate rain forest (tem′pər·it rān fôr′ist) A biome with a lot of rain and a cool climate. (p. 212)

temperature (tem′pər·ə·chər) A measurement of the average energy of particles in an object. (p. 626)

thermal conductivity (thûr′məl kon′duk·tiv′i·tē) The ability of a material to transfer heat. (p. 630)

thermal contraction (thûr′məl kən·trak′shən) The contraction of matter caused by a change in heat. (p. 524)

thermal expansion (thûr′məl ek·span′shən) The expansion of matter caused by a change in heat. (p. 524)

threatened species (thret′ənd spē′shēz) A species that is in danger of becoming endangered. (p. 199)

thunderstorm (thun′dər·stôrm) A rainstorm with both lightning and thunder. (p. 394)

tide (tīd) The regular rise and fall of the water level along a shoreline. (p. 436)

tissue (tish′ü) A group of similar cells that work together at the same job. (p. 28)

topographical map (top′ə·graf′i·kəl map) A map that shows the elevation of an area of Earth's surface using contour lines. (p. 245)

topsoil (top′soil′) The dark, top layer of soil, rich in humus and minerals, in which organisms live and most plants grow. (p. 317)

tornado (tôr·nā′dō) A rotating funnel-shaped cloud with wind speeds up to 500 kilometers per hour. (p. 398)

translucent (trans·lü′s ənt) Blurring light as it passes through. (p. 654)

transpiration (tran′spə·rā′shən) The loss of water through a plant's leaves. (p. 54)

tropical rain forest (trop′i·kəl rān fôr′ist) A hot, humid biome near the equator, with much rainfall and a wide variety of life. (p. 212)

troposphere (trop′ə·sfîr′) The atmospheric layer of gases that is closest to Earth's surface. (p. 366)

tsunami (tsü· nä′mē) A huge wave caused by an earthquake under the ocean. (p. 277)

tundra (tun′drə) A large, treeless plain in the arctic regions, where the ground is frozen all year. (p. 210)

unbalanced forces (un·bal′ənst fôrs′əz) Forces that do not cancel each other out and that cause an object to change its motion. (p. 588)

unicellular (ū′nə·sel′yə·lər) One-celled organism. (p. 23)

vacuum (vak′ū·əm) A space which contains little or no matter. (p. 640)

vascular (vas′kyə·lər) Plant tissue through which water and food move. (p. 38)

vegetative propagation (vej′i ·tā′tiv prop′ə·gā′shən) Asexual reproduction in plants that produces new plants from roots, leaves, or stems. (p. 93)

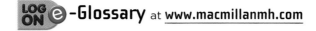

velocity (və·los′i·tē) The speed and direction of a moving object. (p. 575)

vertebrate (vûr′tə·brāt′) An animal that has a backbone. (p. 36)

volcano (vol′kā·nō) An opening in Earth's crust through which lava may flow. (p. 262)

volume (vol′ūm) A measure of how much space the matter of an object takes up. (p. 481)

water cycle (wô′tər sī′kəl) The continuous movement of water between Earth's surface and the air, changing from liquid to gas to liquid. (p. 184)

watershed (wô′tər·shed′) Area from which water is drained; a region that contributes water to a river or river system. (p. 184)

wavelength (wāv′lengkth) The distance from one peak to the next on a wave. (p. 652)

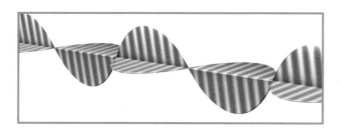

weather (weth′ər) What the troposphere is like at any given place and time. (p. 366)

weather map (weth′ər map) A map that shows the weather in a specific area at a specific point in time. (p. 388)

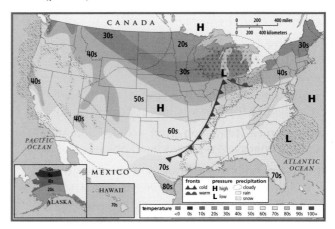

weathering (weth′ər·ing) The process through which rocks or other materials are broken down into smaller pieces. (p. 284)

weight (wāt) A measure of how gravity pulls on an object. (p. 481)

white dwarf (hwīt dwôrf) A star that can no longer turn helium into carbon; it cools and shrinks, becoming very small and dense. (p. 459)

work (wûrk) The use of force to move an object a certain distance, or to change an object. (p. 598)

xylem (zī′ləm) The tissue through which water and minerals move up through a plant. (p. 53)

Index

Note: Page references followed by an asterisk indicate activities.

T

Illustration Credits: All Illustrations are by Macmillan/McGraw Hill (MMH) except as noted below:

6: John Megahan. 24: John Megahan. 26: John Megahan. 28: John Megahan. 50: John Megahan. 52: John Megahan. 54: John Megahan. 55: Virge Kask. 56: John Megahan. 74: (overlay) Jennifer Fairman. 75: Jennifer Fairman. 76: (overlay) Jennifer Fairman. 78: Jennifer Fairman. 79: Linda Nye. 80: (overlay) Jennifer Fairman. 100: Emil Huston. 101: Emil Huston. 102: Sam Tomasello. 103: Sam Tomasello. 104: Emil Huston. 106: Sam Tomasello. 107: Sam Tomasello. 108: Emil Huston. 110: Sam Tomasello. 116: John Megahan. 144: Yuan Lee. 145: Yuan Lee. 146-147: Yuan Lee. 148-149: Sharon & Joel Harris. 152: Terry Kovalcik. 153: GGS Design. 170: (inset) Laurie O'Keefe. 185: John Kaufman. 186-187: John Kaufman. 188: John Kaufman. 195: Emil Huston. 198: Gary Raham. 200: Stephen Durke. 208: XNR Productions. 209: XNR Productions. 210: XNR Productions. 211: XNR Productions. 212: XNR Productions. 214: XNR Productions. 223: Yuan Lee. 224: Yuan Lee. 240-241: John Kaufman. 242: Steve Weston. 244: XNR Productions. 245: XNR Productions. 246: Howard Friedman. 252-253: XNR Productions. 255: John Edwards Illustration. 256: Cindy Shaw. 258: John Edwards Illustration. 259: XNR Productions. 262-263: Steve Weston. 263: (t) XNR Productions. 264: Steve Weston. 266: Steve Weston. 272: Cindy Shaw. 274: (bl) Jeff Grunewald. 275: XNR Productions. 277: John Edwards Illustration. 289: (insets) Sebastian Quigley. 297: XNR Productions. 304: Jeff Grunewald. 306-307: Steve Weston. 317: Frank Ippolito. 318: XNR Productions. 327: Jeff Grunewald. 328: (tr) Frank Ippolito. 334: (inset) Argosy. 336: (b) John Kocon. 342-343: John Kaufman. 345: John Edwards Illustration. 346: Jeff Grunewald. 347: John Kocon. 364-365: John Edwards Illustration. 365: (insets) Jeff Grunewald. 366: Stephen Durke. 367: John Edwards Illustration. 368: Jeff Grunewald. 369: John Edwards Illustration. 369: (b) Jeff Grunewald. 370: (b) Jeff Grunewald. 371: (t) Jeff Grunewald. 371: (b) Argosy. 372-373: John Kaufman. 374: (tr) John Kocon. 382-383: John Megahan. 384: (tr) John Kaufman. 384: (cr) John Kaufman. 384: (b) XNR Productions. 386-387: (insets) Jeff Grunewald. 388: XNR Productions. 394: John Kaufman. 395: John Kaufman. 408: (bl) Jeff Grunewald. 408: (br) XNR Productions. 409: John Megahan. 410: XNR Productions. 411: (tl) Stephen Durke. 412: John Kaufman. 416: (br) Bruce Van Patter. 423: Jeff Grunewald. 424-425: Jeff Grunewald. 426: John Edwards Illustration. 428-429: Jeff Grunewald. 433: Jeff Grunewald. 434: Jeff Grunewald. 436: Jeff Grunewald. 438: Eric Larsen. 439: John Edwards Illustration. 444-445: John Kaufman. 458-459: Stephen Durke. 460: Jeff Grunewald. 471: (br) Bruce Van Patter. 482: (insets) Argosy. 484: (overlays) Argosy. 491: (inset) Argosy. 492: Argosy. 493: Argosy. 494-495: GGS Design. 500-501: Steven Stankiewicz. 520: Argosy. 522: Jeff Grunewald. 524: Argosy. 533: (inset) Argosy. 534: Jeff Grunewald. 535: John Kocon. 543: (insets) Argosy. 544: (insets) Argosy. 545: Argosy. 555: Argosy. 557: Argosy. 558: Argosy. 571: (cr) Jeff Grunewald. 572: John Edwards Illustration. 574: Laurie O'Keefe. 575: John Edwards Illustration. 576-577: John Edwards Illustration. 580: (insets) Jeff Grunewald. 587: Jeff Grunewald. 598: John Edwards Illustration. 600: (tr) Stephen Durke. 608: Jeff Grunewald. 609: (t) Jeff Grunewald. 609: (cr) Laurie O'Keefe. 612-613: John Edwards Illustration. 616: Stephen Durke. 619: (tl) Jeff Grunewald. 626: Jeff Grunewald. 628-629: John Edwards Illustration. 630: (overlay) Argosy. 646: John Edwards Illustration. 652: (overlay) Stephen Durke. 656: (tr) John Kocon. 657: (t) John Kocon. 660: Jeff Grunewald. 665: (cr) Jeff Grunewald. 666: (b) Steven Stankiewicz. 668: Jeff Grunewald. 669: (bl) Argosy. 670: Jeff Grunewald. 671: Jeff Grunewald. 674-675: John Edwards Illustration. 678: Argosy. 679: Jeff Grunewald. 680: Stephen Durke. 681: Stephen Durke. 682: (tr) John Edwards Illustration. 682-683: (cr) Steven Stankiewicz. 684: (inset) Stephen Durke.

Photography Credits: All photographs are by Macmillan/McGraw-Hill (MMH) and Ken Cavanagh, Jacques Cornell, Ken Karp and Joseph Polilio except as noted below:

vi: (tl) Siede Preis/Getty Images; (tr) Burke/Triolo Productions/Brand X Pictures/Getty Images; (bl) American Images/Getty Images; (br) Burke/Triolo Productions/Getty Images. vii: (tr) Alan and Sandy Carey/Getty Images; (b) Corbis/PunchStock. viii: Royalty-Free/Corbis. ix: (tr) Comstock Images/Alamy; (b) NASA/JPL-Caltech. x: (tr) WidStock/Alamy; (bl) CMCD/Getty Images; (b) Andrew Lambert Photography/Photo Researchers, Inc. xi: (tr) PhotoLink/Getty Images; (l) Purestock/PunchStock; (b) Robert Glusic/Getty Images; (br) Photodisc/Getty Images. 1: Eye of Science/Photo Researchers, Inc. 2-3: Nature Picture Library/Alamy. 3: (l, r) Courtesy American Museum of Natural History. 4-5: MdeMS/Alamy. 5: (tl) Roger Eritja/Alamy; (r) Claus Meyer/Minden Pictures. 6-7: Garry Black/Masterfile. 7: Courtesy American Museum of Natural History. 8: (tr, b) Courtesy American Museum of Natural History. 9: (tl, c, br) Courtesy American Museum of Natural History. 10: (tr) LMR Group/Alamy; (bl) Courtesy American Museum of Natural History. 10-11: NHPA/Martin Harvey. 11: (cl) Frans Lanting/Minden Pictures; (br) Courtesy American Museum of Natural History. 12: (l to r) Mark A. Schneider/Photo Researchers, Inc.; Andrew J. Martinez/Photo Researchers, Inc.; Mark A. Schneider/Photo Researchers, Inc. 15: Royalty Free/CORBIS. 16: Shin Yoshino/Minden Pictures. 17: Gunter Ziesler/Peter Arnold, Inc. 18-19: M I (Spike) Walker/Alamy. 19: (t to b) Juniors Bildarchiv/Alamy; Phototake Inc./Alamy; B&B Photos/Custom Medical Stock Photo. 20-21: STEVE GSCHMEISSNER/Photo Researchers, Inc. 21: (t to b) Kevin & Betty Collins/Visuals Unlimited; B&B Photos/Custom Medical Stock Photo. 22: L. S. Stepanowicz/Visuals Unlimited. 23: (t) Digital Vision/Getty Images; (tc) Phototake Inc./Alamy; (tr) B&B Photos/Custom Medical Stock Photo. 28: Juniors Bildarchiv/Alamy. 29: (t to b) Digital Vision/Getty Images; Juniors Bildarchiv/Alamy. 30: Ed Reschke/Peter Arnold, Inc. 32-33: Chris Newbert/Minden Pictures. 34: Stuart McClymont/Getty Images. 35: (l to r, t to b) Dynamic Graphics Group/IT Stock Free/Alamy; BRUCE COLEMAN INC./Alamy; Alan & Sandy Carey/Getty Images; Creatas/PunchStock; Frans Lanting/Minden Pictures; Peter Coombs/Alamy; Comstock Images/Jupiter Images; BRUCE COLEMAN INC./Alamy; Alan & Sandy Carey/Getty Images; Creatas/PunchStock; Frans Lanting/Minden Pictures; Peter Coombs/Alamy; Comstock Images/Jupiter Images; Alan & Sandy Carey/Getty Images; Creatas/PunchStock; Frans Lanting/Minden Pictures; Peter Coombs/Alamy; Comstock Images/Jupiter Images; Creatas/PunchStock; Frans Lanting/Minden Pictures; Peter Coombs/Alamy; Comstock Images/Jupiter Images; Frans Lanting/Minden Pictures; Peter Coombs/Alamy; Comstock Images/Jupiter Images; Peter Coombs/Alamy; Comstock Images/Jupiter Images; Comstock Images/Jupiter Images. 36: (l to r, t to b) Ryan McVay/Getty Images; IT Stock/PunchStock; Royalty-Free/CORBIS; Creatas Images/PictureQuest; Imageshop/Alamy; Gail Shumway/Getty Images. 38: (l to r, t to b) David Sieren/Visuals Unlimited; L. Mellichamp/Visuals Unlimited; Bryan & Cherry Alexander Photography/Alamy; Nature Picture Library/Alamy; Henry W. Robison/Visuals Unlimited; Adam Jones/Visuals Unlimited. 39: (bl) John Mielcarek/Dembinsky Photo Associates; (bc) Jack M. Bostrack/Visuals Unlimited; (br) Scott Camazine/Alamy. 40: (l to r) Wolfgang Baumeister/Photo Researchers, Inc.; PHOTOTAKE Inc./Alamy; Dr. David Phillips/Visuals Unlimited; Stem Jems/Photo Researchers, Inc.; (bl) VEM/Photo Researchers, Inc. 41: (tr) Dr. Dennis Kunkel/Visuals Unlimited; (bl) Wim van Egmond/Visuals Unlimited/Getty Images; (bc) Peter Arnold, Inc./Alamy; (br) Eye of Science/Photo Researchers, Inc. 42: (b) Dr. George Chapman/Visuals Unlimited; (br) Dr. Harold Fisher/Visuals Unlimited. 43: (t to b) Stuart McClymont/Getty Images; Nature Picture Library/Alamy; Dr. Dennis Kunkel/Visuals Unlimited. 44: (l) Courtesy of the American Museum of Natural History; (bc) Duncan Smith/Getty Images; (br) Jim Wehtje/Getty Images. 45: (t) Digital Vision; (inset) Peter Weber/Getty Images. 46-47: George H. H. Huey/CORBIS. 48: (bc) LifeFile Photos Ltd/Alamy; (br) Ed Reschke/Peter Arnold, Inc. 49: (tr) Burke/Triolo Productions/Getty Images; (l to r) B. Rondel/zefa/Corbis; John Emker/Animals Animals-Earth Scenes; JTB Photo/Alamy; Markus Botzek/zefa/CORBIS. 50: Eisenhut & Mayer/FoodPix/Jupiter Images. 50-51: Dwight Kuhn. 51: (cl) D. Hurst/Alamy; (r) Nicole Duplaix/Omni-Photo Communications; (br) Richard Carlton/Visuals Unlimited. 52-53: Siede Preis/Getty Images. 54: Melanie Acevedo/Jupiter Images. 57: (t to b) LifeFile Photos Ltd/Alamy; Richard Carlton/Visuals Unlimited; Melanie Acevedo/Jupiter Images. 58: (r) Photodisc/Getty Images; (bl) Patricio Robles Gil/Nature Picture Library. 59: (l) Siede Preis/Getty Images; (bkgd) Siede Preis/Getty Images. 60-61: Kjell B. Sandved/Photo Researchers, Inc. 61: (t to b) Royalty-Free/Corbis; Comstock Images/Jupiter Images; Creatas Images/PictureQuest; Steve Hopkin/Ardea London Ltd.; Jeremy Woodhouse/Masterfile. 62: Heather Perry/National Geographic Image Collection. 62-63: David Wrobel/Visuals Unlimited. 63: (tl) Ken Lucas/Visuals Unlimited; (tr) Reinhard Dirscherl/Visuals Unlimited. 64: (tr) Fred Bavendam/Peter Arnold, Inc.; (bl) Jean Michel Labat /Ardea London Ltd. 65: (tl) Jean Paul Ferrero /Ardea London Ltd.; (tr) Jean Michel Labat /Ardea London Ltd.; (bee) Stefan Sollfors/Alamy; (ladybug) Westend61/Alamy; (cricket) Photodisc/Getty Images. 66: (tr) Creatas/PunchStock; (b) Stephen Frink/CORBIS. 67: (cl) Michael & Patricia Fogden/Minden Pictures; (b) Craig K. Lorenz /Photo Researchers, Inc. 68: (tr) Klein/Peter Arnold, Inc.; (cr) Reg Morrison/Auscape/Minden Pictures; (bl) Arco Images/Alamy. 69: (t to b) David Wrobel/Visuals Unlimited; Fred Bavendam/Peter Arnold, Inc.; Creatas/PunchStock. 70: (t) RAYMOND MENDEZ/Animals Animals/Earth Scenes; (c) Juniors Bildarchiv/Alamy; (bkgd) Don Farrall/Getty Images. 71: (c) Brand X Pictures/PunchStock; (br) Burke/Triolo Productions/Brand X Pictures/Getty Images. 72-73: Pete Oxford/Minden Pictures. 74: Andrew Darrington/Alamy. 75: Jim Occi, FUNDAMENTAL PHOTOGRAPHS, NEW YORK. 76: Juniors Bildarchiv/Alamy. 77: George G. Lower/Photo Researchers, Inc. 78: (tr) Andre Seale/Alamy; (b) blickwinkel/Alamy. 80: Juniors Bildarchiv/Alamy. 81: (t to b) Andrew Darrington/Alamy; Juniors Bildarchiv/Alamy. 83: Zoran Milich/Getty Images. 86-87: Thomas & Pat Leeson/Photo Researchers, Inc. 87: Digital Vision Ltd.; Science Pictures Limited/Photo Researchers; Mitsuhiko Imamori/